Lecture Notes in Civil Engineering

Volume 252

Lecture Notes in Civil Engineering (LNCE) publishes the latest developments in Civil Engineering - quickly, informally and in top quality. Though original research reported in proceedings and post-proceedings represents the core of LNCE, edited volumes of exceptionally high quality and interest may also be considered for publication. Volumes published in LNCE embrace all aspects and subfields of, as well as new challenges in, Civil Engineering. Topics in the series include:

- Construction and Structural Mechanics
- Building Materials
- Concrete, Steel and Timber Structures
- Geotechnical Engineering
- Earthquake Engineering
- Coastal Engineering
- Ocean and Offshore Engineering; Ships and Floating Structures
- Hydraulics, Hydrology and Water Resources Engineering
- Environmental Engineering and Sustainability
- Structural Health and Monitoring
- Surveying and Geographical Information Systems
- Indoor Environments
- Transportation and Traffic
- Risk Analysis
- Safety and Security

To submit a proposal or request further information, please contact the appropriate Springer Editor:

- Pierpaolo Riva at pierpaolo.riva@springer.com (Europe and Americas);
- Swati Meherishi at swati.meherishi@springer.com (Asia - except China, and Australia, New Zealand);
- Wayne Hu at wayne.hu@springer.com (China).

All books in the series now indexed by Scopus and EI Compendex database!

More information about this series at https://link.springer.com/bookseries/15087

Marcello Biocca · Eugenio Cavallo ·
Massimo Cecchini · Sabina Failla ·
Elio Romano
Editors

Safety, Health and Welfare in Agriculture and Agro-food Systems

Ragusa SHWA 2021

Springer

Editors
Marcello Biocca
CREA
Rome, Italy

Eugenio Cavallo
CNR - STEM
Torino, Italy

Massimo Cecchini
Tuscia University
Viterbo, Italy

Sabina Failla
Di3A
University of Catania
Catania, Italy

Elio Romano
CREA
Rome, Italy

ISSN 2366-2557 ISSN 2366-2565 (electronic)
Lecture Notes in Civil Engineering
ISBN 978-3-030-98091-7 ISBN 978-3-030-98092-4 (eBook)
https://doi.org/10.1007/978-3-030-98092-4

This Springer imprint is published by the registered company Springer Nature Switzerland AG
The registered company address is: Gewerbestrasse 11, 6330 Cham, Switzerland

Preface

Dear Colleagues and dear Friends,

The 6th edition of Ragusa SHWA, the International Conference devoted to the Safety, Health and Welfare in Agriculture and Agro-food systems, was held during September 15–16, 2021.

The previous editions of Ragusa SHWA took place in September 2008, 2010 and 2012 in Ragusa, Italy. The 4th SWHA edition was held in 2015 in Lodi, Italy, together with the International Congress on Rural Health (ICRH). The two International Conferences adopted together the Lodi Declaration on Rural Health.

Due to the Covid-19 pandemic, Ragusa SHWA 2020 and Ragusa SHWA 2021 were celebrated online, with prestigious scientists reporting their studies highly topical issues.

The Ragusa SHWA conferences have two distinctive characteristics. First, it is celebrated in the same place and shows a high return rate of the participants. Those who participate in one edition are likely to return to the following. This means that the opportunities offered by the location are regarded as authentic strengths. In particular, Sicily and its southeastern part are a real laboratory for actions related to agriculture, agro-food and well-living, in urban, coastal and rural spaces. All factors that promote conviviality, well-being and ultimately the formation of pleasant, profitable and long-lasting partnerships among scientists are present. Moreover, Southeast Sicily, with its lovely and attractive countryside, wonderful beaches, gorgeous Late Baroque UNESCO sites, is a great location fitted for fruitful scientific meetings.

The Ragusa SHWA conferences took place under the aegis of the Presidency of the Italian Republic and were supported by International Commission of Agricultural and Biosystems Engineering (CIGR), Italian Society of Agricultural Engineering (AIIA) and the Department of Agriculture, Food and Environment of the University of Catania, Italy.

During the 2021 6th Ragusa SHWA, 2 Lectio Magistralis, 4 main communications, 25 oral and 40 poster contributions were presented online by more than 240 authors.

The conference has hosted a special session on "Innovation for Smart Dairy Farming" (CowBhave project, PRIN 2017 project), meeting Chair Prof. Claudia Arcidiacono (University of Catania, Italy) and session Chair Prof. Daniele Torreggiani (University of Bologna, Italy) with two lectures on "Innovation and sustainability for smart dairy farming COWBHAVE final results" and eight oral contributions.

The number and the quality of the contributions confirm that Ragusa SHWA conference has once again achieved its three main objectives: updating, contamination and promotion of interdisciplinary research on safety, health and welfare.

This volume collects the 52 reviewed contributions presented at the conference.

Ragusa SHWA loves contamination, collaboration and sharing dreams and doubts with the aim to make results from research actual solutions able to improve life for people and environment.

Our studies are faced up following the "One Health" approach, which recognizes that the human health is tightly connected to the health of the animals and environment.

Finally, the Organizing Committee would like to thank the authors and the presenters for their contributions. A special thanks to Prof. Sabina Failla for her outstanding job as Scientific Secretariat of the conference and as Managing Editor of this volume and to Prof. Massimo Cecchini for the digital management of the online conference tools.

Together with Prof. Remigio Berruto, CIGR president, Prof. Danilo Monarca past president of Italian Association of Agricultural Engineering (AIIA), the 6th International Conference Ragusa SHWA Steering Committee invite you to the next edition of the conference, in September 2023, in the late Baroque city of Ragusa Ibla, in the southeastern Sicily, Italy.

January 2022

Marcello Biocca
Eugenio Cavallo
Massimo Cecchini
Sabina Failla
Elio Romano
Giampaolo Schillaci

Organization

Conference Scientific Committee

Remigio Berruto	University of Turin, Italy
Marcello Biocca	CREA-Consiglio per la ricerca in agricoltura e l'analisi dell'economia agraria, Italy
Eugenio Cavallo	CNR-Consiglio Nazionale delle Ricerche, Italy
Massimo Cecchini	University of Tuscia, Italy
Sabina Failla	University of Catania, Italy
Pierluigi Febo	University of Palermo, Italy
Gioia Gibelli	Italian Society of Landscape Ecology, Italy
Celsino Govoni	Department of Public Health-Local Health Unit, Modena, Italy
Martina Jakob	Prevention and Research Unit (EU-OSHA), Bilbao, Spain
Peter Lundqvist	Swedish University of Agricultural Sciences, Sweden
Danilo Monarca	University of Tuscia, Italy
Domenico Pessina	University of Milan, Italy
Elio Romano	CREA-Consiglio per la ricerca in agricoltura e l'analisi dell'economia agraria, Italy
John Rosecrance	Colorado State University, USA
Giuseppe Santucci	Sapienza University, Italy
Giampaolo Schillaci	University of Catania, Italy

Editorial Committee

Marcello Biocca	CREA-Consiglio per la ricerca in agricoltura e l'analisi dell'economia agraria, Italy
Eugenio Cavallo	CNR-Consiglio Nazionale delle Ricerche, Italy
Massimo Cecchini	University of Tuscia, Italy

Sabina Failla	University of Catania, Italy
Elio Romano	CREA-Consiglio per la ricerca in agricoltura e l'analisi dell'economia agraria, Italy
Giampaolo Schillaci	University of Catania, Italy

Convener

Giampaolo Schillaci	University of Catania, Italy

Contents

WMSDs Work Related Musculo-Skeletal Disorders

Machine Milking, Animal Welfare, Sustainable Livestock Farming

Precision Farming and Traceability

Environment Safety, People Health Protection and Welfare

ROPS and Stability Research

Food Safety

WMSDs Work Related Musculo-Skeletal Disorders

Quantification of Trunk Postures Among Fruit and Vegetable Pickers in Sardinia

John Rosecrance[1(✉)], Maria Caria[2], Molly Hischke[1], and Giuseppe Todde[2]

[1] Colorado State University, Fort Collins, CO 80523, USA
John.Rosecrance@colostate.edu
[2] University of Sassari, 07100 Sassari, Italy

Abstract. Throughout the world, fruits and vegetables are grown and harvested through labor-intensive work tasks. The work involved with manually harvesting agricultural produce expose workers to awkward trunk postures that have been associated with the development of musculoskeletal disorders. Row crops, grown close to the ground, are especially task intensive and involve physical labor that has been associated with increased risk of low back injury. The purpose of this study was to quantify trunk postures during manual harvesting of fruits and vegetables.

Vegetable and fruit pickers were recruited from a farm in Sardinia, Italy. Participants wore an accelerometer-based trunk motion logging system for approximately two hours while harvesting eggplants (9 workers) and strawberries (7 workers). The posture data was processed in a custom MATLAB script. The magnitude, frequency, and time spent in four categories of trunk postures were assessed.

Pickers had a mean age of 39.4 years (SD, 13.6) which included 7 females and 9 males. Eggplant pickers had a mean maximum trunk posture of 82° compared to 93° for strawberry pickers. The eggplant pickers spent the majority of their work time between 0° and 30° of trunk flexion, while strawberries pickers spent the majority of work time in 60° to 90° of trunk flexion. The frequency of the trunk postures were also different between workers harvesting eggplant and strawberries. Eggplant workers spent an average of 27 s in trunk flexion during picking as compared to 49 s for strawberry workers. In addition, there were significant differences in the time spent in the four defined posture categories.

The results of the study have demonstrated that an accelerometer-based trunk motion logging system was successful in quantifying trunk postures among fruit and vegetable pickers. Quantifying posture data among fruit and vegetable pickers can be used to establish a baseline to assess, develop and manage improved work designs that reduce the risk of low back stress during manual harvesting.

Keywords: Low back injury · Agriculture · Exposure assessment · Fruit and vegetable picking

M. Biocca et al. (Eds.): SHWA 2020, LNCE 252, pp. 3–10, 2022.
https://doi.org/10.1007/978-3-030-98092-4_1

1 Introduction

Low back disorders a major occupational health problem facing the agricultural workforce [1, 2]. Manual harvesting of fruit and vegetables often requires workers to maintain a flexed or stooped posture, which place agricultural workers at a higher risk of sustaining a low back disorders [3–5]. Row crops, grown close to the ground, require workers to sustain stooped or awkward work postures during planting, weeding, and harvesting. The physical risk factors associated with these work activities have been strongly associated work-related musculoskeletal disorders of the low back [6].

Field-based health and safety assessments are often used to identify work tasks that involve high physical loads and anatomical postures that expose workers to high risk of work-related low back disorders. Beyond the identification of these risks, it is important to also quantify the magnitude and temporal components of the physical exposures. Although field-based assessments within agriculture have been conducted, the risk assessment tools often lack high levels of accuracy and precision when conducting work-related musculoskeletal risk assessments completed over the duration of a workday.

Advancements in technology have resulted in low cost, miniaturized, wearable sensors that detect both physical and physiological signals of the body. Wearable technologies are a novel method to conduct physical risk-assessments for musculoskeletal hazards in the workplace. Although wearable devices have been used to measure occupational physical and physiological workloads, these technologies are not often used in the agricultural sector. Thus, it is necessary to demonstrate the usefulness of wearable sensor systems in terms of data produced, simplicity in the field, and acceptance among agricultural workers when performing field-based risk assessments.

Currently, researchers and practitioners frequently use observational tools that are easily administered but often only quantify the presence of exposures at a single point in time, potentially overestimating or underestimating the magnitude, frequency, and duration of the exposure. The latest wearable technology, however, has the capacity to collect and store physical (and physiological) data from large samples of workers and during an entire work shift (8 h) to accurately quantify the components of exposures of interest. The primary purpose of the present study was to demonstrate the quantification of trunk postures during fruit and vegetable harvesting among agricultural works in Sardinia, Italy.

The aims of the present study were to 1) characterize trunk postures of agricultural workers performed during vegetable and fruit pickings using wearable technologies, and 2) to assess the usefulness of wearable microsensors to assess trunk postures during agricultural field research.

2 Methods

An agricultural organization that produced several varieties of fruits and vegetables in Sardinia, Italy and their farmworkers were recruited for this study. The research methodology was reviewed and approved by the university Institutional Review Board and farmworkers that agreed to participate provided informed consent. The farm that was involved in the study was familiar with the university research team prior to their participation. Participants wore an accelerometer-based trunk motion logging system (Zephyr Bioharness™ 3, Zephyr Technology Corporation, USA) (see Fig. 1) for approximately two hours while harvesting eggplants (9 workers) or strawberries (7 workers). Demographic data collected included age, height, weight, sex, produced harvested, and work experience. Each of the participants wore a chest mounted logging device during the initial calibration procedure (Fig. 3a), during 120 min of picking (Fig. 3a and b), and during the post-calibration procedure.

The trunk posture data from the Bioharness™ 3 was processed in a custom MATLAB script. The magnitude, frequency, and time spent in four categories of trunk postures were assessed. The trunk posture categories assessed included Category 1, trunk extension to upright (<0° to 0°), Category 2, (>0° to <30°), Category 3 ($\geq$30° to $\leq$60°), and Category 4, ($\geq$60°), see Fig. 4. The total time picking, frequency of picking task (expressed as cycles in trunk flexion and extension per hour), average cycle time, and average maximum trunk flection per cycle were quantified.

The study design was descriptive involving one farm, two crops, and a limited number of subjects. Thus, statistical differences between the dependent variables as not assessed (Fig. 2).

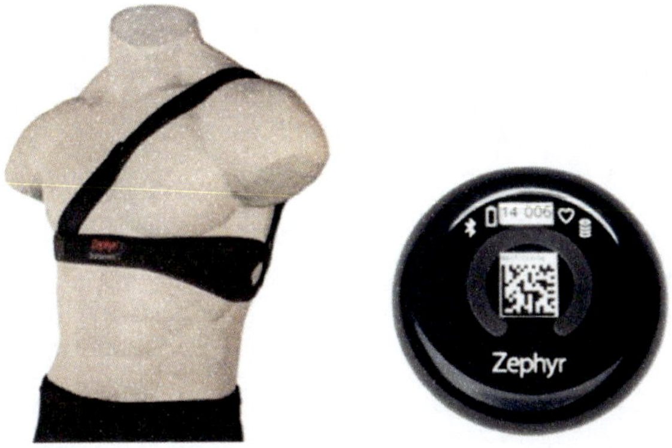

Fig. 1. Zephyr Bioharness™ 3 with the closeup of the module housing the datalogger and tri-axial accelerometer.

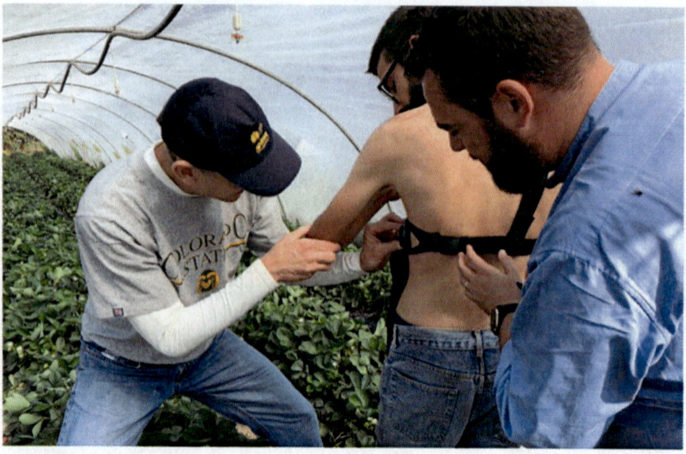

Fig. 2. Research team attaching Bioharness™ 3 to strawberry picker.

(a) (b) (c)

Fig. 3. a, b, c. Study participant (3a.) shown during calibration procedure to establish a neutral 0° posture with plumb line through ear, shoulder, hips and ankle. Typical stooped working postures are shown for eggplant (3b.) and strawberry (3c.) pickers during data collection.

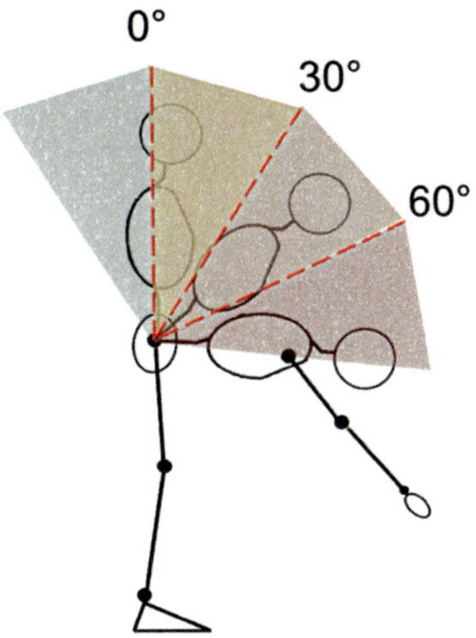

Fig. 4. Four posture categories used in the analyses of posture risk.

3 Results

Pickers had a mean age of 39.4 years (SD, 13.6) which included 7 females and 9 males. The mean time spent picking during the data collection period was 102.1 min (SD, 5.6) for eggplant pickers and 99.5 min (SD, 8.5) for strawberry pickers Eggplant pickers had a mean maximum trunk posture of 82° compared to 93° for strawberry pickers. The eggplant pickers spent the majority of their work time between 0° and 30° of trunk flexion, while strawberries pickers spent the majority of work time in 60° to 90° of trunk flexion (Fig. 5). The frequency of the trunk postures were also different between workers harvesting eggplant and strawberries. Eggplant workers spent an average of 27 s in trunk flexion during picking as compared to 49 s for strawberry workers. In addition, there were large differences in the mean task frequencies with eggplant workers averaging 132 cycles per hour while strawberry workers completed 73 cycles per hour. One task cycle was defined as the duration workers were stooped while picking, that is, from beginning of picking to end of cycle when the worker deposited the product in the bin or basket.

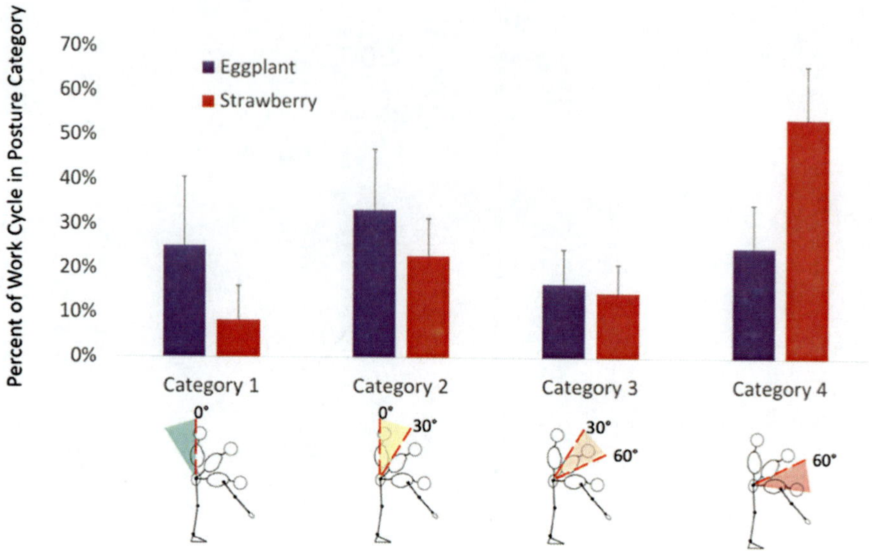

Fig. 5. Percent of work cycle spent in each posture category for eggplant and strawberry pickers.

4 Discussion

The present study described the quantification of trunk postures during fruit and vegetable harvesting among agricultural works in Sardinia, Italy. The study results identified clear differences in the trunk postures between eggplant and strawberry pickers with regard to the magnitude of trunk flexion, duration of time picking in high risk posture categories, average cycle times and mean time for a picking cycle. The results were expected given that the tasks involved with picking strawberries are quite different than that of eggplants.

Strawberry picking involves a smaller product that is typically grown closer to the ground than eggplant which is larger and hangs on a taller plant. Additionally strawberries were picked until the worker was able to fill a small plastic carton that held approximately 20–25 pieces of fruit. Once the plastic carton was filled, the worker would stand upright and place the plastic bin on a wheeled cart. The total cycle time to fill the plastic carton was approximately 50 s. In contrast, eggplants were picked by searching the plant for ripe produce and cutting them with a hand snip. With the cutting snips in one hand and holding eggplants in the other, workers were limited to hold four eggplants before standing upright to dump them into a large bin pulled in a wheeled cart. Thus, the mean cycle time was only approximately 25 s; half that of strawberry harvesting. The differences in these picking methods resulted in a greater magnitude of trunk flexion in a stooped posture for a greater percentage of the task cycle for strawberry workers as opposed to eggplant pickers.

The literature has consistently indicated that stooped postures expose a large segment of agricultural workers to a high risk of developing a low back disorder as well as the need to address these issues through better work methods [3]. Studies assessing the effects of stooped work tasks on the muscle activity and kinematics of the low back, suggest that short, frequent rest breaks during sustained stooped work tasks may be beneficial to the recovery of passive musculoskeletal tissues [5]. Additionally, in a study of rest break interventions among commercial strawberry pickers, researchers reported significantly less severe musculoskeletal symptoms among workers with 5-min breaks every hour versus control workers not receiving the breaks [7].

A secondary aim of this descriptive study was to assess the usefulness of wearable microsensors to assess trunk postures during agricultural field research. In the past decade, there has been greater use of wearable measurement systems to assess occupational physical and physiological workloads [8]. A *useful* wearable measurement system for the researcher or practitioner may be characterized as requiring minimal time and resources to operate and one that obtains quantitative data. The measurement system should be easy to attach to the individual, have a short, uncomplicated calibration procedure, and elementary data processing techniques. The Bioharness™ 3 physical and physiological monitoring system demonstrated these characteristics and was relatively easy to operate in our field conditions. Setup and calibration required approximately 10 min per subject and the harness was reported to be comfortable by workers during the two-hour data collection period. In addition to quantifying trunk posture, the Bioharness™ 3 system also measures heart rate and activity levels, which could provide beneficial information to health and safety professionals interested in a physiologic metrics related to work-environment interactions [9]. The ease of use of a wearable measurement system enables researchers and practitioners the ability to quantify trunk postures pre- and post-interventions to evaluate effectiveness in reducing extreme trunk postures during work tasks. A recent study of forestry loggers assessed the magnitude and duration of trunk postures during various logging tasks to assess biomechanical loads using the same measurement system described in the present study [10].

The data generated from the present field-based descriptive study of trunk postures among fruit and vegetable pickers in Sardinia will be used in the development of work design solutions to reduce potentially harmful physical exposures experienced during manual harvesting tasks. The quantification of frequency, duration, and magnitude of awkward trunk postures during stooped work activities is an important step in the development of safer and effective work design solutions.

Acknowledgements. The authors wish to thank Prof. Lelia Murgia and Dr. Gabriele Sara for their valuable assistance in data collection during on-farm activities. Sincere thanks also goes out to the farm management and the farmworkers for their willingness participate in these research activities. This research was supported by the High Plains Intermountain Center for Agricultural Health and Safety CDC/NIOSH Grant No. U54OH008085 and the Mountain and Plains Education and Research Center CDC/NIOSH Grant No 254-2012-M-52941. Its content is solely the responsibility of the authors and does not necessarily represent the official views of CDC/NIOSH.

References

1. Osborne, A., et al.: Prevalence of musculoskeletal disorders among farmers: a systematic review. Am. J. Ind. Med. **2**(55), 143–158 (2012)
2. Rosecrance, J., Rodgers, G., Merlino, L.: Low back pain and musculoskeletal symptoms among Kansas farmers. Am. J. Ind. Med. **49**(7), 547–556 (2006). https://doi.org/10.1002/ajim.20324. PMID: 16685722
3. Fathallah, F.A., Miller, B.J., Miles, J.A.: Low back disorders in agriculture and the role of stooped work: scope, potential interventions, and research needs. J. Agric. Saf. Health. **14**(2), 221–245 (2008). https://doi.org/10.13031/2013.24352. PMID: 18524286
4. Fathallah, F.: Musculoskeletal disorders in labor-intensive agriculture. Appl. Ergon. **41**, 738–743 (2010)
5. Miller, B.J., Fathallah, F.A.: The effects of a stooped work task on the muscle activity and kinematics of the lower back. Proceed. Human Factors Ergon. Soc. Ann. Meeting **50**(13), 1284–1288 (2006). https://doi.org/10.1177/154193120605001306
6. Putz-Anderson, V., et al.: musculoskeletal disorders and workplace factors: a critical review of epidemiologic evidence for work-related musculoskeletal disorders of the neck, upper extremity, and low back. In: Bernard, B.P., Cincinnatic, O.H. (eds.). National Institue for Occupational Safety and Health (1997)
7. Faucett, J., Meyers, J., Miles, J., Janowitz, I., Fathallah, F.: Rest break interventions in stoop labor tasks. Appl. Ergon. **38**(2), 219–226 (2007). https://doi.org/10.1016/j.apergo.2006.02.003
8. Chaffin, D., et al.: The promise and perils of wearable sensors in organizational research. Organ. Res. Methods **20**(1), 3–31 (2017). https://doi.org/10.1177/1094428115617004
9. Cheng, T., Teizer, J., Migliaccio, G.C., Gatti, U.C.: Automated task-level activity analysis through fusion of real time location sensors and worker's thoracic posture data. Autom. Constr. **29**, 24–39 (2013). https://doi.org/10.1016/j.autcon.2012.08.003
10. Masci, F., Spatari, G., Giorgianni, C.M., Bortolotti, S., Rosecrance, J., Colosio, C.: A Wearable Device to Assess the Spine Biomechanical Overload in a Sample of Loggers. In: Black, N.L., Neumann, W.P., Noy, I. (eds.) IEA 2021. LNNS, vol. 223, pp. 162–170. Springer, Cham (2022). https://doi.org/10.1007/978-3-030-74614-8_19

Image Analysis for Ergonomic Risk Assessment for Rope Arborists

Massimo Cecchini(✉), Andrea Colantoni, Danilo Monarca, Filippo Cossio, Francesco Mazzocchi, and Leonardo Bianchini

Department of Agriculture and Forest Sciences (DAFNE), University of Tuscia, Via San Camillo De Lellis, 01100 Viterbo, Italy
cecchini@unitus.it

Abstract. The work of rope arborist involves multiple risks, as falling from above, the risk of cutting for the use of chainsaw or physical risks (noise and vibrations) that cause professional diseases.

Ergonomic risk also plays an important role. The tree climbers perform a work getting not vertical postures to maintain a stable position and performing repetitive movements due to the use of the saws and chainsaw.

This work aimed to assess the ergonomic risk for rope workers on trees by analyzing the image and applying indices.

The assessment were carried out during pruning activities in fruit chestnut trees in Viterbo province and the postures of 3 operators have been analyzed from an ergonomic point of view.

To obtain this evaluation, 100 observations were examined, and image analysis was done using an Ergofellow 3.0 ergonomic analysis software applying two indices for postural analysis: REBA (Rapid Entire Body Assessment) and RULA (Rapid Upper Limb Assessment).

From the data analysis, the diversity of the calculation process emerged; the two indices are not interchangeable and return a different level of risk for the same task: MEDIUM with action required for REBA and "further study required and rapid changes applied" for RULA.

The two indices REBA and RULA were not designed to analyze rope work.

In conclusion, a specific index for rope workers is desirable, which considers postures and work equipment. This work introduces the foundations for the development of this rope worker's risk index.

Keywords: Tree climbing · Ergonomic · Work positions · REBA Index · RULA Index

1 Introduction

The profession of the arboriculturist on ropes involves many risks, from the most intuitive ones such as the risk of falling from a height [1] and the risk of being cut during the use of the chainsaw [2], to the more subtle ones such as physical risks (noise and vibrations) that cause occupational diseases.

More and more attention is being given to this area, due to the increase in the number of operators who are engaged in this work activity. This phenomenon is due to

M. Biocca et al. (Eds.): SHWA 2020, LNCE 252, pp. 11–18, 2022.
https://doi.org/10.1007/978-3-030-98092-4_2

the increasing attention to the planning and management of trees in the city both for safety and ecosystem services issues. This increase in operators and interest also corresponds to many accidents, given the dangerousness of the activity [3–6].

In the scientific field more and more works are dealing with this sector, going from the interaction between the equipment/operator and the tree, especially regarding the anchorage point [7–9], to certain operations such as rigging [4], to safety with the study of knots used during the same practices [10, 11] or those for the operator's insurance, friction knots, or even the study of the compatibility of certain PPE [12], up to safety standards and surveys among operators to define, for example, the degree of training, equipment used or injuries suffered [6, 13, 14].

In this panorama and given the multitude of risks, the ergonomic one is also a very important factor [2] in this working sector, as the operator performs a demanding physical work and often in positions that are not perfectly vertical, having to rely on ropes and other equipment to maintain a stable position (citing tree climbing questionnaire). The arboriculturist is found cutting with saws or chainsaws, to carry out pruning and controlled felling (rigging) repeating movements and postures that have never been evaluated objectively to quantify the associated risk.

The aim of this work was to evaluate the postures assumed with the technique of tree climbing, during pruning operations of fruit chestnut trees, through image analysis and using two established indices: REBA [15] and RULA [16], both valid for the analysis of postures at work but with different methodological approaches: the first refers to the whole body, while the second mainly to the upper limbs.

2 Materials and Methods

The work sites where the pruning activities were carried out are two chestnut groves located on Mount Cimino, in the province of Viterbo. Three operators were required for the survey: one on the ground for the assistance and possible rescue of the worker (therefore qualified to climb trees and trained to recover the injured person) and an operator in the canopy equipped with a video camera in charge of filming the third operator while he was carrying out the pruning operations. The equipment used is the usual tree climber's equipment, which refers to the provisions of the Italian Consolidated Law on Safety at Work D.lgs 81/08 and consists of access and positioning devices, individual protection and hand and power cutting tools such as hacksaws, telescopic poles, and chainsaws. The SRT (Single Rope Technique) ascent technique was used. Once the operator reaches the highest position, he secures himself with a safety lanyard and installs the work rope, a DdRT (Doubled Rope Technique) system; then, he releases himself from the single rope system with which he ascended and is ready to move inside the canopy, reaching the branches to be eliminated or on which to make the return cut. Each time he has to make a cut, the operator secures himself with the lanyard, for double safety [17]. The second operator on the plant performs the same access and positioning procedure and then reaches a suitable position to film the cutting operator. The position should be as much as possible at the same height as the other operator and positioned laterally, so that the working angles can be best interpreted during image analysis. Pruning operations were performed with hand and powered tools.

The assessment of ergonomic risk for workers can be easily carried out using objective indices that place the operator within risk categories, from null to high (depending on the type of index). The choice of index must be made according to the type of work, so that it considers the limbs more subject to effort or repetitive work rather than impulsive. After accurate bibliographic research, no specific index emerged for the assessment of the ergonomic risk of ropeway workers, so it was decided to use two well-established indices for postural analysis: REBA and RULA. Both indices consist of calculation matrices that, depending on the postures assumed by the operator, return a score [15, 16] that increases according to the lack of ergonomics during work. The video analysis was carried out with Ergofellow 3.0, linera test version, with which the ergonomic indices REBA Index (Rapid Entire Body Assessment) and RULA Index (Rapid Upper Limb Assessment) were calculated. The REBA index makes it possible to carry out a postural assessment of the risk of muscular-skeletal pathologies of the upper limb and neck using the REBA method (Rapid Entire Body Assessment). The REBA method is an internationally used method that allows a quantitative analysis of the main risk factors to which a worker may be subject. The method is to be used in a context of general postural assessment; the fundamental output of the REBA method is to identify situations/conditions of work that could lead to serious diseases in workers. The RULA index was developed to investigate upper extremity musculoskeletal risk exposure during work. Part of the development took place in the garment industry, where the assessment was made for workers who performed tasks including cutting while standing at the workstation, using a variety of sewing machines, trimming, and performing inspection and packing tasks. RULA was developed by evaluating the postures adopted and the forces required and for operations while working [16]. Images were the most important tool for posture and ergonomic risk analysis. They provided all the information for calculating the angles of the limbs with respect to the body, torso torsion, head tilt, and weight bearing on the limbs. It was necessary to find them exactly at the same height as the operator to ensure that the perspective from below did not alter the perception of the angles, since the software requires very detailed information, and the input must be as accurate as possible. In order to avoid gross errors in the evaluation it was preferred to film the tree climber through a second operator positioned in the canopy. The images were taken with a Panasonic Lumix tz 80 digital camera (1,920 × 1,080 pixel, 50 fps). 100 observations were taken. The REBA and RULA indexes were calculated for each of them, and the Pearson correlation and covariance were observed for the comparison of the indicators.

3 Results

From an ergonomic point of view, the postures of three operators engaged in the pruning of tall trees using the technique of tree climbing were analyzed. Ropeway operators generally use equipment that is not very heavy, except for particular controlled felling tasks, which were not taken into consideration because they cannot be classified as repetitive works that require operators to repeat the same movement many times during the working day. Controlled felling is reserved for experienced and well-trained operators who, above all for their own safety, cannot and should never put the speed of execution of the work ahead of their own safety and the tranquility of the entire team assisting them. In pruning work, on the other hand, it is possible to find good procedures which, while always requiring the utmost attention, lead the operator to repeat the same gesture many times, albeit with slight variations in angle and positioning. In particular, the use of the pruning saw is the most frequent working situation (Fig. 1) and valid for analyzing the ergonomic risk for the operator. Various postures have also been considered for the use of the chainsaw, but given the particular attention that its use requires, the work rhythms and therefore the repetition of movements are not comparable to the previous type of tool (Fig. 2).

Fig. 1. Example of posture during pruning with handsaw.

Fig. 2. Example of processed image for identifying limb angles during pruning.

From the analysis of the data, it emerges the diversity between the indices both in the calculation process and in giving a different weight to the various postures according to the risk class. Even though the maximum values are different, the final figure places the operator in two different risk classes: MEDIUM with action required for REBA and "further investigation and rapid application of changes required" for

RULA. The distribution of the values obtained is also substantially different, which confirms what is underlined by the Pearson and Covariance coefficients, both close to zero (Table 1): the two indices are not interchangeable but return a different level of risk for the same task.

Table 1. Average, Correlation and Covariance of the REBA RULA indices

Index	Average ± SD	Correlation	Covariance
REBA	7.01±1.59	0.10	0.13
RULA	6.27±0.78		

It is evident that REBA index, which takes into consideration the whole body, returns for the type of work of pruning on ropes average values in its risk scale, while RULA, which considers only the upper part of the body, returns a risk value that is almost always high (6.27 on a scale of 7- Fig. 3, 4).

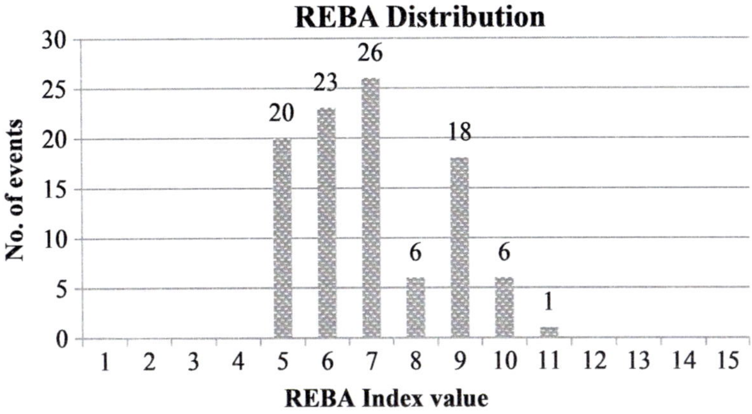

Fig. 3. Distribution of risk classes calculated for the REBA index.

In order to have a subjective term of comparison on which of the two indices is more adequate, a third sample was introduced, i.e., all the images were evaluated by 6 expert operators who gave a judgment by assigning a value from 1 to 11 according to their personal judgment of ergonomics of postures (Fig. 5). This sample is considered a useful judgment of comparison, as it is expressed by people who are experts in the type of work.

To compare the three data sets, they were normalized, and covariance was calculated between the REBA, RULA, and values provided by the expert operators. The analysis of covariance returns a more similar value between RULA index, and the values assigned by the expert operators, although the risk level is more similar to the REBA index values (Table 2).

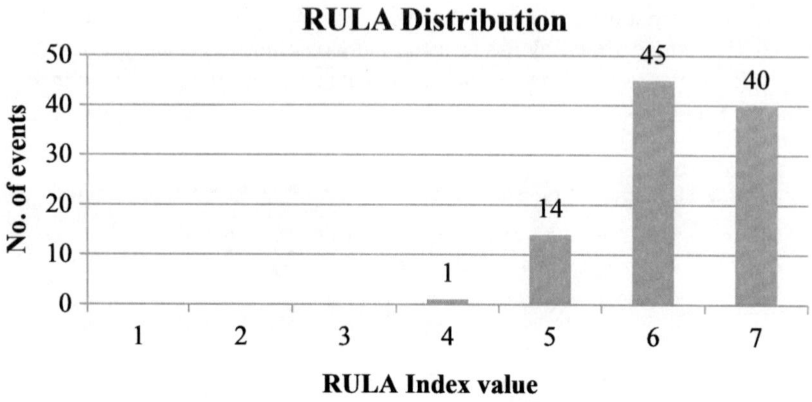

Fig. 4. Distribution of risk classes calculated for the RULA index.

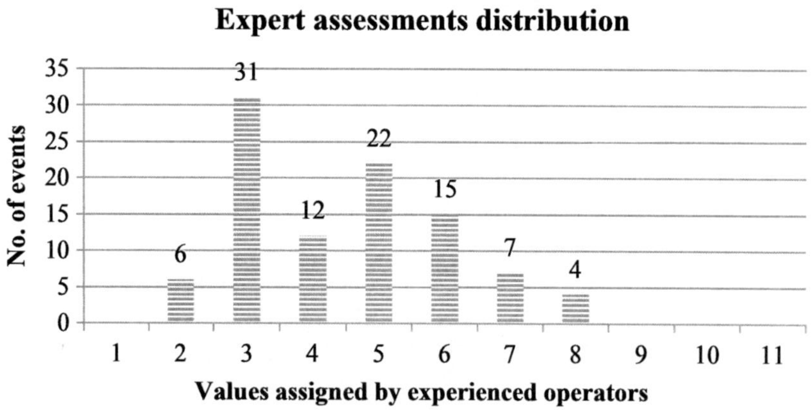

Fig. 5. Distribution of risk classes expressed by experienced operators.

Table 2. The covariance between REBA, RULA values and those provided by experienced operators.

Covariance REBA-expert operators	Covariance RULA-expert operators
0.20	0.39

4 Conclusions

Image analysis has proved to be extremely useful to analyze data on the postures of a worker on a rope intent on pruning. Thanks to this system, it was possible to collect a good amount of postures and analyze them later thanks to dedicated software that allows to detect with greater precision the angles and postures of the operator under examination, which otherwise would be analyzed quickly and in the field [18].

The results confirm that the two indices REBA and RULA were not designed to analyze rope work. In fact, the risk values are neither similar nor covariant. For a more conservative evaluation, it would be suggested to use RULA, which returns a higher value and suggests to look for solutions to improve the ergonomics of the operator's work.

On the other hand, the distribution of values within quartiles is much more similar between REBA and experienced operators (Fig. 6).

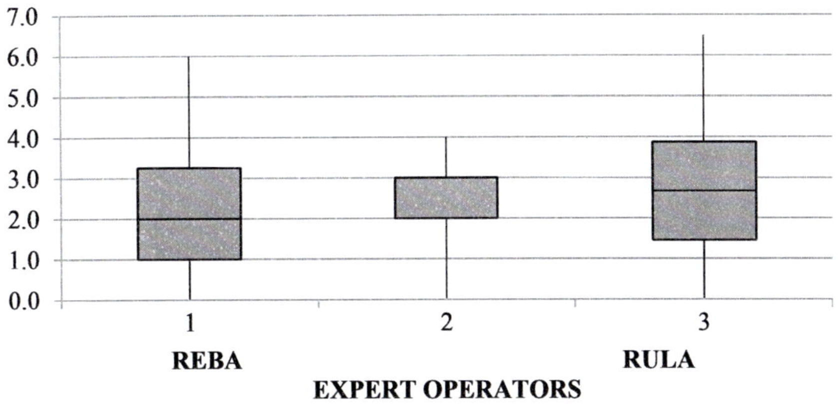

Fig. 6. Box Plot values REBA, RULA, experienced operators

In conclusion it is desirable to have a specific index for ropeway workers, taking into account working postures and equipment, which is necessarily influenced, even positively, by the harness and positioning ropes this work lays the groundwork for the development of such an index of risks for ropeway workers.

References

1. Chen, H., Luo, X.: Severity prediction models of falling risk for workers at height. Procedia Eng. **164**, 439–445 (2016). https://doi.org/10.1016/J.PROENG.2016.11.642
2. Doyle, Y.G., Conroy, R.M.: Prevention of timber felling and chainsaw-related accidents in the Republic of Ireland. Accid. Anal. Prev. **21**, 529–534 (1989). https://doi.org/10.1016/0001-4575(89)90068-7
3. Kane, B.: Forces generated in rigging trees with single and co-dominant stems. Urban For. Urban Green. **24**, 14–18 (2017). https://doi.org/10.1016/j.ufug.2017.03.016
4. Kane, B., Brena, S., Autio, W.: Forces and stresses generated during rigging operations. Arboric. Urban For. **35**, 68–74 (2009)
5. Lim, J., Kane, B., Bloniarz, D.: Arboriculture safety standards: consistent trends. Urban For. Urban Green. **53** (2020). https://doi.org/10.1016/j.ufug.2020.126736

6. Proto, A.R., et al.: A survey on occupational injuries in works on trees in italy. Procedia Soc. Behav. Sci. **223**, 435–441 (2016). https://doi.org/10.1016/j.sbspro.2016.05.266
7. Kane, B.: Loads borne by a tie-in point during ascents and descents on a basal-anchored stationary rope system. Urban For. Urban Green. **51**, 126687 (2020). https://doi.org/10.1016/j.ufug.2020.126687
8. Kane, B.: Loads borne by a tie-in point (TIP) during arboricultural climbing operations. Urban For. Urban Green. **49**, 126625 (2020). https://doi.org/10.1016/j.ufug.2020.126625
9. Kane, B.: Loading experienced by a tie-in point during ascents. Urban For. Urban Green. **34**, 78–84 (2018). https://doi.org/10.1016/j.ufug.2018.06.001
10. Kane, B.: Breaking load of hitches and ropes used in rigging. Arboric. Urban For. **38**, 1–5 (2012)
11. Kane, B.: Frictional properties of arborist rigging blocks. Urban For. Urban Green. **42**, 31–38 (2019). https://doi.org/10.1016/j.ufug.2019.05.003
12. Kane, B.: Compatibility of toothed ascenders with arborist climbing ropes. Arboric. Urban For. **37**, 180–185 (2011)
13. Mazzocchi, F., et al.: An overview of risk assessment for tree climber arborists. Contemp. Eng. Sci. **8**, 1171–1177 (2015). https://doi.org/10.12988/ces.2015.56175
14. Bianchini, L., Cecchini, M., Gallo, P., Biocca, M.: A survey on rope-based ascending techniques and materials of professional arborists in Italy. Environ. Sci. Proc. **3**, 23 (2020). https://doi.org/10.3390/iecf2020-07881
15. McAtamney, L., Hignett, S.: Rapid entire body assessment (REBA). Appl. Ergon. **31**, 201–205 (2000). https://doi.org/10.1016/S0003-6870(99)00039-3
16. Matamney, L., Corlett, N.: RULA: a survey method for the investigation of work-related upper limb disorders. Appl. Ergon. **24**, 91–99 (1993). https://doi.org/10.1016/0003-6870(93)90080-S
17. Jepson, J.: The tree climber's companion: a reference and training manual for professional tree climbers: Amazon.com: Books (2002)
18. Corlet, E.X., Bishop, R.P.: A technique for assessing postural discomfort. Ergonomics **19**, 175–182 (1976). https://doi.org/10.1080/00140137608931530

Biomechanical Risk for the Hand-Arm System During Work Activities on Peach Fruit Tree

Massimo Cecchini(✉), Andrea Colantoni, Danilo Monarca, Filippo Cossio, Leonardo Bianchini, and Simone Riccioni

Department of Agriculture and Forest Sciences, University of Tuscia, Viterbo, Italy
cecchini@unitus.it

Abstract. Biomechanical disorders of the upper limb represent an important social and economic burden in Italy. Every year a significant number of occupational diseases are reported, the annual average in the period 2015/2019 is 59,552, of which 65%, or 38,704, are attributable to occupational diseases of the osteomuscular system and connective tissue; if we then look at the occupational sector, about 22.8% are attributable to the agricultural sector. These significantly high figures result in a total economic burden estimated at 104,844 million euros, or 6.3% of the national GDP. The result is an expenditure that burdens the coffers of the State, causing a "change of course of funds" that, instead of being used for prevention and improvement of the quality of working life, are unfortunately used for the treatment of accidents and occupational diseases. These data show that there is a greater need to carry out a risk assessment, in accordance with current legislation, in order to identify the exact level of exposure of workers and thus outline appropriate prevention and protection measures that significantly reduce the risk of accidents but above all the occurrence of occupational diseases. The aim of this research was to assess the risk of exposure to biomechanical overload of the upper limbs for workers employed in the manual thinning of peach fruits, a seasonal operation necessary to control the productive load of the crop.

Keywords: Farms · Safety · Musculoskeletal disorders · Risk assessment · Ocra index

1 Introduction

Currently, the most commonly reported occupational diseases as a result of exposure to occupational hazards are musculoskeletal system disorders [1–3]. Exposure to the risk of biomechanical overload, especially of the upper limbs (WMSDs), is one of the possible causes of increased probability of occurrence of such occupational diseases. The disorders associated with excessive exposure to different kinds of risk sources, mainly affect the tendons, joints, and nerves of the anatomical parts of the hands, wrists, forearms, elbows and shoulders [4]. The pathologies that can be attributed to the handling of low loads at high frequency are: carpal tunnel, epicondylitis, epitrocleitis, tendinitis of the wrists and shoulders, scapulohumeral periarthritis and trigger finger

M. Biocca et al. (Eds.): SHWA 2020, LNCE 252, pp. 19–28, 2022.
https://doi.org/10.1007/978-3-030-98092-4_3

[5]. The movements, which could lead to these effects, are normally within normal human capabilities and, if carried out in an ordinary manner, do not in themselves represent a source of risk. According to numerous epidemiological studies carried out by the National Institute for Occupational Safety and Health (NIOSH), disturbances caused by workplace-related biomechanical overload may be associated with various factors such as: physical strain related to lifting loads, postures assumed during lifting and handling operations (e.g. twisting and bending), exposure to both hand-arm vibration (HAV) and whole-body vibration (WBV), exposure to unfavourable micro-climatic factors (cold, humidity, draughts, etc.). Handling of low loads at high frequency can generate more evident effects, where the work tasks are carried out in the absence of basic preventive measures, such as sufficient and distributed breaks in the shift, and with high frequency of acts per minute, insufficient or absent recovery time, control of work rhythms. A correct risk assessment [6–11] allows to significantly reduce the risk level, taking into account that the main ergonomic recommendations are provided by the technical standard UNI EN ISO 11228-3 [12] concerning the handling of low loads at high frequency. This technical standard is also the main reference, together with the application guide ISO/TR 12295 [13], which offers the preeminent methodology for risk assessment on the health of the exposed working population. In the agricultural and agro-industrial sector, numerous work activities that involve exposure of the working population to risks from biomechanical overload of the upper limbs are present. The pome fruit and stone fruit cultivation sector involves work tasks, such as pruning, thinning and harvesting, which expose the operators involved to risk factors from biomechanical overload of the upper limbs.

2 Materials and Methods

The OCRA mini-checklist methodology [14] was used to assess the exposure to risk from biomechanical overload of the upper limbs of workers involved in fruit thinning operations in a peach orchard located in Tuscia Viterbese (Latium, Italy). The orchard, in its eighth year of production, is structured with an espalier system on an area of about 25 ha and the operators who participated in the thinning operations were 12. All operators were right-handed. In order to be able to carry out this assessment, an interview of the operators involved was firstly carried out to obtain the information necessary for the correct application of the OCRA methodology, so as to understand in detail: the work organisation, the number of workers employed for this task, the working times and hours, the experience of the staff employed, the distribution of breaks and their duration. Subsequently, observations of the workers employed in the thinning process were carried out in order to establish the correct observation time and the existence or otherwise of cycles. Once the correct observation time had been defined, a series of three video clips were recorded for each operator, taking care to frame in a clear and distinguishable manner the postures assumed by the torso and above all the anatomical structures comprising the upper limbs. In order to limit errors in the assessment and to capture as many postural details as possible, the films were taken in front of the operator from three different angles. The acquired videos were then analysed on a PC with the aid of the Kinovea software (San Francisco, USA), which not only allowed the frames to be slowed down as necessary to count the number of technical actions performed in the cycles, but also to perform summary calculations of the postural angles assumed by the column, wrists, elbows, shoulders and hand. This type of task involves the use of both limbs in a non-symmetrical manner, which is why it was necessary to make separate assessments for each limb. The application of the OCRA mini-checklist requires the insertion, on a special excel spreadsheet set up by the Epm International Ergonomics School (Milan, Italy) [15], of a series of data to be introduced in appropriate highlighted areas of the mini-checklist itself. The structure and functioning of the applied methodology is then illustrated.

2.1 Company Data, Total Exposed Workers and Identification of Repetitive Tasks

The first part of the document allows you to enter the various company data, the presence or absence of homogeneous groups and the number of employees in the repetitive task analysed divided by gender (Fig. 1).

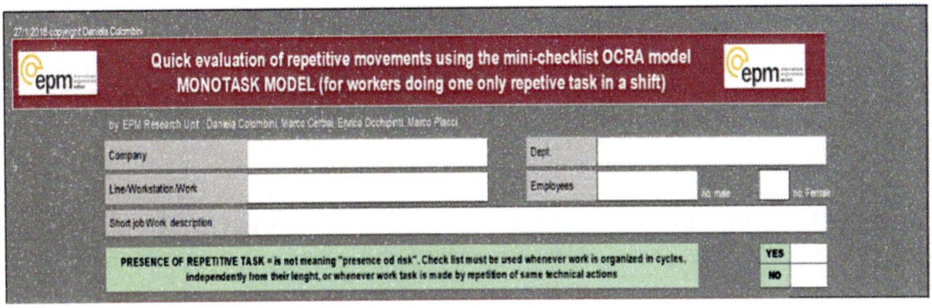

Fig. 1. Company description with repetitive tasks carried out by the worker(s) employed in the repetitive task.

2.2 Net Working Time Calculation

In order to obtain this value (Fig. 2), the mini checklist requires a series of information such as:

- Identification of tasks that are inclusive of duration, that do not involve the repetition of the same work gestures for more than half of the time and are therefore not classifiable as repetitive (e.g. procurement, preparation, cleaning, transport and other occasional tasks carried out for no more than one hour in the work shift);
- Number of breaks present (excluding the meal break, which must be counted separately), as long as they are no less than 8 min long and are not distributed either at the last hour of the shift or at the hour before the meal break;
- Duration of meal break, if paid time;
- Number of interruptions of activities of more than 30 min as a canteen break outside working hours or transfers;
- Presence or absence of cycles within the shift.

The Excel spreadsheet, once the required data has been entered, will automatically provide the values of the recovery and duration multipliers according to the contents of the tables contained within ISO/TR 12295.

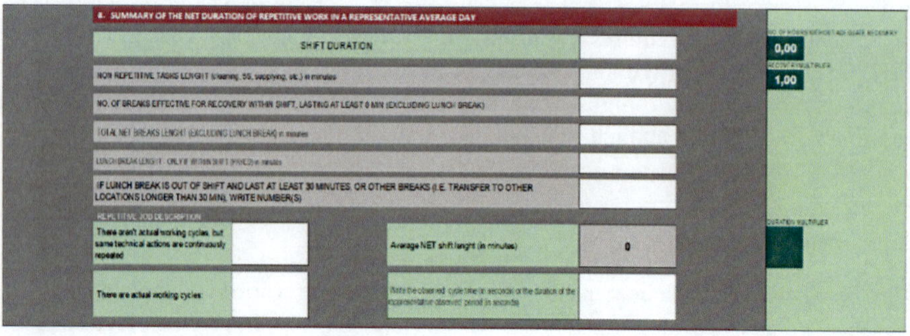

Fig. 2. Part of the mini-checklist to identify the representative day and the net duration of repetitive tasks.

2.3 Evaluation of Frequency and Technical Actions

Before starting the assessment of the various risk factors, it is necessary to indicate which limb the specific checklist refers to; the options available in the mini checklist are three: right limb, left limb, or bilateral if the engagement of both limbs is equivalent or symmetrical. In order to carry out the assessment correctly, it is necessary to calculate the frequency, i.e. the number of technical actions carried out in a given (observed or representative) time interval or cycle.

$$Frequency = No.\,of\,actions * 60/Time\ (s) \tag{1}$$

In the case of repeated cycles, the technical actions will be observed in a single representative cycle, whereas in the case of repetitive work where the same gestures are repeated most of the time, the representative period may be set at 1 to 2 min. By technical action we mean the set of elementary manual actions involving muscular-tendon activity, necessary for the completion of work operations. The choice of technical actions to be considered for their quantification, depends on certain criteria that each gesture must possess are indicated in UNI ISO 11228-3. Knowledge of the technical actions performed by each observed limb is therefore fundamental. As shown in Fig. 3, in the OCRA mini checklist the calculation of frequency is simplified by providing only two options of choice A, C plus an intermediate one defined between A and C, which to simplify the explanation we will call option or scenario B, which is used when the previous options are not verified. Option A must be marked in the case of very low frequencies, which occur when the number of actions observed does not exceed thirty actions per minute (less than one every 2 s). Scenario C should be marked when the frequency is high because the actions performed during the observations are very fast (more than one actions per second). Scenario B, on the other hand, is marked alternatively when neither option A nor C are fulfilled, or when the object is held statically for a large part of the time.

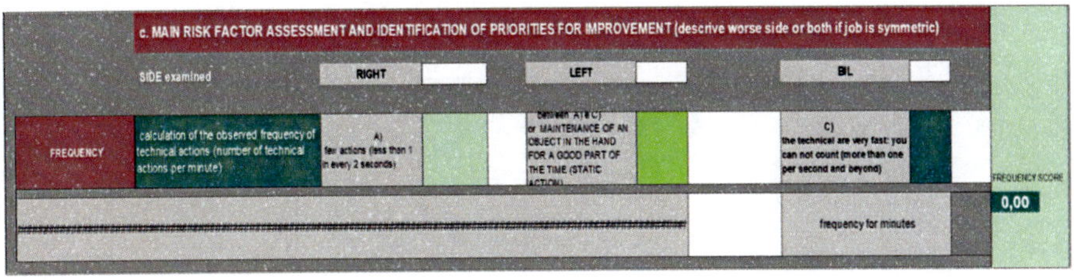

Fig. 3. Part of the checklist devoted to the choice of the examined limb and to the definition of the frequency by applying the choice either of the predetermined scenarios or by indicating the number of technical actions detected in the cycle observation or in the representative period.

For each scenario the spreadsheet automatically assigns a score according to the criteria indicated in the technical standard UNI EN ISO 11228-3, ISO/TR 12295: option A corresponds to a score of 1, option C corresponds to a score of 9 while B corresponds to a score of 5. In order to make the result more precise, especially in intermediate cases, the spreadsheet also allows you to enter the number of technical actions counted within the cycle or representative time unit.

2.4 Assessment of Incongruous Postures

In addition to the technical actions counted during the performance of repetitive tasks, the postures assumed by the hands, wrists, elbows and shoulders must be observed and described over the same period of time, as they are considered elements that contribute significantly to the increased risk of musculoskeletal disorders. In the classic OCRA checklist, a posture is defined as incongruous when the anatomical part of the arms analysed is forced to assume angular excursions of more than 50% of the maximum foreseen. For each of the four anatomical segments observed, increasing multipliers are assigned, the more the incongruous posture is maintained over time within the cycle (1/10, 1/3, 2/3, 3/3). Incongruous postures with the relative index are indicated both in UNI EN ISO 11228-3, in ISO/TR 12295, and in the updated procedures and application criteria of the OCRA checklist.

With regard to the shoulder, it must be verified that the arms:

- Are not held or lifted at shoulder height for more than 2 actions per minute or for more than 10% of the observed cycle;
- Are not maintained between 45° and 80° for more than 10 actions per minute or more than 1/3 of the cycle duration.

In the case of the elbow, it is necessary to check for bending and extension (flexion-extension) or rotation of the forearm (prone-supination) of $\geq 60°$. For the wrist it is necessary to check whether or not there are any deviations of the wrist from its axis in terms of flexion-extension ($\geq 45°$) or along the ulnar radial axis (≥ 15–$20°$). Another thing to check is the presence or not of sockets and their type: hook or palmar. Among the values obtained in the 4 blocks that make up the area of incongruous postures, the highest one must be taken as reference. With respect to the above, the OCRA mini checklist provides for simplifications as shown in Fig. 4.

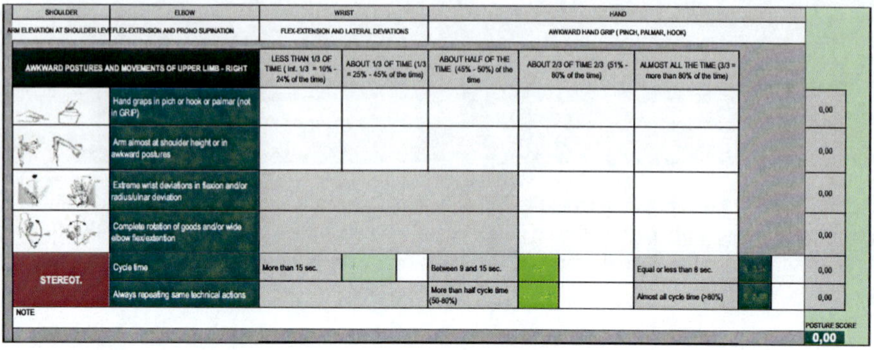

Fig. 4. Scheme for the assessment of incongruous postures with the calculation of the relative score.

The images of the main incongruous postures related to the anatomical parts of the upper limbs are always shown; however, for the part of the deviations of the wrist and arm, it is possible to highlight their presence only if they are present for more than 2/3 of the time. As with the method of the classic OCRA checklist, even in the mini checklist it is possible to indicate whether or not stereotypy is present, i.e. the presence of repetition of identical gestures repeated in the observation time (from 15 s to less than 8 s) or in the cycle; the presence of stereotypy involves the introduction of an index that is added to the highest index found in the analysis of the four blocks of incongruous postures of the upper limbs. Stereotypy may be present regardless of the presence or absence of incongruous postures.

2.5 Strength Risk Factors

Every technical action performed involves the use of a certain force. The quantification of force is difficult because it can take on different values depending on the characteristics of the person performing the biomechanical task. In order to make a representative assessment, it is necessary to interview workers involved in repetitive tasks and to try to translate the information obtained into a numerical quantity. The scale used to do this is divided into 10 points and is the Borg scale, which aims to quantify the subjective muscular effort perceived during the execution of repetitive tasks. The value of the Borg scale, obtained after the interview, is necessary for the attribution of the relative multiplier, whose entity to be attributed, depends on the careful observation of repetitive tasks and their duration in time. The use of the OCRA mini-checklist, as indicated in Fig. 5, requires the ticking of some boxes, according to the function:

- The presence or absence of moderate, strong or very strong force if also present as peaks;
- Of the time span of application of the perceived force.

The values, of the relevant multiplier indices, are attributed directly by the spreadsheet; this methodology considers perceived light values as negligible.

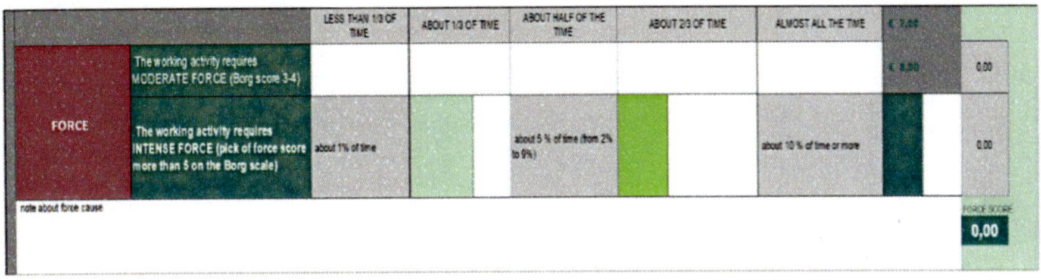

Fig. 5. Scoring scheme following the force detected.

2.6 Complementary Risk Factors

In order to increase the effectiveness and truthfulness of the assessment of the biomechanical overload of the upper limbs, the methodology also envisages taking into account other risk factors defined as "Complementary" which, together with repeated movements, contribute significantly to increasing the probability of the occurrence of pathologies affecting the musculoskeletal system. These multiple factors can be grouped into three main categories:

- Mechanical factors (generated by the use of tools, etc.);
- Environmental factors (temperature, etc.);
- Organisational factors (pace imposed by the machine, etc.).

Each factor has a certain index assigned to it. In the application of the OCRA mini-checklist, as shown in Fig. 6, these factors are restricted to a limited number of cases and, in any case, each one can only be considered if it occurs with a certain timing or frequency depending on the factor analysed. Even if the presence of each factor leads to the addition of an index equal to 2, the presence of several factors at the same time will still lead to the attribution of a maximum value of 3.

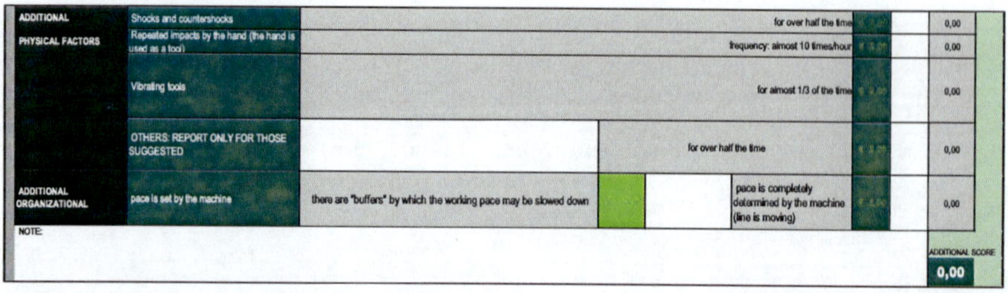

Fig. 6. A number of complementary risk factors are considered in the OCRA mini-checklist, which are useful for calculating the relevant score.

Table 1. Assignment of the risk level according to the score obtained from the compilation of the mini OCRA checklist.

Scores			
OCRA index	Checklist OCRA	Exposure level	
≤ 2.2	≤ 7.5	Accepted	Green
2.3–3.5	7.6–11.0	Very low risk	Yellow
3.6–4.5	11.1–14.0	Low risk	Orange
4.6–9.0	14.1–22.5	Medium	Red
≥ 9.1	≥ 22.6	High	Violet

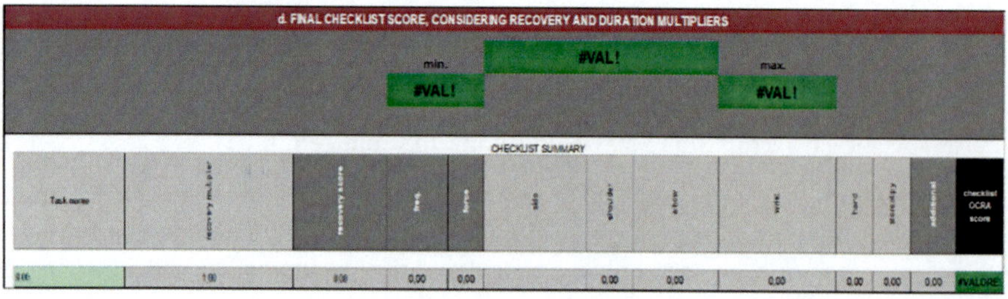

Fig. 7. Final score summary with a summary of the levels obtained by the various multipliers.

2.7 The Final Score of the OCRA Mini-Checklist

The sum of the scores of the various risk factors (frequency, strength, incongruous postures, complementary factors), appropriately multiplied by the corrector due to the recovery and the duration, allows to obtain the final value (Fig. 7). In order to attribute the risk range it is necessary to compare the values obtained with the limits foreseen by the OCRA checklist as indicated in Table 1; the colour obtained is already in itself an indicator of the risk range.

3 Results

The results of the risk assessment, corresponding to the final scores obtained for the right and left limb during peach fruits thinning, are shown in Table 2.

Table 2. Average value of the values obtained from the assessments with the standard deviation.

Average risk index value			
Right limb	Standard deviation	Left limb	Standard deviation
16.82	1.42	12.02	1.51

The scores obtained were appropriately compared with Table 1; for all the assessments the observed task is classified as medium risk. The values found are independent by the strength multipliers and complementary factors, which, with a value of 0, do not contribute to an increasing in risk indices. On the other hand, the factors that make a significant contribution to the risk scores are mainly due to the presence:

- Of incongruous hand and arm postures;
- Of insufficient recovery time.

As far as hands are concerned, the attribution of indexes is linked to the maintenance, even if for about 1/3 of the time in most of the cases analysed, of the pinch grip, a postural attitude necessary to grasp the chosen fruit and remove it from the branch.

The incongruous posture assumed in most of the cases analysed, with the arms placed at a height above the shoulder, occurs instead for less than 1/3 of the time, and is due to a postural defect present in most of the operators, who, although trained, informed and instructed, have succumbed to the unconscious automatism acquired with experience.

The frequency-dependent factor is particularly important, especially for the right limb, due to the high number of technical actions performed in the unit of time.

4 Conclusions

The application of the OCRA mini-checklist showed the presence of a medium risk exposure for operators employed in thinning operations. The values obtained are highly influenced by the frequency, i.e. the number of technical actions carried out during the observation time or cycle. These values tend to be higher the more skilled and "automatic" the workers are in the repetitive operation.

It can be concluded that, by their very nature, thinning operations involve the use of the upper limbs by the operators, however, it is possible to intervene in order to reduce the effects produced. The frequency factor can be reduced by acting on the organisation of work, such as the rescheduling of breaks within shifts. Possible interventions to correct the postures that contribute to the increasing in risk index can be carried out through frequent education and training campaigns, including on-the-job training, exploiting alert technologies, and using classic technical aids such as facilitating carts for the removal of fruit falling at heights above the operators' shoulders.

References

1. Kuta, L., Cież, J., Młotek, M.: Musculoskeletal load assessment of farmers during selected agricultural works. In: 6th International Conference on Applied Human Factors and Ergonomics and the Affiliated Conferences, AHFE (2015)
2. Holmberg, S., Stiernström, E.L., Thelin, A., Svärdsudd, K.: Musculoskeletal symptoms among farmers and non-farmers: a population-based study. Int. J. Occup. Environ. Health (2013)
3. López-Aragóna, L., López-Liria, R., Callejón-Ferreb, A.J., Pérez-Alonso, J.: Musculoskeletal disorders of agricultural workers in the greenhouses of Almería (Southeast Spain). Saf. Sci. **109**, 219–235 (2018)
4. ACGIH: Threshold limit value for chemical substances and physical agents and biological exposure indices. American Conference of Governmental Industrial Hygienists (2001)
5. INAIL: Relazione annuale del Presidente (2019)
6. Occhipinti, E.: Attività lavorative comportanti movimenti ripetitivi a carico degli arti superiori: confronto di diversi metodi di valutazione del rischio. G. Ital. Med. Lav. Erg. **30** (3), 32–38 (2008)
7. Occhipinti, E., Ghibaudo, L.: development of a risk assessment procedure for upper limbs based on combined use of eaws 4th section and ocra high precision checklist. In: Bagnara, S., Tartaglia, R., Albolino, S., Alexander, T., Fujita, Y. (eds.) Proceedings of the 20th Congress of the International Ergonomics Association (IEA 2018). IEA 2018. Advances in Intelligent Systems and Computing, vol. 820. Springer, Cham (2019). https://doi.org/10.1007/978-3-319-96083-8_988
8. CEN. EN: Safety of machinery - Human physical performance -: Part 5: Risk Assessment for repetitive handling at high frequency, 1005–5 (2006)
9. Occhipinti, E., Colombini, D., Occhipinti, M.: Metodo Ocra: messa a punto di una nuova procedura per l'analisi di compiti multipli con rotazioni infrequenti. Med. Lav. **99**, 234–241 (2008)
10. Colombini, D., Occhipinti, E., Fanti, M.: Il metodo OCRA per l'analisi e la prevenzione del rischio da movimenti ripetuti. Collana Salute e lavoro, Franco Angeli Editore (2005)
11. Colombini, D., Occhipinti, E., Cerbai, M.: Appunti di metodo: aggiornamento della specificazione di procedure e criteri nella applicazione della Checklist OCRA. Med. Lav. **102**, 00 (2011)
12. UNI EN ISO, Ergonomics - Manual handling – Handling of low loads at high frequency, 11228–3 (2007)
13. ISO/TR 12295:2014. Ergonomia — Documento per l'applicazione delle norme ISO alla movimentazione manuale di carichi (ISO 11228-1, ISO 11228-2 e ISO 112283) e la valutazione delle posture di lavoro statiche (ISO 11226)
14. Colombini, D., Occhipinti, E.: La valutazione del rischio da sovraccarico biomeccanico degli arti superiori con strumenti semplificati: la minichecklist OCRA. Contenuti, campo applicativo e validazione, Med. Lav. **102**(1), 000 (2011)
15. EPM International Ergonomics School http://www.epmresearch.org/. Accessed 24 Aug 2021

Health and Safety Risks in Hop-Picking Activities: An Analysis of the State of the Art

Giovanni Matranga[1(✉)], Niccolò Pampuro[1], Federica Caffaro[2], Lucia Vigoroso[1], and Eugenio Cavallo[1]

[1] Institute of Sciences and Technologies for Sustainable Energy and Mobility (STEMS), National Research Council of Italy (CNR), Strada delle cacce, 73, 10135 Turin, Italy
giovanni.matranga@stems.cnr.it

[2] Department of Education, University of Roma Tre, Via del Castro Pretorio, 20 – 00185 Rome, Italy

Abstract. In the last decade, hop cultivation has seen greater interest in small and medium-sized farms due to the rising importance of craft beer micro-breweries requiring local raw materials. However, high investments are needed for mechanized harvesters. Small-scale farms cannot face these costs; thus, manual harvesting of hop cones is frequently adopted. The study investigated the risks for the health and safety of operators during manual and mechanized harvesting, also examining musculoskeletal disorders the workers can incur. A literature review and secondary data analysis investigated risk factors associated with the manual and mechanized harvesting process. Furthermore, the operator's workload related to the hop cones picking time was analyzed based on international standards. The results showed that operators face several critical issues during the different tasks of the manual harvesting of hops (e.g. cutting and removing the bines from the field and the manual cones picking). Some of the operations are carried out at height with the risk of falling and assuming bad postures that can pose risks to the operator's health. Furthermore, the study points out major health concerns for the workers because of the repetitiveness of the operations they have to carry out. The mechanization of harvesting involves a significant increase in efficiency by reducing the number of operators involved. This results in an improvement in operators' health conditions. However, the harvesting equipment can represent a safety risk for the operators because of their interaction with the moving parts of the machinery.

Keywords: Hop · Mechanized harvesting · Manual picking · Musculoskeletal disorders · Occupational safety

1 Introduction

In the last two decades, the beer industry has experienced substantial market worldwide growth, recording +36% [1] globally in beer productions. This trend caused major requests in hop production and consequently incremented cultivated areas, especially in Centre-East Europe and the US, where hop cultivation is historically grounded. The Italian brewing industry is recording significant growth (+39%) due to a new trend in

M. Biocca et al. (Eds.): SHWA 2020, LNCE 252, pp. 29–37, 2022.
https://doi.org/10.1007/978-3-030-98092-4_4

the brewing compartment: the craft beer industry [2]. By its legal definition, in Italy, craft beer is produced by small breweries, or microbreweries, able to produce volumes not exceeding 200.000 hectoliters per year. National craft brewing is still being addicted to imported hop: only 1.2% of hop implemented in the Italian brew industry is local, corresponding to about 41 ha of cultivated areas, mainly distributed in the north (ISTAT 2019). In regions where hop cultivation is not traditionally implemented, small-scale farms are commonly the first to tempt new cultivations. For example, the hop cultivated area for each farm in Italy amounts to around 0, 4 ha [3]. However, small-scale cultivation does not allow to justify investments to realize mechanized systems all along with the growing and harvesting cycles. Therefore, small-scale farms usually cope manually with almost the entire production cycle, especially harvest operations.

Hop harvesting time is the most intense of the entire production cycle. A proper harvest has a positive impact on the quality of the final product [4]. Hop cones must be collected, cleaned from the vegetal residue, and dried as fast as possible to ensure high-quality and sanitary standards [5]. Manual hop-picking operations require long times, repetitive gestures, and assuming awkward postures that can expose workers to health risks, especially upper limbs overload.

Different previous studies showed that the agricultural sector causes many health threats due to ergonomic and accidental risks [6–8]. The present study investigates the risks for the health and safety of operators during manual harvesting, examining the possible risks of work-related musculoskeletal disorders (WMSDs) due to static postures, low loads at high frequencies related to manual harvesting of hop florescence, comparing them to the possible health and safety risks related to the worker-machine interaction (hop-picking machines, specifically) in mechanized harvesting. International standards most suitable for risk assessment in hop-picking activities were also identified. The study intends to fill the gap in the literature around this theme, defining state-of-the-art occupational risks in hop-picking activities and putting the ground for future research, with the final aim to identify areas for possible interventions in terms of machinery re-design and training to promote operators' health and safety.

2 Methods

The study was articulated in three parts. The first step consisted of reviewing the literature about each hop harvest method, comparing manual and mechanized operations. Literature about manual and mechanized hop-picking solutions has been revised through a targeted search on scientific journals databases, such as Science Direct, Scopus, and Web of Science. A keyword-based search was used to identify articles, using the following keywords: hop harvest* OR hop-picking AND manual activit* OR manual operation* OR mechanized OR machine*. Specialized magazines and corporate websites were consulted as well. In the second step, an analysis of potential safety risks during all harvesting processes was made to highlight the critical issues. Afterwards, literature on risk assessment and prevention of work-related health and safety risks was reviewed, especially regarding static and awkward postures, low loads manual handling, and repetitive operations. Eventually, because there is no specific literature on the health risks related to hop harvest, an analysis of existing international standards assessing biomechanical overload and risks in human–machine interaction was performed.

3 Results

3.1 Hop Harvesting Methods

The Hop harvesting period corresponds with the inflorescence ripening, when a rich complex of resins and essential oils accumulate inside flower glands. As shown in studies, both in central Europe [4] and North America [9], ripening is gradual and generally begins in late July and ends in early September. In these regions, mature cones are collected from mid-August to September. Accurate harvest timing is crucial to get a high-quality product since it significantly influences the final aroma [4].

Hop harvest can be realized through different solutions, partly depending on the territorial context:

1. Infield manual picking of hop cones. The inflorescences are directly collected from bines by workers through ladders or trailers with elevators [10].
2. Bines cutting and loading on trailers, and cones picking is carried out later by operators. Bines cutting can be arranged manually (Fig. 1) or through specific cutting mechanized systems [11].
3. Cutting and loading mechanized of the bines, transport in shed and load in the hop-picking machines [12].

Manual hop harvesting is the low-tech solution; it consists of manual cutting of bines and manual picking of cones, directly in the field or operating out of the field, from previously harvested bines. The bines manual harvest involves cutting the stem below and then above using shears or sickles, implying elevators and adopting, in some conditions, awkward postures. The time–frequency of such postures depends on the number of workers and harvested area. The hop picking consists of manually collecting cones, thus it can be operated directly in the field through ladders or elevators or collected out of the field, from harvested bines.

The mechanized hop harvesting consists firstly of a bine cutter to collect the stems loaded on a trailer and secondly in a hop picker. There are many types of bine cutter and loader; most of them are devices adjustable into the tractor via the Power Take-Off (PTO) [13]. The hop picking machine sepa-rates hop cones from the stem and the cleaning process to eliminate residues. Its primary energy source is electricity, and typically is located in a shed area. The machine size, the workload and the number of workers depend on the number of bines to process, hence by the harvested area and the hop variety [14].

3.2 Critical Issues in the Hop Harvest

Manual harvesting solutions are associated with some critical issues that could pose health risks: hop cones collection implies working at height, through the use of ladders or elevators, assuming static and dynamic postures by the bines cutting and loading efforts, and operating at high time frequencies for an extended period, due to the amount of handwork [15].

On the other hand, mechanized hop collecting provides more harvest efficiency and reduces manual efforts and their negative consequences for the operator. Nonetheless, human–machine interaction could cause many safety risks if the safety provisions are not assured or observed [16].

Furthermore, few studies have been published on occupational diseases during hop harvest and processing because of hop dust contact or inhalation [17, 18]. Reeb-Whitaker and Bonauto [18] reported that respiratory work diseases in hop cultivation are 30 times greater than the rate of respiratory illnesses diagnosed in open field vegetable crop farms. Principal diagnoses associated with exposure to hop dust are asthma and other respiratory disorders related to bronchial allergy, mainly observed in harvest activities and post-harvest processing.

Fig. 1. Hop harvest activities in northern Italy: manual bines cutting and loading into a trailer.

3.3 Assessment and Prevention of Work-Related Health and Safety Risks

International standardization bodies have substantially contributed to promoting work-related health by providing standards to control and prevent potential MSDs. Specifically, for the risk assessment related to inadequate static postures adopted throughout the harvest activities, we can refer to the ISO 11226: 2019 "Evaluation of Static Working Postures" [19], which defines as a static posture any posture that holds the position for at least 4 s, considering minor variations. This international standard establishes the limits for static working postures, examining externally applied force, body angles and time [20]. It gives recommendations concerning work modalities that guarantee sufficient physical and mental variations. The valuation procedure is used to determine the acceptability of work postures, considering body segments and joints in one or two control tiers. Tier one considers only body angles, analyzing risks due to ligaments, cartilages and intervertebral discs. The evaluation can lead to an acceptable, unrecommended activity. The assessment continues to tier two, where different variables should be considered, such as the holding time of specific body angles.

The ISO 11228-3:2009 "Handling of low loads at high frequency" [21] is another standard that could be suitable to assess the risks associated with specific activities related to manual and mechanical harvesting. It belongs to the ISO 11228 series, which also deals with lifting and transport operations and pushing and pulling of loads. ISO 11228-3:2009 includes a risk assessment concerning handling low loads in repetitive tasks, leading to biomechanical overload pathologies involving the osteoarticular, musculotendinous, and nerve-vascular structures. In addition, the upper limbs stress can cause pain and fatigue, leading to musculoskeletal disorders, reduced productivity, and deteriorated postures and movement coordination. The risk assessment procedure consists of hazard identification, which accounts for frequent repetitive movements, complex postures and combined movements, use of force, duration and insufficient recovery, vibration and impact forces, environmental conditions as the presence of noise or non-optimal climate. Each potential risk factor is evaluated by a checklist, which allows for the classification of the risk by the three-zone model ("no risk", "very low risk" and "risk") and determines the consequences to be acted upon. Additionally, the standard provides the Occupational repetitive action index (OCRA) as a method for detailed risk assessment, the result of the ratio between the number of technical actions and the number of reference technical actions for each upper limb. OCRA index was designed to imply various risk factors: body postures, handhold modalities, action duration, work organization, task complexity, recovery duration, applied force, upper limb and whole-body postures [22].

Concerning working at height risks and prevention, the case was studied from the standard operating procedure, which has referenced to ISO 45001:2018 "Occupational health and safety management systems-Requirements with guidance for use" (clause 8.1) [23]. The document's purpose is to define controls for protection from identified hazards to any modality of working at heights. Working at heights includes all the work activities for which the measure from the soles of worker's feet to the ground is over 2 m, which can be associated with severe injuries in case of falling. Therefore, in assessing the risks, the design and layout of elevated work areas should be considered, as well as the distance of potential falls, the proximity of people to unsafe areas, climate conditions, adequacy of operators training, suitability of proper work clothes and personal protective equipment (PPE).

Regarding the risk stems from human–machine interaction, the international standard ISO 12100:2010 "Safety of machinery—General principles for design—Risk assessment and risk reduction" [24] recommends a method to ensure safety for machine operators, specifying a risk assessment and risk reduction addressing to producers and users. Risk assessment is a series of logical steps that determine the machinery identification limits, consisting of all the machinery life phases, through hazard identification, risk estimation, and risk evaluation. Hazard identification corresponds to the systematic identification of reasonably foreseeable hazards, hazardous situations and hazardous events during all phases of machine use. The risk estimation shall be carried out for each hazardous situation by determining the number of elements of risk such as the severity of harm, hazard exposure, the occurrence of a dangerous event, the number of persons exposed, frequencies and duration of exposure, the relationship between exposure and possible harm and the technical and human possibilities to avoid or limit the harm. Furthermore, a risk evaluation determines if risk reduction measures are required. Consequently, risk reduction can be achieved by eliminating hazards or by reducing the severity of harm from the hazard and the probability of that harm.

Where some risks remain despite safe design measures, residual risks shall be identified by providing specific information to the users through the safety guidelines provided by the machine producer. As an example, into the operation and maintenance manual of the most common hop-picking machine "Wolf WHE 500" in the section "General Safety-Accident Prevention" several safety signs and instructions are provided, also attaching a technical drawing of workspaces (Fig. 2) around the picking

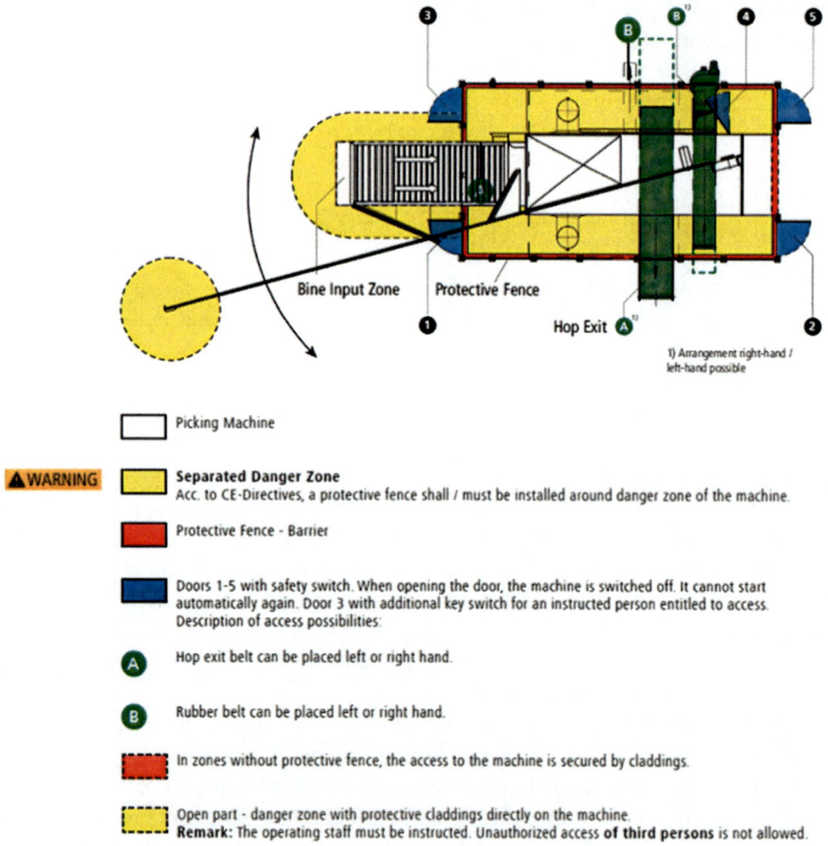

Fig. 2. Technical drawing of safety areas to observe around the hop-picking machine (Operation and Maintenance Manual of WHE 500). https://www.wolf-geisenfeld.de/en/harvesting-technology/whe-picking-machine

machine that points out the danger zones and where to install protective fences. The principal risks reported and highlighted by safety signs are gear entanglement, crushing hazard (due to suspended loads) and slipping and stumbling in the shed environment. Entering is allowed only with PPE as helmets, glasses, ear protectors, gloves, safety clothes etc.

Table 1 shows the health and safety risks and standard references for risk assessment for both manual and mechanical hop harvesting activities.

Table. 1. Summary table of health and safety risks in different hop-picking modalities and standards referred to for risk assessment.

Harvest method	Factor	Risk	Reference
Manual hop picking	Working at height	Falls from ladder or elevator	ISO 45001:2018
			SOP for Working on Height Hazards
	Repetitiveness	Biomechanical overload	ISO 11228-3 2009: "Handling of low loads at high frequency"
			The OCRA index
	Awkward postures	Pain, fatigue and MSDs	ISO 11226: 2019 "Evaluation of Static Working Postures"
			LUBA-OWAS index: posture-related load and handling
Mechanized hop picking	Human–machine interaction	Gear entanglement	ISO 12100:2010 Safety of machinery-general principles for design-Risk assessment and risk reduction
	Suspended loads	Crushing hazard	Machinery 's operation and maintenance manual
	Work shed environment	Slipping and stumbling	

4 Conclusion

This paper has highlighted different critical health and safety risks issues during hop harvesting depending on various harvest methods. Regarding manual harvest, the principal hazard is attributed to uncomfortable work conditions taking unpleasant positions, repetitiveness and duration of activities. Instead, during the mechanized harvesting operations, the risks are mainly attributed to a dangerous environment due to specific features of the machine. Although adopting barriers and devices to secure harvesting machine is crucial to reduce risks, the human factor is essential to prevent incorrect use. In [25] reported empirical evidence that operator's manuals are infrequently read, and warnings reported on them are scarcely noticed. This gives the importance of the operator's manual as a source of information for machine safe operation and maintenance [25]. In [26], additionally, emphasized the limited efficacy of the operator's manual in conveying safety information, stressing the need to revise this instrument to achieve better readability.

Despite the mechanized harvesting equipment can represent a safety risk for the operators, if the safety guidelines established by the manufacturer are observed and if the PPE is properly worn, safety hazards in the working environment are reduced. Thus, mechanized operations lead to an improvement in workers' health conditions. Moreover, although small farms have difficulties making investments to mechanize the harvesting operations, this results in advantage not only for occupational health but also for the production efficiency [27, 28].

Further work needs to be carried out to assess the gravity of risks and define recommendations and precautions required to reduce risks. Under the ISO 11226:2019 and ISO 11228-3:2009, detailed risk assessment could provide crucial information on the operators working conditions and help identify possible solutions to risk reduction. Therefore, the hop-picking machine must be designed following provisions from ISO 12100:2010, to guarantee worker safety. Moreover, the employment of high trained operators and the compliance of safety guidelines specified by the manufacturer will reduce work-related injuries.

References

1. Food and Agriculture Organization of the United Nations. FAOSTAT Statistical Database, Rome, FAO (2018)
2. Cabras, I., Higgins, D., Preece, D. (Eds.). Brewing, beer and pubs: a global perspective. Springer. (2016)
3. Gargani, E., Faggioli, F., Haegi, A.: A survey on pests and diseases of Italian Hop crops. Italus Hortus **24**(24), 1–17 (2018)
4. Bailey, B., Schönberger, C., Drexler, G., Gahr, A., Newman, R., Pöschl, M., Geiger, E.: The influence of hop harvest date on flavor stability in dry-hopped beers. Tech. Quarter. **46**(2), 1–7 (2009)
5. Lafontaine, S., Varnum, S., Roland, A., Delpech, S., Dagan, L., Vollmer, D., Kishimoto, T., Shellhammer, T.: Impact of harvest maturity on the aroma characteristics and chemistry of Cascade hops used for dry-hopping. Food Chem. **278**, 228–239 (2019)
6. Micheletti C.M., Giustetto, A., Caffaro, F., Colantoni, A., Cavallo, E., Grigolato, S.: Risk assessment for musculoskeletal disorders in forestry: a comparison between RULA and REBA in the manual feeding of a wood-chipper. Int. J. Environ. Res. Public Health, **16**(5) (2019)
7. Naeini, H.S., Karuppiah, K., Tamrin, S.B., Dala, K.: Ergonomics in agriculture: an approach in prevention of Work-related Musculoskeletal Disorders (WMSDs). J. Agric. Environ. Sci. **3**(2), 33–35 (2014)
8. Lundqvist, P., Gustafsson, B.: Accidents and accident prevention in agriculture A review of selected studies. Int. J. Ind. Ergon. **10**, 311–319 (1992)
9. Lizotte, E.: The Art and Science of Hop Harvest. Michigan State University Extention. Michigan, USA (2015)
10. Campbell, S. M., Pearson, B. J.: Harvest techniques for hops (Humulus lupulus). EDIS 2019. **6**, 4 (2019)
11. Sirrine, R.J., et al.: Sustainable hop production in the Great Lakes region. Extension Bull. **3083** (2010)

12. Kumhála, F., Kavka, M., Prošek, V.: Capacitive throughput unit applied to stationary hop picking machine. Comput. Electron. Agric. **95**, 92–97 (2013)
13. Zbynek, A.: Method and equipment for harvesting hop vines. European Patent Office (Agrozet Koncern Zemedelskeho Strojirenstvi) (1987)
14. Daniel, K., Charron, E.J.: Hop picking machine. United States Patent (Dk Fab Inc). (2017)
15. Assirelli, A., Carbone, K., Ciccoritti, R.: Mechanical hop-picking solutions in Italian cultivated areas. Sustainability (Switzerland) **12**(12), 1–13 (2020)
16. Caffaro, F., Micheletti Cremasco, M., Roccato, M., Cavallo, E.: It does not occur by chance: a mediation model of the influence of workers' characteristics, work environment factors, and near misses on agricultural machinery-related accidents. Int. J. Occup. Environ. Health **23**(1), 52–59 (2017)
17. Arcangeli, G., et al.: Allergic anaphylactic risk in farming activities: a systematic review. Int. J. Environ. Res. Public Health **17**(14), 1–20. MDPI AG (2020)
18. Reeb-Whitaker, C.K., Bonauto, D.K.: Respiratory disease associated with occupational inhalation to hop (Humulus lupulus) during harvest and processing. Ann. Allergy Asthma Immunol. **113**(5), 534–538 (2014)
19. ISO 11226: 2019 Evaluation of Static Working Postures, International Organization for Standardization, Geneva, Switzerland (2019)
20. Delleman, N.J., Dul, J.: International standards on working postures and movements ISO 11226 and EN 1005-4. Ergonomics **50**(11), 1809–1819 (2007)
21. ISO 11228-3: Handling of low loads at high frequency, International Organization for Standardization, Geneva, Switzerland (2009)
22. Lavatelli, I., Schaub, K., Caragnano, G.: Correlations in between EAWS and OCRA Index concerning the repetitive loads of the upper limbs in automobile manufacturing industries. Work, **41**(1), 4436–4444 (2012)
23. ISO 45001: Occupational health and safety management systems-Requirements with guidance for use, International Organization for Standardization, Geneva, Switzerland (2018)
24. ISO 12100: Safety of machinery—General principles for design—Risk assessment and risk reduction, International Organization for Standardization, Geneva, Switzerland (2010)
25. Tebeaux, E.: Safety warnings in tractor operation manuals, 1920–1980: manuals and warnings don't always work. J. Tech. Writ. Commun. **40**(1), 3–28 (2010)
26. Tebeaux, E.: Improving tractor safety warnings: readability is missing. J. Agric. Saf. Health **16**(3), 181–205 (2010)
27. Callahan C.: Harvest and beyond harvesters, Drying Oasts, and Pelletizing (2014)
28. Neve, R.A.: Hops. In: B.V. Springer-science+business media (Ed.), Euphytica Netherlands J. Plant Breed. (1) (1991)

Machine Milking, Animal Welfare, Sustainable Livestock Farming

Precision Livestock Farming for Mediterranean Water Buffalo: Some Applications and Opportunities from the Agridigit Project

David Meo Zilio(✉), Roberto Steri, Miriam Iacurto, Gennaro Catillo, Vittoria Barile, Antonella Chiariotti, Francesco Cenci, Maria Chiara La Mantia, and Luca Buttazzoni

Consiglio per la ricerca in agricoltura e l'analisi dell'economia agraria, Rome, Italy
david.meozilio@crea.gov.it

Abstract. AGRIDIGIT is a wide spectra project dealing with digital agriculture. It wanders from forestry, to agrometeorology, crop and livestock production for a full range sustainability of the sector. The sub-project PLF4MILK focuses on precision farming of dairy cattle and buffalo, branching out into 5 Working Packages (WPs). This paper refers to buffalo (WP4). The aim of WP4 was to increase efficiency of buffalo farms through the precision farming approach. Specific objectives are the development of breeding and feeding protocols to ensure a higher welfare in buffalo, according in particular to thermal stress and individual enteric methane (CH_4) emissions measurements.

Analysis of behavior, welfare and milk production according to the season, monitored by individual sensors and punctual weather station, was performed from July 2020 to January 2021. Ear tags measuring activity, resting, eating, rumination and ear temperature coupled with a milk meters electronic system were used. The study of enteric CH_4 emission trough individual measurements is based on remote controllable laser detectors.

Data are then integrated each other. Multivariate approach and local and/or non-parametric regressions allow to highlight patterns of (co)variance among data from different sensors.

Data will be used to set up algorithms and models suitable for buffalo. A software interface for decision supporting, will be released. The project addresses the needs for a higher efficiency in buffalo milk supply chain along with the improvement of animal welfare and environmental sustainability.

Keywords: Precision farming · Water buffalo · Heat stress · Methane emissions

M. Biocca et al. (Eds.): SHWA 2020, LNCE 252, pp. 41–50, 2022.
https://doi.org/10.1007/978-3-030-98092-4_5

1 Introduction

There are about 400K buffaloes in Italy (90% of European population) out of 204 Mil heads worldwide, mainly located in disadvantaged areas. Local or marginal spread may have induced an underestimation of its potential and a gap in technology, when compared to dairy cattle that plays a primary role in precision livestock farming applications. Precision livestock farming (PLF) is a system that is used to manage livestock based on process engineering principles and technologies [1], but it is scarcely used for buffalo farming [2]. Obviously, in order to be effective and appropriate, PLF must be tailored on the targeted species whose biological and morphological characteristics must be accurately determined [2] but most of the algorithms and models are developed for dairy cattle on intensive system.

Indeed, some applications are common for all animal species, but some others vary based on specific characteristics [3]. An accurate precision livestock management system must rely on rigorous scientific data, from raw data to processed data, passing through the analysis and interpretation up to the data use [2].

The Agridigit project is a wide ongoing project dealing with digital agriculture and declining into several sub-projects, one of which relates to livestock production. Within the zootechnical sub-project a very important role is played by the digitalization of the buffalo farm, in order to increase its efficiency and to improve welfare and sustainability as well.

Thus, the aim of this work was both to describe the application of some PLF technologies at the CREA's buffalo farm, located in Monterotondo (Rome, IT) research station of the Research Centre for Animal Production and Aquaculture and to provide preliminary results. Specific objectives were to test the ear tag accelerometers for buffalo welfare and behavior analyses and the laser methane pistols as a feasible tool to discriminate animals on the base of their methane emissions.

1.1 The Agridigit Project in Brief

The project is funded by the Italian Ministry of Agriculture, Food and Forestry Policies (MiPAAF) and covers a three-year activity. The sub-project "Digital technologies in the cow and buffalo milk supply chain (PLF4MILK)" accounts for almost 1.5 million euros for research and equipment and is divided into five work packages (WPs). The WP 4-Precision livestock farming in the buffalo species-included the study of production, behavior and welfare of lactating buffaloes according in particular to thermal stress (task 1) and the study of individual methane emissions, also for selective purposes (task 2). The rationale of WP 4 is reported in Fig. 1.

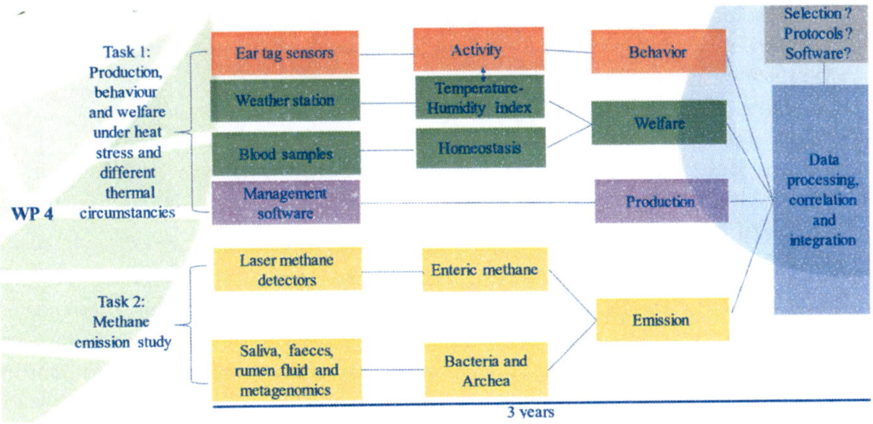

Fig. 1. The WP4 flow chart.

2 Material and Methods

The CREA's Buffalo farm of is equipped with the following technologies:

- Radio Frequency Identifier (RFID) apparatus: each animal is boled by a rumen RFID that identifies it and, through the receiver, continuously transmits data to a management software. This software makes available a great deal of individual and group data as well as production, reproduction and economic indexes.
- Milking parlor with automatic milk flow meters that allow the storage of yield and flow emission data for each experimental subject, twice a day.
- Temperature and humidity data loggers for Temperature Humidity Index (THI) calculation. THI is a bioclimatic index that combines the simultaneous effect of temperature and relative humidity and is used to characterize heat stress in livestock [4]. THI = db°F-{(0.55–0.55 RH) (db°F-58)}, where dB was dry bulb temperature and RH the relative humidity [5].
- Ear tag sensors system for recording activity, resting, rumination and temperature at ear level and to get information on health, nutrition and reproduction. Ear tag accelerometers were used to monitor 40 lactating buffaloes according to seasonal variation from July 2020 to January 2021. The used systems is commercially available and is largely applied to manage dairy cattle herds. At CREA experimental farm this technology has recently been implementing and testing on buffaloes, to improve management and operations but also to check potential correlations between activity related variables and animal behavior and welfare, in the perspective of thermal stress prevention in producing subjects.
- A demo unit laser telemeter for acquisition of remote biometric measurements, such as fattening (body condition score or BCS) and body weight. This is a smartphone dedicated tool operating in Android environment. It allows linear, angular and area calibrated measures, shared via Bluetooth.

- Cameras for video surveillance. They are used to monitor animals during their regular routine. Recorded video will be then used for validation of ear tag sensors classification of activities.
- Laser methane pistols for individual methane emissions detection:

40 animals were monitored twice a day, during the milking session (1' recording each), over three weeks per months in two different seasons (winter and summer). In addition, once per month, a 5' recording was performed by directly pointing the laser beam into the muzzle of haltered animals.

Sensors' data will be correlated with blood analyses (for oxidation and stress markers), objective evaluations (for BCS and internal body temperature), direct and indirect rumen samples (rumen liquor, nose-pharyngeal swabs, metagenomics) for validation and strengthening of results. Descriptive statistics were calculated on sensors data. Following models have been applied to the dataset:

- Local regressions by fitting a cubic spline function with twenty-two knots:

$$Y_t = a + b_1 t + b_2 t^2 + c\left(t - N_j\right)^3 \tag{1}$$

where, Y_t is sensor time dependent variables, and N_j is the position of j-th knot. This approach allows an effective exploratory visualization of the longitudinal data without preliminary assumptions about their trend [6, 7].
- Linear model by fitting a linear mixed model with the animal as a random factor:

$$Y = \mu + \text{Parity} + b(\text{DIM}) + \text{Month} + \text{hour} + \text{Month*hour} + \text{cow} + e \tag{2}$$

where DIM was days in milk. Model's least squares solution allows to reconstruct the temporal variation curve corrected for environmental effects included in the model.

3 Results and Discussion

All data obtained according to the previous scheme (Fig. 1) will be merged and correlated in order to make up breeding and feeding protocols, to release an easy-to-use application software for decision making and to drive selection of the most deserving animals as a function of environmental impact and climate change resilience.

3.1 Ear Tag Sensors

Basic statistics and relative correlations, performed on data streaming out of sensors are displayed in Tables 1 and 2.

Table. 1. Ear tags mean values.

Variable	Mean	S.D.	Range
Ruminating min/h	9.01	9.31	0–60
Eating min/h	3.53	6.72	0–60
Not active min/h	23.55	17.64	0–60
Active min/h	15.16	10.86	0–60
Highly active min/h	8.98	9.19	0–60
Temperature °C	24.36	5.96	1.93–49.75
Moving* min/h	24.14	2.84	0–60

S.D. = standard deviation; *Moving = active + highly active.

All sensors' variables are very wide ranging and characterized by high dispersion around the mean.

Table 2. Ear tags variables correlations.

Variable	Temperature	Ruminating	Eating	Not active	Active	Highly active	Moving*
Temperature	1	0.05	−0.04	0.04	−0.03	−0.05	−0.07
Ruminating	<.0001	1	−0.04	−0.53	−0.07	0.12	0.03
Eating	<.0001	<.0001	1	−0.42	−0.29	0.47	0.09
Not active	<.0001	<.0001	<.0001	1	−0.37	−0.64	−0.77
Active	<.0001	< .0001	<.0001	<.0001	1	−0.19	0.71
Highly active	<.0001	<.0001	<.0001	<.0001	<.0001	1	0.56
Moving*	<.0001	<.0001	<.0001	<.0001	<.0001	<.0001	1

*Moving = active + highly active.

Correlations are low as intensity, apart from activity, high activity and moving. There was a contrasting correlation between eating, activity and high activity. In some case a stronger correlation would be expected (*i.e.,* activity *vs* inactivity and high activity *vs* activity).

A certain correlation between eating and high activity is confirmed by the daily data and their relative trends. Estimate means for all animals over the whole observation period sorted per hour are reported in Fig. 2.

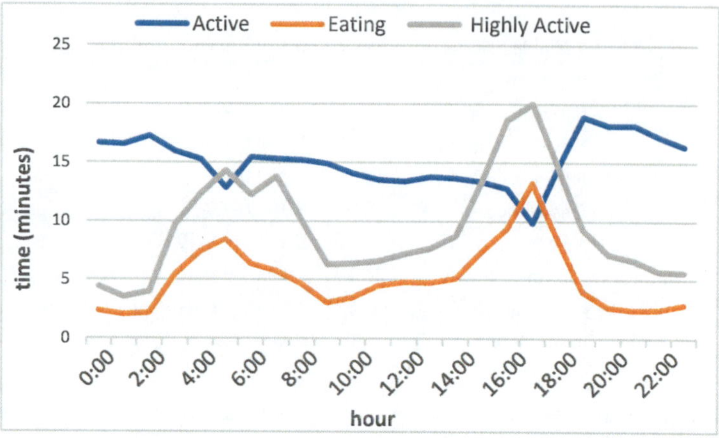

Fig. 2. Daily activity, high activity and eating of animals during the observation period.

Moreover, all considered variables seemed to follow a temperature gradient, from cold to hot season, suggesting a strong effect of climatic condition on buffalo behavior. [9] Reported changes in buffalo behavior after half an hour solar radiation exposure. [10] Stated that neuroendocrinal response to heat stress affects behavior, production and reproduction in buffalo. This trend is reported in Fig. 3, for eating (on the left) and high activity (on the right) respectively, but it was similar for all considered categories of activity. Anyway, there is high variability between animals (Fig. 4). If the proposed models fit well, this variability is basic and may be used to drive selection accordingly, to improve the specific traits as the genetic variation available in sweat glands and heat-tolerance characteristics gives a possibility for selection for heat tolerance [10].

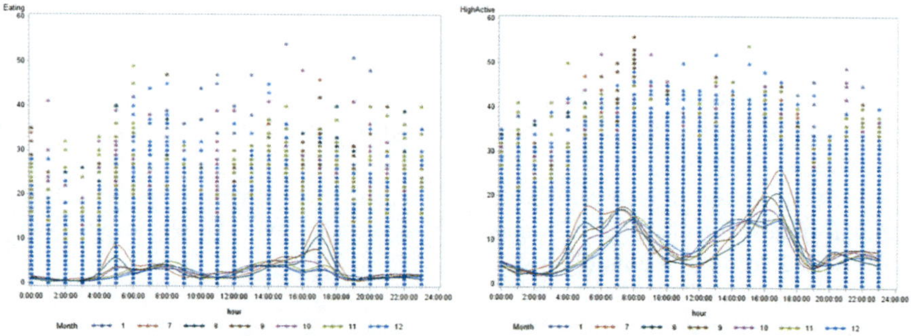

Fig. 3. Trend of eating and high activity according to month.

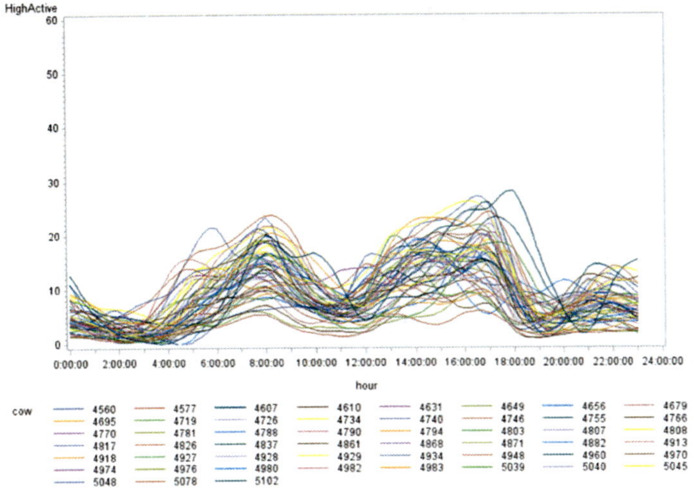

Fig. 4. High activity variability among animals.

Looking in particular at eating behavior, buffaloes showed a strong preference for specific timeslots during the day (*i.e.,* early morning and late afternoon), when temperature was hot (data not shown). Moreover, from July to September, the concentration at the manger was significantly higher twice a day and in absolute terms it decreased according to the changing months. A strong pattern depending on season was found for each considered variable. Based on the weather station located nearby the stable, July was the hottest month (mean temperature 25.8 °C, maximum temperature >40 °C) followed by August and September, in that order. In addition, the temperature/humidity index (THI), calculated on punctual data logger located in the pens, was steadily straddling two zones, the risk zone and the alert zone (*i.e.,* THI >72) from June to September. In July the mean THI was 78.18 (ranging from 72.5 to 81.5 approximatively). The sensors recorded a mean body surface temperature at ear of 31.2 °C in July.

Conversely sensor response gave more regular and flatter curves for the mild or cold season. This evidence would suggest that, even though buffalo is often recognized as a very well adapted species to hot climate condition, it is quite sensitive to temperature and modify its behavior accordingly at our latitude. [11] Reported a thermal optimum between 15 and 30 °C for Mediterranean water buffalo. So, a certain thermal discomfort was expected in the hot period. As a partial confirmation of the effect of different external condition on animals, in a test performed comparing the mean rectal temperature on 20 animals during a mild day (April the 5th 2021) and a hot one (August the 31st 2021), after exposure to direct solar radiation (10:00 to 12:00 A.M.), we found a 1 °C difference (39.02±0.56 °C vs. 40.125±0.83 °C respectively). Moreover, the range and the standard deviation were much greater on the hot day, suggesting a different individual ability to cope with thermal stressors and to thermoregulate, in particular with high temperatures and confirming what found by [12]. As for the cold period, it seems associated to more regular and distributed behavioral pattern, but we must consider that no critical temperatures were reached during observations (T min

−7 °C, average 6.4 °C, ear sensor temperature 17.6 °C, in January). [10] Stressed quite a tolerance to cold stress in buffalo. In mild season (data not reported) values were intermediate between the extremes.

The relationship between the sensors response and some specific activities could be of interest and help for the stakeholders, not only in decision making, but also, in tuning the timing, setting and planning of specific and/or routinary activities (*i.e.,* feed delivery or approaching, grouping and groups' managing and so on).

3.2 Body Condition Score Evaluation

As a side test, we used an electronic device for zoometric measurements to get fattening degree of buffaloes on the base of the equation proposed by [13]. Our provisional results were quite encouraging since the model gave a high determination coefficient and the measures obtained by the electronic tool and the direct observation from expert operators were highly correlated. This data will be object of a further stand-alone publication.

3.3 Laser Methane Pistols

A laser methane detector (LMD) is a system for remote monitoring of methane gas concentrations in ambient air. It is specifically designed for the constant scanning and monitoring gas tanks, storage facilities, pipelines, gas stations *etc*. Nevertheless, different uses and applications are grown over the last decade. An interesting one is its possible utilization for livestock methane emissions measurements [14]. It is based on tunable diode laser technology and allows high-sensitivity real-time detection of methane from a distance. Methane can be quickly revealed by pointing the laser beam toward the source, thus enabling to easily and safely access hard to reach locations. Average methane density between the detector and target is displayed in ppm*m.

There are applications of LMD on cow, sheep and goat enteric methane emissions measurement, but as far as we are concerned not on buffalo. The objective of the activity was to test this technology to qualitative and quantitative assessment of methane emissions in breath, also for selective purposes. The rationale was to discriminate and rank animals on the base of laser methane response, to quantify the real punctual and daily individual methane emissions and to select the more deserving buffaloes for culling and replacement schemes.

Our first results (Fig. 5) clearly highlight and differentiate CH_4 peaks due to the regular respiration and those due to eructation or other events (such as the zero-response due to apnea). Also, in this case a very strong variability came out when comparing different animals (Fig. 6). Nevertheless, from a very preliminary analysis, the most emitting animals, both in terms of peak intensity and in absolute terms, would seem to show a certain repeatability in signal. Moreover, an intermediate heritability (0, 15) was reported for cattle methane emissions measured by the LMD [15]. What above bodes well for the potential use of this approach to select for the lower emitting animals.

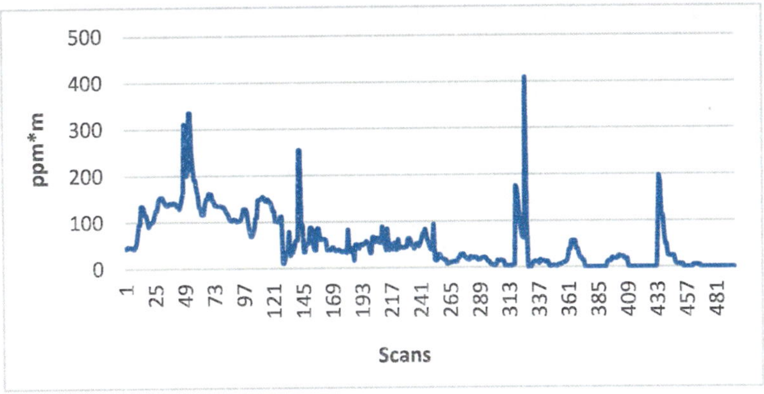

Fig. 5. Portion of LMD spectrum of a single animal (2 scans/sec).

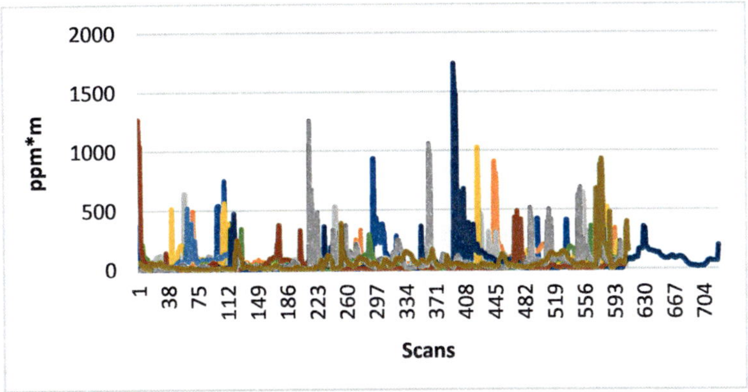

Fig. 6. LMD spectrum (2 scans/sec) variability among different animals.

4 Conclusion

A sensor approach to buffalo farming may lead to on optimization of management (*i.e.*, feeding, milking, grouping etc…) according to specific needs, behaviors and animal-based responses.

Methane emissions seems to be very variable and perhaps moderately heritable and so viable for selection.

Use of technology can help to safely manage the herds (*e.g.*, by remote acquisition of BCS and body weight) and to improve buffalo farming efficiency (*i.e.*, by preventing heat stress, nutrition or fertility problems etc.…), through the development of an easy-to-use application software for stakeholders.

For farming and breeding efficiency improvement it would be important to monitor not only milking but also dry, free ranging, growing, or parturient animals.

By steady setting remote-controlled LMDs in the milking parlor or in the pens a great deal of data from a higher number of animals every day, all year long, would be available.

The numerosity is crucial to further in deep study complex issues and to better understand the mechanism controlling their interactions.

Further challenges are: calibration and validation of sensors for buffalo species suitable models; validation of LMD for quantitative methane emissions measurements in buffalo; selection based on lower emission and higher resilience to climate change.

Aknowledgments. The authors thank Dr. Emanuela Rossi for technical supporting and Mr. Adriano Medori and Gianluca Paris for animals managing.

References

1. Morgan-Davies, C., et al.: Impacts of using a precision livestock system targeted approach in mountain sheep flocks. Livestock Sci. **208**, 67–76 (2018)
2. Ermetin, O.: Precision livestock farming: potential use in water buffalo (Bubalus bubalis) operations. Anim. Sci. Papers Reports **39**(1), 19–30 (2021)
3. Guarino, M., Norton, T., Berckmans, D., Vranken, E., Berckmans, D.: A blueprint for developing and applying precision livestock farming tools: a key output of the EU-PLF project. Anim. Front. **7**, 12–17 (2017)
4. Bohmanova, J., Misztal, I., Coleet, J.B.: Temperature-HumidityIndicesasindicatorofmilk productionlosses duetoheatstress. J. Dairy Sci. **90**, 1947–1956 (2007)
5. LPHSI: Livestock and Poultry Heat Stress Indices. The Livestock and Poultry Heat Stress Indices for Cattle, Sheep and Goats. Agriculture Engineering Technology Guide, Clemson University, Clemson, SC, USA (1990)
6. Gauthier, J., Wu, Q.V., Gooley, T.A.: Cubic splines to model relationships between continuous variables and outcomes: a guide for clinicians. Bone Marrow Transp. **55**, 675–680 (2020)
7. Smith, P.L.: Splines as a useful and convenient statistical tool. Am. Stat. **33**, 57–62 (1979)
8. Littell, R.C., Henry, P.R., Ammerman, C.B.: Statistical analysis of repeated measures data using SAS procedures. J. Anim. Sci. **76**, 1216–1231 (1998)
9. Asker, A.A., Ragab, M.T.: Causes of variation in birth weight of Egyptian cattle and bufftaloes. Indian J. Veterin. Sci. **22**, 265–272 (1952)
10. Marai, I.F.M., Haeeb, A.A.M.: Buffalo's biological functions as affected by heat stress—a review. Livestock Sci. **127**, 89–109 (2010)
11. Zicarelli, L., La bufala da latte. 30 giorni Speciale, **1**(8), 79–82 (2008)
12. Shafie, M.M.: Biological adaptation of buffaloes to climatic conditions. In: Prospects of Buffalo Production in the Mediterranean and the Middle East. CONFERENCE 1993, Cairo. Wageningen Academic Publishers, Wageningen, vol. 62, pp. 176–185 (1993)
13. Negretti, P., Bianconi, G., Bartocci, S., Terramoccia, S., Verna, M.: Determination of live weight and body condition score in lactating Mediterranean buffalo by Visual Image Analysis. Livest. Sci. **113**, 1–7 (2008)
14. Chagunda, M.G.G.: Opportunities and challenges in the use of the Laser Methane Detector to monitor enteric methane emissions from ruminants. Animal **7**(s2), 394–400 (2013)
15. Mühlbach, S., Sorg, D., Kecman, J., Rosner, F., Swalve H.H.: Heritabilities for methane concentrations measured by laser methane detector in the breath of cows. In: 68th Annual Meeting of the European Federation of Animal Science Proceedings. Wageningen Academic Publishers, Wageningen, p. 200 (2017)

Pre-slaughtering Phases and Meat Quality of Highly Profitable Cattle (Piedmontese Fat Ox)

Simone Stella, Silvia Mazzola, Annamaria Costa, Erica Tirloni, Ernesto Beretta, and Francesco M. Tangorra(✉)

Department of Veterinary Medicine and Animal Sciences, Università degli Studi di Milano, Via dell'Università 6, 26900 Lodi, Italy
francesco.tangorra@unimi.it

Abstract. The objective of the study was to evaluate the effect of the pre-slaughtering phases on animal welfare of highly profitable cattle (Piedmontese fat ox) and on meat quality, aged for long time (4–6 months).

The study involved 14 Piedmontese fat oxes from four farms located in Piedmont (Northern Italy). The animals were slaughtered in a family-run slaughterhouse and the carcasses were stored in a static refrigerating room at 2 °C and aged for 4–6 months. Loading and transportation times, waiting times before stunning and travelling times were recorded. Hair and blood cortisol were measured (ELISA essay kit and chemiluminescence), to evaluate animal welfare respectively at farm level and during slaughter. The pH of meat was measured 72 h post-mortem and every week until the end of aging. The meat color was determined at 100–140 days of aging, using a colorimeter. Cooking losses were determined by cooking meat samples at a final internal temperature of 71 °C, and the Warner-Bratzler shear force (WBSF) test was used to assess their tenderness.

The average distance travelled from the farms to the slaughterhouse was approximately 200 km, with an average travel time of 3 h and an interval time between loading and stunning of 8 h. The average hair and blood cortisol concentrations were respectively 1.05 pg/mg and 4.58 µg/dl. The pH of meat ranged between 5.87 and 5.99 72 h post-mortem and reached values close to neutrality at the end of aging period. The color of meat surface was compatible with the long aging time. Cooking losses were on average 21.6% of the samples weight. The meat showed a marked tenderness with low shear force values (1.72 kgf/cm^2). Pre-slaughtering phases did not affect animal welfare quality of meat aged for long time.

Keywords: Pre-slaughtering · Animal welfare · DFD meat · Piedmontese ox

M. Biocca et al. (Eds.): SHWA 2020, LNCE 252, pp. 51–59, 2022.
https://doi.org/10.1007/978-3-030-98092-4_6

1 Introduction

1.1 Influence of Animal Welfare on the Quality of Beef

Animal welfare is increasingly considered by the Food Business Operators, both for ethical purposes, and for its influence on the quality of the obtained meat. The impact of stressful conditions on the final quality of meat and derived products is well known: the main negative effect of animal stress is the depletion of muscle glycogen, leading to an insufficient post mortem acidification and to the occurrence of dark cutting beef. This phenomenon can affect meat quality (e.g. pH, water holding capacity, color), but can also have a negative impact on meat safety (higher diffusion and potential growth of pathogenic microorganisms) [1].

In this light, pre-slaughtering phases are extremely critical, as stressful events can occur mainly during transport and related stages (loading and unloading), and during slaughter lairage [2, 3]. Several factors influence the welfare status of the animals in these phases, such as animal density, climate, distance/duration of the transport, fasting, structure adequacy and personnel expertise [3, 4], but also intrinsic characteristics of the animal, such as age and temperament have an important role. González et al. [4] found that cattle weighing more than 500 kg do not suffer negative effects of transport procedures, thanks to a better body condition and a proper functionality of immune system. The animal temperament plays a major role, influencing both physiological parameters and meat quality characteristics (in particular tenderness) [2, 5].

Among the methods applied for the evaluation of stressful condition of the animals, the quantification of some hormones, and in particular of cortisol, have been suggested as reliable index [1]. Cortisol plays an important role in the response to acute stress, by increasing the efficiency of amino acids and fatty acids metabolism to meet the higher energy demand. The serum cortisol concentration is applied to evaluate acute stress condition, such as that caused by unloading and handling of the animals; a particular care should be taken in blood sampling, as an increase in the hormone concentration could be also induced just by the sampling procedures [6]. When produced, cortisol is accumulated in animal hair; thus, it can be used as chronic stress indicator [7]; hair sampling is simple and not influenced by animal handling during sampling, and doesn't need specific transport and storage requirements [8].

1.2 Effect of Prolonged Aging on Beef Quality

Meat aging is defined as an amicrobial catabolic process involving muscle tissue, applied to achieve optimal characteristics of mature meat (color, tenderness, juiciness, flavor). The mechanisms of the acquisition of sensorial characteristics during aging is complex, not yet cleared completely, even if is well known that it involves proteolytic, lipolitic and oxidative reactions [9]. To obtain high quality meat, the aging period should be extended for some weeks (a minimum of 12 days at 4 °C as suggested by Kinsman et al. [10]), but it generally lasts just for some days/one week, in order to avoid weight loss and possible microbial spoilage. Two main aging methods are usually applied ("wet" and "dry" aging); in particular, the dry aging method is considered the most traditional one, and consists in hanging half carcasses or quarters in cold rooms, at refrigeration temperature (about 4 °C). Such method has some disadvantages, namely a weight loss due to surface evaporation, and the need for surface trimming, but the meat obtained is particularly appreciated by some consumers for its tenderness and for its marked, persistent flavor, defined as "buttery and rich", or "earthy and nutty" [11–13].

Dry aging of beef could be sometimes prolonged for several weeks, but only little information is available on the quality of meat obtained, and on the evolution of chemical-physical characteristics of muscle tissue during the process. Some studies indicate that a prolonged aging (more than three weeks) don't exert a positive effect, as off-flavors can be perceived [13], and meat color is modified, becoming darker and less homogenous during prolonged aging. These effects are balanced by the increased flavor and tenderness of the meat obtained. Data from Phelps et al. [14] indicated a marked tenderization through the first three weeks, followed by a gradual, slow tenderization for the whole aging period (70 days). The long-aged beef is often intended for consumption at restaurants, thus enhancing its quality attributes [15].

1.3 Aim of the Study

This study aimed to evaluate some aspects of the production of high quality beef from highly profitable cattle (Piedmontese oxes): in particular, the work focused on a) the welfare/stress condition of the animals, evaluated by the quantification of serum and hair cortisol, b) the effect of very prolonged dry aging on pH of the quarters obtained, and c) the quality characteristics of very "mature" beef obtained by prolonged aging.

2 Materials and Methods

2.1 Experimental Plan

The study involved a total of 14 Piedmontese oxes coming from four farms located in the provinces of Cuneo and Alessandria (Piedmont region), slaughtered through a nine months' period. Animals aged approximately three years (mean age = 1070 days) and weighing about 900 kg (mean weight: 926 kg) were selected for the study. At the end of the finishing period, the animals were truck-transported at a family-run slaughterhouse with low operating capacity (2–3 animals/day), and slaughtered the same day. The duration of loading, transport and unloading phases, and the time before stunning were recorded, as well as the distance travelled from the farms to the slaughterhouse; the mean distance between the farms and the slaughterhouse was about 200 km, and the mean transport duration was of about 3 h.

The total time spent between animals load on the truck and slaughtering (load, transport, unload, lairage, immobilization and stunning) was of about eight hours. After slaughtering, the carcasses were sectioned in quarters, and put in a dedicated refrigerated chamber (2–3 °C), where they were submitted to a prolonged dry aging and to the weekly evaluation of pH. After 100–140 days of aging, beef samples were submitted to the evaluation of quality characteristics.

2.2 Cortisol Determination

Hair Cortisol Determination. The hair cortisol concentration was evaluated as it could indicate a chronic stressful condition, caused by non-optimal farming procedures. In order to limit variability due to differences among locations, hair sampling was made by a razor from the frontal region of the head of the animals; the sampling was made one week before the scheduled slaughtering. For the determination, hair samples were washed with warm water and air-dried for one day. A 250 mg aliquot was washed with 5 ml of isopropanol for three minutes and left to dry for five days, then ground with a ball mill (30 revolutions per second) for five minutes. The ground samples were sonicated with methanol for 30 min, and incubated for 18 h, at 50 °C in a shaker (100 rpm), then submitted to cortisol quantification by a Cortisol ELISA kit (ENZO Life Science, Pero, I) following the manufacturer's instructions [16].

Blood Cortisol Determination. The blood cortisol content was determined to evaluate the onset of eventual acute stress during pre-slaughtering phases (transport, unload, short lairage). The blood was taken from the caudal vein, immediately after stunning. The cortisol determination was made by a Cortisol chemiluminescent ELISA kit (Creative Diagnostics, London, UK) [17].

2.3 Muscle pH Measurement

The muscle pH was measured in the hindquarter (tri-tip and flank cuts) and in the forequarter (neck muscles) of each carcass by a pH-meter with an infission probe (HI 99163, Hanna Instruments, Ronchi di Villafranca Padovana, I). The determination was made at 72 h post-mortem and was repeated weekly until the end of the experimental period, that lasted for 20 weeks for forequarters, and 32 weeks for hindquarters.

2.4 Aged Beef Quality Characteristics

Measurement of Color Parameters. The meat color was determined on m. longissimus dorsi (at the twelfth-thirteenth rib level) after a prolonged dry aging period, set between 100 and 140 days. The determination was performed by a CR-300 Chroma-Meter (Minolta, Osaka, JPN), after a blooming of 45 min to allow deoxymyoglobin oxygenation; the CIE L^*, a^* and b^* values were taken at 6 locations, and the mean value was calculated. The hue angle (h) was calculated as $h = \arctan(b^*/a^*)*57.32$, and Chroma was calculated as $Ch = \sqrt{(a^*2 + b^*2)}$.

Cooking Loss Determination. The meat cooking loss was deterimined following the method by Honikel [18] with minor modifications: the steaks were weighed and put into 190 × 300 mm, 65-μm thick Polysilk bags (thermal resistance: −40 /+80 °C; Baglight, Interscience, Saint Nom, France) and placed in a thermostatic water bath at 75 °C, until they reached a targeted peak internal temperature of 71 °C. Temperature measurements were obtained by a thermometer (735-2, Testo, Settimo Milanese, Italy) equipped with a probe (PT-100, Testo) inserted into the core of each steak. When the end-point temperature was reached, bags were removed from the water bath, cooled under tap water and then chilled in refrigerator. Then, the steaks were weighed again, and the ratio between the weight measured after and before cooking was calculated.

Warner-Bratzler Shear Force Determination. The cooked meat samples were then submitted to determination of the Warner-Bratzler shear force (WBSF) evaluation by an Instron Universal Testing Machine (model 5542, Instron Engineering Corp., Canton, MA; blade speed = 200 mm/min); the analysis was performed on 6 shares (1.27 cm in diameter) from each sample. The shares were cut parallel to the longitudinal orientation of muscle fibres; the peak shear force was measured (Warner-Bratzler blade speed 200 mm/min), and mean values were recorded.

3 Results and Discussion

3.1 Cortisol Determination

Hair Cortisol Determination. The mean cortisol concentration detected in hair samples was 1.05 ± 1.29 pg/ml; quite all the samples showed values lower than 1 pg/ml (median value equal to 0.52 pg/ml), with only two animals (n° 6 and 14) with higher results (4.05 and 3.39 pg/ml, respectively). The information about cortisol hair concentration in literature data is scarce, but some authors measured this parameter in different cattle categories, observing higher values: various authors evaluated the hair cortisol concentration in cattle (mainly milking cows) [6, 7, 19–22]. These studies showed higher concentrations in all cattle categories: values of 1 to 10 pg/ml were detected in adult cattle (mainly dairy cows), while higher concentrations were found in younger animals such as heifers (more than 10 pg/ml) and especially in calves (more than 100 pg/ml). The hair cortisol concentration is influenced by several factors, such as the health status of cattle and the farming system, but other factors should be considered, such as the sampling site, the hair color length and growth rate and the sample preparation technique (use of ball mill vs. scissors, presence/absence of follicles) [7, 19–22]. Even when taking into account these variables, the values observed in this study were lower than those obtained in previous studies, suggesting the absence of an evident stressful condition of the animals during farming. The age of the animals should also be considered as an important factor, as significantly higher cortisol concentrations have been detected in calves than in adult cattle.

Blood Cortisol Concentration. The mean cortisol content in the blood of Piedmontese oxes was 4.58 ± 2.05 μg/dl, with a range between 0.51 and 7.60 μg/dl, and a median value of 5.1 μg/dl. The values obtained are in agreement with those form previous studies performed on castrated cattle transported for two to three hours [23], while entire males and females showed lower values, between 1 and 3 μg/dl [24–26].

3.2 Evolution of Meat pH During the Prolonged Aging

The meat of Piedmontese oxes used for this study was submitted to a very prolonged aging; the period lasted for 20 weeks (forequarters) and 32 weeks (hindquarters). The mean initial pH value of forequarter muscles was 5.99 ± 0.22, while a value of

5.87 ± 0.20 was found on hindquarter (average of the measurements taken from tri-tip and flank). The muscle pH increased during the trial (Fig. 1), reaching high values (around or above 6.7) in the final part of the study, both in fore and hindquarters. A comparison with literature data is difficult, as information about a so prolonged aging is scarce, but the values observed were higher than those measured in previous studies [15, 27]: this finding could be due to the typology of animals considered in our study (castrated animals, aged about three years), that are characterized by "mature" muscle tissue. It has to be noted that the high pH values detected were not associated to an evident dark cutting of meat; moreover, although the surface of quarters was gradually covered by a layer of moulds, no evident meat defects were detected after surface trimming. An evident difference was detected between the values obtained from forequarters and hindquarters, with significantly higher values in the forequarter ($P < 0.01$) for the whole experimental period considered (20 weeks), whereas the evaluation of separated values from tri-tip and flank (until week 23, when both cuts were available) showed very similar values and trends (Fig. 1).

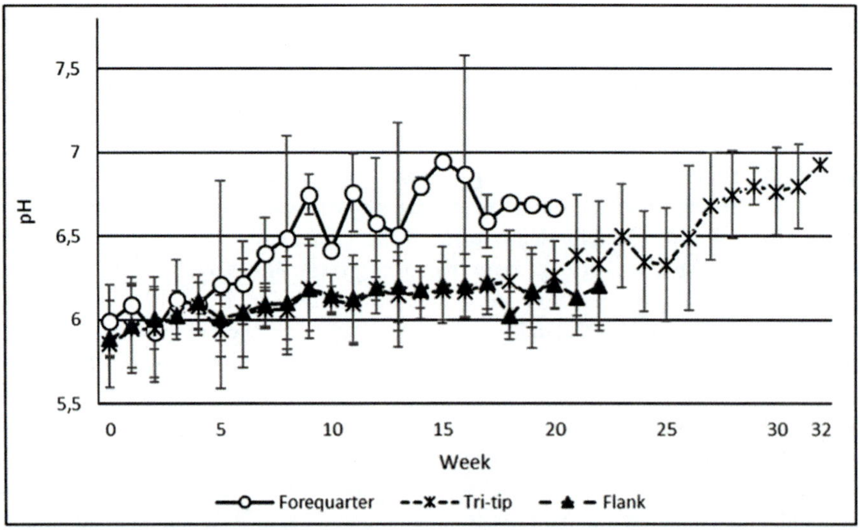

Fig. 1. Evolution of muscle pH (forequarter, tri-tip and flank) during the prolonged dry aging period.

3.3 Quality Characteristics of Long Aged Piedmontese Beef

Determination of Color Parameters. The color parameters determination performed of m. longissimus dorsi gave the following results: the luminosity fell mainly in the range 30–35 (32.2 ± 3.0), the red index showed a mean value of 15.3 ± 2.7 (lower than the value generally considered as normal for fresh meat [28], but typical of mature, aged beef), and yellow index showed low values (blueish tonality, 7.7 ± 0.8). The comparison with literature data is difficult: few studies evaluated the color of beef after prolonged aging [15], with higher values for all the three color parameters, but the results of our study are justified by the age of the animals and the extremely prolonged dry aging period. The hue angle values (27.2 ± 4.2) fell in the normal range for beef (as both red and yellow index had low values), and Chroma (color saturation index) values (17.2 ± 2.5) were near the threshold generally considered for fresh meat (a value of 18 is reported by McDougall [29].

Cooking Loss. The aged beef showed a mean cooking loss of 21.6%, with an evident variability among the samples (range 14.1–24.6%). This value can be considered low if compared to other studies; a decrease in cooking loss after a prolonged dry aging has been already described, but few results are available [15]. The low cooking loss values obtained can be explained by the high level of maturation of beef, owing to both the age of the animals ("mature" beef) and to the prolonged dry aging: a gradual proteolysis occurs during the aging period, resulting in a generally higher water holding capacity by the muscle tissue. Moreover, a continuous, although slow, meat surface drying takes place, thus causing a weight loss of the whole quarters, and consequently limiting the further loss during cooking.

Warner-Bratzler Shear Force. As already shown by other studies, Piedmontese beef is characterized by a marked tenderness, determined by the breed genotype, and this characteristic is well recognized by the consumers [28, 30, 31]. All the beef samples showed a marked tenderness: the mean WBSF value was equal to 1.72 ± 0.93 kgf/cm^2, with the values ranging from 0.90 and 1.84 (median value = 1.53). The values obtained were compared to the threshold set by Belew et al. [32], based on consumers' hardness perception: all the samples fell within the range of "very tender" meat (values below 2.41 kgf/cm^2). Previous studies showed the gradual decrease in WBSF values through the aging period; anyway, the authors evaluated the WBSF for shorter aging periods, and values detected were similar or higher than those observed in our study: Juarez et al. [33] found a value of 1.88 kgf/cm^2 in striploin samples aged for 56 days, while higher values were obtained by Phelps et al. (70 days of aging) and Colle et al. (63 days of aging) [14, 15].

4 Conclusions

This study aimed to evaluate different aspects regarding the production of meat from highly profitable cattle (Piedmontese oxes). First, the evaluation of animal welfare by cortisol determination showed the absence of particularly stressful condition during farming and pre-slaughter phases. This result could be due to the management of small animal groups or of single animals during farming, transport and slaughtering, and to the temperament of the animals. The second aspect investigated in this study was the effect of prolonged aging on the quality of the beef obtained; our data showed the continuous, gradual increase in meat pH, that indicates an evolution in muscle degradation process. The characteristics of beef obtained after a prolonged dry aging process (more than three months) are well recognizable by the consumers, who appreciate this extremely "mature" beef, in particular thanks to its tenderness and taste.

In consumers beef purchase choice, animal welfare and meat quality are nowadays considered as important determinants, in addition to the retail price. In the specific case of Piedmontese beef, a previous study by Merlino et al. [30] highlighted that the purchase decision is mainly influenced by the price. This outcome apparently reduces the importance of meat quality and animal welfare, but, as described by the authors, it really indicates that welfare of Piedmontese cattle and high quality characteristics of the derived meat are so well recognized to be considered as a prerequisite. Thus, experimental data are demanded by the producers to supply objective information for the promotion of this traditional beef production chain.

References

1. Terlouw, E.M.C., et al.: Pre-slaughter conditions, animal stress and welfare: current status and possible future research. Animal **2**, 1501–1517 (2008)
2. Del Campo, M., Brito, G., Soares de Lima, J., Hernández, P., Montossi, F.: Finishing diet, temperament and lairage time effects on carcass and meat quality traits in steers. Meat Sci. **86** (4), 908–914 (2010)
3. Mounier, L., Dubroeucq, H., Andanson, S., Veissier, I.: Variation in meat pH of beef bulls in relation to condition of transfer to slaughter and previous history of the animals. J. Anim. Sci. **84**, 1567–1576 (2006)
4. González, L.A., Schwartzkopf-Genswein, K.S., Bryan, M., Silasi, R., Brown, F.: Relationships between transport conditions and welfare outcomes during commercial long haul transport of cattle in North America1. J. Anim. Sci. **90**(10), 3640–3651 (2012)
5. Burdick, N.C., Carroll, J.A., Hulbert, L.E., Dailey, J.W., Willard, S.T., Vann, R.C.: Relationships between temperament and transportation with rectal temperature and serum concentrations of cortisol and epinephrine in bulls. Livest. Sci. **129**, 166–172 (2010)
6. Burnett, T.A., et al.: Relationship of concentrations of cortisol in hair with health, biomarkers in blood, and reproductive status in dairy cows. J. Diary Sci. **98**, 4414–4426 (2015)
7. Burnett, T.A., et al.: Factors affecting hair cortisol concentrations in lactating dairy cows. J. Dairy Sci. **97**, 7685–7690 (2014)
8. Prandi, A., Comin, A., Peric, T., Montinello, M., Omodeo, S.G.: Nuovo approccio non invasivo per la valutazione del benessere animale. Quaderno SOZOOALP **6**, 183–191 (2010)
9. Smith, A.M., Harris, K.B., Griffin, D.B., Miller, R.K., Kerth, C.R., Savell, J.W.: Retail yields and palatability evaluations of individual muscles from wet-aged and dry-aged beef ribeyes and top sirloin butts that were merchandised innovatively. Meat Sci. **97**, 21–26 (2014)
10. Kinsman, D.M., Kotula, A.W., Breidenstein, B.C. (eds.): Muscle Foods. Chapman & Hall, New York (1994)
11. Khan, M.I., Jung, S., Nam Ki Chang Nam, K., Jo, C.: Postmortem aging of beef with a special reference to the dry aging. Korean J. Food Sci. Anim. Resour. **36**, 159–169 (2016)
12. Savell, J.W.: Dry aging of beef. Executive summary (2008). https://www.beefresearch.org/Media/BeefResearch/Docs/dry_aging_of_beef_08-20-2020-28.pdf
13. Smith, R.D., et al.: Dry versus wet aging of beef: Retail cutting yields and consumer palatability evaluations of steaks from US Choice and US Select short loins. Meat Sci. **79**, 631–639 (2008)

14. Phelps, K.J., et al.: Effect of extended postmortem aging and steak location on myofibrillar protein degradation and Warner-Bratzler shear force of beef M. semitendinosus steaks1. J. Anim. Sci. **94**(1), 412–423 (2016)
15. Colle, M.C., et al.: Influence of extended aging on beef quality characteristics and sensory perception of steaks from the gluteus medius and longissimus lumborum. Meat Sci. **110**, 32–39 (2015)
16. Enzo: Product Manual. Cortisol ELISA kit. Catalog number: ADI-900-071. https://www.enzolifesciences.com/fileadmin/files/manual/ADI-900-071_insert.pdf
17. Creative Diagnostics: Cortisol ELISA KIT Data sheet. https://www.creative-diagnostics.com/Cortisol-EIA-Kit-110975-465.htm
18. Honikel, K.O.: Reference methods for the assessment of physical characteristics of meat. Meat Sci. **49**, 447–457 (1998)
19. Gonzàlez-de-la-Vara, M.R., Valdez, R.A., Lemus-Ramirez, V., Vazquez-Chagoyan, J.C., Villa-Godoy, A., Romano, M.C.: Effects of adrenocorticotropic hormone challenge and age on hair cortisol concentrations in dairy cattle. Can. J. Vet. Res. **75**, 216–221 (2011)
20. Tallo-Parra, O., Manteca, X., Sabes-Alsina, M., Carbajal, A., Lopez-Bejar, M.: Hair cortisol detection in dairy cattle by using EIA: protocol validation and correlation with faecal cortisol metabolites. Animal **9**, 1059–1064 (2015)
21. Comin, A., Prandi, A., Peric, T., Corazzin, M., Dovier, S., Bovolenta, S.: Hair cortisol levels in dairy cows from winter housing o summer highland grazing. Livest. Sci. **138**, 69–73 (2011)
22. Moya, D., Schwartzkopf-Genswein, K.S., Veira, D.M.: Standardization of a non-invasive methodology to measure cortisol in hair of beef cattle. Livest. Sci. **158**, 138–144 (2013)
23. Tennessen, T., Price, M.A., Berg, R.T.: Comparative responses of bulls and steers to transportation. Can. J. Anim. Sci. **64**, 333–338 (1984)
24. Sporer, B.K.R., Weber, P.S.D., Burton, J.L., Earley, B., Crowe, M.A.: Transportation of young beef bulls alters circulating physiological parameters that may be effective biomarkers of stress. J. Anim. Sci. **86**, 1325–1334 (2008)
25. Gupta, S., Earley, B., Crowe, M.A.: Effect of 12-hour road transportation on physiological, immunological and haematological parameters in bulls housed at different space allowances. Vet. J. **173**, 605–616 (2007)
26. Liotta, L., Costa, L.N., Chiofalo, B., Ravarotto, L., Chiofalo, V.: Effects of lairage duration on some blood constituents and beef quality in bulls after long journey. Ital. J. Anim. Sci. **6**, 375–384 (2007)
27. Kim, Y.H., Kemp, R., Samuelsson, L.M.: Effects of dry-aging on meat quality attributes and metabolite profiles of beef loins. Meat Sci. **111**, 168–176 (2016)
28. Destefanis, G., Brugiapaglia, A., Barge, M.T., Lazzaroni, C.: Effect of castration on meat quality in Piemontese cattle. Meat Sci. **64**, 215–218 (2003)
29. MacDougall, D.B.: Changes in the colour and opacity of meat. Food Chem. **9**(1), 75–88 (1982)
30. Merlino, V.M., Borra, D., Girgenti, V., Dal Vecchio, A., Massaglia, S.: Beef meat preferences of consumers from Northwest Italy: analysis of choice attributes. Meat Sci. **143**, 119–128 (2018)
31. Zwambag, A., Kelly, M., Schenkel, F., Mandell, I., Wilton, J., Miller, S.: Heritability of beef tenderness at different aging times and across breed comparisons. Can. J. Anim. Sci. **93**, 307–312 (2013)
32. Belew, J.B., Brooks, J.C., McKenna, D.R., Savell, J.W.: Warner-Bratzler shear evaluations of 40 bovine muscles. Meat Sci. **64**, 507–512 (2003)
33. Juarez, M., et al.: Extended ageing time and temperature effects on quality of sub-primal cuts of boxed beef. Can. J. Anim. Sci. **90**, 361–370 (2010)

Calculation of the Mixing Time as a Function of the Dairy Cow Diet Chemical Homogeneity Inside the Mixing Hopper

Francesco M. Tangorra[1(✉)], Vera Perricone[1], Alessandro Agazzi[1], Aldo Calcante[2], Giovanni Savoini[1], and Annamaria Costa[1]

[1] Department of Veterinary Medicine and Animal Sciences, Università degli Studi di Milano, Via dell'Università 6, 26900 Lodi, Italy
francesco.tangorra@unimi.it

[2] Department of Agricultural and Environmental Sciences - Production, Landscape, Agroenergy, Università degli Studi di Milano, Via G. Celoria 2, 20133 Milan, Italy

Abstract. The objective of the study was to develop and implement a Near InfraRed Spectroscopy (NIRS) based system to determine the optimal mixing time of the Total Mixed Ration inside the hopper of a mixer wagon. To do this, a diode array spectrometer (950–1650 nm range), with a calibration model for predicting the main chemical parameters of the ration, was mounted at the bottom of the hopper of a horizontal self-propelled mixer wagon, close to the unloading belt conveyor, and was integrated in an on-line and real-time measurement system that was activated during the mixing procedure. The optimal mixing time was assumed to be achieved when the stability ratio of the chemical parameters measurement process was < 1.5. Student's t-tests were used to compare the distribution of the main nutrients in the ration inside the mixing hopper and in the delivered ration.

Results showed that the chemical homogeneity inside the mixing hopper was reached on average at 3.31 ± 0.87 min after the last ingredient was loaded.

The ration inside the hopper reflected the delivered ration for Dry Matter, Neutral Detergent Fiber and starch (P > 0.05), while exceeded the delivered ration for Crude Protein, Acid Detergent Fiber and ash (P < 0.01) and underfed Ether Extract (P < 0.01). A fine-tuning of the algorithm implemented as well as an adjustment of the NIR calibration for Crude Protein, Ether Extract, Acid Detergent Fiber, and ash is required to improve the capability of the system in evaluating the ration chemical homogeneity.

Keywords: Total mixed ration · Near-infrared spectroscopy · Mixing wagon

1 Introduction

The optimal mixing time for a dairy cow Total Mixed Ration (TMR) depends on many factors such as the type of ingredients used and their loading order, mixing equipment features and maintenance, as well as the inclusion of liquid feeds in the ration [1].

M. Biocca et al. (Eds.): SHWA 2020, LNCE 252, pp. 60–66, 2022.
https://doi.org/10.1007/978-3-030-98092-4_7

Adequate mixing time leads to a more homogenous TMR and less feed sorting contributing to optimize dry matter and nutrients intake and reduce metabolic disorders [2].

For the ration to be defined homogeneous, it must contain the individual ingredients uniformly distributed without any undesirable or harmful deficiencies or excesses. Non-homogeneous rations lead to a separate ingestion of nutrients due to the easy selection operated by animals. Alteration of ruminal and production efficiencies as well as an increase in variability among individuals are the consequences [3].

Homogeneity of TMR can be measured by sampling along the barn feeding-line and subjecting TMR samples to chemical analysis and sieving by particle size Penn State Particle Separator [4]. The use of portable Near InfraRed Spectroscopy (NIRS) instruments to detect the homogeneity of the TMR was experimented for the first time by Serva et al. [5] who developed an algorithm capable to estimate the homogeneity of TMR for beef and dairy cattle. To the authors' knowledge no other studies were carried out on the use of portable NIRS on the mixer wagon to detect the TMR homogeneity in real-time.

The objective of the study was to develop and implement a near infrared spectroscopy (NIRS) based system to determine the TMR optimal mixing time evaluating the TMR homogeneity inside the hopper of a mixer wagon.

2 Materials and Methods

2.1 Spectrometer

A diode array spectrometer (Corona Extreme, Carl Zeiss AG, Oberkochen, Germany) operating in the range 950–1650 nm with a calibration model for predicting the TMR main chemical parameters was used. Its main technical characteristics are reported in Table 1.

Table 1. Main technical characteristics of the NIR spectrometer used

Parameter	Value
Measurement range	950–1650 nm
Mean spectral pixel pitch	3 nm
Spectral resolution (half width at 1/10 max.)	≤ 10 nm
Wavelength accuracy	≤ 1 nm
Wavelength reproducibility	≤ 0.1 nm
Light source	Halogen
Dimensions	256 × 191 × 253 mm
Weight	10 kg
Range of operating temperatures	−15 °C to +50 °C
Power supply voltage	9–36 V

2.2 Integrated Measurement System

The NIR spectrometer was installed outside the hopper of a horizontal self-propelled mixer wagon (Gulliver 6014, Sgariboldi srl, Codogno, Italy). By positioning the spectrometer measuring head at the bottom of the hopper and close to the unloading belt conveyor, a continuous flow of material over the spectrometer lens was guaranteed during both the mixing phase operated by the mixing reel blades, and the TMR unloading (Fig. 1).

The NIR spectrometer was integrated in an on-line measurement system designed to measure in real-time and continuously the TMR main chemical parameters during the mixing and unloading phases (Fig. 2). For this purpose, front-end electronics and control units have been implemented using the know-how acquired by PTM srl (Visano, Brescia, Italy) in its decades of experience in the sector.

The following devices composed the system:

- PC Windows: Trough the InProcess Software (Carl Zeiss AG, Oberkochen, Germany) it allows to control the spectrometer, to setup and execute measurement procedures, to visualize and archive spectral and numerical data from the spectrometer. The PC also executes the algorithm for estimating the TMR homogeneity and manage the communication with the control unit.
- WLAN: This data exchange module allows the control unit connecting to a Wi-Fi network and receiving the weights of TMR ingredients to be loaded from the cloud management software PTM MANAGEMENT. Once the operational phase is over, the module transmits to the management software the loaded weights of each TMR ingredient and the total weight of the TMR distributed to the various dairy cows feeding groups.
- ADVANCE SUPER USB: The control unit is responsible for estimating the TMR homogeneity during the mixing phase and for driving the operator, by means of a TFT display installed in the driver's cab of the mixer wagon, through the different phases of preparation and distribution of the TMR along the feeding manger. When the loading phase of the TMR ingredients is ended, the control unit starts the NIR analysis communicating with the PC and executes a specific algorithm during the next mixing phase measuring the TMR homogeneity in real time and suggesting when proceeding with the unloading phase.

Fig. 1. NIR sensor mounted outside the mixing hopper (on the left); spectrometer measuring head at the bottom of the hopper and close to the unloading belt conveyor (on the right).

2.3 TMR Homogeneity Evaluation

The TMR homogeneity was assessed through the Stability Ratio (SR) calculated as:

$$\left(\frac{Overall\,sigma}{Within\,sigma}\right)^2 \tag{1}$$

where the *Overall sigma* is the estimate of the standard deviation based on all observations, while *Within sigma* is the estimate of the standard deviation based on within subgroup variation. *Within sigma* is calculated based on a moving range as:

$$\hat{\sigma} = \frac{\overline{MR}}{d_2(2)} \tag{2}$$

where $\overline{MR}$ is the mean of non-missing moving ranges computed as $(MR2 + MR3 + \ldots + MRN)/(N-1)$ with $MR_i = |y_i - y_{i-1}|$. $d_2(2)$ for moving range with length 2 is 1.128.

A stability ratio that exceeds 1.5 indicates that the process is unstable.

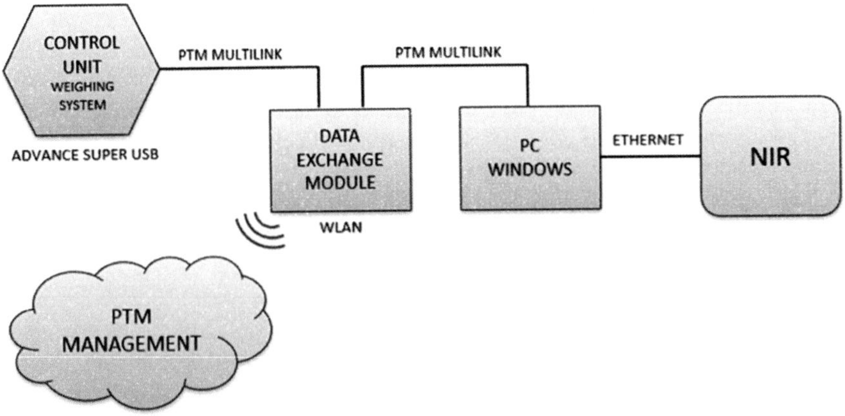

Fig. 2. Block diagram of the real-time measurement system implemented.

2.4 Field Tests and Statistical Analysis

Field tests were carried out at the University of Milan experimental farm (Landriano, Pavia, Italy). During May 2019 the corn and soy silage-based TMR at 50.5% DM for lactating cows was prepared and distributed once a day along the feeding manger.

The TMR was unloaded when the real-time measurement system implemented advised the operator that the TMR homogeneity was reached. TMR samples (200 g each) were collected for each day of the trial in six different points from the beginning to the end of the feeding manger and analyzed to determine dry matter (DM), crude protein (CP), ether extract (EE), ash, and starch content following the relative Association of Analytical Communities official methods of analysis [6]. Neutral detergent fibre (NDF) and acid detergent fibre (ADF) were measured through the Van Soest method.

Student's t-tests were used to compare the distribution of the main nutrients in the TMR inside the mixing hopper, as measured by NIR, and in the TMR delivered. Statistical significance was declared for $P < 0.05$.

3 Results and Discussion

The TMR homogeneity real-time measurement system reported the achievement of the ration chemical homogeneity on average at 3.31 ± 0.87 min after the last ingredient was loaded. This value agrees with the manufacturer's recommended mixing times ranging from 3 to 6 min.

Mixing time is a key area to watch when preparing a TMR, affecting the integrity of the feed ingredients. Mixing time has been proven to significantly alter particle size distribution [7]. Under mixing a TMR decreases uniformity along the feeding manger whereas overmixing reduces the particle size of forages [8, 9]. Analyzing samples from the mixer unloading chute after 4, 8, 16 and 32 min of mixing time, Heinrichs et al. [10] found that increasing the mixing time over 8 min reduced the fraction of particles longer than 18 mm. As particle length declines, cows spend less time chewing, decreasing the volume of saliva produced that acts to buffer ruminal contents [11]. This factor allows ruminal pH to fall, depressing activity of cellulolytic bacteria and causing a prolonged period of low ruminal pH [12].

On average, the TMR inside the hopper reflected the delivered ration for DM, NDF and starch ($P > 0.05$), while exceeded the delivered ration for CP, ADF and ash ($P < 0.01$) and underfed EE ($P < 0.01$), as reported in Table 2.

These results confirm the good predictive capabilities of the NIRS based system to detect in real-time the DM, NDF, and starch content of the TMR inside the mixing hopper as observed by Serva et al. [5] in a study aimed to calculate the homogeneity index of the distributed ration through the comparison between chemical-physical analysis and portable NIRS. A fine-tuning of the algorithm implemented as well as an adjustment of the NIR calibration for the others parameters (CP, EE, ADF, and ash) are required to improve the capability of the system in evaluating the TMR chemical homogeneity.

Table 2. Dry matter (DM) content (%), crude protein (CP), ether extract (EE), ash, neutral detergent fibre (NDF), acid detergent fibre (ADF) and starch (% of DM) contents of TMR. Values are expressed as means ± SD

TMR nutrients	TMR delivered	TMR inside the hopper	*P*
DM	48.6 ± 2.1	48.8 ± 2.8	>0.05
CP	14.6 ± 0.5	15.9 ± 1.0	<0.01
EE	3.7 ± 0.2	3.1 ± 0.6	<0.01
Ash	6.0 ± 0.5	6.5 ± 0.9	<0.01
NDF	34.9 ± 2.5	35.4 ± 3.1	>0.05
ADF	18.2 ± 1.3	23.0 ± 2.5	<0.01
Starch	26.1 ± 1.8	25.8 ± 0.8	>0.05

4 Conclusions

A NIRS based system to calculate the TMR optimal mixing time through the evaluation of the TMR homogeneity inside the hopper of a mixer wagon was developed, implemented and tested. Results highlighted the good predictive capabilities of the NIRS based system to detect in real-time the DM, NDF, and starch content of the TMR inside the mixing hopper. A fine-tuning of the algorithm implemented as well as an adjustment of the NIR calibration for the other parameters (CP, EE, ADF, and ash) will improve the capability of the system to measure the TMR homogeneity in real-time avoiding to under or over mix the TMR.

The research was co-funded by Regione Lombardia in the framework of POR FESR 2014–2020 (Axis I, Objective 1.b.1). Project P.L.U.S. - Precision Livestock Unifeed System.

References

1. Schingoethe, D.J.: A 100-year review: total mixed ration feeding of dairy cows. J. Dairy Sci. **100**(12), 10143–10150 (2017)
2. Marchesini, G., et al.: Effect of total mixed ration processing time on ration consistency and beef cattle performance during the early fattening period. Anim. Feed Sci. Technol. **262**, 114421 (2020)
3. Stauder, A., Humer, E., Neubauer, V., Reisinger, N., Kaltenegger, A., Zebeli, Q.: Distinct responses in feed sorting, chewing behavior, and ruminal acidosis risk between primiparous and multiparous Simmental cows fed diets differing in forage and starch levels. J. Dairy Sci. **103**(9), 8467–8481 (2020)
4. Moallem, U., Lifshitz, L.: Accuracy and homogeneity of total mixed rations processed through trailer mixer or self-propelled mixer, and effects on the yields of high-yielding dairy cows. Anim. Feed Sci. Technol. **270**, 114708 (2020)
5. Serva, L., et al.: Uso di uno strumento NIR portatile per la valutazione dell'omogeneità dell'unifeed distribuito in mangiatoia. In: Atti del Simposio SISNIR, Proceedings of the 7th Simposio Italiano di Spettroscopia NIR, Milano, Italy (12–14 October 2016); SISNIR: Lodi, Italy, pp. 138–142 (2016)

6. AOAC. Official methods of analysis, 18th edn. Gaithersburg (MD): Association of Official Analytical Chemists, International (2005)
7. Buckmaster, D.R., Heinrichs, A.J., Thomas, R.S.: Particle size reduction during total mixed ration preparation and delivery. Paper No. 97-AETC104. American Society of Agricultural Engineers, St. Joseph (1997)
8. Dahlke, G., Strohbehn, D.R.: Feed particle separation due to feed delivery and time in feed bunk and effects on cattle performance. Iowa State University Animal Industry Report: AS 655, ASL R2422 (2009)
9. Costa, A., et al.: Influence of different loading levels, cutting and mixing times on total mixed ration (TMR) homogeneity in a vertical mixing wagon during distribution: a case study. Ital. J. Anim. Sci. **18**(1), 1093–1098 (2019)
10. Heinrichs, A.J., Buckmaster, D.R., Lammers, B.P.: Processing, mixing, and particle size reduction of forages for dairy cattle. J. Anim. Sci. **77**, 180–186 (1999)
11. Nasrollahi, S.M., Imani, M., Zebeli, Q.: A meta-analysis and meta-regression of the impact of particle size, level, source and preservation method of forages on chewing behavior and ruminal fermentation in dairy cows. Anim. Feed Sci. Technol. **219**, 144–158 (2016)
12. Owens, F.N., Basalan, M.: Ruminal fermentation. In: Millen, D.D., De Beni Arrigoni, M., Lauritano Pacheco, R.D. (eds.) Rumenology, pp. 63–102. Springer, Cham (2016). https://doi.org/10.1007/978-3-319-30533-2_3

Using of NMR Milk Metabolomics to Evaluate Mammary Gland Health Status in Dairy Cows

Cinzia Marchitelli[1(✉)], Daniele Pietrucci[2], Marco Milanesi[2], Luca Casadei[2], Fernando Porcelli[2], Giovanni Chillemi[2], and Alessandra Crisà[1]

[1] Centro di ricerca Zootecnia e Acquacoltura, Consiglio per la Ricerca in Agricoltura e l'Analisi dell'Economia Agraria (CREA), Rome, Italy
cinzia.marchitelli@crea.gov.it

[2] Dipartimento per la Innovazione nei sistemi Biologici, Agroalimentari e Forestali (DIBAF), Tuscia University, Viterbo, Italy

Abstract. Objectives. Milk of mammal origin is an important food in the diet during infancy as well as in adulthood. It contains lipids, proteins, carbohydrates, essential nutrients, bioactive compounds (immunoglobulins, oligosaccharides), organic acids, nucleotides, micronutrients, and metabolites. Milk metabolites may originate from multiple cell types or metabolisms. The milk quality is affected by different factors, such as lactation stage, feed, genetic background, and cow health. Somatic cell count (SCC) is normally used as an indicator of mastitis. Here we proposed to use milk metabolites as indicators of mammary gland health status.

Methods. A total of 44 dairy cows of four different breeds (Holstein, Simmenthal, Simmenthal x Holstein crossbred, and Podolica) were enrolled for this study. Immediately after morning milking, milk samples (around 60 days after calving) were collected and placed on ice for transport to the laboratory. Once in the laboratory, the milk samples were aliquoted, and frozen at −80 °C. Nuclear magnetic resonance (NMR) spectroscopy was used to determine milk metabolites. Samples were classified as "high" ($\geq$400.000) or "low" ($\leq$400.000) SCC content. The metabolite profiles in the two classes were analysed to identify differences. A General Linear Model (GLM) procedure was used to identify the statistically significant differences of metabolites content.

Results. NMR spectra confirmed milk metabolites associated with a high number of SCC, already identified in previous studies, where many milk samples were analysed. In this study, four new metabolites (N-acetyl-lysine, N-acetyl-glutamine, creatinine, phosphocreatine,) were associated to high somatic cell count and could be used as biomarkers of mammary gland health status.

Keywords: Dairy cattle · NMR · Milk metabolites · Mammary gland · Somatic cell count

1 Introduction

Cows' milk consumption varies worldwide, with an average from 10 to 212 kg per-person per year (World Health Organization 2018). Milk contains 18 of 22 essential nutrients (Zhang et al. 2021), including a various of bioactive peptides and fatty acids

M. Biocca et al. (Eds.): SHWA 2020, LNCE 252, pp. 67–75, 2022.
https://doi.org/10.1007/978-3-030-98092-4_8

(*i.e.* alpha-linolenic (ALA), conjugated linoleic (CLAs), ω3), caseins, whey proteins, lactose, oligosaccarides and minor costituents (*i.e.* calcium, phosphorous, magnesium) which have an important impact on human and cow health (Garcia et al. 2019). Milk, as a source of nutrition, is especially important for infants and children, who need nutrient and energy-rich foods for growth, but also for adults, because is an excellent source of nutrients (Breda et al. 2020). A range of different factors, *i.e.* diets, genetics, number and stage of lactation, season of calving, management and cow health, influence the milk composition. These factors influence milk quality (type of nutrients and their quantity) and the technological properties of milk (Taničin et al. 2018). One of these factors is represented by the milk somatic cells, composed by different kinds of cells: epithelial, neutrophils, lymphocytes, macrophages. Somatic cell count (SCC) is a useful predictor of mastitis, an intramammary infection (IMI) caused by the introduction and multiplication of pathogenic microorganisms (*i.e. Escherichia coli*, *Staphylococcus aureus*, *S. uberis*). Physiologically, mastitis can be classified into subclinical and clinical type depending on the severity of inflammation (Maréchal et al. 2011). The subclinical mastitis is not easily detected, as no visual indicators of the inflammation exist; however, SCC is elevated, and the milk quality is affected (Sundekilde et al. 2013). The clinical mastitis is normally associated with inflammation signs of the udder (redness, swelling) together with obvious abnormalities in the milk (flakes, clots), compositional changes and elevated somatic cell count (Luangwilai et al. 2021). Dairy cow mastitis is the most expensive infectious disease with significant economic implications in the dairy industry (Guimarães et al. 2017). The annual losses are caused by a poor milk quality and quantity, the cost of veterinary treatment and the increasing cost of farm management (Hu et al. 2021). Management practices and selection for mastitis-resistant phenotypes are conventional methods to reduce the incidence of mastitis within a dairy herd (Alhussien et al. 2018). SCC is the primary parameter used to determine the usefulness of milk for human consumption. The accepted threshold for SCC in bulk milk in EU is 400.000 cells/ml, while in United States and Canada is 200.000 cells/ml (Sundekilde et al. 2013). Mammary gland inflammation elevated SCC and the presence of microorganisms influence protein and lipid metabolism, innate immune response and milk metabolites (Alnakip et al. 2014). Milk metabolites reflect the metabolic activity of the mammary gland and, during an inflammation, they may be secreted by microorganisms, immune cells (dead or alive) and mammary epithelial cells (Luangwilai et al. 2021). This study aimed to analyse milk samples of different cattle breeds to determine metabolites by nuclear magnetic resonance (NMR) and to explore possible associations with different milk SCC. In conclusion, we propose some milk metabolites as indicators of mammary gland infections.

2 Materials and Methods

2.1 Animals and Milk Samples

A total of 44 dairy cows of four different breeds (Holstein - HO, Simmental - SM, Simmental x Holstein crossbred - SMxHO, and Podolica - POD) were involved for this study. Podolica breed is a local cow breed characterized by an extensive production system, while the other three breeds are part of CREA cross herd and they are reared in an intensive production system. Immediately after morning milking, individual milk samples (around 60 days after calving) were collected and placed on ice to be transported to the laboratory. Once in the laboratory, the milk samples were analyzed for the SCC content and frozen at −80 °C for NMR.

Milk samples were classified in two health status using as SCC threshold value 400.000 cells/ml: 37 healthy (nine HO, ten SMxHO, nine SM, and nine POD) and 7 no-healthy (one HO, two SMxHO, and four SM). These groups were used for the subsequent analysis.

2.2 NMR

For sample preparation and NMR analysis a modified protocol was used (Sundekilde et al. 2013).

2.2.1 Sample Preparation

Prior to NMR spectroscopy the milk samples were thawed, thoroughly shaken to homogenize the sample and stored at 4 °C for at most 12 h before preparation. The metabolite profiling consisted of mixing 700 μl of milk with 700 μl of dichloromethane (CH_2Cl_2). The mixture was homogenized by vortexing and then was incubated for 15 min at room temperature. The mixture was then centrifuged at 14000 rpm for 30 min and the supernatant was recovered. Then, 500 μl of the supernatant were mixed with 400 μl of sodium phosphate buffer ay pH 7.4, containing 0.025% of $2H_2O$ and 0.9% of sodium trimethylsilyl [2,2,3,3-$2H_4$] propionate (TSP), to avoid chemical shift drifts due to pH effects.

2.2.2 NMR Analysis

About 600 μl of each sample were transferred into 5 mm NMR tube for the analysis. One-dimensional 1H NMR spectra were acquired on a Bruker Avance spectrometer operating at a 400 MHz at a temperature of 25 ± 1 °C. Tuning, matching and shimming was per-formed before each sample and the 90° proton pulse length was calibrated on each sample. The spectral width was 11 ppm and 32 scans (4 dummy scans) were collected with 32k points. Free induction decays were multiplied by an exponential function equivalent to a 1.0 Hz line-broadening factor before applying Fourier transform. Transformed spectra were automatically corrected for phase and base-line distortions and calibrated (TSP peak at 0.00 ppm) using TopSpin (Bruker). For multivariate analysis, each 1D spectrum in the range between 0.02 and 10.00 ppm was segmented into 0.02-ppm chemical shifts bins, and the corresponding areas were integrated using AMIX software (Bruker BioSpin).

2.3 Statistical Analysis

Data mining analysis tool in STATISTICA© 12.0 package (StatSoft Inc., Tulsa, OK, USA) was used to reduce the total dataset. This tool explores the data searching systematic relationships between variables, and individuating variables as best predictors for interested phenotype, in this case, SCC. Equation (1) describe the general linear model (GLM) procedure on the reduced dataset:

$$Y_{ijklnopq} = b_0 + B_i + P_j + S_k + b_1(d_1) + b_2(milk_m) + b_3(FAT_n) \\ + b_4(PROTEIN_o) + b_5(LACTOSE_p) + e_{ijklmnopq} \tag{1}$$

Where $Y_{ijklmnopq} =$ metabolites content expressed as proton concentration (calculated as integral of a bin, corresponding to the concentration of the metabolite behind the signal), $b_0 =$ was the overall average of the metabolite; B_i = fixed effect of breed (i = 1–4, 1 = HO, 2 = SMxHO, 3 = SM and 4 = POD), P_j = fixed effect of lactation number (j = 1–3, 1 = first, 2 = second, 30third), S_k = fixed effect of season of sampling (k = 1–4, 1 = spring, 2 = summer, 3 = autumn, 4 = winter), $SCC_l =$ fixed effect of SCC class, $b_1 =$ the regression coefficient of the days in milk; $b_2 =$ the regression coefficient of the milk yield, $b_3 =$ the regression coefficient of the %FAT, $b_4 =$ the regression coefficient of the %PROTEIN, $b_5 =$ the regression coefficient of the % LACTOSE, $e_{ijklmnop}$ = was the random error. The significance of the metabolites content was determined using Tukey's test in the GLM procedure.

3 Results and Discussion

3.1 NMR Metabolites Assignment

Nuclear magnetic resonance spectra of 44 milk samples were represented by 400 bins. Each bin corresponded to a range of chemical shifts expressed as part per million (ppm). The initial dataset was analyzed with a data mining analysis tool. We aimed to identify metabolites that could be predictive of two SCC classes: healthy vs no-healthy. The data mining analysis identified 40 bins as best predictors of SCC (p-value < 0.001), and consequently of mammary gland healthy status.

The assignment of each of the 40 bins to a metabolite was obtained by both a consultation of available scientific literature and HUMAN/BOVINE metabolome databases (Wishart et al. 2007; Foroutan et al. 2020). Only 14 bins were assigned to 13 known metabolites, in particular: acetate (1.91..1.94), N-acetyl-glutamine (2.04..2.05 ppm), N-acetyl-lysine (1.80..1.82 ppm), N-acetyl-glucosamine (2.05..2.08), alanine (1.46..1.49), caprate (1.24..1.26 ppm), creatinine (2.90..3.00), phosphocreatine/creatinine (3.03..3.06), 2-hydroxyvalerate (1.41..1.43), isoleucine (0.91–0.94), lactate (1.31–1.34; 4.10–4.13), lactose (3.80..3.83), and valine (1.03–1.06).

3.2 Analysis of Metabolites

Among the 13 identified metabolites, lactose had the highest concentration, as expected and confirmed by other NMR milk studies (Sundekilde et al. 2013; Kim et al. 2021). On the other hand, creatinine (when we considered without phosphocreatine) and N-acetyl-lysine were the least concentrated (Table 1). The metabolites here identified were involved in protein metabolism (alanine, isoleucine, valine), acetyl-carbohydrates and aminoacids metabolism (N-acetyl-glucosamine, N-acetyl-glutamine, N-acetyl-lysine), glycolysis (lactate), fatty acid metabolism (acetate, caprate), urea cycle (creatine-P/creatinine).

Table 1. Descriptive statistics of 13 milk metabolites.

Metabolites	Mean	SE	MIN	MAX	Lower quartile	Upper quartile
Lactose	480.30	2.07	437.00	510.00	472.50	487.50
N-acetyl-glucosamine	26.02	0.66	16.50	37.80	23.15	28.65
Isololeucine	16.41	0.60	10.70	27.60	13.40	18.10
Creatine-P/creatinine	14.23	0.37	9.67	19.90	12.65	16.40
Lactate	9.03	0.41	4.67	14.90	6.82	11.70
Caprate	9.00	0.21	6.26	11.40	8.04	10.10
2-hydroxyvalerate	8.23	0.23	5.22	11.90	7.19	9.16
Acetate	8.04	0.22	5.30	11.50	7.09	8.90
Alanine	7.72	0.20	5.01	10.80	6.86	8.64
Valine	6.34	0.18	3.51	8.46	5.56	7.24
N-acetyl-glutamine	5.84	0.18	4.34	9.56	5.02	6.32
N-acetyl-lysine	4.58	0.13	2.73	6.38	4.03	5.18
Creatinine	2.84	0.10	1.46	4.47	2.52	3.31

SE: Standard Error, MIN: minimum, MAX: maximum

In this study we confirmed acetate, N-acetyl-glucosamine, alanine, caprate, 2-hydroxyvalerate, isoleucine, lactate, lactose, associated with somatic cell count, corroborating similar studies (Sundekilde et al. 2013; Kim et al. 2021; Zhu et al. 2021). While N-acetyl-glutamine, N-acetyl-lysine, creatinine, phoshocreatine, were just identified in other studies where different cattle breeds were compared (Buitenhuis et al. 2013), or milk samples of early and late lactation were analysed (Klein et al. 2010).

The multivariate data analysis revelead statistically significant differences between SCC classes, where only lactose showed a lower concentration in milk samples with high SCC (classified no-healthy), while all other metabolites had a higher concentration in no-healthy milk samples (Figs. 1 and 2).

Lactose is the principal metabolite that predominated in NMR spectra, and we confirmed that lactose content is lower when a intramammary infection is present (Sundekilde et al. 2013; Leitner et al. 2019; Kim et al. 2021). When the number of somatic cell is high, membrane of mammary cells is damaged, blood costituents flow in milk, so lactose is decreased to maintain osmotic pressure. It has been shown that acetate and caprate fatty acid derive from the microorganisms that cause mastitis (Alhussien et al. 2018). Lactate is the final product of anaerobic glycolysis, and it can be produced both by milk microorganisms and by mammary ephitelial cells during a mastitis (Blum et al. 2020). The alanine, valine, and isoleucine metabolic pathways are altered by subclinical *S. agalactiae* mastitis in dairy cattle (Tong et al. 2019). The production of 2-hydroxyvalerate (a short chain fatty acid - SCFA) is influenced by rumen bacteria fermentation. It has been reported that SCFAs have a protective effect on mammary gland inflammatory responses (Hu et al. 2019). It has been shown that N-acetylglucosamine derived by bacteria is able to counteract mammary immune response, during a chronical mastitis caused by *S. aureus* (Marbach et al. 2019).

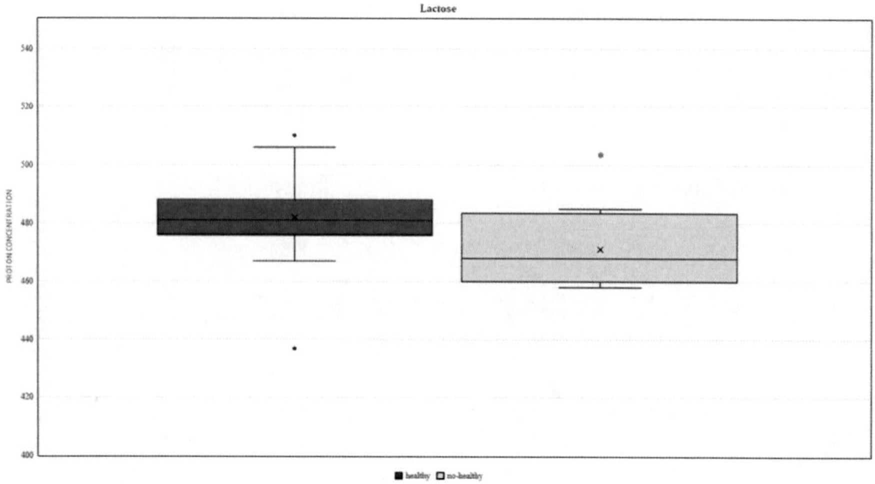

Fig. 1. Boxplot comparing the amount of lactose in the healthy (37 samples with SCC lower than 400.000 cells/mL – dark grey) and no-healthy (7 samples with SCC greater than 400.000 cells/mL – light grey) evaluated using NMR. The healthy samples have a significant ($p < 0.05$) higher value of lactose, than no-healthy samples.

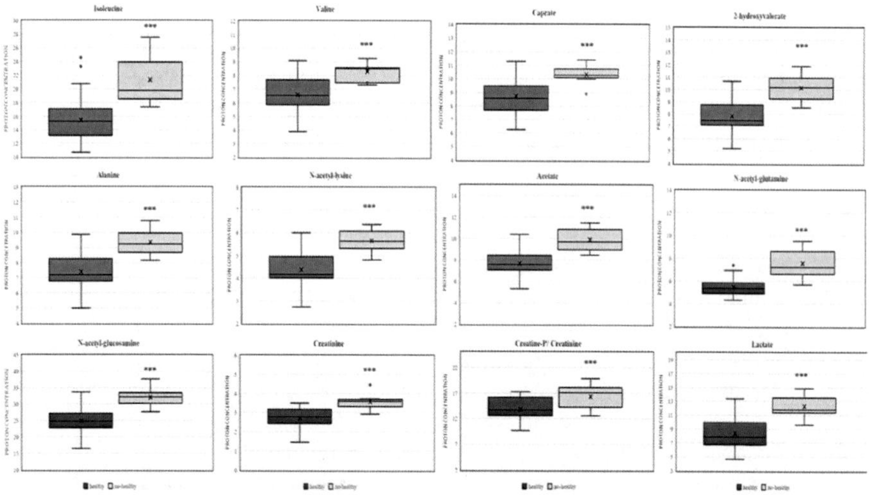

Fig. 2. Boxplots comparing the amount of isoleucine, valine, caprate, 2-hydroxyvalerate, alanine, N-acetyl-lysine, acetate, N-acetyl-glutamine, N-acetyl-glucosamine, creatinine, phosphocreatine/creatinine and lactate in the healthy (37 samples with SCC lower than 400.000 cells/mL – dark grey) and no-healthy (7 samples with SCC greater than 400.000 cells/mL – light grey) evaluated using NMR. In all the 12 metabolites, the healthy samples have a highly significant ($p < 0.001$) lower value, compared with the no-healthy samples.

The process of protein aceylation that produce N-acetylaminoacids in mitochondria has been found associated with immune response during fatty liver disease in dairy cattle (Le-tian et al. 2020).

Phosphocreatine is used as a source of inorganic phosphorus in the conversion of ADP to ATP in the skeletal muscles (Zhang et al. 2021). Creatinine is a waste product produced in the catabolism of phosphocreatine in the muscles and is filtered mostly by the kidneys and excreted in the urine. An higher concentration of creatinine was observed in blood of cow with retained placenta. Cows with retained fetal membranes are at increased risk of metritis, displaced abomasum, and mastitis (Dervishi et al. 2016). In this study we observed a high concentration of both in milk samples with a SCC > 400.000cells/ml, so probably derived from mammary gland with an infection. In comparison with Zhang et al. (2021) study, we could speculate that in presence of a mammary gland infection, there is an altérated urea cycle.

4 Conclusion

This study shows an application of NMR metabolomics for discrimination of dairy milk samples collected from healthy and no-healthy mammary glands. The highest concentration of free amino acids and organic acids in milk were significantly associated with elevated milk SCC level. Some metabolites pointed to the presence of bacteria that caused mastitis in dairy cows (*S. uberis, S. agalactiae, S. aureus, E. coli*). This result could useful to identified cows with subclinic mastitis, so to prevent chronic mastitis. Beside the existing biomarkers reported in literatures, we found new metabolites, such as creatinine, phosphocreatine, and n-acetylaminoacids associated with infected milk samples. Traditional detection method for mastitis-resistant phenotypes don't elucidate the clear pathogenesis of this disease, and it is time-consuming, low sensitivity, low specificity and accuracy (Hao et al. 2018; Li et al. 2019). While these results confirm that metabolomic technology could be a powerfull tool to detect specific metabolites for both a healthy mammary gland and with different grade of inflammation (Mansor et al. 2013). Metabolomic could help researcher to prevent and control mastitis, positively influencing high management costs of this important disease (Hu et al. 2021).

Acknowledgement. This research was funded by MIPAAF in the national research project MIQUALAT (D.M. 16844/7100/2019).

References

Alhussien, M.N., Dang, A.K.: Milk somatic cells, factors influencing their release, future prospects, and practical utility in dairy animals: an overview. Vet. World **11**(5), 562–577 (2018)

Alnakip, M.E., Quintela-Baluja, M., Böhme K., Fernández-No, I., Caamaño-Antelo, S., et al.: The immunology of mammary gland of dairy ruminants between healthy and inflammatory conditions. J. Vet. Med. Article ID 659801, 31 p (2014)

Blum, S.E., Heller, D.E., Jacoby, S., Krifuks, O., Merin, U., et al.: Physiological response of mammary glands to *Escherichia coli* infection: a conflict between glucose need for milk production and immune response. Sci. Rep. **10**, 9602 (2020)

Breda, J., et al.: Towards better nutrition in Europe: evaluating progress and defining future directions. Food Policy **96**, 101887 (2020)

Buitenhuis, A.J., Sundekilde, U.K., Poulsen, N.A., Bertram, H.C., Larsen, L.B., Sørensen, P.: Estimation of genetic parameters and detection of quantitative trait loci for metabolites in Danish Holstein milk. J. Dairy Sci. **96**, 3285–3295 (2013)

Dervishi, E., Zhang, G., Hailemariam, D., Dunn, S.M., Ametaj, B.N.: Occurrence of retained placenta is preceded by an inflammatory state and alterations of energy metabolism in transition dairy cows. J. Anim. Sci. Biotechnol. **7**, 26 (2016)

Foroutan, A., Fitzsimmons, C., Mandal, M., Piri-Moghadam, H., Zheng, J., et al.: The bovine metabolome. Metabolites **10**, 233 (2020)

Garcia, S.N., Osburnb, B.I., Cullora, J.S.: A one health perspective on dairy production and dairy food safety. One Health **7**, 100086 (2019)

Guimarães, J.L.B., Brito, M.A.V.P., Lange, C.C., Silva, M.R., Ribeiro, J.B., et al.: Estimate of the economic impact of mastitis: a case study in a Holstein dairy herd under tropical conditions. Prev. Vet. Med. **142**, 46–50 (2017)

Hao, J.F., Li, X.W., Li, X.B., Zhao, C.X., Liu, G.W., et al.: New progress of research on diagnosis technology in cow recessive mastitis. Heilongjiang Anim. Sci. Vet. Med. **2018**, 63–5, 69 (2018)

Hu, X., Li, S., Fu, Y., Zhang, N.: Targeting gut microbiota as a possible therapy for mastitis. Eur. J. Clin. Microbiol. Infect. Dis. **38**(8), 1409–1423 (2019)

Hu, H., Fang, Z., Mu, T., Wang, Z., Ma, Y., et al.: Application of metabolomics in diagnosis of cow mastitis: a review. Front. Vet. Sci. **8**, 747519 (2021)

Kim, H.S., et al.: Exploration of metabolite profiles in the biofluids of dairy cows by proton nuclear magnetic resonance analysis. PLoS ONE **16**(1), e0246290 (2021)

Klein, M.S., et al.: Nuclear magnetic resonance and mass spectrometry-based milk metabolomics in dairy cows during early and late lactation. J. Dairy Sci. **93**, 1539–1550 (2010)

Leitner, G., Lavon, Y., Merin, U., Jacoby, S., Blum, S.E., Krifucks, O., et al.: Milk quality and milk transformation parameters from infected mammary glands depends on the infecting bacteria species. PLoS ONE **14**(7), e0213817 (2019)

Le-tian, Z., Cheng-zhang, H., Xuan, Z., Zhang, Q., Zhen-gui, Y., et al.: Protein acetylation in mitochondria plays critical functions in the pathogenesis of fatty liver disease. BMC Genom. **21**, 435 (2020)

Li, G.D., Lu, D.Y., Tian, X.Z., Ji, P.Y., Guo, J.P., et al.: Research progress of omics technologies in cow mastitis. Sci. Agric. Sin. **52**, 350–358 (2019)

Luangwilai, M., Duangmal, K., Chantaprasarn, N., Settachaimongkon, S.: Comparative metabolite profiling of raw milk from subclinical andclinical mastitis cows using1H-NMR combined with chemometric analysis. Int. J. Food Sci. Tech. **56**, 493–503 (2021)

Mansor, R., Mullen, W., Albalat, A., Zerefos, P., Mischak, H., et al.: A peptidomic approach to biomarker discovery for bovine mastitis. J Proteomics. **85**, 89–98 (2013)

Marbach, H., Mayer, K., Vogl, C., Lee, J.Y.H., Monk, J.R., et al.: Within-host evolution of bovine *Staphylococcus aureus* selects for a SigB-deficient pathotype characterized by reduced virulence but enhanced proteolytic activity and biofilm formation. Sci. Rep. **9**, 13479 (2019)

Maréchal, C.L., Thiéry, R., Vautor, E., Loir, Y.L.: Mastitis impact on technologicalproperties of milk and quality of milk products: a review. Dairy Sci. Technol. **91**, 247–282 (2011)

Sundekilde, U.K., Poulsen, N.A., Larsen, L.B., Bertram, H.C.: Nuclear magnetic resonance metabomics reveals strong association between milk metabolites and somatic cell count in bovine milk. J. Dairy Sci. **96**, 290–299 (2013)

Taničin, V., Miklás, Š, Mačuhová, L.: Possible physiological and environmental factors affecting milk production and udder health of dairy cows. A review. Slovak J. Anim. Sci. **51**, 32–40 (2018)

Tong, J., Zhang, H., Zhang, Y., Xiong, B., Jiang, L.: Microbiome and metabolome analyses of milk from dairy cows with subclinical *streptococcus agalactiae* mastitis—potential biomarkers. Front. Microbiol. **10**, 2547 (2019)

Wishart, D.S., Tzur, D., Knox, C., Eisner, R., Guo, A.C., et al.: HMDB: the human metabolome database. Nucleic Acids Res. **35**, D521–D526 (2007)

World Health Organization. Global and regional food consumption patterns and trends (2018). http://www.who.int/nutrition/topics/3_foodconsumption/en/index4.html

Zhang, G., Tobolski, D., Zwierzchowski, G., Mandal, R., Wishart, D.S., Ametaj, B.N.: A targeted serum metabolomics GC-MS approach identifies predictive blood biomarkers for retained placenta in holstein dairy cows. Metabolites **11**, 633 (2021)

Zhang, X., Chen, X., Xu, Y., Yang, J., Du, L., Zhou, Y.: Milk consumption and multiple health outcomes: umbrella review of systematic reviews and meta-analyses in humans. Nut. Metab. **18**, 7 (2021)

Zhu, C., Tang, K., Lu, X., Tang, J., Laghi, L.: An Untargeted Metabolomics Investigation of Milk from Dairy Cows with Clinical Mastitis by 1H-NMR. Foods **10**, 1707 (2021)

Spatial Variability of Ammonia Concentrations in an Open-Sided Dairy Barn

Provvidenza Rita D'Urso(✉), Claudia Arcidiacono, and Giovanni Cascone

Department of Agriculture, Food and Environment (Di3A), Building and Land Engineering Section, University of Catania, Via Santa Sofia, 100, 95123 Catania, Italy
provvidenza.durso@phd.unict.it

Abstract. The variability of daily concentrations of ammonia (NH_3) was assessed at different sampling locations (SLs), horizontally and vertically distributed at the centre of an open-sided dairy barn in Mediterranean climate. Concentrations of NH_3 were continuously measured at 6 SLs in a cubicle free-stall dairy barn located in the province of Ragusa (Italy). Data acquired were organised in a dataset and statistical analyses were performed to delineate potential differences in gas concentration distribution. Application of a one-way analysis of variance (ANOVA) tested the equality of the mean values of gas concentrations for each group of the parameters analysed for groups of SLs. In the post-hoc analysis the mean values were separated by Tukey's honestly significant difference at $P < 0.05$. NH_3 concentrations changed in time and with SLs during the day inside the barn. There was a significant difference ($P < 0.001$) for NH_3 concentrations at different SLs. In detail, the highest values of NH_3 concentrations were recorded at 0.40 m from the floor where the trend of the gas showed the highest variability, whereas the lowest values of NH_3 were measured at 1.55 m from the floor. At 0.40 m and 2.70 m from the floor, concentrations decreased in the prevailing air direction due to the dispersion and dilution effects. At 1.55 m from the floor, this was not confirmed due to the effect of the fans on the air flux.

Keywords: Livestock buildings · Ammonia measurements · Sampling distribution · Monitoring · Dairy cows

1 Introduction

Ammonia (NH_3) emissions from livestock barns have been the focus of research around the world for many years [1]. It is widely acknowledged that the estimation of emissions from naturally-ventilated (NV) buildings is a challenging task due to the difficulties in the determination of ventilation rate [2]. According to Ogink et al. [3], the NH_3 emissions can be estimated with the tracer gas ratio method, based on the product of the ventilation rate and the difference between indoor and outdoor ammonia concentrations. Since this method requires accurate measurement of tracer release rate and a representative mixing ratio estimation between NH_3 and tracer gases, knowledge on the spatial variability of pollutants and tracer gases can contribute in defining areas within the barn where mixing ratios are considered to be representative for the whole barn [4].

M. Biocca et al. (Eds.): SHWA 2020, LNCE 252, pp. 76–84, 2022.
https://doi.org/10.1007/978-3-030-98092-4_9

The barn structure and the ventilation system are relevant feature in the NH_3 estimation. Several studies [5–7] were carried out in NV barns with large openings, protected by nets. Curtains were installed at the side walls to protect the cows during windy and very cold weather. The NV system was integrated by a ventilation system. In the analyses performed in a dairy farm located in the south of Sweden [8, 9] the computation of gaseous emissions was performed in a NV barn where the NV system was automatically regulated by ventilation flaps mounted on the sidewalls below the eaves and also at the ridge. Other studies [10, 11] were carried out in NV buildings with an open ridge vent and curtains located at the openings to control the inlet air flux. In the study carried out by Bjerg et al. [12], the structures had an open ridge and large openings in side walls.

Most of the current research has computed emissions from NV dairy barn located in temperate oceanic climate zone (Cfb in the Koppen classification), where the animals are usually kept in buildings with mechanical or natural ventilation. Few studies have been performed on very open dairy barn characterized by the absence of perimeter walls in Mediterranean climate. These structures are typical in hot summer Mediterranean climate (Csa in Koppen classification) where climate is characterised by temperatures above 22 °C in the warmest month and precipitation values lower than 30 mm in the driest month in summer. In these dairy buildings, the natural ventilation is generally combined with a cooling system (e.g., fans and sprinklers) to reduce heat stress of the cows [13–16].

Recently, D'Urso et al. [17] investigated the horizontal distribution of NH_3 concentrations in an open sided dairy barn in Mediterranean area, but no information was provided about the vertical distribution of the gas.

In order to fill this gap, multi-location measurements were carried out at different sampling locations (SLs) in a open NV barn. In this direction, the main objectives of this study were related to the acquisition of knowledge on gas concentration variability and distribution at different sampling points.

2 Materials and Methods

2.1 Barn Description and Management

The site of the experiment was located in the province of Ragusa (Southern Italy), which is a geographical area with the highest number of dairy barns in the Mediterranean basin. Data were acquired in a cubicle free-stall dairy barn from 11th to 18th June 2021 during a period characterized by homogeneous climatic conditions as well as barn management.

The open structure (Fig. 1) was defined by three sides completely open and one side closed by a continuous wall with four small openings. The rectangular plan (i.e., 55.50 m long and 20.80 m wide) was covered by a concrete floor with a roof made of fibre-reinforced concrete corrugated panels. The heights at the ridge vent and at the eave was 7 m and 4 m, respectively.

Fig. 1. Indoor view of the barn with two sides completely opened on the edge.

Different functional areas can be distinguished in the plan (Fig. 2): the feeding area, subdivided in the feeding alley, the manger, and the feeding passage; the resting area with 64 head-to-head cubicles in two rows and three pens; the service alley for herd management; offices and boxes for calves, located in the south-west side.

Fig. 2. Plan of the barn with the localization of the functional areas, the position of the fans and the two sampling poles for the measurement of gas concentrations.

Beside the natural ventilation of indoor space due to the open structure, the barn was equipped with sprinklers and fans in the feeding alley and a fogging system with fans in the resting area. The fans in both the resting area and feeding alley had a tilt angle of 20° from the horizontal. In detail, fans in the resting area had a rotation axis located at 2.75 m from the floor along the longitudinal axis of the barn, whereas fan in the feeding alley had a rotation axis at 2.70 m from the floor and parallel to the longitudinal axis of the feeding alley.

Fifty-seven Friesian cows (i.e., 21 in the first pen, 14 in the second pen and 21 in the third pen) were milked twice a day at about 5:15 a.m. and 5:00 p.m. The floor of the barn was cleaned by a tractor with a scraper every day at about 7:40 a.m. A mixed ratio was delivered at 12 a.m. for at libitum consumption. Daily feed consumption per cow was 7 kg grass silage, 9 kg protein-based concentrate, 6.5 kg corn silage (i.e., flour and flakes), 0.1 kg sodium bicarbonate, 0.2 kg fat, 0.08 kg salt, 0.05 kg urea, 0.03 kg yeast, 30 kg citrus pulp and 0.2 kg buffer package.

The cooling system in the feeding alley was operated from 8.30 a.m. to 9.30 a.m., from 12.00 a.m. to 2.00 p.m. and from 4.00 p.m. to 5.00 p.m., whereas the cooling system in the resting area was operated from 9.30 a.m. to 12.00 a.m., 2.00 p.m. to 4.00 p.m. From 7.00 p.m. to 5.00 a.m., only fans in the resting area were operated.

2.2 Experimental Set-Up

Measurements of NH_3 concentrations were continuously recorded by six portable devices (SKY2000-M2, Digitron Italia, Ferentino, Frosinone, Italy), organized in two sampling poles located in the central area of the barn (Fig. 1). In detail, the sampling poles are placed one below the fan and one further away in order to study the two typical conditions in the barn.

For each sampling pole, three devices were located at 3 different vertical distances from the floor: 0.40 m from the floor, 1.55 m from the floor and 2.70 m from the floor (Fig. 3). Each portable device had an internal sampling pump linked to a sampler tube in PTFE (polytetrafluoroethylene). An air-filter was attached to the end of each sampling tube to keep the sampler free of particles. The measurement range was 0–100 ppm, resolution 0.01 ppm and precision 2%FS. The measurement frequency was 2 min and 30 s. The calibration of each instrument was performed by the manufacturer just before the measurements.

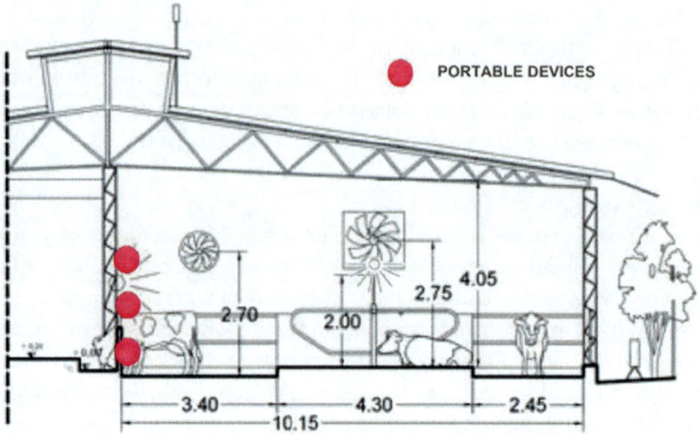

(a)

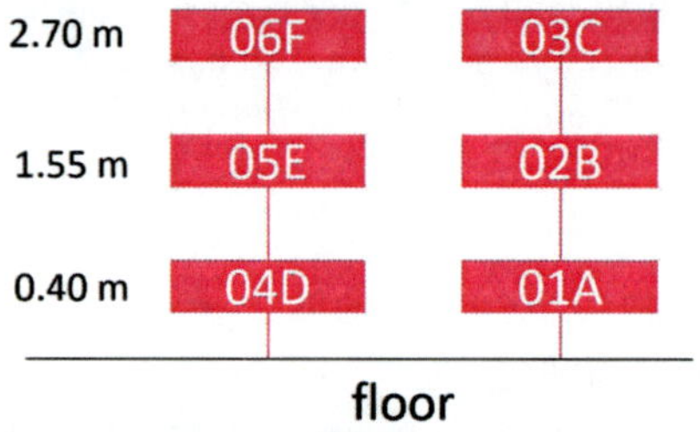

(b)

Fig. 3. Section of the barn with vertical distribution of the sampling locations (SLs) in each sampling pole (a); nomenclature of each SLs (b).

2.3 Data Analysis

Data recorded during the measurement campaign were statistically analysed in order to study differences in the gas distribution in the barn. The one-way analysis of variance (ANOVA) was carried out by using NH_3 concentrations acquired at six SLs. In the post hoc analyses, the mean values were separated by Tukey's honestly significant difference at $p < 0.05$. Then, the daily variability of the gas was studied for each SLs.

3 Results and Discussion

3.1 Spatial Distribution of NH_3 Concentrations

In the literature [8, 17, 18], it was found that the NH_3 concentrations were not uniform in the barn plan. Besides the heterogeneous distribution in the plan, this study proved vertical variability of NH_3 concentrations. The result of the one-way ANOVA showed a significant difference ($p < 0.001$) between NH_3 concentrations recorded at different SLs in the barn. The highest NH_3 concentrations were recorded in the SLs near the floor, whereas NH_3 decreased in the SLs at 1.55 m and 2.70 m. The results of this study appear tightly related to the effect of the action of fans that dilutes and flush air at about 1.55 m from the floor. Moreover, the gaseous production occurs mainly at the floor level in the AOZ, and then NH_3 rises upward very fast because it is lighter (bulk density of 0.66 kg/m^3 in standard conditions) than air (bulk density of 1.293 kg/m^3 in standard conditions). The air flux of the fans increases the dilution effect on NH_3 and its transport toward the outdoor environment. In fact, the fans were proved to smooth the concentration gradients inside the barn [6]. Based on the results, concentrations decreased at 0.40 m and 2.70 m from the floor in the prevailing air direction due to the dispersion and dilution effects as it was previous found at the floor level in previous studies [17, 20]. At 1.55 m from the floor, this was not confirmed and higher NH_3 concentrations were found at SL 02B than those at SL 05E. A relevant factor that explains this result is the position and the tilt angle of the fans near the SLs. In fact, the fan above the SL 02B did not have a direct effect on NH_3 measured at that location, whereas NH_3 concentrations at SL 05E were diluted and flushed from the air of the fan located between the two pens. Moreover, there was not a significant difference between NH_3 concentrations at SL 02B and SL 03C.

Conversely, in a NV barn without fans, such as that reported by Mendes et al. [4], the results are different from those obtained in this study. Specifically, NH_3 concentrations were more stable at 3 m from the floor above the animal occupied zone (AOZ). This observation proves the importance of carrying out specific experiments in the different operative conditions of a barn to provide a suitable basis for effective gas measurement guidelines (Table 1).

Table 1. Results of the statistical analyses for indoor temperature at different days

NH_3 concentrations (ppm)							
SLs	Mean value	SD	Tukey Post Hoc Test*				
04D	10.46	3.2	A				
01A	7.90	2.81		B			
06F	4.24	2.39			C		
03C	2.49	3.19				D	
02B	2.48	1.78				D	
05E	1.16	1.33					E

*Groups of values that do not share a letter (A, B, C, D, E) are significantly different.

3.2 Daily Variation of NH_3 Concentrations

The variability of NH_3 concentrations during the day was analysed for each SLs. Figure 4 showed that the trend of the gas is far more defined for SLs near the floor. In detail, the variability of NH_3 was characterised by four peaks (i.e., 7 a.m., 11 a.m., 15 a.m., 18–19 a.m.) during the day for both SL 01A and SL 04D. According to previous studies carried out in this barn typology, among the main influencing factors it was found that these peaks are related to the cleaning of the barn [17], the number of the milking sessions [19], the microclimatic conditions and management of the cooling system [13].

In the other SLs, information on gas variability was reduced. In fact, the trend of NH_3 was almost constant and showed a weak increase at 12 a.m. for SL 02B, SL 05E, SL 06F and two peaks at 11 a.m. and 3 a.m. Therefore, the dilution effect of fans on the gas measurements affected also measurements on a daily basis.

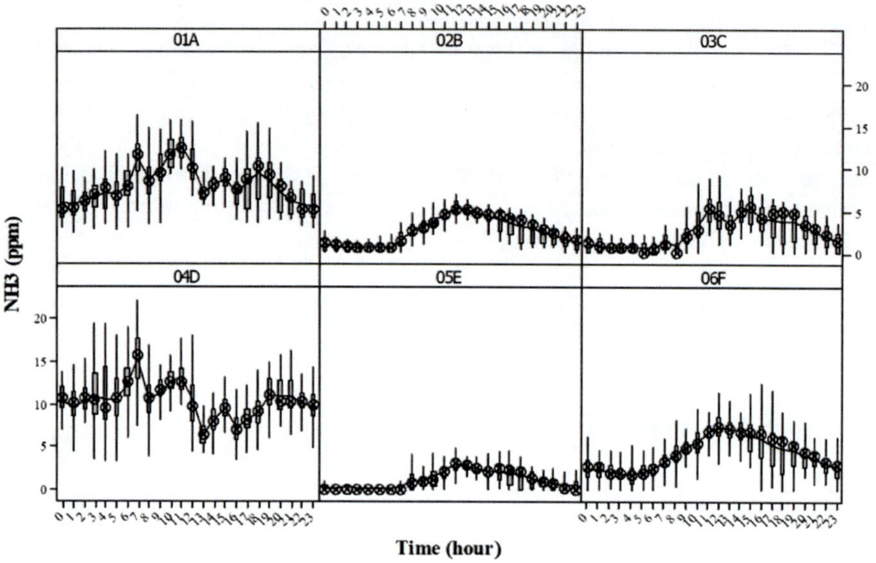

Fig. 4. Variability of NH_3 concentrations (ppm) during the day at six SLs.

4 Conclusion

A study of the variability of NH_3 concentrations in an open dairy barn in a Mediterranean area was presented. By applying a statistical analysis, it was found a heterogeneous distribution of the gas at different heights from the floor. In the SLs located near the floor, the gas concentrations were the highest and the trend of the gas varied expressed different peaks. The gas concentrations measured at 1.55 m and 2.70 m from the floor were influenced by environmental conditions due to the semi-open envelope of the building and the presence of the cooling system with fans typical of this type of barns. Further research is in progress to perform specific experiments in the different operative conditions of the barn and during different seasons in order to contribute to the construction of a suitable basis for effective gas measurement guidelines.

Acknowledgements. The authors are grateful to the farm ALPA S.S. for providing the opportunity of carrying out the tests.

Funding. The research study was funded by the University of Catania through the 'Piano incentivi per la ricerca di Ateneo 2020–2022-Linea 2′ project on 'Engineering solutions for sustainable development of agricultural buildings and land—LANDSUS' (ID: 5A722192152) coordinated by Prof. Claudia Arcidiacono. The INNOVA device was funded by the project "Centro per l'innovazione dei sistemi di qualità tracciabilità e certificazione dell'agroalimentare"—AGRIVET (ID: G46D15000170009).

References

1. Zhang, G., et al.: Emission of ammonia and other contaminant gases from naturally ventilated dairy cattle buildings. Biosyst. Eng. **92**, 355–364 (2005)
2. Calvet, S., et al.: Measuring gas emissions from livestock buildings: a review on uncertainty analysis and error sources. Biosys. Eng. **116**, 221–231 (2013)
3. Ogink, N.W.M., Mosquera, J., Calvet, S., Zhang, G.: Methods for measuring gas emissions from naturally ventilated livestock buildings: developments over the last decade and perspectives for improvement. Biosyst. Eng. **116**, 297–308 (2013)
4. Mendes, L., Edouard, N., Ogink, N., Van Dooren, H.J.C., FerreiraTinôco, I.D.F., Mosquera, J.: Spatial variability of mixing ratios of ammonia and tracer gases in a naturally ventilated dairy cow barn. Biosyst. Eng. **129**, 360–369 (2015)
5. Fiedler, M., et al.: Spatial distribution of air flow and CO_2 concentration in a naturally ventilated dairy building. Environ. Eng. Manag. J. **13**, 2193–2200 (2014)
6. König, M., Hempel, S., Janke, D., Amon, B., Amon, T.: Variabilities in determining air exchange rates in naturally ventilated dairy buildings using the CO_2 production model. Biosyst. Eng. **174**, 249–259 (2018)
7. Janke, D., et al.: Calculation of ventilation rates and ammonia emissions: comparison of sampling strategies for a naturally ventilated dairy barn. Biosyst. Eng. **198**, 15–30 (2020)
8. Ngwabie, N., Jeppsson, K.-H., Nimmermark, S., Swensson, C., Gustafsson, G.: Multi-location measurements of greenhouse gases and emission rates of methane and ammonia from a naturally-ventilated barn for dairy cows. Biosyst. Eng. **103**, 68–77 (2009)
9. Ngwabie, N.M., Jeppsson, K.-H., Gustafsson, G., Nimmermark, S.: Effects of animal activity and air temperature on methane and ammonia emissions from a naturally ventilated building for dairy cows. Atmos. Environ. **45**, 6760–6768 (2011)
10. Hempel, S., et al.: Non-linear temperature dependency of ammonia and methane emissions from a naturally ventilated dairy barn. Biosyst. Eng. **145**, 10–21 (2016)
11. Schmithausen, A.J., et al.: Quantification of methane and ammonia emissions in a naturally ventilated barn by using defined criteria to calculate emission rates. Animals **8**, 75 (2018)
12. Bjerg, B., Zhang, G., Madsen, J., Rom, H.B.: Methane emission from naturally ventilated livestock buildings can be determined from gas concentration measurements. Environ. Monit. Assess. **184**(10), 5989–6000 (2012). https://doi.org/10.1007/s10661-011-2397-8
13. D'Urso, P.R., Arcidiacono, C., Cascone, G.: Environmental and animal-related parameters and the emissions of ammonia and methane from an open-sided free-stall barn in hot mediterranean climate: a preliminary study. Agronomy **11**, 1772 (2021)
14. Honig, H., et al.: Performance and welfare of high-yielding dairy cows subjected to 5 or 8 cooling sessions daily under hot and humid climate. J. Dairy Sci. **95**, 3736–3742 (2012)
15. Porto, S.M., D'Emilio, A., Cascone, G.: On the influence of the alternation of two different cooling systems on dairy cow daily activities. J. Agric. Eng. **48**, 21–27 (2017)
16. Drwencke, A.M., et al.: Innovative cooling strategies: Dairy cow responses and water and energy use. J. Dairy Sci. **103**(6), 5440–5454 (2020)

17. D'Urso, P.R., Arcidiacono, C., Valenti, F., Cascone, G.: Assessing influence factors on daily ammonia and greenhouse gas concentrations from an open-sided cubicle barn in hot mediterranean climate. Animals **11**, 1400 (2021)
18. Fiedler, A., Müuller, H.-J.: Emissions of ammonia and methane from a livestock building natural cross ventilation. Meteorol. Z. **20**, 59–65 (2011)
19. D'Urso, P.R., Arcidiacono, C.: Effect of the milking frequency on the concentrations of ammonia and greenhouse gases within an open dairy barn in hot climate conditions. Sustainability **13**, 9235 (2021)
20. Arcidiacono, C., Porto, S.M.C., Cascone, G.: On ammonia concentrations in naturally ventilated dairy houses located in Sicily. Agric. Eng. Int. CIGR J. **2015**, 294–310 (2015)

The Effect of Microclimatic Conditions on Ammonia Emissions from an Open-Sided Dairy Barn During Spring

Provvidenza Rita D'Urso(✉), Claudia Arcidiacono, and Giovanni Cascone

Department of Agriculture, Food and Environment (Di3A) Building and Land Engineering Section, University of Catania, Via Santa Sofia, 100, 95123 Catania, Italy
provvidenza.durso@phd.unict.it

Abstract. The aim of the study was to assess the influence of microclimatic conditions on ammonia (NH_3) emissions in an open dairy barn during spring in Mediterranean climate. Measurements were carried out in a cubicle free-stall dairy barn located in the province of Ragusa (Italy) from 7th to 21st June 2016. Concentrations of NH_3 as well as carbon dioxide (CO_2) were continuously measured by a photo-acoustic analyser. Measurements of microclimatic variables (i.e., temperature, relative humidity, wind speed, and wind direction) were acquired by sensors installed inside the barn. The CO_2 mass balance method was applied in order to estimate NH_3 emissions.

Data collected were organised in a dataset to carry out statistical analyses on gas emissions for groups of days characterised by different climatic parameters. The first group (G1) was characterised by higher values of the parameters compared to the second group (G2). The one-way analysis of variance (ANOVA) and correlation analyses were applied in order to identify significant differences for tested groups and relations among them. In the post-hoc analysis the mean values were separated by Tukey's honestly significant difference at $P < 0.05$.

NH_3 emissions in G1 were significantly different from those in G2 with higher NH_3 emissions in G2. An almost constant trend was found for NH_3 emissions in G1. During G2, the correlation between emissions and air temperature was equal to −0.67 ($p < 0.001$), whereas NH_3 emissions showed a different trend based on air velocity. The results of this study highlighted that NH_3 emissions in open structures were influenced by both microclimatic variables and barn management.

Keywords: Livestock buildings · Ammonia measurements · Microclimatic variables monitoring · Dairy cows · Emission evaluation

1 Introduction

The gaseous emissions from dairy houses not only have a relevant impact on dairy cows' production but affect the environment [1]. Among the main harmful gases, ammonia (NH_3) is an atmospheric pollutant responsible for acidification and

M. Biocca et al. (Eds.): SHWA 2020, LNCE 252, pp. 85–94, 2022.
https://doi.org/10.1007/978-3-030-98092-4_10

eutrophication of several ecosystems [2]. In the atmosphere, it reacts with other compounds leading to the formation of atmospheric particulate matter [3] with harmful effects on both welfare and health of humans and animals [4].

Along the milk supply chain, livestock barns are the main responsible of NH_3 emissions to the atmosphere due to their substantial footprints [5]. NH_3 is mainly produced from the urine of the cows deposited on the barn floor. Specifically, the manure originated from the mix of urine and faeces can lead to volatilisation and emission of NH_3 by a series of enzymatic reactions (e.g., bacteriological degradation, and urea mineralization) [6].

Differences in the diet of the animals, the cow behaviour, the barn management, the climatic conditions, the dairy barn structure, the ventilation system, the flooring system, the characteristic of the manure and the manure removal system affect the NH_3 emissions [3, 7–10]. In naturally ventilated structures, the production of polluting gases is strongly affected by the continuously changing climatic conditions. Through a correlation analysis, Zou et al. [1] and Ngwabie et al. [11] found a high relation between temperature and NH_3 emissions. Moreover, the analyses carried out by Hempel et al. [12] on NH_3 proved its exponential dependency on temperature. In the study of Li et al. [2], the high NH_3 emission rate found after the noon time was attributed to the higher temperature compared to the other hours of the day. In addition, air relative humidity, wind speed and direction, hour of the day and day of the year were proved to significantly affect NH_3 emissions by Saha et al. [13], and specifically wind speed and interactions of external wind speed and direction [14].

To the authors' knowledge, few studies have been performed in open dairy buildings in Mediterranean regions, where the barn structure is generally characterised by wide openings and cooling systems (e.g., sprinklers, and fans). This specific building design is adopted to increase the ventilation rate in the barn and reduce the heat stress risk for cows.

The present study aims at assessing the influence of microclimatic parameters in open dairy barns. The specific objectives are to identify groups of days characterised by similar climatic conditions and to study the variability of NH_3 emissions at different groups of variables.

2 Materials and Methods

2.1 Barn Description and Management

The study was carried out in a cubicle free-stall dairy barn located in the province of Ragusa (Southern Italy), which is a geographical area with the highest density of dairy barns in the Mediterranean basin. Data were acquired from 7th to 21st June 2016.

The barn facility (Fig. 1) had an open structure with three sides completely open and one side (i.e., the south-west side) closed by a continuous wall with four small openings. The rectangular plan was about 55.50 m long and 20.80 m wide and covered by a concrete floor, whereas the heights at the ridge vent and at the eave was 7 m and 4 m, respectively.

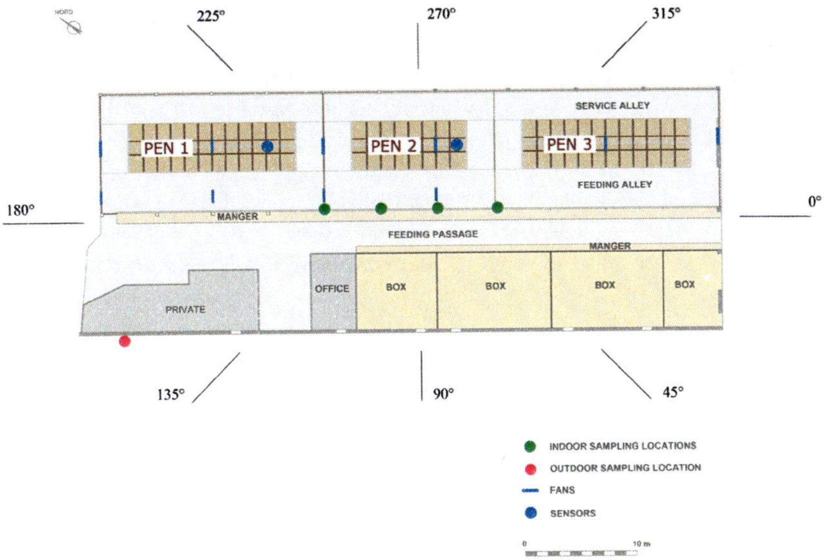

Fig. 1. Plan of the barn containing the description of the functional areas, the location of the sampling points and sensors, and the air velocity directions.

The dairy barn had different functional areas: the feeding area, located longitudinally along the central area of the barn, was composed of the feeding alley, the manger, and the feeding passage; the resting area had 64 head-to-head cubicles organised in two rows and subdivided in three pens; a service alley, in the north-east side, was used for herd management; offices and boxes for calves, located in the south-west side.

The open structure of the barn ensured a natural ventilation of indoor space. In addition, sprinklers with fans and a fogging system with fans were installed in the feeding alley and in the resting area, respectively, in order to reduce heat stress in hot climatic conditions.

With regard to the barn management, the milking occurred twice a day at about 5:00 a.m. and 5:00 p.m.; the cleaning of the floor was performed by a tractor with a scraper every day at about 7:30 a.m.; and the feed was delivered after the cleaning of the barn.

2.2 Experimental Set-Up

Data of gas concentrations and climatic parameters were monitored during the experimental period. An infrared photo-acoustic analyser (INNOVA) composed of a Multigas Monitor mod 1412i and a multipoint sampler 1409/12 (Lumasense Technology A/S) continuously recorded concentrations of NH_3 and CO_2 at five sampling locations (SLs) with a sampling frequency of 15 min. Each inlet channel had a PTFE (polytetrafluoroethylene) tubing to reduce the adsorption of samples and an air filter was located at the end of each sampling tube to keep the multipoint free of particles. The plan distribution of the SLs consisted of four indoor SLs in the central area of the barn at 20 cm from the barn floor and an outdoor SL located at 3 m above the floor and upwind at 2 m from the front of the barn (Fig. 1). The calibration of the instrument was performed by the manufacturer just before the measurements.

Air temperature, relative humidity, indoor air and wind speed and direction were acquired by sensors located at 2.0 m above the floor inside the barn (Fig. 1). These variables were measured with a sampling frequency of five seconds by a data-logger CR10X (Campbell, UK) that, every five minutes, calculated the average values and stored them in memory locations. In detail, the sensors had the following characteristics: air temperature was measured by platinum thermo-resistances (Pt 100 Ω 0 °C) (Rotronic Italy s.r.l., Milano, Italy) with a measurement range between −40 and +60 °C and a precision of ±0.2 °C (at 20 °C); air relative humidity was measured by an hygrometer (Rotronic Italy s.r.l., Milano, Italy) with a sensitivity of ±0.04%RH/°C and a precision of ±2% (at 20 °C); the anemometers (WindSonic, Gill instruments Ltd., Lymington, UK) had two-dimensional sonic sensors with a velocity measuring interval of 0–60 m s^{-1} (i.e., a precision of ±2% (at 12 m s^{-1}), a resolution of 0.01 m s^{-1} and a threshold of 0.01 m s^{-1}) and a direction measuring interval of 0–359° (i.e., a precision of ±3% (at 12 m s^{-1}) and a resolution of 1°).

2.3 Estimation of NH_3 Emissions

Since the gas concentration is not uniform in the barn and changes with time [15], in this study an ideal mix of the gas in the building was hypothesised in order to obtain an estimation of the NH_3 emissions in the barn.

The NH_3 emissions were estimated using the following equation (Eq. 1):

$$E_t = Q\,(C_{in} - C_{out}) \tag{1}$$

where E_t represents the emission rate of gas (g h^{-1}), Q is the ventilation rate (m^3 h^{-1}), and C_{in} and C_{out} (g m^{-3}) represent the average concentrations of the gas inside the barn (i.e., average of the measurements at the 4 central SLs) and outside the building, respectively. The Q was estimated by applying the CO_2 mass balance method, where the CO_2 produced from animal respiration and manure was used as a tracer gas. In detail, the Eq. 2 applied for Q was the following:

$$\mathrm{Q} = \mathrm{P_{CO2}} \times \mathrm{N}/(\mathrm{C_{CO2in}} - \mathrm{C_{CO2out}}) \tag{2}$$

where P_{CO2} is the excretion rate of CO_2 by one cow (g cow^{-1} h^{-1}), N represents the number of cows housed inside the building, and $\mathrm{C_{CO2in}}$ and $\mathrm{C_{CO2out}}$ are the hourly average concentrations of the gas inside (i.e., average of the measurements at the 4 central SLs) and outside the building in the prevailing air direction (135°–225°), respectively (g m^{-3}). The P_{CO2} was estimated following the CIGR instructions [16] and neglecting the CO_2 production from manure.

2.4 Data Analysis

The data recorded during the measurement campaign were statistically analysed in order to identify homogeneous days of microclimatic conditions in the barn. Comparisons between different groups of days were carried out by applying the one-way analysis of variance (ANOVA) on climatic variables. In the post hoc analyses, the mean values were separated by Tukey's honestly significant difference at $p < 0.05$. Then, two homogeneous groups of climatic variables were identified, and the influence of climatic variables on NH_3 emissions was statistically assessed. Differences between groups of NH_3 emissions were studied through the one-way ANOVA and the Tuckey's post hoc test for $p < 0.05$.

3 Results and Discussion

3.1 Microclimatic Conditions

The microclimatic conditions were not uniform during the observation periods. The results of statistical analyses showed significant differences ($p < 0.001$) for indoor temperature (Table 1) and indoor air velocity (Table 2) at different days in June.

Table 1. Results of the statistical analyses for indoor temperature at different days.

Indoor temperature (°C)							
Day	Mean value	SD	Tukey post hoc test				
16	31.7	6.3	A				
17	30.7	4.7	A				
18	27.5	4.1		B			
15	26.2	6.3		B	C		
19	25.2	3.6			C	D	
12	23.2	3.8				D	E
21	23.2	3.6				D	E
13	23.1	3.1				D	E
10	22.8	3.5					E
7	22.7	5.1					E
8	22.6	4.5					E
11	22.4	4.5					E
9	22.1	3.6					E
14	21.7	3.4					E
20	21.7	4.1					E

*Groups of values that do not share a letter (A, B, C, D, E) are significantly different.

Table 2. Results of the statistical analyses for indoor air velocity at different days

Indoor air velocity ($m\ s^{-1}$)							
Day	Mean value	SD	Tukey post hoc test*				
16	1.21	0.72	A				
17	0.86	0.54		B			
18	0.83	0.41		B	C		
19	0.82	0.42		B	C	D	
13	0.63	0.51			C	D	E
14	0.61	0.40			C	D	E
21	0.60	0.41				D	E
7	0.60	0.45				D	E
20	0.58	0.41					E
11	0.58	0.41					E
15	0.57	0.46					E
8	0.55	0.37					E
9	0.54	0.39					E
10	0.51	0.46					E
12	0.50	0.39					E

*Groups of values that do not share a letter (A, B, C, D, E) are significantly different.

Based on the Tukey post hoc tests, two groups of days having homogeneous climatic conditions were identified: G1 was the period composed of days with higher indoor air temperature and velocity (i.e., 16th, 17th, 18th, and 19th June), whereas G2 was a period with days characterised by a lower indoor air velocity and temperature (i.e., 8th, 9th, 10th, and 11th June).

3.2 Ammonia Emissions

The mean value of NH_3 emissions during G1 was 5.25 g $LU^{-1}\ h^{-1}$, whereas the mean value of NH_3 emissions during G2 was 7.04 g $LU^{-1}\ h^{-1}$. The one-way ANOVA highlighted that NH_3 emissions were significantly affected by microclimatic conditions ($p < 0.001$). In detail, the emissions during G2 were higher than those recorded during G1. In the literature, it was found that a higher air velocity reduces gas concentrations [15] with effects on gas emissions. Moreover, a high wind speed reduces gas emissions

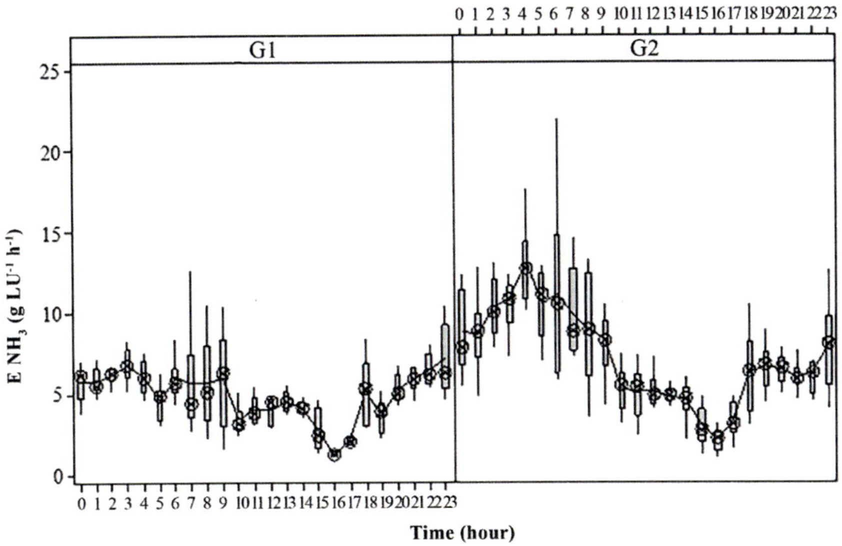

Fig. 2. Daily NH_3 emissions (g LU^{-1} h^{-1}) for two groups of days (i.e., G1 and G2) with different indoor climatic parameters. G1 represents a group of days with high indoor air velocities and temperatures, whereas G2 is characterised by low indoor air velocities and temperatures.

[8]. In fact, the trend of the daily NH_3 emissions (Fig. 2) was almost constant during G1 compared to that during G2.

Although a positive correlation was found in the literature between air temperature and hourly NH_3 emissions [2, 12], in this study the correlation *r* between NH_3 concentrations and air temperature was equal to −0.59 ($p < 0.001$) as well as the correlation between emissions and air temperature that reached −0.67 ($p < 0.001$) during G2. The negative relation between NH_3 and air temperature (Fig. 3a) is due to the activation of the cooling system in the central hours of the day when air temperature had the highest values. In fact, the water on the floor diluted urine on the floor with a reduction effect on the conversion of ammonium (NH_4^+) into NH_3. Moreover, the fans increased the dilution effect on the gas in the central hours of the day.

A deeper analysis on the data showed a specific effect of air velocity on gas emissions. Figure 3b showed that NH_3 emissions increased for air velocities lower than 0.5 m s^{-1}, whereas the trend was almost constant at air velocities higher than 1 m s^{-1}.

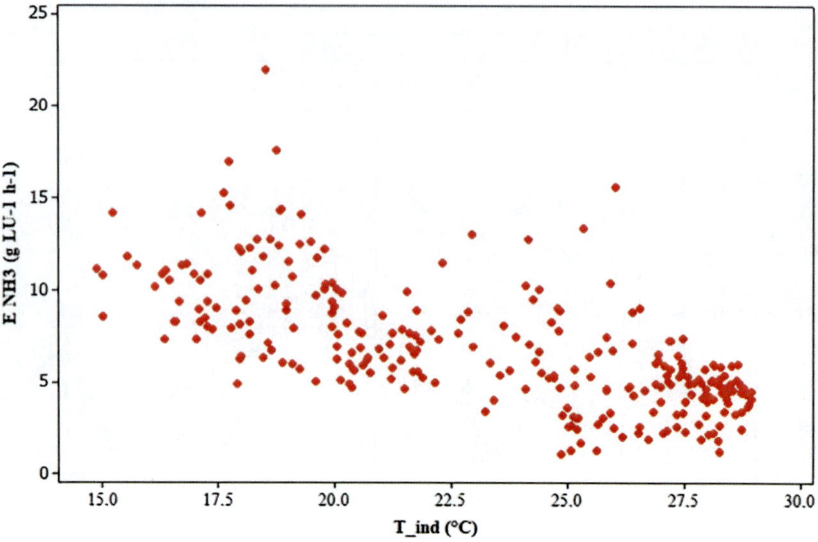

(a)

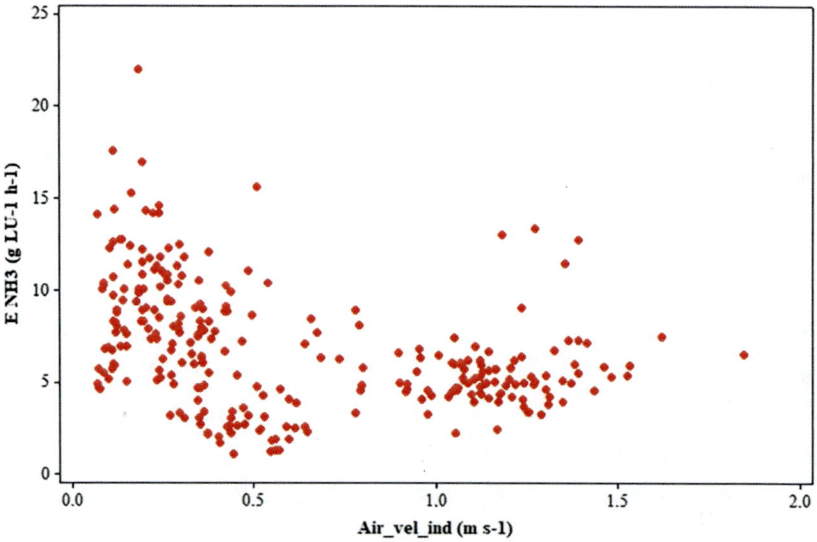

(b)

Fig. 3. Scatter plots of NH_3 emissions related to air temperature T_ind (a) and air velocity Air_vel_ind (b) during G2 characterised by days with low indoor air velocities and temperatures.

4 Conclusion

The outcomes of this research work proved with statistical evidence that environmental parameters (i.e., air velocity and temperature) have a significant influence on NH_3 emissions. During days with high air velocity, the NH_3 emissions are reduced, and generally there is a greater difficulty in studying relations between variables. Both the barn typology with open structure and the management of the cooling system affects microclimate with effects on the variability of gaseous emissions.

Knowledge on how gas concentrations and emissions are affected by microclimate are highly valuable to improve the design of mitigations strategies. However, more in-field measurements are necessary to provide robust suggestions for reducing environmental impacts.

Acknowledgements. The authors are grateful to the farm ALPA S.S. for providing the opportunity of carrying out the tests.

Funding. The research study was funded by the University of Catania through the 'Piano incentivi per la ricerca di Ateneo 2020–2022-Linea 2' project on 'Engineering solutions for sustainable development of agricultural buildings and land—LANDSUS' (ID: 5A722192152) coordinated by Prof. Claudia Arcidiacono. The INNOVA device was funded by the project "Centro per l'innovazione dei sistemi di qualità tracciabilità e certificazione dell'agroalimentare"—AGRIVET (ID: G46D15000170009).

References

1. Zou, B., Shi, Z., Du, S.: Gases emissions estimation and analysis by using carbon dioxide balance method in natural-ventilated dairy cow barns. Int. J. Agric. Biol. Eng. **13**(2), 41–47 (2020)
2. Rong, L., Liu, D., Pedersen, E.F., Zhang, G.: Effect of climate parameters on air exchange rate and ammonia and methane emissions from a hybrid ventilated dairy cow building. Energy Build. **82**, 632–643 (2014)
3. Baldini, C., Borgonovo, F., Gardoni, D., Guarino, M.: Comparison among NH3 and GHGs emissive patterns from different housing solutions of dairy farms. Atmos. Environ. **141**, 60–66 (2016)
4. Cambra-López, M., Aarnink, A.J.A., Zhao, J., Calvet, S., Torres, A.G.: Airborne particulate matter from livestock production systems: a review of an air pollution problem. Environ. Pollut. **158**(1), 1–17 (2010)
5. Wang, X., et al.: Reliable low-cost devices for monitoring ammonia concentrations and emissions in naturally ventilated dairy barns. Environ. Pollut. **208**, 571–579 (2016)
6. Mendes, L.B., Pieters, J.G., Snoek, D., Ogink, N.W.M., Brusselman, E., Demeyer, P.: Reduction of ammonia emissions from dairy cattle cubicle houses via improved management- or design-based strategies: a modeling approach. Sci. Total Environ. **574**, 520–531 (2017)
7. Arcidiacono, C., Porto, S.M.C., Cascone, G.: On ammonia concentrations in naturally ventilated dairy houses located in Sicily. Agric. Eng. Int. CIGR J. **2015**, 294–310 (2015)

8. D'Urso, P.R., Arcidiacono, C., Cascone, G.: Environmental and animal-related parameters and the emissions of ammonia and methane from an open-sided free-stall barn in hot mediterranean climate: a preliminary study. Agronomy **11**, 1772 (2021)
9. D'Urso, P.R., Arcidiacono, C.: Effect of the milking frequency on the concentrations of ammonia and greenhouse gases within an open dairy barn in hot climate conditions. Sustainability **13**, 9235 (2021)
10. Zhang, G., et al.: Emission of ammonia and other contaminant gases from naturally ventilated dairy cattle buildings. Biosyst. Eng. **92**, 355–364 (2005)
11. Ngwabie, N.M., Jeppsson, K.-H., Gustafsson, G., Nimmermark, S.: Effects of animal activity and air temperature on methane and ammonia emissions from a naturally ventilated building for dairy cows. Atmos. Environ. **45**, 6760–6768 (2011)
12. Hempel, S., et al.: Non-linear temperature dependency of ammonia and methane emissions from a naturally ventilated dairy barn. Biosyst. Eng. **145**, 10–21 (2016)
13. Saha, C., et al.: Seasonal and diel variations of ammonia and methane emissions from a naturally ventilated dairy building and the associated factors influencing emissions. Sci. Total Environ. **468–469**, 53–62 (2014)
14. Saha, C.K., et al.: The effect of external wind speed and direction on sampling point concentrations, air change rate and emissions from a naturally ventilated dairy building. Biosyst. Eng. **114**, 267–278 (2013)
15. D'Urso, P.R., Arcidiacono, C., Valenti, F., Cascone, G.: Assessing influence factors on daily ammonia and greenhouse gas concentrations from an open-sided cubicle barn in hot mediterranean climate. Animals **11**, 1400 (2021)
16. CIGR. Climatization of animal houses. In: Pedersen, S., Sallvik, K. (eds.) 4th Report of Working Group: Heat and Moisture Production at Animal and House Levels. CIGR: Liege, Belgium (2002)

The COWBHAVE System: An Open-Source Accelerometer-Based System to Monitor Dairy Cows' Behavioural Activities

Claudia Arcidiacono[1(✉)], Massimo Mancino[1], Simona M. C. Porto[1], Giovanni Cascone[1], Victor Bloch[2], and Matti Pastell[2]

[1] Department of Agriculture Food and Environment, Building and Land Engineering Section, University of Catania, Via S. Sofia 100, 95123 Catania, Italy
claudia.arcidiacono@unict.it
[2] Natural Resources Institute Finland (LUKE), Latokartanonkaari 9, 00790 Helsinki, Finland

Abstract. The main objective of the COWBHAVE project originates from the idea of making available a low cost open-source cow behaviour monitoring system to farmers, based on a novel accelerometer firmware version. A GUI and open APIs complete the system.

Keywords: Cows' behaviours · Acceleration thresholds · Monitoring system · Graphical user Interface · Application programming interface

1 Introduction

Inertial sensor-based technology has spread in recent years in livestock housing and accelerometer systems have been developed for health and welfare assessment [1] for commercial or scientific purposes. Various applications in dairy cattle housing, ranging from behaviour recognition and monitoring [2–4], to cow's status detection [5–7], have been proposed in various countries.

Utilisation and spread of this kind of technologies among farmers and stakeholders is connected with the capability of the system to fulfil the requirements and have an adequate cost. Basic knowledge of cows' behavioural activities in the housing area would be valuable for barn management, facilitating the control of the herd and reducing the presence of operators in the breeding area. When economic development of the country affects the application of modern technologies, the availability of automated monitoring systems at low cost would be crucial [8].

From these issues the basic idea of ICT-Agri COWBHAVE project has been conceived. Specifically, the main objective of the project is "the design and development of a basic low-cost automated monitoring system based on accelerometers and open-source software tools with the aim of discriminating dairy cow behavioural activities". A monitoring system of this kind, based on low-cost hardware, open-source software, and exchange of firmware/software in open repositories (e.g., GitHub) would enhance potential for utilisation among researchers and dairy farmers.

M. Biocca et al. (Eds.): SHWA 2020, LNCE 252, pp. 95–105, 2022.
https://doi.org/10.1007/978-3-030-98092-4_11

2 The System Architecture

The components of the COWBHAVE system for cows' behaviour discrimination are summarised in Table 1. Ruuvitag devices, developed by Ruuvi Innovations LTD – Finland, acquired data by using a specifically developed firmware [9], named COWBHAVE firmware mode C hereafter, that handled aggregated data thus being suitable for reducing packet loss and increase battery duration. Data were received by a single board computer, the Raspberry Pi by Raspberry Foundation (United Kingdom). Other specific information on the acquisition system is reported by Arcidiacono et al. [9]. The development of an automated data analysis system was performed by using a PostgreSQL DBMS, located on board of a Raspberry.

Table 1. System components of the COWBHAVE accelerometer system.

	Components type	System components
Hardware components	*Wearable device*	Ruuvi Ruuvitag
	Communication module	Bluetooth Low Energy
	Single Board Computer	Raspberry Pi 3 Model B/B+
	Data storage	SD card 32 GByte
Software components	*Firmware*	CowBhave firmware Mode C
	Operative System	Raspbian OS
	Data collection software (.csv files)	Python scripts based on BLUEZ and Ruuvitag libraries
	Automated Data Analysis	Raspberry Pi on board data analysis (DBMS)
	Farmer User Interface	Graphical User Interface to manage reference data and to observe the results

2.1 The COWBHAVE Firmware, the Acceleration-Based Variable, and the Threshold-Based Algorithm

With the aim of improving the system performance in terms of reduction of battery power consumption, minimisation of packet loss, and the overall improvement of data collection in the barn, a new firmware for the Ruuvitag and a special variable based on acceleration along the x axis (Fig. 1), was defined as follows (Eq. 1) [9]:

$$sum\_acc\_x_h = \sum_{i=1}^{50} |acc\_x_i| \text{ right shifted of 3 bits} \quad (1)$$

This definition allowed computing the variables derived from sensor raw data straight on-board of the device by designing and implementing a specific firmware. The Mode C of the COWBHAVE firmware is suitable for discriminating the position of the tag ('down' or 'up'), during a period of 5 s, by considering the acceleration measurements along the x axis (Fig. 1).

The acceleration values are collected in the FIFO buffer of the Ruuvitag, then a routine extracts them every 2.5 s and sums their absolute value in a temporary variable. Finally, every 5 s the last value of the 2.5-s temporary variable is summed to the previous one, and the new result, right shifted of 3 bits, constitutes the value of the $sum\_acc\_x_h$ variable. Therefore, the accelerometer acquisition frequency is 10 Hz whereas the communication is every 5 s. Moreover, the previous values of the variable are stored in the same message, thus solving the problem of packet loss [9].

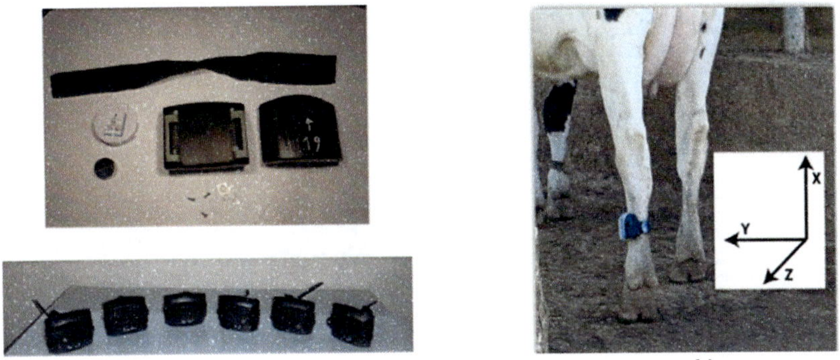

a) b)

Fig. 1. The tag and the additional shelter with leg band (a); the orientation of the axes in the tag fixed at the cow's hind leg (b).

Thresholds for the acceleration variable were determined in specific experiments [4, 7, 9, 10] for lying, standing, and walking behaviours. Moreover, the system allows defining specific thresholds for the group of cows considered, therefore facilitating the farmer when analysing different groups (e.g., for parity, age, breed, and new heads in the dairy cattle herd).

2.2 The Database

In the Raspberry, a module was implemented in Python to handle the packets received by the Ruuvitag devices, fixed at the cows' legs in the barn, through the Bluetooth Low Energy connection of the Raspberry; it allows obtaining the values of the acceleration-based variable during time for each cow. These values and the related timestamp and MAC address of the device are registered in .csv text files. Each file contains 10 min of data related to the same cow. The scheduling of the data import is performed by using the crontab of the Raspbian; specifically, the .csv files are imported every 15 min into the PostgreSQL database and data are elaborated.

In Fig. 2a, the Conceptual model of the database is reported. The entity-relation model is described by the entities with their attributes and the relations among them. In this model, there are dynamic data, such as Cow_behaviours, connected to reference data (e.g., cows). In Fig. 2c, an example of four records (every 5 s) related to the behaviour detected for the cow identified by the number 64 are reported.

(a)

(b)

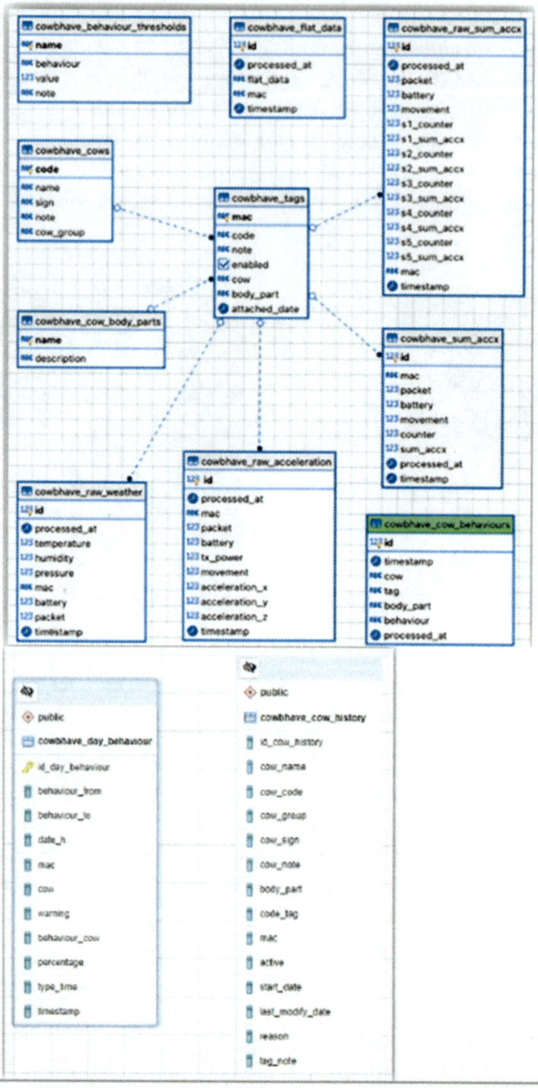

	id [PK] integer	timestamp timestamp without time zone	cow character varying (20)	tag character varying (20)	body_part character varying (20)	behaviour character varying (20)
1	18687	2020-10-22 18:15:53	64	FD:A2:BE:3F:8C:6F	Hind leg	Lying
2	18688	2020-10-22 18:15:58	64	FD:A2:BE:3F:8C:6F	Hind leg	Lying
3	18689	2020-10-22 18:16:03	64	FD:A2:BE:3F:8C:6F	Hind leg	Lying
4	18690	2020-10-22 18:16:08	64	FD:A2:BE:3F:8C:6F	Hind leg	Lying

(c)

Fig. 2. Conceptual model of the database (a); additional tables of the database for GUI enhancement (b); example of records and fields of the database (c).

The choice of designing and developing a relational database was founded on the need to perform complex data queries from the .csv files and relating them with the reference information of the barn (e.g., cow groups, which cow wears the tag and in which body part). The PostgreSQL DBMS offers many advantages: it adopts a client-server architecture; it is a cross-platform thus the server side can be managed on the Raspberry Pi and the client side on a different device, and also the server side could be moved on a different device when the Raspberry's workload increases with the number of connected Ruuvitags; it uses the stored procedures, i.e., SQL procedures that process the data and can be scheduled as well as the Python scripts.

The Relational database contains static data useful to provide information that does not change with time, the 'reference data' (e.g., cows, groups, tags, behaviours, and acceleration thresholds), which are connected through join operators. For instance, 'Cow behaviour' is a dynamic data connected to 'Cow' that is a reference data.

The data in .csv format are loaded and elaborated at scheduled times, i.e., at device reboot and every 15 min as default. In detail, programmable scheduled processes were established by using crontab. For instance, the variable $sum\_acc\_x_h$ is computed and compared with the thresholds, and consequently the classification of the behaviours is obtained. Therefore, automated recognition of cow's behavioural activities, such as lying and standing or walking, is executed.

Implementation of a Graphical User Interface (GUI) for the farmers was carried out by using the PHP language. With the aim of allowing data display by the GUI independently from data being processed in real time, the Database was enriched with two tables and four functions both to store any user event on the elements of a generic observation, and to aggregate the data (by day and by hour of each observation day), during the night. Specifically, the table 'cowbhave_cow_history' was added. In this table, information about RuuviTag-Cow associations and bodypart specific features (beginning and ending dates of association) is historicised, as well as variations of information on notes for RuuviTag and on specific signs, notes, and group for the cows (Fig. 2b).

2.3 The Graphical User Interface (GUI) and the Application Programming Interface (API)

The programming language PHP (version: 8.0.2) was used for the design of the GUI. The PHP interpreter is a free software and, therefore, a web-oriented interface was created. To use the application, it is necessary to install an Apache web server and the PHP interpreter on the Raspberry device dedicated to publishing web pages. Access to the Web application is immediate and, therefore, does not require user registration, login and logout actions.

Through the GUI, all the changes made to any modifiable element of an association in progress are logged, including several RuuviTags for the same association.

The design is user experience oriented (that is to say, it aims at providing the user with what he expects from the application) and consists of a simple interface (in tabular mode) with simple and easy-to-understand labels. The layout is purpose-oriented and each element performs a clear (application) function; the colours, text, fonts and font sizes have been designed to be appropriate to the function they perform and easily recognisable. Furthermore, the GUI informs about any functional or system errors.

The publishing service of the Web pages, making up the GUI, is provided by the Raspberry Pi itself; it is fixed at the centre of the considered pen in the stable and is connected to the router, both via Wi-Fi and via Ethernet (optional). Therefore, any device (e.g., computer, smartphone, tablet, or notebook) connected to the same LAN or WLAN, and with a web browser installed, can view the web pages related to the application.

The GUI has input, reporting, and service functions, with sections for cows and RuuviTags. Furthermore, it allows definition of threshold values for the accelerometer variable, suitable for adapting the default values for lying-standing-walking to the specific cow or group of cows, and for fixing specific thresholds for other conditions, such as oestrus detection or lameness. Through the creation of an associative table, the GUI of the CowBhave project stores all the associations made between cattle, RuuviTag and detected data, providing tools for the creation, modification, deletion and table display of the associations and any modifications thereof.

Application Programming Interface (API) functions were implemented. They run through a client, by using the cow input parameters and the time interval of observation, and return a JavaScript Object Notation (JSON) file.

3 The Experimental Tests in the Lab and in a Free-Stall Barn to Assess System Functioning

Laboratory tests were performed to assess the firmware compared to the original one of the Ruuvitag. Results demonstrated how a high increase of battery estimated duration (from 65 days to 246 days) and an almost complete reduction of packet loss (0.14%) were achieved, besides a lower CPU usage and storage memory occupation, compared to the original firmware [9].

The aim of in-barn experimental tests was to assess the behaviour recognition capability of the system. During these tests, Ruuvitag devices were fixed to the cow's hind legs and were connected to a Raspberry 3 model B [9].

For each behaviour (lying, standing, and walking) samples were randomly selected from the accelerometer data, using an equal number of samples for each cow, and retrieved in the video-recordings by the timestamp. The videos were acquired through cameras, installed in the pen of the barn where the considered cows were housed, and utilised for validation of cow behaviour detection. This video-recording system was composed of IP video cameras (Kon.El.Co, Italy) connected to a personal computer by two switches and Ethernet cables [7].

The values of the $sum\_acc\_x_h$ variable were used for cow's behaviour prediction by comparing them with the thresholds for lying, standing, and walking, equal to 3150 mg and 6250 mg [9, 10]. Specifically, a sample was considered as lying if the value of $sum\_acc\_x_h$ was less than the threshold 3150 mg, else it was considered as standing, whereas the value 6250 mg discriminated the standing from walking.

Classifier accuracy was assessed by using the following indicators: Misclassification Rate (MR), Sensitivity, Precision, Specificity, Quality Percentage (QP), Branching Factor (BF), and Miss Factor (MF). The results demonstrated that a very high accuracy (100%) was achieved in behaviour classification.

4 GUI Sections and API Functions of the Case Study

In the "Home" section (Fig. 3), the GUI provides the direct display of the data related to the associations in progress, highlighting the behaviour of the cow both through thresholds of values (that can be customised by the user through a special section), and through a distribution, in percentage, of the standing, lying and walking behaviours of the active association.

On the home page, the anomalous behaviours of cows in the previous 24 h are highlighted in red; the colour of the report button also highlights the anomalies. By clicking on it, you can immediately access the details about hourly percentages of behaviours in the last 24 h with evidence, also, of the daily anomalies. The report shows, in cyan color, the cow with an observation in progress, the buttons for accessing all the observation days, and the fields for the time interval search for valid observations. Therefore, the user can analyse the behaviour of the cow by watching the reports of the previous days.

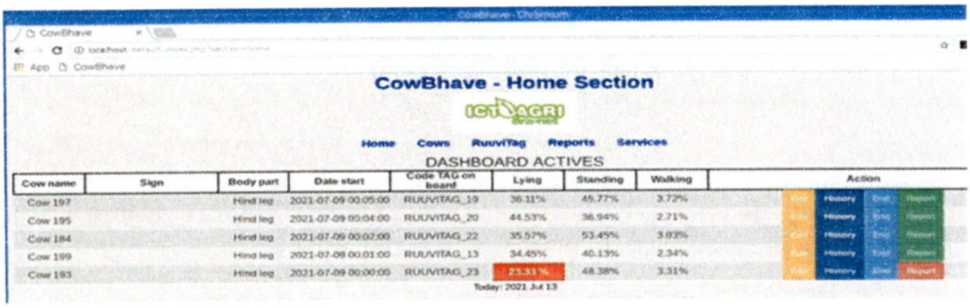

Fig. 3. Home section of the GUI. (Color figure online)

At the top of the "Reports" section (Fig. 4), a list of all the cows involved, at least once, in any association is shown. By selecting a cow, you will be able to access the data related to the last seven days, if present, otherwise, the data related to the last valid association. It is possible to select a time interval and, if present, the data related to the association or associations involved in the selected interval will be displayed, each with the option of selecting the daily data for the entire duration of the association.

In the Services section (Fig. 5), the average of the behaviours is provided by considering the last observation (completed or in progress) and evidence is given of the days, and the time period, over which it is calculated. For each behaviour, a maximum daily value (or a minimum value) can be defined. When above (or below) this value, a warning will be logged with evidence of the threshold values related to the day of data processing.

In the cow subsection (Fig. 6), you can easily perform the following tasks: insert a new cow; delete an existing cow; edit the notes; modify the signs; start a new association; terminate an association in progress; view the history of the associations (History button); modify the associated tag (History -> Edit buttons); modify the part of the body on which the tag is positioned (History -> Edit buttons); and delete association. Any changes made to existing associations are historicised. Similar tasks can be carried out for tags in a specific subsection of the GUI.

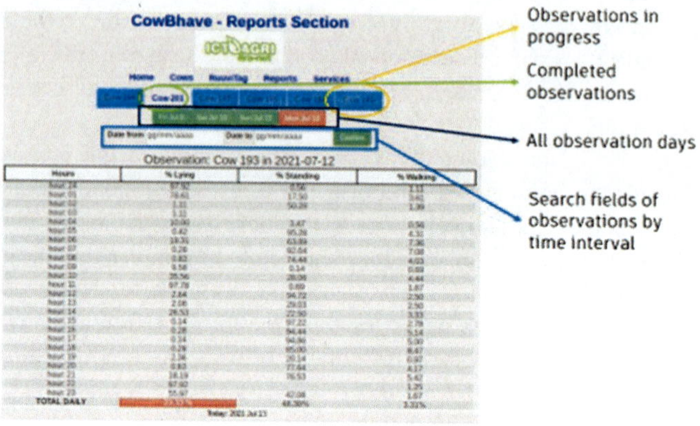

Fig. 4. Reports section of the GUI.

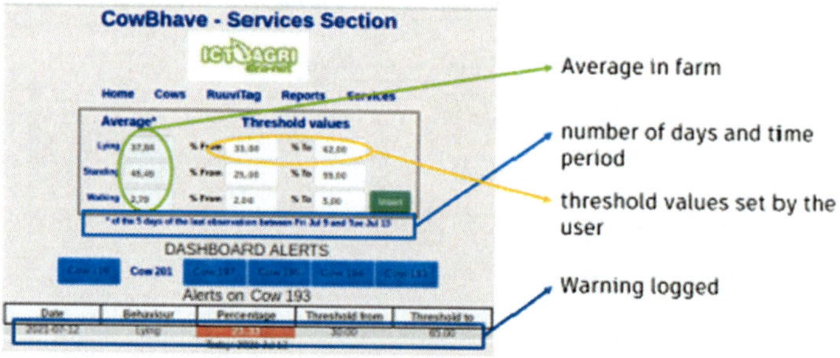

Fig. 5. Services section of the GUI.

In order to export data to other systems, Application Programming Interface (API) functions were provided. In detail, any implemented API can be run through a client, by using the cow input parameters and the time interval of observation, and it returns a JSON file containing:

- the hourly aggregation of the percentages of lying, standing and walking activities, related to the individual observation days (read_rep.php);
- the values of the $sum\_acc\_x_h$ acceleration variable with maximum granularity of five seconds (read_sum.php).

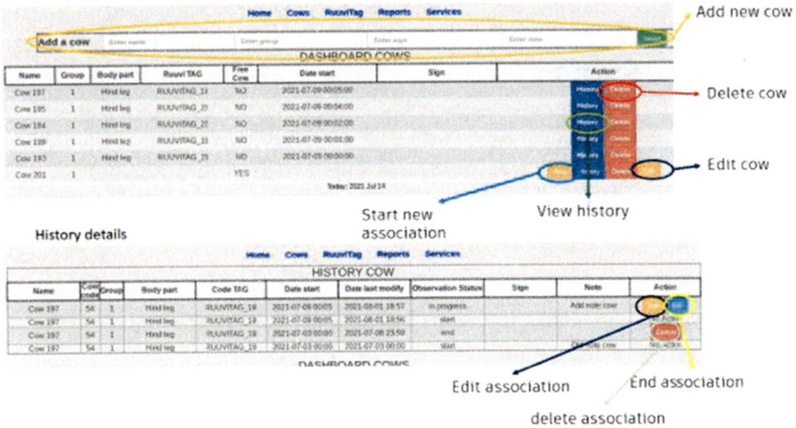

Fig. 6. Editing section of the GUI: example of functions for cow subsection.

For instance, in the browser URL the input parameters Cow 199 from 2021-07-01 to 2021-07-20 are inputted as follows:

http://hostname/CowBhaveAPI/read/read_rep.php?cow=Cow 199&start=2021-07-01 00:00:00 &end=2021-07-20 23:59:59

and the response will be a JSON file of the type reported in Fig. 7.

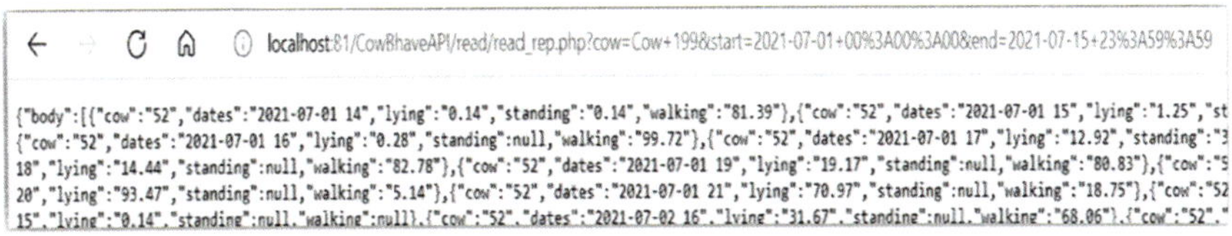

Fig. 7. Output of the API function related to Cow199 and the observation interval from 2021-07-01 to 2021-07-20.

5 Conclusions

The main objective of the COWBHAVE project, financed within the ICT-AGRI Call 2017 'Farm Management Systems for Precision Farming', originated from the idea of making available a low cost open-source cow behaviour monitoring system to the farmers. This system would facilitate herd management by discriminating dairy cow behavioural activities and deviating behaviour.

A novel firmware version of a Micro Electro-Mechanical Systems (MEMS) device (Ruuvitag by Ruuvi, Finland) and behavioural classification algorithms were developed in order to build a real-time system based on an accelerometer fixed at the cows' leg and a threshold-based algorithm. A Graphical User Interface developed in php completes the system and open API for the system have been implemented to export the data. Applications of the system were carried out in a cubicle free-stall dairy barn located in the province of Ragusa (Italy).

The results showed the suitability of the system to perform discrimination of the behaviours with minimum data loss, thanks to the specific firmware developed for the Ruuvitag device. Moreover, many advantages derived from the adoption of a client-server architecture and the cross-platform feature connected to the use of Python scripts and PostgreSQL. The GUI Web application facilitates interaction between users and the system, tags handling, data retrieval and analysis; it was developed into sections including cows' information, tag information, association between cows and tags, and related editing tools, data analysis, parameters settings (e.g., cow behaviour thresholds), and warnings on behaviours.

Key advantages of the COWBHAVE system encompasses the following characteristics: it is not a proprietary System since it is an open System based on open-source software; it uses low-cost hardware; it is robust, thanks to the improvements on the firmware; and it allows detection of various cow behaviours (lying, standing, walking, anomaly detection) based on thresholds.

Furthermore, the operating system, the applications, the firmware, all the software produced and the settings, will be released, in a single repository on GitHub and other platforms for code sharing, as an image to be copied into an SD memory.

Acknowledgements. This study was carried out in the framework of the ICT-Agri2 project "CowBhave – a basic low-cost open-source automated monitoring system for discrimination of dairy cow behavioural activities", Coordinator Prof. Claudia Arcidiacono.

The authors are grateful to the farm ALPA S.S. for giving the opportunity of carrying out the tests.

The authors express their thanks to Dr. R. Faella for his contribute to the GUI development.

References

1. Chapa, J.M., Maschat, K., Iwersen, M., Baumgartner, J., Drillich, M.: Accelerometer systems as tools for health and welfare assessment in cattle and pigs – a review. Behav. Processes **181**, 104262 (2020)
2. Martiskainen, P., Jarvinen, M., Skon, J.-P., Tiirikainen, J., Kolehmainen, M., Mononen, J.: Cow behaviour pattern recognition using a three-dimensional accelerometer and support vector machines. Appl. Anim. Behav. Sci. **119**, 32–38 (2009)
3. Darr, M., Epperson, W.: Embedded sensor technology for real time determination of animal lying time. Comput. Electron. Agric. **66**(1), 106–111 (2009)
4. Arcidiacono, C., Porto, S.M.C., Mancino, M., Cascone, G.: Development of a threshold-based classifier for realtime recognition of cow feeding and standing behavioural activities from accelerometer data. Comput. Electron. Agric. **134**, 124–134 (2017)
5. Pastell, M., Tiusanen, J., Hakojarvi, M., Hanninen, L.: A wireless accelerometer system with wavelet analysis for assessing lameness in cattle. Biosys. Eng. **104**, 545–551 (2009)
6. Chanvallon, A., et al.: Comparison of three devices for the automated detection of estrus in dairy cows. Theriogenology **82**(5), 734–741 (2014)

7. Arcidiacono, C., Mancino, M., Porto, S.M.C.: Moving mean-based algorithm for dairy cow's oestrus detection from uniaxial-accelerometer data acquired in a free-stall barn. Comput. Electron. Agric. **175**, 105498 (2020)
8. Arcidiacono, C.: Engineered solutions for animal heat stress abatement in livestock buildings. CIGR e-Journal 2018 Special issue: Animal Housing in Hot Climate, pp. 1–22 (2018). http://www.cigrjournal.org/index.php/Ejounral/article/view/4705/2734
9. Arcidiacono, C., Mancino, M., Porto, S.M.C., Bloch, V., Pastell, M.: IoT device-based data acquisition system with on-board computation of variables for cow behaviour recognition. Comput. Electron. Agric. **191**, 106500 (2021). https://doi.org/10.1016/j.compag.2021.106500. ISSN 0168-1699
10. Arcidiacono, C., et al.: Feasibility study on the development of a real-time inertial sensor-based system for dairy cow behaviour detection in free stall barns. In: 7th European Conference on Precision Livestock Farming (EC-PLF), Milano 15–18 September 2015, pp. 782–791 (2015)

On the Determination of Acceleration Thresholds for the Automatic Detection of Cow Behavioural Activities in Extensive Livestock Systems

Simona Maria C. Porto[1], Giulia Castagnolo[2(✉)], Massimo Mancino[1], Dominga Mancuso[1], and Giovanni Cascone[1]

[1] Department of Agriculture, Food and Environment, University of Catania, Catania, Italy
simona.porto@unict.it
[2] Department of Electrical, Electronic and Computer Engineering, University of Catania, Catania, Italy
giulia.castagnolo@phd.unict.it

Abstract. In recent years the use of wearable sensors has proved to be valid tools in the field of livestock farming. They have played a crucial role in the automation and evolution of animal health and welfare monitoring processes, especially in the case of extensive farms where there is an infrequent farmer-to-animal contact. Accelerometers allow the measurements of movements of a body in space and are currently very popular sensors in the zootechnical field for monitoring livestock as they can be worn without being invasive for animals. Therefore, the aim of this research activity was to prove the feasibility of using acceleration thresholds for the automatic detection of grazing cow behavioural activities. In this study, a customized electronic device equipped with a triaxial accelerometer was built and placed in a collar of a grazing cow. The accelerometer data acquired at 4 Hz from the device were sent to a cloud server one time an hour. Then they were subsequently analysed to be able to establish thresholds for the discrimination of cow behaviours. As stated in previous studies, algorithms based on statistically defined acceleration thresholds could be easily implemented in a firmware for real-time detection of cow behaviour, therefore, they could be a valuable low-cost tool for monitoring livestock in extensive farms.

Keywords: Sensor-based systems · MEMS · Cow welfare · Automated monitoring systems

1 Introduction

Over the last few years there has been an increasing use of IoT sensors in precision livestock farming techniques [1, 2]. The most used category of sensors is non-invasive i.e., sensors that can be used inside or outside the barn, in feeding systems or worn by animals such as accelerometers and GPS. Smart sensors allow the collection of a large volume of data, also they can process the raw data and, under specific range of values,

M. Biocca et al. (Eds.): SHWA 2020, LNCE 252, pp. 106–114, 2022.
https://doi.org/10.1007/978-3-030-98092-4_12

alert farmers if the cattle behaviour is abnormal. Automated monitoring process of livestock has been directed towards various tasks, such as control of welfare, animal's health status and early detection of diseases [3], localization and tracking [4], behaviour detection [5] and animal's identification, all aimed at reducing the need for human intervention and consequently improving the management of livestock [6, 7]. In the state of the art, various systems have been designed in accordance with the development of increasingly precise and miniaturized sensors and IoT devices, some of them are briefly described below. Arcidiacono et al. [8] proposed an open-source behaviour discrimination tool using accelerometer data in free-stall barns, Arcidiacono et al. [9] proposed an inertial sensor-based systems to perform real-time cow step counting in free-stall barns; Maroto et al. [10] proposed a low-cost IoT based system to localize a whole herd, Shen et al. [11] developed a sensor-based automatic recognition system to detect ingestive-related behaviour. Another automatic detection sensor-based system is RumiWatch (*Itin + Hoch GmbH, Liestal, Switzerland*) [12], developed for the detection of rumination, feed intake, water intake and locomotion in barn and is composed of a pedometer, which contains a triaxial-accelerometer, and noseband sensor. RumiWatch was subsequently used for the evaluation of grazing cow behaviour [13]. The continuous automatic analysis of behaviour is a very important task, especially in the case of extensive pastures, since farmer-animal contact is certainly less frequent than in the barns. The farms, to date, are more and more extensive therefore it is not always possible to monitor, through human control, the behavior of animals, especially nutrition: it is of fundamental importance for the farmer to know if the animal has fed. As described above, in literature there are several different systems proposed for animal housed in barns, however in the case of extensive farms the number of proposed solutions is significantly reduced. Probably, this is related to the difficulty of using automatic monitoring systems in extensive farms where the areas of the pastures are not always covered by efficient telecommunication networks. Moreover, often the use of sensors equipped with batteries could imply costs of management and maintenance if the batteries run out quickly and if the proposed systems are not optimized for energy savings [14, 15]. The main objective of this paper is to prove the feasibility of using accelerometers for the automatic detection of grazing cow behavioural activities. As stated in previous studies, algorithms based on statistically defined acceleration thresholds could be easily implemented in a firmware for real-time detection of cow behaviour. This feature could allow the processing of algorithms in the devices, avoiding the intensive use of the telecommunications network and therefore extending life-time battery.

2 Materials and Methods

2.1 The Herd Considered in the Research Study

The experimental trial was carried out in an existing 180 hectares semi-natural pasture characterized by good availability of meadow and cultivated grazing areas. The breeding system adopted is that of the cow-calf line, that involves keeping calves with mothers during the lactation period until weaning (6/8 months). It can take place entirely outdoors, with cows living in pastures all year. During the summer, from 7:00 to 18:00, cows stayed near the farmer house in a large enclosure of about 2 ha (Fig. 1) where there are several watering tanks while in the rest of the day and night they are moved to the pasture. In this study a 19-month-old cow, which is part of a group of 10 Limousines replacement heifers, was monitored.

Fig. 1. Large enclosure near the farmer house where cows stayed from 7:00 to 18:00

2.2 Device and Data Acquisition

A customized device (Fig. 2a) was built to carry out the experimental activities of this research study and it was equipped with: triaxial MEMS accelerometers, omnidirectional antennas, 32-bit Cortex Microcontroller, GSM/GPRS quad band modules, Li-SOCL2 high-capacity battery and flash memories. This device, protected by a rigid case, was put to the collar of the monitored cow (Fig. 2b) adopting a leather-reinforced mesh collar. The case containing the electronic device was fixed to the collar by using electrical clamps and duct tape. To limit the rotation of the device around the cow's neck, a weight of 1 kg was attached into the collar. The distance of the device from the weight was chosen to detect the accelerations coming from the jaw oscillations during the rumination phase.

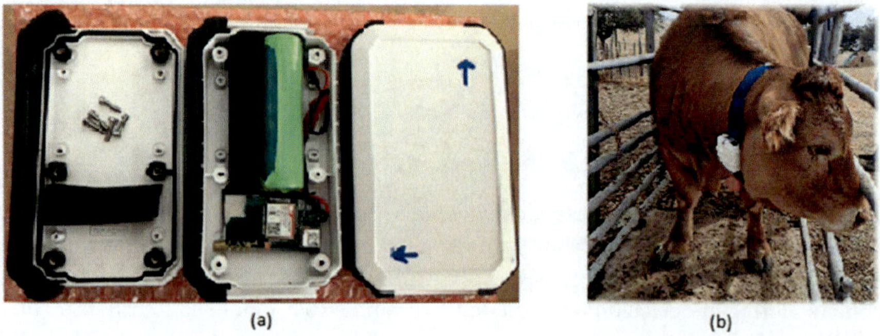

Fig. 2. a) Customized device equipped with triaxial accelerometers, 32-bit Cortex Microcontroller, GSM/GPRS modules, high-capacity battery; b) Device attached to the cow collar.

The acceleration components along the x, y, and z axes were acquired with a frequency of 4 Hz and recorded in the internal memory of the devices. Through the GSM communication module, the data collected were sent to a cloud, one time an hour, for their processing and display through a Web App (Fig. 3) developed for the management of the dataset used in the subsequent labelling phase.

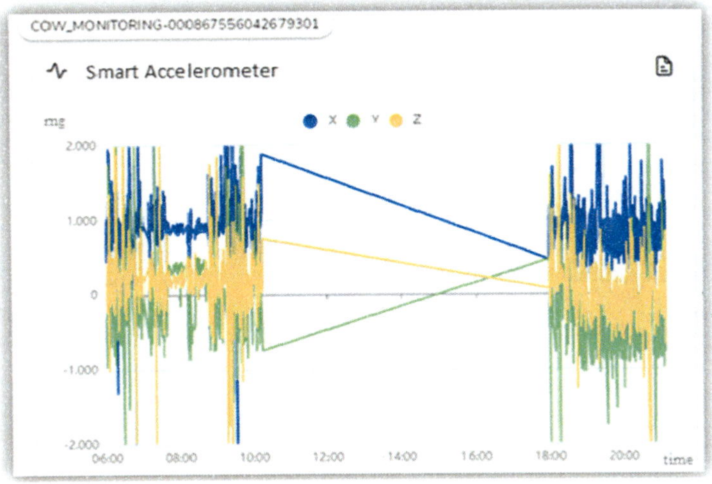

Fig. 3. Graph of the WebApp that was developed to collect and manage the acquired data. To increase battery life the device entered in sleep mode from 10:00 a.m. to 17:59.

2.3 Dataset and Labelling

The behavioural activities analysed in the present study were rumination in standing position (R), lying (L), lying with rumination (L-R), walking (W), feeding (F), feeding and walking (F-W), feeding to the manger (FM) and drinking (D). These behavioural activities of the monitored cow were observed in two different time intervals: the first in May 2021, for 5 days and the second in June 2021 for 4 days. It was not possible to extend the duration of data acquisition for each time interval because the battery was discharged. The device used in this study is an experimental prototype that use the GSM network, which is highly energy-intensive in the phase of sending accelerometer data to the cloud. The acquired acceleration data were labelled by using video-labelling technique, that is the recognition of the behavioural activities carried out by the cow through the analysis of video recordings acquired during the observation periods. Video recordings of the monitored cow were taken daily in the coolest hours of the day, in the morning from 6:00 to 09:59 and in the afternoon from 18:00 to 21:00. To increase battery life the device entered in sleep mode from 10:00 to 17:59 (Fig. 3). A free third-part app was used to apply timestamp in each frame. This made it possible to synchronize the video recordings with the accelerometer data that the device periodically sent to the cloud.

1	Time	X	Y	Z	W	L	R	F
35235	2021/06/27 09:30:27:000	995,52	-312,32	160,064	W			
35236	2021/06/27 09:30:27:250	925,248	-269,376	199,104	W			
35237	2021/06/27 09:30:27:500	1.034,56	-230,336	230,336	W			
35238	2021/06/27 09:30:27:750	1.011,14	-234,24	242,048	W			
35239	2021/06/27 09:30:28:000	1.085,31	-58,56	124,928	W			
35240	2021/06/27 09:30:28:250	1.022,85	-222,528	199,104	W			
35241	2021/06/27 09:30:28:500	1.026,75	-245,952	386,496	W			
35242	2021/06/27 09:30:28:750	1.026,75	-191,296	144,448	W			
35243	2021/06/27 09:30:29:000	999,424	15,616	355,264	W			
35244	2021/06/27 09:30:29:250	1.206,34	-355,264	66,368	W			

Fig. 4. Example of the acquired dataset. In the first column there is the timestamp, in the next three columns there are the accelerometer data (mg) measured along the three axes and then the label such as W (Walking), F (Feeding), R (Rumination) etc. The behaviour label is associated for each sample.

By carrying out the visual analysis of the acquired video recordings, it was possible to label the accelerometer data by indicating the observed behavioural activities of the cow (Fig. 4). During the visual analysis of the video recordings, and the related labelling phase, occurrences of minor behavioural activities were observed such as brusque movements of the ears to ward off flies, brusque movements of the head backwards to scratch or lick itself, sniff the ground before positioning itself in decubitus. Therefore, to reduce the incidence of outliers in the acceleration dataset, it was necessary to delete the acceleration values related to these minor behavioural activities.

2.4 Data Analysis

Data analysis was initially carried out by using descriptive statistics, e.g., the mean, maximum, minimum and standard deviation values of the accelerations detected along the x, y and z axes, grouped by behavioural activities and observation day. Subsequently, the ANOVA test was conducted to compare the data relating to the groups of behaviour considered, to highlight statistically significant differences, and define the range of accelerations for each axis. The ANOVA test was performed by calculating the median of the accelerations acquired in one second (4 Hz sampling) for each behaviour considered. Finally, the Tukey test was carried out to make a comparison between the behavioural activities and to establish, therefore, any overlaps and which axes could discriminate the behaviours.

3 Results and Discussion

From the analysis conducted it was possible to identify the acceleration components required in order to define thresholds to be able to discriminate cow behavioural activities. Table 1 shows, for the x-axis, the results of the ANOVA test conducted for each behaviour. Table 4 and Fig. 5a show the results of Tukey's test for the x axis, and as it can be observed, when considering only the x axis, some behavioural classes overlap, such as: rumination (R) and feeding to manger (FM), lying with rumination

(L-R) and walking (W), drinking (D) and lying (L). Acceleration along x-axis made it possible to discriminate only the feeding activity, either in standing or walking position. The ANOVA test results for the y-axis are shown in Table 2 while Table 4 and Fig. 5b show the results of Tukey's test for y-axis. Unlike x-axis component, there are no overlapping areas among the behavioural activities. The same analysis was carried out for the z-axis. Table 3 shows the results of to the ANOVA test, while the results of the Tukey test are shown in the Table 4 and in the Fig. 5c. Similarly, to the case of the x-axis, there are some overlapping values among the behavioural classes when considering only the z-axis and this occurred for lying with rumination (L-R) and rumination (R) and feeding to manger (FM), walking (W) and drinking (D). Acceleration along z-axis made it possible to discriminate lying (L) and the feeding activity, either in standing (F) or walking position (F-W). The Table 5 summarize the analysis carried out and it does not show the drinking (D) behaviour because the number of samples (N) of these behavioural activities in the whole dataset is strongly unbalanced with respect to the other behaviours. The study showed (Table 5) that the components to be considered in terms of accelerations for the detection of cattle behaviour are: x and y axes for rumination (R), y for feeding to manger (FM), y and z (uncertain) for lying with rumination (L-R), y and z for walking (W), all three axes for lying (L), feeding (F) and feeding with walking (F-W). The results of this experimental activity are a first step forward the determination of acceleration thresholds for the automatic detection of the behavioural activities of grazing livestock.

Table 1. Results of the Anova test for x-axis, grouped by behaviour

Median label	N	Mean	StDev	95% CI
L	977	907.10	115.07	(902.05; 912.15)
L-R	1408	934.31	51.01	(930.10; 938.52)
D	94	908.39	43.62	(892.11; 924.66)
F	1504	803.96	103.85	(799.89; 808.03)
FM	1234	952.31	39.77	(947.82; 956.80)
R	2127	956.875	30.309	(953.45; 960.29)
W-F	4073	782.31	93.00	(779.83; 784.78)
W	1540	933.18	87.76	(929.15; 937.20)

Table 2. Results of the Anova test for y-axis, grouped by behaviour

Median label	N	Mean	StDev	95% CI
L	977	−103.06	194.07	(−122.58; −93.55)
L-R	1408	−40.12	174.94	(−48.50; −32.30)
D	94	−322.6	141.0	(−353.30; −291.90)
F	1504	−538.20	163.71	(−545.87; −530.53)
FM	1234	17.32	143.26	(8.85; 25.78)
R	2127	−6.15	112.68	(−12.60; 0.30)
F-W	4073	−585.11	121.33	(−589.77; −580.45)
W	1540	−151.12	203.91	(−158.70; −143.54)

Table 3. Results of the Anova test for z-axis, grouped by behaviour

Median label	N	Mean	StDev	95% CI
L	977	169.91	232.18	(162.41; 177.41)
L-R	1408	196.06	131.22	(189.82; 202.31)
D	94	119.82	93.90	(95.64; 144.00)
F	1504	−72.15	129.37	(−78.19; −66.10)
FM	1234	192.78	77.84	(186.10; 199.45)
R	2127	195.67	66.95	(190.58; 200.75)
F-W	4073	−105.62	98.16	(-109.29; -101.95)
W	1540	152.60	130.39	(146.63; 158.58)

Table 4. Results of the Tukey test for all axes

Median label	Grouping X	Grouping Y	Grouping Z
R	A	A	A
FM	A	B	A
L-R	B	C	A
W	B	D	B
D	B C	E	C
L	C	F	C
F	D	G	D
F-W	E	H	E

Table 5. Accelerometer axes to be considered for the detection of cattle behaviour.

Behaviour	Acceleration component		
	x	y	z
R	Required	Required	Not required
L-R	Not required	Required	Uncertain
W	Not required	Required	Required
L	Required	Required	Required
F	Required	Required	Required
F-W	Required	Required	Required
FM	Not required	Required	Not required

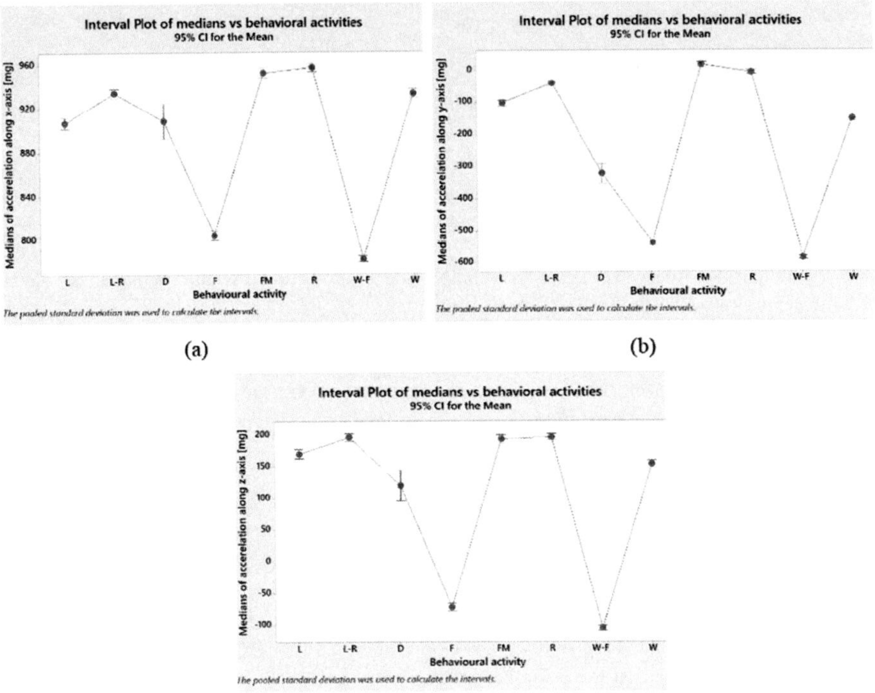

(a) (b) (c)

Fig. 5. Interval plot of medians for x-axis (a), y-axis (b), z-axis (c), grouped by behaviour.

4 Conclusions

Indicators based on cow behavioural activities are relevant to evaluate physiological and physical status of animals, especially in the case of extensive farms where there is an infrequent farmer-to-animal contact. Feeding, rumination, lying and walking are the main daily activities of grazing cattle that should be monitored to ensure the maintenance of animal welfare as well as to improve herd management.

The use of motion sensors can provide a valuable method for monitoring animal activities and determining the time budget spent in daily activities, such as for example the evaluation of the feeding and ruminating period, the early detection of abnormalities in walking, the onset of oestrus. Given the promising results of this work, future developments will regard the analyses of multiple data coming from a larger group of animals and the use of machine and deep learning techniques to be able to establish, in a completely automatic process, the thresholds for the discrimination of cow behavioural activities.

Acknowledgment. The research study was funded by PRIN 2017 project "Smart dairy farming: innovative solutions to improve herd productivity" (ID: E64I18002270001) coordinated by Prof. Simona M.C. Porto for the University of Catania. The authors are grateful to Massimo Pollino and Paolo Vasta of Trecastagni s.r.l., developers of the device and the WebAPP.

References

1. Akhigbe, I., Munir, K., Akinade, O., Akanbi, L., Oyedele, L.O.: IoT technologies for livestock management: a review of present status, opportunities, and future trends. Big Data Cogn. Comput. **5**(1), 10 (2021)
2. Zhang, M., Wang, X., Feng, H., Huang, Q., Xiao, X., Zhang, X.: Wearable Internet of Things enabled precision livestock farming in smart farms: a review of technical solutions for precise perception, biocompatibility, and sustainability monitoring. J. Clean. Prod. **312**, 127712 (2021)
3. Neethirajan, S.: Recent advances in wearable sensors for animal health management. Sens. Bio-Sens. Res. **12**, 15–29 (2017)
4. Bloch, V., Pastell, M.: Monitoring of cow location in a barn by an opensource, low-cost, low-energy bluetooth tag system. Sensors **20**, 1–20 (2020)
5. Arcidiacono, C., Porto, S.M.C., Mancino, M., Cascone, G.: Development of a threshold-based classifier for real-time recognition of cow feeding and standing behavioural activities from accelerometer data. Comput. Electron. Agric. **134**, 124–134 (2017)
6. Neethirajan, S., Kemp, B.: Digital livestock farming. Sens. Bio-Sens. Res. **32**, 100408 (2021)
7. Schillings, J., Bennett, R., Rose, D.C.: Exploring the potential of precision livestock farming technologies to help address farm animal welfare. Front. Anim. Sci. **2**, 13 (2021)
8. Arcidiacono, C., Porto, S.M.C., Mancino, M., Valenti, F., Cascone, G.: New open-source software tools using accelerometer data for the discrimination of cow behavioural activities in free-stall barns. In: Precision Livestock Farming 2017 - Papers Presented at the 8th European Conference on Precision Livestock Farming, ECPLF 2017, pp. 276–287 (2017)
9. Arcidiacono, C., Porto, S.M.C., Mancino, M.: Cascone G A threshold-based algorithm for the development of inertial sensor-based systems to perform real-time cow step counting in free-stall barns. Biosys. Eng. **153**, 99–109 (2017)
10. Maroto-Molina, F., et al.: A low-cost IOT-based system to monitor the location of a whole herd. Sensors **19**(10), 2298 (2019)
11. Shen, W., Cheng, F., Zhang, Y., Wei, X., Fu, Q., Zhang, Y.: Automatic recognition of ingestive-related behaviors of dairy cows based on triaxial acceleration. Inf. Process. Agric. **7**, 427–443 (2020)
12. Zehner, N., et al.: Validation of a new health monitoring system (RumiWatch) for combined automatic measurement of rumination, feed intake, water intake and locomotion in dairy cows (2012)
13. Werner, J., et al.: Evaluation of the RumiWatchSystem for measuring grazing behaviour of cows. J. Neurosci. Methods **300**, 138–146 (2018)
14. Kwong, K.H., et al.: Practical considerations for wireless sensor networks in cattle monitoring applications. Comput. Electron. Agric. **81**, 33–44 (2012)
15. Arcidiacono, C., et al.: Smart dairy farming: innovative solutions to improve herd productivity. In: Coppola, A., Di Renzo, G.C., Altieri, G., D'Antonio, P. (eds.) MID-TERM AIIA 2019. LNCE, vol. 67, pp. 265–270. Springer, Cham (2020). https://doi.org/10.1007/978-3-030-39299-4_30

Instrumentation, Equipment, Periodic Procedures and Tests

Contribution of Inspection Methods to Monitoring Operator Comfort During Agricultural Operations

Elio Romano[1(✉)], Carlo Bisaglia[1], Aldo Calcante[2], Roberto Oberti[2], Alberto Zani[2], Denis Vinnikov[3], Andrea Marconi[4], Ermanno Vitale[4], Massimo Bracci[5], and Venerando Rapisarda[4]

[1] Consiglio per la ricerca in agricoltura e l'Analisi dell'Economia Agraria, Centro di Ricerca Ingegneria e Trasformazioni agroalimentari, Via Milano 43, 24047 Treviglio, BG, Italy
elio.romano@crea.gov.it

[2] Department of Agricultural and Environmental Sciences-Production, Landscape, Agroenergy, Università di Milano, 20133 Milan, Italy

[3] Department of Epidemiology, Biostatistics and Evidence-Based Medicine, Al-Farabi Kazakh National University, Almaty 050040, Kazakhstan

[4] Occupational Medicine, Department of Clinical and Experimental Medicine, University of Catania, 95123 Catania, Italy

[5] Occupational Medicine, Department of Clinical and Experimental Medicine, Polytechnic University of Marche, 60126 Ancona, Italy

Abstract. In agricultural activities, the tractor driver is subjected to many stresses from physical and attentional points of view. Driving tractors requires the operator to pay great awareness both during the agricultural operation and when handling the vehicle. The seat and the driving place play an important role by representing the interface between the man and the machine. In recent years, various investigation techniques have been used with the aim of studying the operator's comfort during the mechanized operations. The techniques most aimed at observing posture and postural commitment examine the positions of specific body parts and their variation over time, the pressure measured between the body and the seat and the muscle electromyographs of the limbs and back. In this study, three driving modes have been considered: i) manual driving, ii) assisted driving and iii) semi-automatic driving in the execution of an agricultural practice of harrowing soil following plowing. Ten operators were hired to carry out the task on the same field alternating the three different driving methods. The procedure of investigation of the postures assumed by the body consisted in the application of a low-cost 3D camera (Microsoft Kinect), a continuous reading tool of the posture and of the crucial nodes of the body. For the evaluation of the pressures of the operator's body on the seat, a sensitized carpet consisting of a matrix of sensors was used. Finally, portable electromyographs were placed in the shoulders, left arm and back of the operator for the assessment of muscle engagement. The values collected by the three

M. Biocca et al. (Eds.): SHWA 2020, LNCE 252, pp. 117–126, 2022.
https://doi.org/10.1007/978-3-030-98092-4_13

methodologies were processed with multivariate analysis, to verify the contribution of each methodology in the characterization of the driving methods and in the measure of the physical correlated effort.

Keywords: Pressure sensors · Electromyography · Kinect · Postures · Soil tillage

1 Introduction

Soil working, in the conventional management of agricultural production, still forms the basis of preparing a field for subsequent cultivation. The main stresses to the operator driving a tractor are caused by tillage operations, both in the more intense ones of plowing and in the less heavy ones of harrowing for refining the soil for sowing.

Therefore, tractor drivers, especially when they are engaged in tillage such as plowing and harrowing, are exposed to noise, vibrations and possible incorrect postures that can generate musculoskeletal disorders [1, 2].

Surveys conducted on farms have shown that strain on the forearms and neck during a few hours of work as a carrier can be associated with discomfort and pain in the upper limbs and neck [3].

Back pain and sitting disorders are the main complaints reported by drivers of agricultural machinery [4–6]. It was found [7] that 24% of agricultural tractor drivers reported symptoms in the neck area and 96% of them reported bending or twisting their necks during various stages of work; 64% of tractor operators reported back symptoms, mainly reporting pain and stiffness problems.

The assessment of driver comfort during tillage operations can be studied with an ergonomic approach. The goal of ergonomics is not only to improve work performance, but also to improve human comfort and safety [8].

During agricultural operations the tiring of the operator could be due not only to the physical workload and uncomfortable postures, but also to the complexity of the task, the overload of information to manage, the pressure of time [9]. This wearying can be measured in terms of the stress on the tractor driver [10]. The most common tests typically include electrocardiography (for monitoring heart rate), electromyography (for monitoring muscle activity through their electrical potentials), pneumography (for monitoring breathing) or skin conductivity (to measure sweat activity) [11, 12].

Reliable information on people's real-time stress levels could be provided by many physiological cues [13]. The approach with different measuring instruments, including Motion Capture for reading movements, has been applied to experimental tests for the design of ergonomic dashboards [14].

The new technologies, based on GNSS and mechatronic components, aim to help the operator to drive tractors and self-propelled machines along predefined lines. This, from an ergonomic point of view, allows to simplify operations, reducing the number of incorrect or uncomfortable postures and improving operator awareness. The most popular GNSS-controlled guidance devices are the assisted and semi-automatic guidance systems.

A previous study [15] was carried out by comparing the same agricultural practice of harrowing conducted with manual guidance or guidance assisted by satellite guidelines or semi-automatic guidance. Comfort was observed through three methods, barometric, electromyographic and postural ones.

The purpose of the present study is to evaluate the sharing of these methods in verifying the overall assessment of operator comfort.

2 Materials and Methods

The study was performed through experimental tests that considered 10 operators, with similar body size (height: 1.80 ± 0.02 m, weight: 80 ± 1 kg, BMI: 24 ± 0.7 kg m^{-2}), who had to carry out the same agricultural operation harrowing with the same tractor, but in three different steering modes.

The tractor, a 120kW nominal power JOHN DEERE 6920S (John Deere, Moline, IL, USA) variable transmission tractor was equipped with an S-Guide RTK (Star-Fire6000, John Deere, Moline, IL, USA) automated steering system. Therefore, the harrowing operation had to be carried out by each operator, manually, trying to follow the route following the previous pass, in an assisted way, with the support of the satellite signal for the guidelines, or in a semi-automatic way, through the automatic management of the steering in the straight path and resumption of the command by the operator in the turning phase only.

The tests were carried out on a single field, located in Treviglio (Bergamo) near the CREA headquarters (Council for Agricultural Research and Economics), on which the operators had to perform four harrowing passages 150 m long, with a harrow (MAINARDI 400, Mainardi Machine Agricole, Cremona, Italy) following the path indicated in Fig. 1.

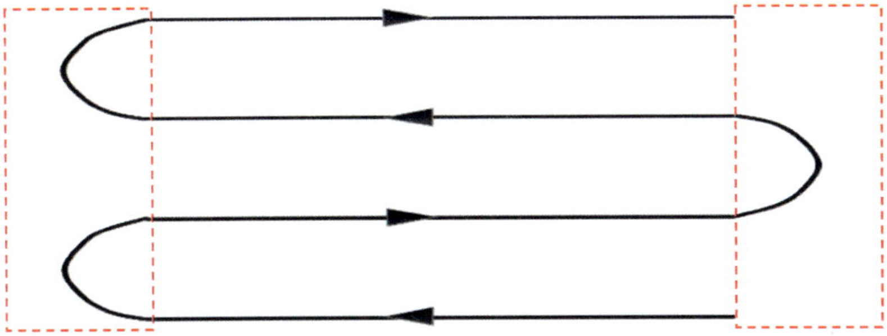

Fig. 1. Track travelled in each drive mode. The turnings (dashed lines) were not considered in the measurements.

The tests were repeated in three consecutive days characterized by the same meteorological conditions.

The soil was homogeneous for physical characteristics: sand = 68%; silt = 24%; clay = 8%. Stone represent 1/3 of the test field. During the tests, the median soil humidity was 18%, and thc median resistance to penetration was 0.98 MPa.

All procedures performed in present study were in accordance with the 1964 Helsinki Declaration and its later amendments or comparable ethical standards. All participants gave their signed consent to participate in the study. The study was approved by the ethics committee of University of Catania number code 321/18.

During the tests, the measurements involved the application of the following instruments:

- four electromyographs for the assessment of muscular engagement at the shoulders (SH_LX, SH_RX), at the lumb (LUMB) and at the right arm (AR_RX);
- two barometric mats for the assessment of body-weight distribution on the seat;
- one postural camera for the assessment of body position on the driving place able to read four nodes: neck (J3), the head (J4), the left shoulder (J5) and the right shoulder (J9).

The details of the three acquisition systems and the relevant methods used were presented in a previous article [15].

Data analyses were conducted with the Comprehensive R [18] Archive Network (CRAN) software (Institute for Statistics and Mathematics, Wien-Umgebung, Austria). Data from the electromyographic, barometric, and postural acquisition systems, as well as those related to geolocation, were processed through descriptive statistical analysis and the contribution of each evaluation method, barometric, electromyographic and postural, was observed through the analysis of the principal components (PCA) [19] developing the *factoMineR* and *factoshiny* packages [20], using the measured variables from each measuring instrument, after standardizing the values.

3 Results

The development of the Principal Component Analysis (PCA) on the dataset which contained 30 individuals and 11 variables, did not detect any outlier.

The inertia of the first dimensions shows if there are strong relationships between variables and suggests the number of dimensions that should be studied.

The first two dimensions (Fig. 2) of analyse express 52.87% of the total dataset inertia; that means that 52.87% of the individuals (or variables) cloud total variability is explained by the plane. This percentage is relatively high and thus the first plane well represents the data variability. This value is greater than the reference value that equals 39.35%, the variability explained by this plane is thus significant (the reference value is the 0.95-quantile of the inertia percentages distribution obtained by simulating 2849 data tables of equivalent size based on a normal distribution).

From these observations, it should be better to also interpret the dimensions greater or equal to the third one.

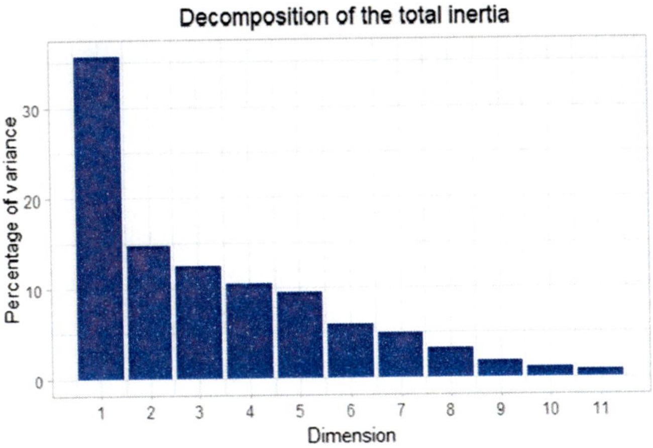

Fig. 2. Decomposition of the total inertia.

An estimation of the right number of axes to interpret suggests restricting the analysis to the description of the first 2 axis. These axes present an amount of inertia greater than those obtained by the 0.95-quantile of random distributions (52.87% against 39.35%). This observation suggests that only these two axes are carrying a real information. The distribution of the points is shown in the biplot of Fig. 3.

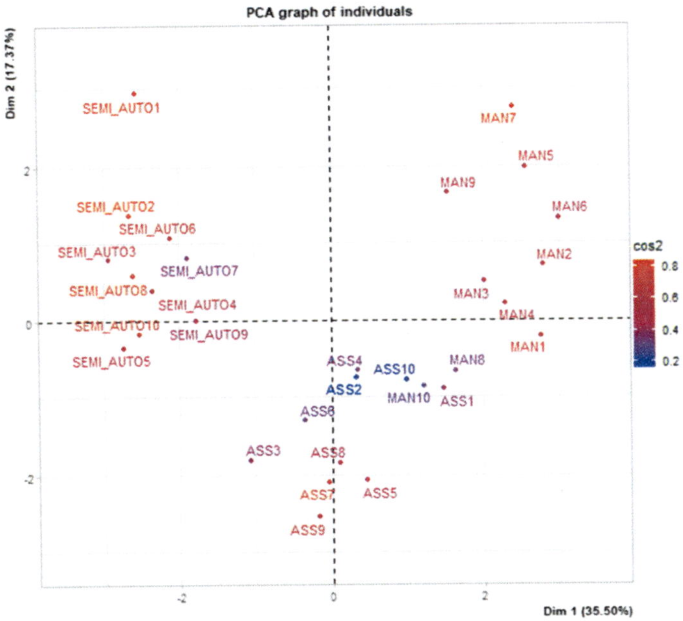

Fig. 3. Individuals factor map (PCA). The labelled individuals are those with the higher contribution to the plane construction. (cos2 shows the significance of a principal component for a given observation).

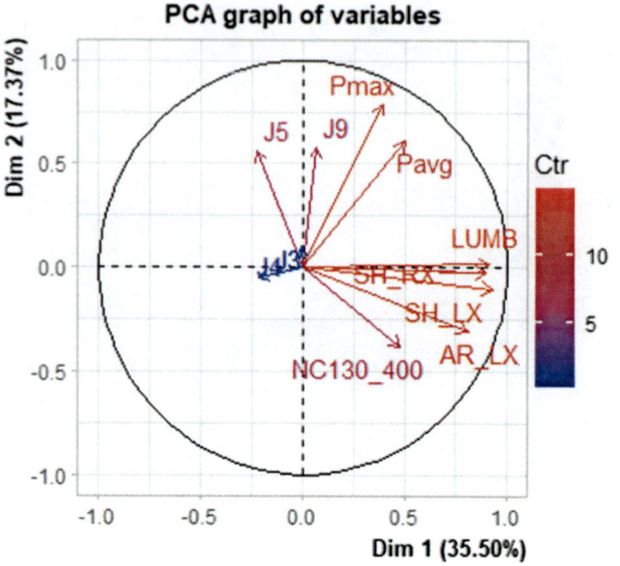

Fig. 4. Variables factor map (PCA) The labelled variables are those the best shown on the plane. Ctr: Contribution of variables to explain the variability.

The dimension 1 (Figs. 4 and 5) opposes individuals such as MAN7, MAN1, MAN5 and MAN2 (to the right of the graph, characterized by a strongly positive coordinate on the axis) to individuals such as SEMI_AUTO2, SEMI_AUTO1, SEMI_AUTO8, SEMI_AUTO10, SEMI_AUTO3 and SEMI_AUTO5 (to the left of the graph, characterized by a strongly negative coordinate on the axis).

The group in which the individuals MAN7, MAN1, MAN5 and MAN2 stand (characterized by a positive coordinate on the axis) is sharing:

- high values for the variables LUMB, SH_LX, SH_RX, AR_LX, Pmax, Pavg and NC130_400 (variables are sorted from the strongest).

The group in which the individuals SEMI_AUTO2, SEMI_AUTO1, SEMI_-AUTO8, SEMI_AUTO10, SEMI_AUTO3 and SEMI_AUTO5 stand (characterized by a negative coordinate on the axis) is sharing:

- low values for the variables SH_LX, AR_LX, LUMB, SH_RX and NC130_400 (variables are sorted from the weakest).

The dimension 2 (Figs. 4 and 6) particularly distinguishes individuals such as ASS7ASS9 (to the bottom of the graph, characterized by a strongly negative coordinate on the axis). These individuals form a group sharing: - low values for the variables Pavg, J9 and J5 (variables are sorted from the weakest).

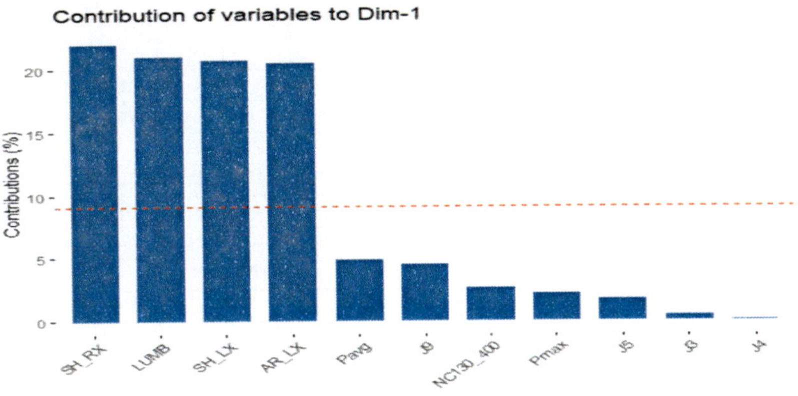

Fig. 5. Contribution of variables to explain the variability on the Dim 1.

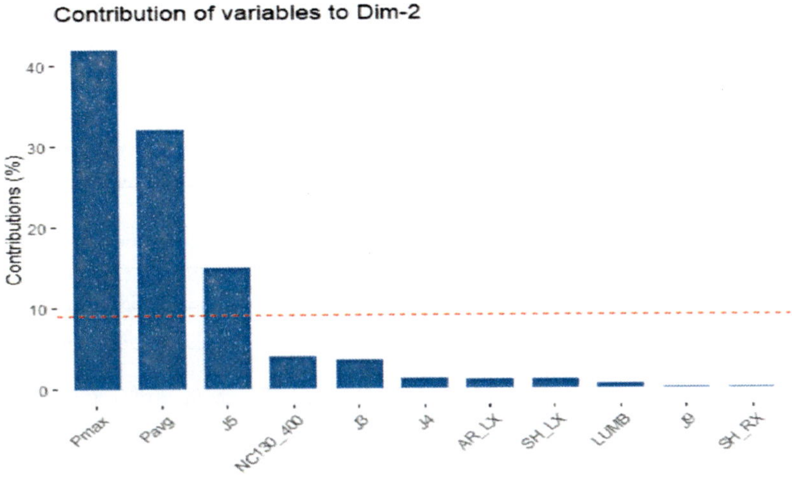

Fig. 6. Contribution of variables to explain the variability on the Dim 2.

The developed PCA is characterized by the contribution on the first dimension (Fig. 5), by the variables LUMB, SH_RX, AR_LX, SH_LX, which correspond to those relating to the electromyographic survey, and by J9 which corresponds to the joint of the right shoulder in the postural evaluation. The second dimension (Fig. 6) has greater contribution from the variables Pmax, Pavg, and NC130_400, which correspond to the measures of the barometric system and from J5 and J3 which correspond respectively to the left shoulder and the neck measures of the postural system.

4 Discussion

The agricultural operations of working the land involve attention, muscular and postural commitment. The technology based on the possibility of using satellite geolocation can modify the commitment required of the operator. The experimentation conducted in this work has allowed us to observe the contribution of three methods of investigating the operator's comfort during the conduct of a harrowing operation. The barometric method was intended to investigate the pressures exerted between the operator's body and the seat and backrest. The electromyographic method had the objective of measuring the muscular commitment of the shoulders, back and one arm. The postural method evaluated the change in the position of corresponding knots in the shoulders, neck and back over time.

The statistical analysis conducted in this article had the main objective of observing the contribution of the three methods in the evaluation of the comfort obtainable if the agricultural practice was conducted with manual guidance or assisted by guidelines indicated by the satellite detection system or finally from a semi-automatic type of driving, that is with an automatic steering wheel activated in the straight path and resumption of the command in the turn.

The multivariate statistical analysis, with an explanation percentage of the overall variance greater than 50%, showed a clear discrimination of the three driving conditions based on the descriptive variables observed.

The variables constituted by the measurements through the postural system showed agreement with the barometric ones, characterizing one dimension more, while the variables relating to the electromyographies participated in the distribution of the data on the plane in another direction, affecting more along the other dimension.

The set of 30 observed cases was then distributed and characterized in a complementary manner by the three types of vectors, thus showing the complementarity of the three investigation tools.

The muscular effort described by the electromyographic measurements has identified a crescendo of values, with the semi-automatic driving characterized by the minimum effort between the three guides, up to the manual which has a significantly greater muscular effort than the upper left. Assisted driving has values very close to the manual in terms of electromyographic measurements, probably because the muscular effort is to be attributed to very similar gestures. In this type of investigation all the observed variables and therefore all the observation points of the electromyographs had an effect in the description of the phenomenon under study.

The barometric variables showed that manual driving generated higher values than the other driving modes. These pressures have been reduced with assisted driving but increase with semi-automatic driving, probably because this driving allows higher speeds. The variables that best described the dataset were those relating to the average and maximum pressures.

The variables relating to postural observations show that the most characteristic nodes were those of the shoulders with higher values in the semi-automatic and manual guides. Probably since in manual driving the body moves a lot due to the commitment to drive, while in semi-automatic driving the operator, by taking his hands off the steering wheel, leaves the body freer and more subject to stress in the cabin. The variables of this survey method that had the greatest effect in the distribution of the points were J5 and J9 which corresponded to the shoulders.

5 Conclusions

Technology, introduced with the aim to increase performance of operators in the agricultural sector, offers new problems regarding safety risks in the short and long terms. The present study, conducted using three measuring methods (electromyographic, barometric and postural) applied to different driving modes on the same agricultural practice (harrowing), has shown that the observation of muscular engagement, of the stresses on the body and of the operator's posture, are three complementary aspects to characterize comfort of tractor driver during the work phases. This suggests the opportunity to consider the contextual use of the three methods to observe and qualify the operator's comfort allowing to intervene during the structural design phases or through corrective tools aimed to reduce the muscular, pressure or postural stresses.

References

1. Krüger, J.: Influence of posture on the deviation of measured acceleration values for tractor operators. Biosyst. Eng. **2019**(185), 56–63 (2019)
2. Brookhuis, K.A., DeWaard, D.: Monitoring drivers' mental workload in driving simulators using physiological measures. Accid. Anal. Prev. **42**, 898–903 (2010)
3. Tuure, V.-M.: Determination of physical stress in agricultural work. Int. J. Ind. Ergon. **10**, 275–284 (1992)
4. Fairley, T.E.: Predicting the discomfort caused by tractor vibration. Ergonomics **1995**(38), 2091–2106 (1995)
5. Futatsuka, M., Maeda, S., Inaoka, T., Nagano, M., Shono, M., Miyakita, T.: Whole-body vibration and health effects in the agricultural machinery drivers. Ind. Health **36**, 127–132 (1998)
6. Morgan, L., Mansfield, N.J.: A survey of expert opinion on the effects of occupational exposures to trunk rotation and whole-body vibration. Ergonomics **57**, 563–574 (2014)
7. Mayton, A., Kittusamy, N., Ambrose, D., Jobes, C., Legault, M.: Jarring/jolting exposure and musculoskeletal symptoms among farm equipment operators. Int. J. Ind. Ergon. **38**, 758–766 (2008)
8. Ergonomics of human system interaction - Part 210: Human-centered design for interactive systems, International Organization for Standardization (ISO), ISO 9241-210 (2009)
9. Zhang, J.Y., Qiu, W.W., Fu, H.J., Zhang, M.T., Ma, Q.G.: Review of techniques for driver fatigue detection. Appl. Mech. Mater. **433–435**, 928–931 (2013)
10. Peruzzini, M., Grandi, F., Pellicciari, M., Campanella, C.E.: User experience analysis based on physiological data monitoring and mixed prototyping to support human-centre product design. In: Rebelo, F., Soares, M.M. (eds.) AHFE 2018. AISC, vol. 777, pp. 401–412. Springer, Cham (2019). https://doi.org/10.1007/978-3-319-94706-8_44

11. Boubaker, K., Colantoni, A., Allegrini, E., Longo, L., Di Giacinto, S., Monarca, D.: A model for musculoskeletal disorder-related fatigue in upper limb manipulation during industrial vegetables sorting. Int. J. Ind. Ergon. **44**, 601–605 (2014)
12. Healey, J.A., Picard, R.W.: Detecting stress during real-world driving tasks using physiological sensors. IEEE Trans. Intell. Transp. Syst. **6**(2), 156–166 (2005)
13. Ergonomic principles related to mental workload — Part 3: Principles and requirements concerning methods for measuring and assessing mental workload, International Organization for Standardization (ISO), ISO 10075-3 (2005)
14. Grandi, F., Peruzzini, M., Campanella, C.E., Pellicciari, M.: Application of innovative tools to design ergonomic control dashboards. Adv. Transdisciplinary Eng. **12**, 193–200 (2020). https://doi.org/10.3233/ATDE200077
15. Romano, E., et al.: Assessment of comfort variation among different types of driving agricultural tractors: traditional, satellite-assisted and semi-automatic. Int. J. Environ. Res. Public Health **17**, 8836 (2020). https://doi.org/10.3390/ijerph17238836
16. Valentino, M., et al.: A new method for the experimental assessment of finger haemodynamic effects induced by a hydraulic breaker in operative conditions. J. Occup. Health **46**, 253–259 (2004)
17. Otte, K., et al.: Accuracy and reliability of the kinect version 2 for clinical measurement of motor function. PLoS ONE **11**, e0166532 (2016)
18. R Core Team: R: A language and environment for statistical computing. R Foundation for Statistical Computing, Vienna, Austria (2021). https://www.R-project.org/
19. Abdi, H., Williams, L.J.: Principal component analysis. WIREs Comput. Stat. **2**(4), 433–459 (2010). https://doi.org/10.1002/wics.101
20. Kassambara, A.: Practical Guide to Principal Component Methods in R. STHDA (2017)

Assessment of External Sprayer Cleaning Efficiency by Comparing Different Cleaning Devices, Sprayer Tank Materials and Operators

Marco Grella[1(✉)], Davide Allochis[2], Paolo Marucco[1], and Paolo Balsari[1,2]

[1] Department of Agricultural, Forest and Food Sciences (DiSAFA), University of Turin (UNITO), Largo Paolo Braccini, 2, 10095 Grugliasco, TO, Italy
marco.grella@unito.it

[2] National Interuniversity Consortium of Materials Science and Technology (INSTM), Via G. Giusti, 9, 50121 Firenze, Italy

Abstract. The external sprayer cleaning is an essential procedure to minimize possible chemical contamination for both the environment and operators. The cleaning efficiency can be affected by the type of cleaning device, the sprayer architecture, and its component materials but also by the skills of operators. For these reasons it is important to investigate the effect of different variables on the potential decontamination of external sprayer surfaces. A new method aimed to evaluate the external sprayer cleaning efficiency in a simple and cost-effective way was tuned. It foresees the i) contamination of the sprayer applying copper oxychloride solution uniformly on the sprayer under controlled laboratory conditions and then ii) the external sprayer cleaning using a fixed volume of pure water. The cleaning efficiency was obtained measuring the amount of copper removed compared to that initially applied. Three different cleaning devices were compared (spray gun, long-range spray gun and lance) and then the operating parameters, namely nozzle size (1.0, 1.2, 1.5 and 2.2 mm in diameter) and spray pressure (0.3, 1.0 and 1.5 MPa) were also investigated for their possible effect on cleaning efficiency. Finally, the effect of tank materials (fiberglass vs. polyethylene) and operator were also tested. Results underlined the effect of nozzle size on the cleaning efficiency while the cleaning devices did not show significant differences. Furthermore, the cleaning efficiency resulted correlated to the roughness of tank materials while no effect of operator was detected applying the same procedure for cleaning.

Keywords: Chemical contamination · Spray gun · Spray lance · Sprayer tank · External surfaces

1 Introduction

Before being authorized for use, a pesticide is evaluated to ensure that it does not pose any undue risk to the spray operator or other workers. The scenarios assessed includes loading/mixing of the concentrated product, application of the diluted product, re-

M. Biocca et al. (Eds.): SHWA 2020, LNCE 252, pp. 127–136, 2022.
https://doi.org/10.1007/978-3-030-98092-4_14

entry/handling of a treated crop and inadvertent exposure to the public [1]. However, during application, a proportion of the pesticides is deposited on the external surfaces of the sprayer [2]. Exposure to these residues does not form any part of the risk assessment procedure possibly because it is assumed that, compared with other tasks one spray operator performs, the risks associated with external residues are relatively low [1], particularly as it is known that exposure during mixing and loading is much greater than that during spraying [3]. However, Ramwell et al. [1] demonstrated that operator contamination attributable to external residues compared with occupational exposure limits indicated that external sprayer contamination is not negligible. Concurrently, the chemical residues of the external sprayer surfaces represent not only a hazard for operators but also for the environment [3], e.g. from washing off after a rainfall. Even small quantities of the concentrated spray solution may represent a serious contamination risk [4]. A drip of formulated preparation can easily contain 1g of active ingredient, and 10,000 m^3 of water is needed to dilute this amount to an acceptable a concentration below 0.1 μg L^{-1} water [5], the threshold stated in EU Water Framework Directive (2000/60/EC).

Sprayer external deposits on bush/tree crop sprayers are generally higher than those found on field crop sprayers due to the turbulence generated by the air support used to convey spray droplets to the target during application [3]. Deposits can be quantified in the range 82.5–207.0 g of active ingredients on the external sprayer surface in one spray season [6]. Ramwell et al. [1] found residues accumulated on external sprayer surfaces higher than 1 g m^{-2}. Concurrently, Balsari et al. [2] found external contamination ranging between 0.33–0.83% of the amount applied per application. Authors agreed that the amount of spray mixture applied which deposits on the external surfaces of the spraying equipment depends on the sprayer type, the operating parameters and the environmental conditions, and is also affected by the agrochemicals' formulations [1–3, 7].

The data described above regarding the amount of external sprayer contamination arises the importance of effective decontamination procedure to minimize both operator exposure and environmental risks, especially in bush/tree crop sprayers. It is therefore important to develop more efficient methods and devices for the external decontamination of sprayers through experimental research finalized to quantify both the actual sprayer external contamination and efficiency of tools/devices used for cleaning, in order to provide reliable information useful at regulatory level and guidelines for farmers. The evaluation of cleaning efficiency in relation to the different tools used for the decontamination will become more and more important considering that it is now recognized that pesticide residues on the external surfaces of sprayers can represent a significant route of exposure for both the spray operator and the environment, and these residues exist despite sprayers being washed [8].

In this context, a further point of relevance is that the amount of clean water available to carry out the external cleaning of sprayers directly in the field right after the Plant Protection Product (PPP) application, as recommended in TOPPS guidelines (www.topss-life.org), is generally very limited. The capacity of the clean water tank is typically just 10% of the nominal mixture tank one, and its volume shall be used both for internal and external cleaning. It is therefore important that the devices available on the sprayers to carry out the external cleaning are efficient and used at adequate pressures in order to allow at least a first effective decontamination of the sprayer surfaces. ISO 22368-2:2004 defines the methodology to assess the efficiency of external cleaning devices for bush and tree crop sprayers contaminating the machine with a water solution of copper oxychloride by driving in circle at 5.0 km h^{-1} for 10 min with all the nozzles activated at 1.0 MPa pressure and the fan rotation speed set according to the manufacturer instructions. This procedure, however, does not provide an even contamination of the sprayer and the amount of copper oxychloride deposited on the external surfaces of the machine in the test replicates is very variable depending on the environmental conditions.

To simplify the test procedure and to work under controlled and stable environmental conditions, a new indoor methodology was proposed and tested using an air-assisted sprayer for vineyards. Different experimental trials were carried out to test the effect of different cleaning devices and operating parameters like nozzle orifice sizes and operating pressures on sprayer cleaning efficiency. Moreover, additional trials were carried out to verify the influence of the tank material, characterized by different roughness, and of the operator on the effectiveness of sprayer external cleaning.

2 Materials and Methods

2.1 Experimental Arrangement and Sprayer Used

A mounted conventional airblast sprayer Dragone Athos (Dragone S.n.c., Castagnole delle Lanze, Asti, AT, Italy) for bush/tree crops equipped with 200 L polyethylene tank was used for experimental purposes (Fig. 1).

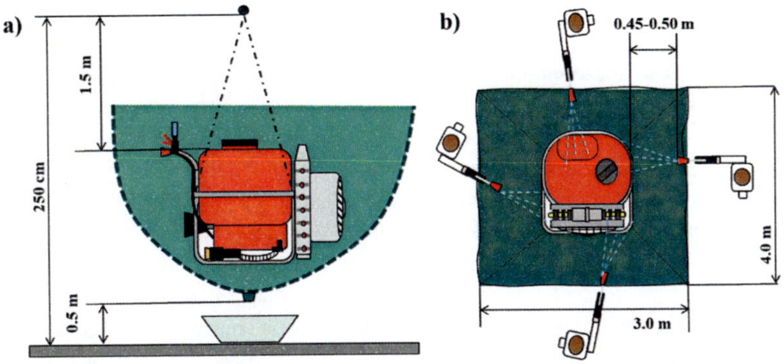

Fig. 1. Schematic of a) sprayer positioning, suspended in the laboratory test area, and of the waterproof sheet used to collect washing liquid (lateral view) and b) positions used by the operator equipped with a knapsack sprayer to contaminate the external surfaces of the suspended sprayer (aerial view).

Laboratory trials were set up at DiSAFA, Crop Protection Technology research group facilities. An ad hoc indoor test area was prepared to suspend the sprayer at a height of 0.5 m from the ground using chains (Fig. 1a). In this way, it was possible to place a waterproof sheet underneath and around it to collect the liquid used for the cleaning (Fig. 1). The sheet had a drain hole at the lowest point, underneath of which a 50 L container was placed. It was fixed so that it would surround the sprayer without preventing the operator from carrying out the cleaning operations.

2.2 Experimental Trials Procedure and Laboratory Analysis

The procedure used was common for all tests and can be divided into five phases: i) sprayer contamination, ii) washing of the collection sheet, iii) drying of the oxychloride solution on the sprayer, iv) sprayer cleaning, v) deep-cleaning of the sprayer using the high-pressure washer.

Phase 1: to carry out the contamination, a 1% copper oxychloride solution was used as required by ISO 22368-2:2004. The solution was applied using a Jacto PJB-20 (JACTO, Pompeia, SP, Brazil) knapsack sprayer equipped with one single hollow cone nozzle Albuz ATR 80 blue (Albuz® CoorsTek, Evreux, France) mounted on a lance with a flow rate of 1.92 L min^{-1} and working at a pressure of 0.3 MPa. To ensure uniform contamination, 4 L of solution were applied, equally distributed by the operator from four positions around the machine (front, back and both lateral sides) (Fig. 1b). It derives that the spray was activated 0.31 s per position with operator ensuring to cover the whole surface accounting for the intended sprayer side. The amount of 4 L was defined based on preliminary trials as it guarantees an even contamination of surfaces.

Phase 2: the copper oxychloride solution dripped from the sprayer during its contamination and deposited on the waterproof sheet was then removed. This operation was carried out by rinsing each of the sheet's walls with clean water at a pressure of 0.3 MPa, taking care not to rinse the sprayer as well. The washings were collected in the container placed under the drain hole of the waterproof sheet. The total amount of water used to rinse the contaminated surfaces of the waterproof sheet was measured and averaged 21 L.

Phase 3: a pre-defined time was set up waiting for the copper oxychloride solution natural-air drying on the sprayer surfaces.

Phase 4: the sprayer external surfaces were cleaned for the removal of dried copper oxychloride using the devices to be tested and 5 L of clean water. The washing water was collected by the waterproof sheet previously mentioned (Phase 2) and described in Fig. 1.

Phase 5: at the end of each test a deep-cleaning of the sprayer surfaces using a high-pressure washer was performed to guarantee the complete removal of copper oxychloride and evaluate the complementary fraction not removed by tested devices during Phase 3. The high-pressure washer was used with clean water at 15 MPa pressure providing a flow rate equal to 10 L min^{-1}. The average amount of water used for this operation was 24 L.

For the determination of copper oxychloride concentration in each phase of the above-described procedure, a total of 11 samples 500 ml each were taken using glass-lab bottles (two initial samples during Phase 1, to check the exact concentration of the copper oxychloride solution at the beginning, and three samples per each cleaning phase namely Phases 2, 4 and 5) which were weighed with a precision balance (full scale 1,000 g and resolution 0.005 g). Then samples were placed in laboratory oven at 105 °C for 48 h. Finally, the dry weights (g ml^{-1}) of the copper oxychloride deposits inside the glass lab-bottles were measured.

The air temperature (°C) and relative humidity (%) were monitored throughout the trials using a thermo hygrometer Testo 625. Average air temperature resulted 25.0 °C, ranging from 23.9 to 26.0 °C, while the average relative humidity resulted 65%, ranging from 60 to 71%.

2.3 Variables Examined for Their Possible Influence on Sprayer Cleaning Efficiency

Three devices were selected and compared for their possible effect on sprayer external cleaning efficiency: i) spray gun Braglia Turbo 400, ii) long range spray gun (miter) Braglia 22.901.161 and iii) spray lance Braglia Spot 600 (Braglia S.r.l., Reggio Emilia, Italy) (Fig. 2).

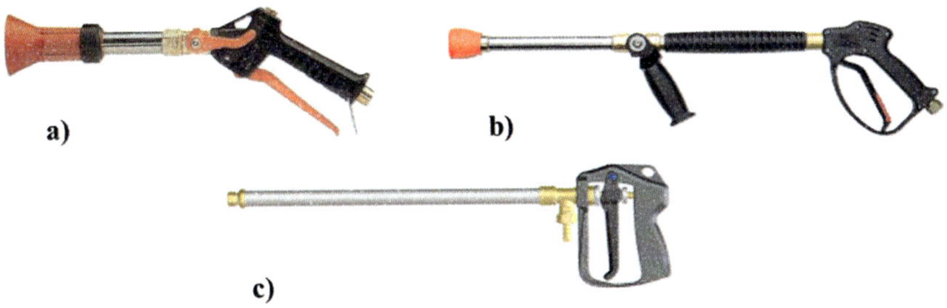

Fig. 2. Cleaning devices compared in the experimental trials: a) spray gun, b) long rang spray gun (miter) and c) spray lance (Braglia S.r.l.).

Each device was tested combined with two drying time, namely 1 and 24 h to possibly simulate a sprayer cleaning made just after the spray application and after long time from spray application with chemical residues on sprayer surfaces completely dried to the natural air. The three devices were used with the same operating modes (1.5 mm nozzle orifice diameter, operating pressure of 1.0 MPa, distance between device and sprayer: 45–50 cm), and therefore with a fix amount of water used for the cleaning equals to 5 L. This latter volume was defined by referring to: i) the overall capacity that, according to the Machinery Directive (Directive 2009/127/EC) and the related harmonized reference standard ISO 16119:2013, a clean water tank mounted on a sprayer should have: 10% of the capacity of the mixture tank of the sprayer; ii) the amount of water used for external cleaning of the sprayer should not exceed 50% of the water present in this tank.

A second set of experiments were carried out to assess the influence of operating parameters on the cleaning efficiency. For these tests only the spray gun device was used with four nozzles characterized by different orifice sizes (1.0, 1.2, 1.5, 2.2 mm in diameter) and combined with three operating pressures, namely 0.30, 1.00 and 1.50 MPa. As expected, nozzles were characterized by different droplets size spectra, with Volume Median Diameter (VMD) raging between 100 and 200 μm. The size of the droplets generated by the twelve different configurations (combination of 4 nozzle sizes and 3 spray pressure) were measured using a Malvern Spraytec laser diffraction system STP5342 (Malvern Instruments Ltd, Malvern, UK) as detailed by Grella et al. [9]. For this experiments the drying time corresponded to 1 h. The selection of cleaning devices was carried out through an inquiry made by a personal interview that was conducted to the more representative spraying machinery manufacturers and spraying machinery retailers. In the inquiry, it was taken into consideration not only the most common types and models of cleaning equipment normally supplied to sprayers, but also the standard configurations and usual practices of using these devices. Based on preliminary measurements, it was identified that for conventional airblast sprayers 70–80% of sprayer completely exposed surface is accounted by the tank. Concurrently the roughness index (RzD, μm) of different sprayer components, characterized by different materials, was measured on different sprayer types per the UNI EN ISO 4288:2000 and an overview is given in Fig. 3. Based on these premises a third experiment was set up to investigate the effect of tank materials on the sprayer external cleaning efficiency. So, two tanks of 200 L nominal capacity, made of polyethylene and fiberglass respectively, were compared as characterized by different roughness. For these tests the spray gun cleaning device equipped with a 1.5 mm diameter nozzle orifice and operated at 1.0 MPa pressure was used. Tests were also conducted in combination with two different drying times for the copper oxychloride solution applied (1 and 24 h).

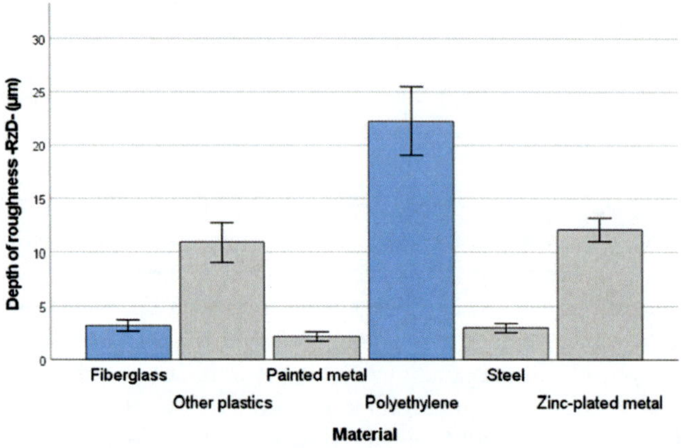

Fig. 3. Depth of roughness (RzD, μm) measured on sprayer's components made by different materials. Light-blue bars identify the materials tested in experimental trials for their effect on cleaning efficiency.

Finally, the fourth experiment consisted in comparing the effect of different operators on the sprayer external cleaning efficiency. Three operators selected among DiSAFA workers, after an appropriate training about the procedure to be used, were compared in making the cleaning of the Dragone Athos sprayer. In this case the spray gun device with 1.5 mm diameter nozzle orifice and operated at 1.0 MPa pressure was used too. A fixed drying time of 1 h was waited before cleaning.

2.4 Data Processing and Statistical Analysis

The cleaning efficiency *E* (%) was then calculated according to Eq. (1) as follows:

$$E = (R_{dev} * 100) / (R_{dev} + R_{spr}) \tag{1}$$

where: R_{dev} (g) is the amount of copper oxychloride found in the samples taken after cleaning the sprayer with the tested device (Phase 4) and R_{spr} (g) is the amount of oxychloride copper in the sample taken after the deep-cleaning with the high pressure washer (Phase 5).

All statistical analyses were performed using IBM SPSS Statistics (Version 26) predictive analytics software for Windows. In order to evaluate the effect of a) different cleaning devices, b) operating parameters used to clean the sprayers c) drying times and d) the tank materials on the sprayer external cleaning efficiency, a two-way Analysis of Variance (ANOVA) was used. Instead for the evaluation of operator effects one-way ANOVA was used. In all cases the means were compared using a FREWG post-hoc test for multiple comparison ($p < 0.05$).

3 Results and Discussion

Figure 4 displays the cleaning efficiency of the three cleaning devices tested waiting for two different drying time of copper oxychloride before the cleaning.

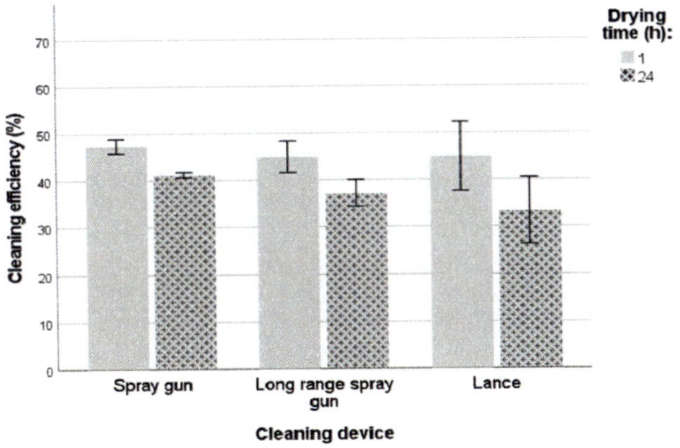

Fig. 4. Cleaning efficiency of the three devices, with 1 and 24 h drying time periods. The bars show the mean ± standard error of the mean.

No statistical effect of cleaning device [F(2,12) = 0.643, p = 0.543] on the cleaning efficiency was detected, while significant effect of drying time [F(1,12) = 5.161, p = 0.042] was noticed. The percentages of cleaning efficiency were very similar especially with the short drying time with values between 44 and 49% while for the 24 h drying time values between 33 and 41% were found. The results are in line with those reported by Ramwell et al. [7] investigating the removal capability of different pesticides from sprayer external surface. It is noticeable that the lance showed the highest variability with coefficient of variation (CV) equal to 36%. Limited variability was detected for the spray gun and long range spray gun with CV equal to 5 and 13%, respectively.

Figure 5 displays the cleaning efficiency obtained using different nozzle orifice sizes (1.0, 1.2, 1.5 and 2.2 mm in diameter) mounted on a spray gun and combined with three spray pressures (0.3, 1.0 and 1.5 MPa). The cleaning efficiency values at 0.3 and 1.0 MPa with all types of nozzles ranged between 32 and 49% while when the pressure was increased up to 1.5 MPa the cleaning efficiency also increased achieving the highest values equal to 59%. Even if in some cases results suggest to increase the liquid pressure in order to increase the cleaning efficiency, no statistical effect of spray pressure on cleaning efficiency increment was detected [F(2,24) = 1.903, p = 0.171]. On the contrary the nozzle size showed significant effect on cleaning efficiency [F (3,24) = 9.651, p = 2.31E-04] even if no clear trend according to the orifice size increment or decrement was observed (Fig. 5).

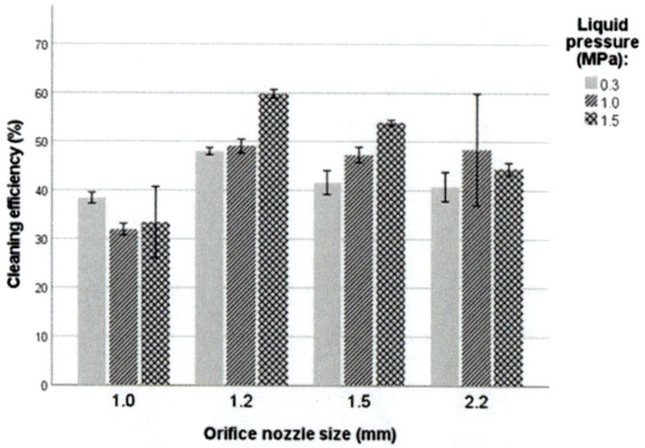

Fig. 5. Cleaning efficiency of the spray gun, obtained with 4 different nozzle types, 3 pressures and a drying time of the sprayer after the contamination of 1 h. The bars show the mean ± standard error of the mean. Different letters denote significant differences among orifice nozzle size (REGWF post hoc test, $p < 0.05$).

Figure 6 showed the cleaning efficiency related to the two different tank materials examined, namely polyethylene and fiberglass, combined with two different times occurred between the contamination and the cleaning, namely 1 and 24 h. Both tested variables, namely material [F(1,8) = 79.662, p = 2.00E-05] and drying time [F (1,8) = 114.928, p = 5.21E-06] showed a significant effect on the cleaning efficiency confirming results obtained with different cleaning devices (Fig. 4). As expected, the effect of roughness on cleaning efficiency was confirmed. Indeed, fiberglass characterized by a lower roughness (Fig. 3) obtained higher cleaning efficiency both with the short and long drying time reaching values of 58 and 43%, respectively. The cleaning efficiency for the polyethylene tank was 46 and 31% according the short and long drying time, respectively. Very low variability was noticed in all cases with CV values between test replicates always lower than 8%.

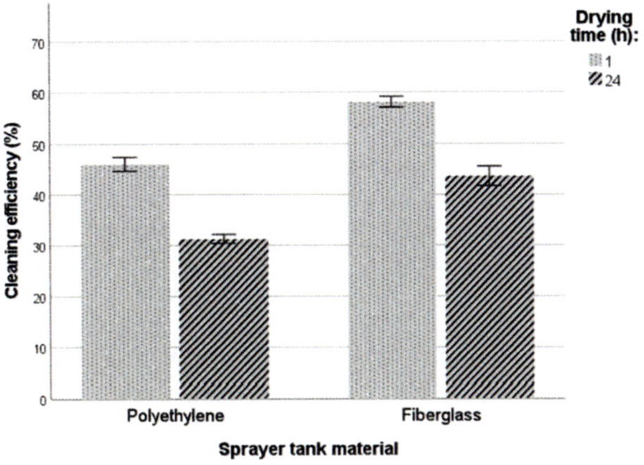

Fig. 6. Cleaning efficiency obtained testing polyethylene and fiberglass tanks, with 1 and 24 h drying time periods. The bars show the mean ± standard error of the mean.

Finally, no significant effect of operator [F(2,6) = 4.579, p = 0.62] on cleaning efficiency was observed, with values ranging between 43 and 47%. Again, a very low variability was measured with CV values between test replicates, always lower than 6%.

4 Conclusions

Even if the results of the tests highlighted the limits that still characterize the sprayer external cleaning procedure and the difficulties in obtaining a good cleaning efficiency, with cleaning efficiency values generally below 50%, the data and information collected enabled to acquire important and significant indications for improving the washing procedure. The time that occurs between the external sprayer contamination and the cleaning plays a key role for the decontamination success of sprayer external surfaces. Indeed, when the cleaning was carried out within 1 h from the sprayer surfaces contamination the higher levels of efficiency were obtained. This demonstrates the importance to clean the sprayer just after the PPP treatment. Despite in the tests carried out with the different cleaning devices no significant differences in efficiency emerged, slightly better values were generally found out using the spray gun. While the variation of the operating pressure had no significant effect in increasing the cleaning efficiency, on the contrary the different orifice nozzle sizes had a significant effect on it, even if a clear trend was not observed. So, the best combination between the liquid outlet hole diameter and the operating pressure value was the one with the 1.2 mm nozzle orifice and 1.5 MPa pressure corresponding to 59% of cleaning efficiency. The results obtained with the two tank materials pointed out that working with the same cleaning device and operating parameters, a better cleaning efficiency was reached on the fiberglass, which is characterized by a lower roughness level, compared to polyethylene. Recently, Balsari et al. [10] tested various cleaning agents for the decontamination of sprayer surfaces. Authors obtained promising preliminary results showing consistent increment in the cleaning efficiency, especially for those agents characterized by acid compounds, and further trials are ongoing.

References

1. Ramwell, C.T., Johnson, P.D., Boxall, A.B.A., Rimmer, D.A.: Pesticide residues on the external surfaces of field-crop sprayers: environmental impact. Pest Manag. Sci. **60**, 795–802 (2004). https://doi.org/10.1002/ps.870
2. Balsari, P., Marucco, P., Oggero, G.: External contamination of sprayers in vineyard. Asp. Appl. Biol. **77**(1), 215–221 (2006)
3. Balsari, P., Marucco, P.: Internal and external contamination of sprayers: causes and strategies to minimise negative effects on the environment. Chem. Eng. Trans. **58**, 793–798 (2017). https://doi.org/10.3303/CET1758133
4. De Wilde, T., et al.: Characterizing pesticide sorption and degradation in microscale biopurification systems using column displacement experiments. Environ. Pollut. **157**(2), 463–473 (2009). https://doi.org/10.1016/j.envpol.2008.09.008
5. De Wilde, T., Spanoghe, P., Debaer, C., Ryckeboer, J., Springael, D., Jaeken, P.: Overview of on-farm bioremediation systems to reduce the occurrence of point source contamination. Pest Manag. Sci. **63**(2), 111–128 (2007). https://doi.org/10.1002/ps.1323
6. Debaer, C., et al.: Volumes of residual of sprayers and their International Standards: impact on farm water treatment systems. Asp. Appl. Biol. **84**, 193–199 (2008)
7. Ramwell, C.T., Johnson, P.D., Corns, H.: Transferability of six pesticides from agricultural sprayer surfaces. Ann. Occup. Hyg. **50**(3), 323–329 (2006). https://doi.org/10.1093/annhyg/mei074
8. Ramwell, C.T., Johnson, P.D., Boxall, A.B.A., Rimmer, D.A.: Exposure to pesticide residues on agricultural spraying equipment, 1st edn. Health and Safety Executive, Caerphilly (2010)
9. Grella, M., Miranda-Fuentes, A., Marucco, P., Balsari, P.: Field assessment of a newly-designed pneumatic spout to contain spray drift in vineyards: evaluation of canopy distribution and off-target losses. Pest Manag. Sci. **76**, 4173–4191 (2020). https://doi.org/10.1002/ps.5975
10. Balsari, P., Allochis, D., Marucco, P., Grella, M.: Assessment of external sprayer cleaning efficiency from different devices and cleaning agents. Asp. Appl. Biol. **143**, 81–89 (2021)

Safety Health and Welfare in Building

The Therapeutic Value of a Green Roof in a Prison Facility

A Case Study in Central Italy

Francesca Piccioni[1(✉)], Massimo Cecchini[1], Chiara Grazini[2], and Leonardo Bianchini[1]

[1] Department of Agriculture and Forest Sciences, University of Tuscia, Viterbo, Italy
francesca.piccioni@unitus.it
[2] Department of Economics, Engineering, Society and Business Organization, University of Tuscia, Viterbo, Italy

Abstract. Does a green roof built in a penitentiary context play an important role in preserving environmental resources and naturalness, with positive effects on inmates' health and well-being? In this setting, the study aims at investigating the potential benefits of a "green roof" for the inmates of a Women's Correctional Institution in Rome, Italy, as well as demonstrating the advantages for the Public Administration: Ministry of Justice - Department of Prison Administration. Though the overall benefits could be different, from easy maintenance to environmental sustainability, from low energy consumption to safety education, from training and employment opportunities for disadvantaged women to the therapeutic impact and general well-being for the inmates, this study focuses on the last two topics. This experimentation is unprecedented and unique in Italy. A questionnaire focusing on the individual relationship with green and the potential effect on the personal well-being was administered to a specific panel of inmates, while a second questionnaire concerning the attitude towards safety and green maintenance was administered only to specially trained convicts.

Results show the significant restorative effects of larger green areas in the penitentiary, especially on some specific groups of prisoners, corroborating the hypothesis of a positive cost-benefit analysis relative to the intervention. We conclude by discussing the implications of the present study on prison design in the context of sustainability.

Keywords: Therapeutic green · Social workers · Penitentiary · Restorativeness · Occupational safety and health

1 Introduction

The aim of this research is to carry out a cost-benefit analysis for the Penitentiary Administration regarding the renovation of a prison roof with a green roof. The cost-benefit analysis, taking for certain the numerous advantages in terms of energy efficiency and increase of environmental sustainability, focuses on the assessment of the regenerative capacity of the green roof in this particular context.

M. Biocca et al. (Eds.): SHWA 2020, LNCE 252, pp. 139–148, 2022.
https://doi.org/10.1007/978-3-030-98092-4_15

Referring to the evidence of positive health effects that access to nature generates in healthcare facilities [1], and to comparative research conducted in the UK and the Nordic region that shows evidence that access to nature generates the same effects in the custodial context as are recognized in healthcare facilities [2, 3], in this paper we explore the potential increase of well-being and restorativeness obtained by the views of and access to green spaces declined as a green roof inserted in the penitentiary context.

The experimentation was conducted in the "G. Stefanini" Women's prison House in Rebibbia, Rome, and the green roof constructed on the library building.

Firstly, the preliminary loading tests on the slab of the building were carried out; following their positive outcome, the design criteria were adopted for ease of maintenance and cost containment: hence the choice of a green roof of the extensive type.

Finally, a training was provided for ordinary maintenance of the green roof to strengthen the involvement of women prisoners in the project.

2 Materials and Methods

2.1 Study Area

Being the largest women's prison facility in Europe, the "G. Stefanini" Women's Prison House in Rebibbia Penitentiary Institute, Rome was selected for this case study. The director of the Institute welcomed the research project which also benefited of the prison technical staff of agronomists and of a prison farm. The prison facility is located in an urban area which proves important for visibility testing and positive effects in terms of energy efficiency. In addition, the facility is a women's institute with a specific attention to gender. Therefore, it proved particularly suitable for the evaluation of the restorative effects of the intervention in a population subject to a high level of stress caused by imprisonment, and of the possibilities of future work reintegration. These were all key factors in the choice.

As mentioned above, the building chosen to construct the green roof was the library, both for its central position, visible from the most populated detention sections and for its symbolic value as a place of culture with a special 'cover' that would uniquely identify it.

2.2 Design of the Green Roof

The design of the green roof results from two objectives: creating a structure with a strong distinctive identity and making it easier to maintain the site. This is why the design of the central path allocates the largest surface area to the plant parts providing a single central path covered with gravel offering ready access to the cultivated parts. A curved design gave the installation an "organic" language of rupture with respect to the regularity of the surrounding buildings. The choice of the plant species has been oriented on perennial officinal herbs which, due to their robustness and rusticity only need low maintenance. Furthermore, they could be used by the Institute that is planning the creation of a laboratory for the production of essential oils in the coming future, thus achieving a virtuous path of circular economy.

The layout of the plants was carried out to produce, on a visual level, an increasing view of green volumes, with staggered blooming and different colours.

2.3 Green Roof Construction

Following the positive result of the load tests, the Institute requested a quotation from a company specialized in waterproofing and certified in the construction of green roofs according to UNI 11235:2015, then it directly employed the company as building contractor to do the work.

The work expected completion period was of 60 days, with the following steps: installation of an aluminium parapet to secure the site; construction works: scaffolding and delimiting areas; demolition, site clearance and transport; support preparation with self-levelling fiber-reinforced mortars; installation of a new waterproofing system with non-woven geotextile materials, synthetic PVC coating, 100% root barrier and accessories; installation of the multilayer system: water-retention and protection felt, aeration and control of the vapor flow, filter cloth and substrate according the requirements of UNI 11235:2015; formation of a gravel path confined between two drainage profiles of aluminium containment; planting of the vegetative layer: carried out by the prisoners involved in the training course.

2.4 Training for Workers in the Maintenance of a Green Roof

The inmates suitable for the training had specific psychological characteristics and aptitudes, and were facing long-term sentences with a planned reintegration programme. Once selected, a test was administered to verify their command of the Italian language. Twenty participants were selected to attend the 13-h classroom training, but due to the long lockdown caused by the Covid-19 pandemic, only 6 inmates completed the course. The inmates' skills and knowledge were assessed with a test consisting of 20 multiple-choice questions.

2.5 Data Collection

The term "Restorativeness" indicates a process of recovery of cognitive and psychological resources compared to conditions of previous mental deficit [4]. Scientific research in this area is based on Stress Recovery Theory (SRT) [5] and on Attention Restoration Theory [6]. ART suggests that some environments can encourage the regeneration of voluntary attention, thus they are defined as regenerative. According to ART a regenerative environment must be characterized by four properties: being-away, extent, fascination, and compatibility. Environments with a higher degree of naturalness generally possess all these properties and a stronger regenerative potential compared to artificial environments [7]. There is currently only one instrument suitable for measuring the regenerative qualities and potential of an environment: Perceived Restorativeness Scale [8]. Some authors evaluated the validity and reliability of the PRS [9–11]. In particular, Pasini et al. demonstrated that the "Italian version" of the PRS (PRS/IT) is reliable and can discriminate between the four environmental properties [12].

On this basis, a questionnaire of 50 questions was designed to measure the regenerative capacity of the project: the first 25 questions, investigated the profile of prisoners and their 'green' attitude, while the last 25 questions were structured on a 5-point Likert scale: exploring the effect of green from a physical, emotional and social point of view in the prison context and outside, including the impact of the green roof project. The paper forms were distributed to the population resident in the main sections "Camerotti" and "Cellulare", both with direct views of the green roof: the first mainly hosts detainees waiting for the first degree of judgment, while the second one houses the inmates with a final sentence. During the survey, the Institute was accommodating 307 people, while the above Sections were hosting 113 people: the sample obtained is considered representative, being 36.81% of the total.

The risk of inaccessibility of the questionnaire was reduced by educating the inmates with greater cultural skills to help the less-skilled ones. A total of 113 questionnaires were distributed, 69 of which in the "Camerotti" and 44 in the "Cellulare" Sections; although a few people did not answer all questions, all questionnaires were considered reliable.

Based on the abridged Italian version of the PRS, 8 questions were used within the questionnaire (2 questions for each property) to obtain, from the arithmetic mean, the regenerative properties of this project: Being-Away (BAWAY), Extent (EXT), Fascination (FASC) and Compatibility (COMP). The Restorativeness variable (REST), relating to the regenerative potential of a site, was calculated as the average of the regenerative properties as suggested by several studies on the subject [13–15].

3 Results

3.1 Profile of the Interviewees and Descriptive Statistics

A total of 113 women between the ages of 20 and 60 participated in the collection of data, with an average age of 45 for Italian women and 35 for foreigners, of which 69 (61% of the total) were resident in the "Camerotti" Section and 44 (39% of the total) in the "Cellulare" Section. 54% of the total are Italian, while the rest are of 23 different nationalities. In the "Camerotti" Section, hosting detainees awaiting trial, there is a greater presence of foreigners (34% of the total sample against 13% of the "Cellulare" Section). The survey confirmed that the average age of foreign inmates is lower than that of the Italian ones, and that those awaiting trial are younger than those with final conviction. 29% of the respondents have no children and 28% have only one child. 70% of the respondents were born in a city, while only 3.5% were from the countryside; 70% did not have a family background in farming.

Considering the educational background, 35% of the sample barely completed compulsory schooling, while 7% have no qualifications at all. On the other hand, nearly 6% of the interviewed have a degree, with a predominance of foreigners. The level of schooling, as confirmed by DAP (Department of Prison Administration) statistics, is therefore rather low, reaching the middle school certificate for most of the sample.

The vast majority of the convicted declared that at the time of imprisonment they had a regular job, a lower number was either unemployed or had an undeclared job, while among the prisoners awaiting trial the most representative sample consisted of undeclared workers. By the way, the definition of "regular work" can convey a special meaning in prison, in which precariousness and marginality are normal and not exceptional [16].

In addition, the questionnaire included, a detailed item on the working situation at the time of arrest: probably, for some of the people who replied, "regular" (27%) also meant "non-criminal". As many as 65% of the interviewed believe that they will be able to resume the job they had before entering prison, but the percentage drops dramatically when the prison sentence, or the time spent in prison exceeds 3 years, while 22% would like to be able to do so, but they think it will not be possible for them.

As for the "green" profile of the sample, intended as affinity with nature and the habit to adopt environmentally friendly behaviour, a first-rate result regarding the preferred gardening activities compared with "at home, I have" highlighted that among those who had a garden at home it was a priority to take care of the plants, followed by walking and being outdoors. Considering the whole sample, plant care totals 46% of the choices, followed by walking and being outdoors (35% of the total). "Picking flowers and fruits" has only 15% of preferences.

More than a third of the sample (35%) has neither a garden, nor a vegetable garden or potted plants at home.

3.2 Evaluation of the Regenerative Capacity of the Intervention

In the current work, the regenerative capacity of the intervention was evaluated on the responding sample using the PRS on a 5-point Likert scale, from "strongly disagree" = 1 to "strongly agree" = 5. The related items of the questionnaire and the variables used are reported in Table 1.

Table 1. Items of the survey and related regenerative properties to calculate Restorativeness

Questionnaire items	Regenerative properties
Spending time in the green gives me a break from my day-to-day routine	BAWAY_1
A green area is a place to get away from things that usually demand my attention	BAWAY_2
It seems that green sites are limitless	EXT_1
There is a clear order in the physical arrangement of the garden	EXT_2
Gardens are fascinating places	FASC_1
In a garden there is much to explore and discover	FASC_2
Being in a garden fits with my personal inclinations	COMP_1
In a garden it is easy to do what I want	COMP_2

Figure 1 shows the average values of the four environmental properties: Being Away, Fascination, Extent, Compatibility and Restorativeness (REST). The average value of REST is very high, being 4,52. In the «Cellulare» Section, hosting the convicts under a final judgement, the benefit of the intervention is greater, being 4.62 vs. 4.45 (Fig. 2).

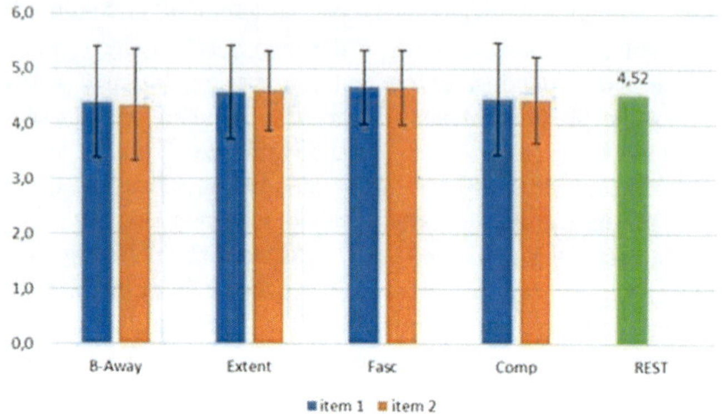

Fig. 1. Average values of regenerative properties and Restorativeness

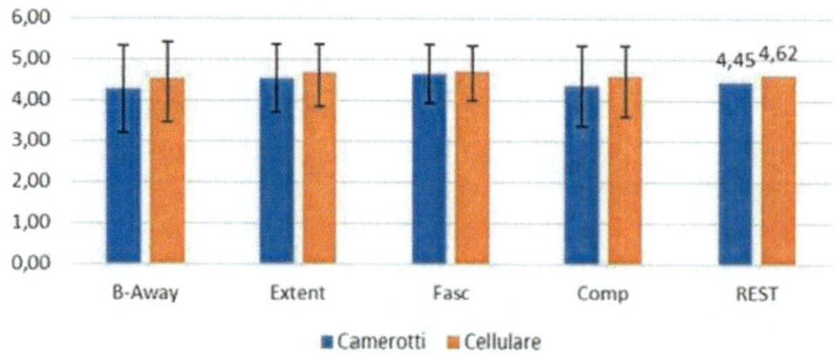

Fig. 2. Average values of regenerative properties and REST in the two Sections

As regards the educational background, the less educated seem to have a greater benefit: the higher value of REST for uneducated or poorly educated people being 4,6, but it decreases as the level of education increases, as shown in Fig. 3. The relationship between Restorativeness and the level of education is finally inversely proportional.

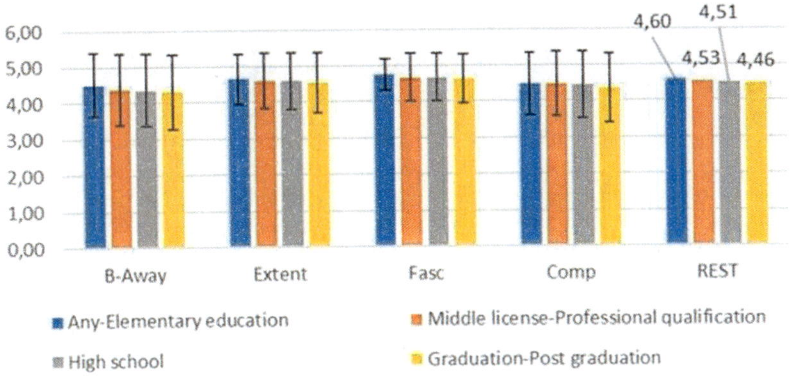

Fig. 3. Average values of regenerative properties and REST for levels of education

Concerning the impact of the intervention on different age groups, older women express a higher Restorativeness value: according to the percentage, the group 40–49 years is the one that benefits most from the intervention, as shown in Fig. 4.

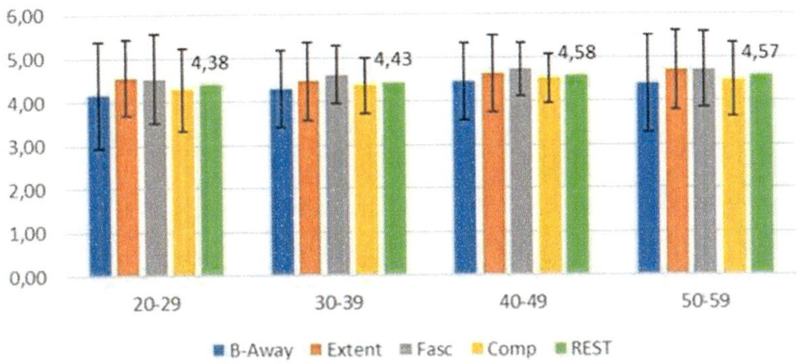

Fig. 4. Average values of regenerative properties and REST for each age group

As shown in Figs. 5 and 6, women who are involved in a working activity in prison and who have been detained from 1 up to 3 years are those with the highest values of Restorativeness, nearly the same as the largest group (jailed from 6 months to 1 year), this is probably related to a positive acceptance of punishment and a wish for social reintegration.

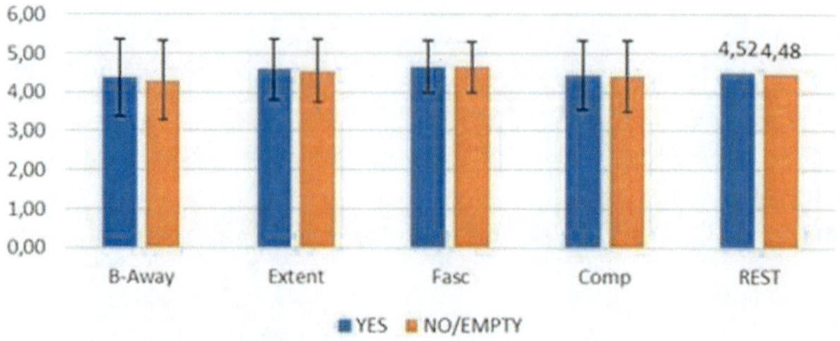

Fig. 5. Average values of regenerative properties and Restorativeness related to the question "current activity status"

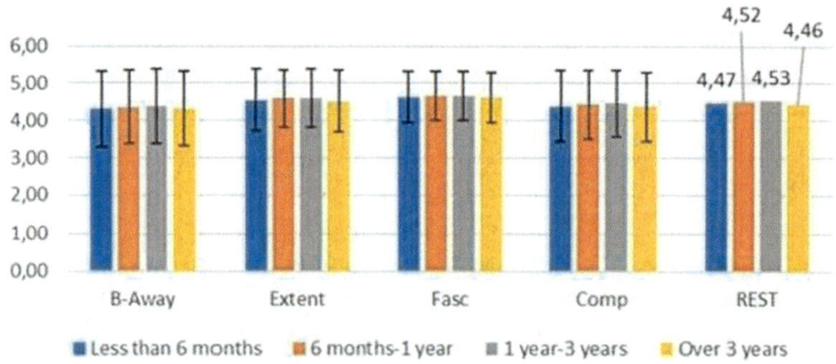

Fig. 6. Average values of regenerative properties and Restorativeness related to the question "how long have you been in prison?"

4 Discussion and Conclusions

The first objective of this study was to assess the regenerative potential of a green roof in a prison context on the occasion of the remaking of a degraded cover. The overall results are encouraging: the average values of Restorativeness (REST) are always higher than 4. Considering that the measuring system adopted for the evaluation of the four regenerative properties is based on a five-level scoring scale (1 = strongly disagree; 5 = strongly agree) we can say that the choice of increasing the green area is, overall, highly regenerative. In the period considered, in front of a standard capacity of 268 inmates, there were, as mentioned, 307 detainees, with a relatively low overcrowding index (+15%) compared to the average of Italian Institutes, representing a medium stress factor. In the section "Cellulare", housing mainly the prisoners with a final judgment, the benefit of the intervention was greater, which is probably due to the specificity of the place; since in the other section, "Camerotti", where are housed the prisoners waiting for judgment, spatial and temporal perception are different. This perception translates into a different relationship with the architectural spaces, as it is recognized that the condemned have greater need for privacy and independence, establishing a deeper and a personal connexion with the environments where they live [17]. Hence the appreciation for an intervention that, even just visually, represents an improvement in their everyday life. The level of REST is also higher among Italian women than among foreign ones: a data to be related to the larger presence of Italians in the "Cellulare" Section. The female prison population has a lower level of education than the entire prison population, and it is precisely for the less educated that the effect of regeneration is higher; this fact is confirmed in a constant higher level of regenerative property "Fascination" compared to the others; the most spontaneous approach to green and nature can indeed impose itself on rational awareness and denounce, on the other hand, on the negative phenomenon of prisoners' infantilization. In the group of women between 40 and 60 years of age the impact of the intervention was more significant, this could be related to a greater maturity and ability to metabolize their condition, supported by the fact that the benefits appear to be greater even among people detained between one and three years; while it decreases in a period of imprisonment of more than three years, probably coinciding with the loss of hope and interest in external life in general. Higher values of REST are found in the group of women who had an occupation, which demonstrates the absolute centrality of work in terms of rehabilitation and reintegration opportunities. Results, consequently, corroborate the hypothesis of a positive cost-benefit analysis relative to the intervention as a valuable long term investment for the Department of Prison Administration. It is important to note that the period under consideration was marked by the outbreak of the Covid-19 pandemic, which inevitably conditioned the fruition and training of the prisoners about the ordinary maintenance of the green roof. It is considered that further research in this area is still needed to validate these results in other categories of prison facilities (for example, male or juvenile institutions) or to follow the evolution over time of the newly constructed green roof and the consequent attitude of the detained population.

This study, which is the result of the interaction between architects, agronomists and educators working in the difficult context of the prison administration, has provided useful information in this field of research, both in the development of study methods and because it provided applicable results in the maintenance of damaged roofs or the design of new pavilions. The designers' awareness of the regenerative properties of green spaces, even in the penitentiary, could be an important step in the direction of urban sustainability, with the parallel non-negligible result, of the production of well-being within a population, such as that jailed, subject to strong stress, which is often cause of intemperance and self-injurious behaviour. The contradiction between the current legislation, which indicates rehabilitation as the goal of punishment, and the condition of the spaces where imprisonment is carried out, is clear. The research applied to the design of detention spaces, whether closed or open, would be useful to realize, in the places frequented daily by the detained population, the true purpose of the norm, allowing greater welfare for both the inmates and the employees in the prison administration.

References

1. Ulrich, R.S.: View through a window may influence recovery from surgery. Science **224** (4647), 420–421 (1984)
2. Moran, D., Turner, J., Schliehe, A.K.: Conceptualizing the carceral in carceral geography. Prog. Hum. Geogr. **42**(5), 666–686 (2017)
3. Moran, D., Turner, J.: Turning over a new leaf: the health-enabling capacities of nature contact in prison. Soc. Sci. Med. **231**, 62–69 (2019)
4. Hartig, T., Staats, H.: Guest Editors'introduction: restorative environments. J. Environ. Psychol. **23**, 103–107 (2003)
5. Ulrich, R.S.: Biophilia, biophobia, and natural landscapes. Biophilia Hypothesis **7**, 73–137 (1993)
6. Kaplan, S.: The restorative benefits of nature: toward an integrative framework. J. Environ. Psychol. **15**, 169–182 (1995)
7. White, K.E.: The Role of Nature in Physiological Recovery from Stress: A Critical Examination of Restorative Environments Theory. Graduate Theses and Dissertations (2013)
8. Hartig, T., Korpela, K., Evans, G.W., Gärling, T.: A measure of restorative quality in environments. Scand. Hous. Plan. Res. **14**, 175–194 (1997)
9. Korpela, K.M., Hartig, T., Kaiser, F.G., Fuhrer, U.: Restorative experience and self-regulation in favorite places. Environ. Behav. **33**, 572–589 (2001)
10. Purcell, T., Peron, E., Berto, R.: Why do preferences differ between scene types? Environ. Behav. **33**, 93–106 (2001)
11. Purcell, A., Berto, R., Mainardi Peron, E.: Restorativeness, preference and the perceived naturalness of places. Medio Ambiente y Comportamiento Humano **3**,19–34 (2002)
12. Pasini, M., Berto, R., Scopelliti, M., Carrus, G.: Measuring the restorative value of the environment: contribution to the validation of the Italian version of the perceived restorativeness scale. Boll. di Psicol. Appl. **257**, 3–11 (2009)
13. Berto, R.: Exposure to restorative environments helps restore attentional capacity. J. Environ. Psychol. **25**, 249–259 (2005)
14. Ivarsson, C.T., Hagerhall, C.M.: The perceived restorativeness of gardens – assessing the restorativeness of a mixed built and natural scene type. Urban For. Urban Green. **7**, 107–118 (2008)
15. Dentamaro, I., Lafortezza, R., Colangelo, G., Carrus, G., Sanesi, G.: Assessing the restorative potential of different types of urban and periurban green spaces. For. Riv. di Selvic. ed Ecol. For. **8**, 162–178 (2011)
16. Molteni, A., Naldi, A. (a cura di): Indagine Sulle Condizioni Sociali, Economiche e Abitative Delle Persone Detenute a Milano e Delle Loro Famiglie. Caritas Ambrosiana, Milano (2007)
17. Giofrè, F., Posocco, P.: Donne in carcere Ricerche e progetti per Rebibbia. Lettera Ventidue Edizioni Srl, Siracusa (2020)

Lighting of Milking Parlours: Results from a Field Study

Francesco M. Tangorra(✉) and Annamaria Costa

Department of Veterinary Medicine and Animal Sciences, Università degli Studi di Milano, Via dell'Università 6, 26900 Lodi, Italy
francesco.tangorra@unimi.it

Abstract. The aim of the study was to investigate the lighting of milking parlours at pit and milking stall levels. The field study was carried out in ten milking parlours with natural light inlets on the sides. The average illuminance (E_a) of the milking pit and milking stalls was calculated according to the CIBSE method. The measurements occurred on July 2018 before and during the afternoon milking session using a lux meter for assessing the impact of the operator and the animals on the illuminance of milking stalls. A k-means cluster analysis was performed using the JMP 15.2 software (SAS, Cary, NC, USA). Clustering analysis divided the milking parlours into three groups. Groups 1 (n = 5) and 2 (n = 3) included herringbone milking parlours respectively with less and more than 12 milking stalls. Group 3 included two parallel milking parlours with 40 milking stalls. On average in all groups of milking parlours the lighting power density (LPD) was below the recommended 6.0 W/m^2. The light fixtures were positioned over the milker's pit and the cow stances respectively at 2.3–3.3 m height above the pit floor and at 1.4–2.4 m height above the milking stall floor. Only Group 3 showed E_a values at milking stall before and during milking above that recommended by the reference technical standards ($E_a > 200$ lx).

Keywords: Dairy cattle farm · Milking parlour · Lighting · Illuminance

1 Introduction

Mechanical milking is one of the most important and critical processes in managing a dairy cattle farm. Typically, it takes place in milking parlours where the animals are led in groups at fixed times, two or three times a day. Some authors proposed minimum lighting for the milking center, suggesting a light intensity of 100 lx in the waiting room before entering the milking parlour and 200 lx in the milking parlour [1]. Rajaniemi et al. [2] recommended illumination intensity of the milking parlour of 200–250 lx, in line with lighting level recommendations by Harner and Smith [3], who suggested illumination levels of 215 lx. The illumination level recommended by the Italian National Unification Body (UNI) for the milking parlour is 200 lx [4]. The lighting level of the milking pit should also be considered, whereby ASABE [5] recommend an illumination level of 500 lx to ensure adequate lighting for controlling hygiene and the attachment of milking clusters.

M. Biocca et al. (Eds.): SHWA 2020, LNCE 252, pp. 149–154, 2022.
https://doi.org/10.1007/978-3-030-98092-4_16

Workplaces in the milking parlour are often badly lit, if compared with those in industry. In a recent work of Celozzi et al. [6] aimed to assess milking performances, cows' behaviour and microclimatic condition in farm areas, the milking parlour emerged as the area with the worst light intensity during all day over different seasons (winter and summer) and in all the farms analysed. This result is worrying, because a bad illumination of the milking parlour can compromise the efficiency and effectiveness of the milking routine, the chance to check disease-related changes in udder or milk, the cow flow towards the milking platform increasing the risk of injuries to cows, the well-being and health of the workers.

Abrosimova [7] found that adequate light intensity during milking could shorten the process by 8–12% assisting the milk let-down in the parlour. Proper lighting could influence oxytocin release and hence milk let-down [8].

The aim of the study was to investigate the lighting of milking parlours and to assess the impact of the operator and the animals on the illuminance of milking stalls during a regular milking session.

2 Materials and Methods

The field study was performed in ten milking parlours, with natural light inlets supplemented by artificial lighting, located in Lombardia (Northern Italy). The measurements of the illuminance levels at milking pit and milking stalls were in accordance with CIBSE method [9]. The basis of this method is to divide the working plane into a number of equal areas, which should be as nearly square as possible. The illuminance at the centre of each area is measured with a lux meter and the mean value calculated, giving an estimate of the average illuminance.

The method first involves calculating the Room index (K_r) with the following formula (1):

$$K_r = \frac{L \times W}{h(L + W)} \tag{1}$$

where:

L = room length (m)
W = room width (m)
h = height of lighting above horizontal reference plane

Table 1 relates the room index to the number of measurement points, while the sides of each area are calculated by applying the following formulas:

$$L_1 = \frac{L}{\sqrt{n}} \tag{2}$$

$$W_1 = \frac{W}{\sqrt{n}} \tag{3}$$

where:

L_1 = length of measurement areas (m)
W_1 = width of measurement areas (m)
n = number of measurement points

Table 1. Relationship between room index and the minimum number of measurement points (n) in CIBSE method

Room index (K_r)	n
$K_r \leq 1$	4
$1 < K_r \leq 2$	9
$2 < K_r \leq 3$	16
$K_r > 3$	25

The average illuminance (E_a) was calculated by the arithmetic mean of the illuminance values measured at the centre of each area with a lux meter:

$$E_a = E_1 + E_2 + \ldots + E_n \quad (4)$$

where:

E_a = average illuminance (lx)
$E_1 \div E_n$ = illuminance values of the measured points (lx)
n = total number of measurement points

The illuminance uniformity (U) was obtained as follows:

$$U = \frac{E_{min}}{E_a} \quad (5)$$

U = illuminance uniformity
E_{min} = minimum illuminance value (lx)
E_a = average illuminance (lx)

The measurements were carried out on July 2018 considering daylighting and artificial lighting together, as this is the actual use condition of the milking parlours. Measurements were performed with a lux meter (PCE-174 Data Logger Light Meter, PCE Instruments, Capannori, Lucca, Italy) with accuracy of 5% of reading + 10 digits and scale of 40, 400, 4000 lx, before and during an afternoon milking session (2–4 p. m.) for assessing the impact of the operator and the animals on the illuminance of milking stalls. A k-means cluster analysis was performed using the JMP 15.2 software (SAS, Cary, NC, USA).

3 Results and Discussion

Clustering analysis divided the milking parlours into three groups. Groups 1 (n = 3) and 2 (n = 5) included herringbone milking parlours respectively with less and more than 12 milking stalls/side. Group 3 included two parallel milking parlours with 20 milking stalls/side.

All the milking parlours used fluorescent lamps and the average lighting power density (LPD), defined as watts per square meter of floor area (W/m^2), increased with increasing the installation height of light fixtures. On average in all groups of milking parlours LPD presented values below the recommended 6.0 W/m^2 [10].

The light fixtures were positioned over the milker's pit and the cow stances respectively at about 2.3 to 3.3 m height above the pit floor (h_1) and at about 1.4 to 2.4 m height above the milking stall floor (h_2) as reported in Table 2. According to Winter and Wagner [11] for the excellent lighting of the udder area lights should be positioned over each side of the cow stances, while over the milker's pit two rows of lights should be positioned around 0,4 m from the mid-axis at about 2.2–2.5 m height above the pit floor to give 500 lx illuminance. A mounting height of less than 3.5 m is suggested by Chastain and Nicholai [12] to give adequate lighting for the pit.

Table 2. Lighting power density (LPD) and height of light fixtures from the pit floor (h_1) and milking stall floor (h_2) in the groups of milking parlours involved in the field study. Values are reported as means ± SD

Group	Milking parlour type	Milking stalls per side (n)	Lamp type	LPD (W/m^2)	h_1 (m)	h_2 (m)
1	Herringbone	<12	Fluorescent	5.7 ± 1.0	3.28 ± 0.46	2.41 ± 0.50
2	Herringbone	≥ 12		2.7 ± 1.8	2.33 ± 0.27	1.41 ± 0.24
3	Parallel	20		3.6 ± 0.3	2.70 ± 0.08	1.68 ± 0.11

Only Group 3 showed satisfactory results at pit floor before milking and at milking stall floor before and during milking because the illuminance presented values above that recommended by the reference technical standards (Table 3), due to the contribute of large natural light inlets on the sides of the milking parlours. Group 1 and 2 substantially met the reference technical standard at pit floor before milking with respectively 243 ± 79 and 197 ± 57 lx, but not at milking stall floor where the light intensity emerged as a critical aspect both before and during milking (Table 3). These results could depend by many factors, considering that much of the variability in illuminance as a function of LPD can be explained based on variations in light fixtures position, room size, luminaire efficiency, scheduled maintenance, and location of the task relative to the light fixtures [13].

The illuminance should be uniform throughout the milking parlour. With an illuminance uniformity of 0.8 (Table 3), only milking parlours belonging to Group 3 allow a correct execution of the milking related tasks. Good lighting can make milking easier and more efficient, improve cow flow and reduce the risk of injuries to cows and the workers [14].

Table 3. Average illuminance at pit floor before milking (Ea_1) and at milking stall floor before (Ea_2) and during (Ea_3) milking. Average illuminance uniformity at milking stall during milking (U). Values are reported as means ± SD

Group	Milking parlour type	Milking stalls per side (n)	Lamp type	Ea_1 (lx)	Ea_2 (lx)	Ea_3 (lx)	U
1	Herringbone	< 12	Fluorescent	243 ± 79	161 ± 45	83 ± 31	0.5
2	Herringbone	≥ 12		197 ± 57	159 ± 64	94 ± 60	0.4
3	Parallel	20		514 ± 157	559 ± 305	234 ± 53	0.8

4 Conclusions

This study aimed to investigate the lighting of milking parlours at pit and milking stall levels. Results showed that in most of the milking parlours the light intensity at milking stall is a critical aspect both before and during milking, highlighting the importance in lamps positioning for proper lighting of the udder area.

Future research needs to address the relation between light intensity in milking parlours and milking routine efficiency, udder health, and milk quality.

References

1. Clarke, S., House, H.: Energy efficient dairy lighting. Agricult. Eng. Ontario, Order **06**, 007 (2016)
2. Rajaniemi, M., Turunen, M., Ahokas, J.: Direct energy consumption and saving possibilities in milk production. Agron. Res. **2015**(13), 261–268 (2015)
3. Harner, J.P., Smith, J.F.: Lighting Low Profile Cross Ventilated Dairy Houses (2008). https://www.asi.k-state.edu/doc/dairy/lighting-low-profile-cross-ventilated-dairy-houses.pdf
4. UNI EN 12464-1:2011
5. ASAE EP344.4 JAN2014. Lighting Systems for Agricultural Facilities
6. Celozzi, S., et al.: The use of integrated data collection system to evaluate milking performance, microclimatic condition and cows' behaviour. Ital. J. Anim. Sci. **19**(1), 856–864 (2020)
7. Abrosimova, R.: Effect of lighting on gaseous exchange and productivity in large ruminants. Anim. Sci. (RU) **2**, 87–88 (1978)
8. Macuhova, J., Bruckmaier, R.M.: Diurnal changes of oxytocin release during au-tomatic milking. In: Meijering, A., Hogeveen, H., de Koning, C.J.A.M. (eds.). Auto-matic Milking A Better Understanding. Wageningen Academic Publishers, Wageningen, The Netherlands, p. 502 (2004)
9. Bean, A.R., Esterson, D.M.: Average illumination measurement— a preliminary investigation. Light Lighting **59**(7), 204–205 (1966)
10. Sangiorgi F.: Fabbricati e impianti per l'allevamento dei bovini da latte. In: Il manuale dell'allevatore, Granlatte, pp. 9–75 (2003)
11. Winter, A., Wagner, T.: Lighting of cowhouse and milking area. LANDTECHNIK **57**(2), 82 (2002)

12. Chastain, J., Nicolai R.: Dairy Lighting System for Free Stall Barns and Milking Centers. University of Minnesota AEU-12 (1996)
13. Collins, B.L., Fisher, W., Gillette, G., Marans, R.W.: Second-level post occupancy evaluation analysis. J. Illum. Eng. Soc. **19**(2), 21–44 (1990)
14. Phillips, C.J.C., Morris, I.D., Lomas, C.A., Lockwood, S.J.: The locomotion of dairy cows in passageways with different light intensities. Anim Welf. **9**(4), 421–431 (2000)

Spatial Analyses to Assess the Availability of Sheep Wool as Potential Eco-friendly Material

Monica C. M. Parlato(✉), Francesca Valenti, and Simona Maria C. Porto

Department of Agriculture, Food and Environment (Di3A), University of Catania, Via Santa Sofia, 100, 95123 Catania, Italy
monica.parlato@unict.it

Abstract. The main objective of this work was to investigate the yearly availability of an agricultural co-product, i.e., sheep wool, suitable as eco-friendly material in building sector. Currently, sheep wool is an agricultural special waste contaminated with impurities, with high disposal costs for breeder, and often illegally disposed (e.g., buried or burned), with a strong environmental impact. On the other hand, by considering the green building sector, sheep wool responds to the requirements of green building components because is an eco-friendly material, a surplus, yearly renewable, and totally recyclable. If used instead of common insulation materials (e.g., fiberglass, rock wool, polyurethane foam, polystyrene), or as reinforcement fibre in bio composite materials, sheep wool offers significant benefits for sustainability such as a reduction for both cost of breeding and environmental pollution. In this study, a Geographical Information System (GIS) - based model to locate and quantify the yearly amount of livestock co-production, i.e., sheep wool, coming from dairy sheep breeding, was put forward and was applied in a study area located in Southern Italy highly characterized by this kind of breeding. The GIS-based model results provided the localization and the quantification of the yearly amount of sheep wool in order to evaluate its availability as new eco-friendly material. Moreover, these results could represent basic information for further analysis of the environmental impact related to the logistics and supply phase of sheep wool to a future collection centre.

Keywords: SHW in building · Natural material · Sheep wool · Natural insulation materials · GIS-based model

1 Introduction

Agricultural co-products, by-products and wastes (AW) if valorized, could reduce pollution and cost production, by playing an important role in bio-architecture [1, 2].

A potential re-use of AW is based on a properly disposal management with the aim to move from a linear economy to a circular one in accordance with New Green Deal Agenda [3].

M. Biocca et al. (Eds.): SHWA 2020, LNCE 252, pp. 155–164, 2022.
https://doi.org/10.1007/978-3-030-98092-4_17

In the Mediterranean area, rural buildings are usually built without a specific temperature control system [4]. The use of natural insulation systems, i.e., sheep wool-based materials, could improve their thermal efficiency by optimizing the building passive conditioning systems [5]. The employ of sheep wool as thermal insulating material represents a consistent advantage for some types of rural buildings where the control of microclimatic conditions is of relevant importance regarding their main functional destination (i.e., buildings for intensive animal breeding, buildings for agroindustry). Sheep wool is suitable as thermal and acoustical insulation material thanks to its capacity to absorb the exceeding moisture and to isolate from vibrations. Moreover, sheep wool neutralizes harmful and odorous substances such as Nitrogen Dioxide, Sulphur Dioxide, Toluene, and Formaldehydes. By using sheep wool insulation indoor air quality (IAQ) and human and/or animal wellness could be improved [6–10].

Currently, sheep wool is an agricultural special waste contaminated with impurities, and with high disposal costs for breeder. In some case, wool is buried or burned, with a strong impact on soil and air pollution.

In the context of a circular economy, the reuse of greasy wool could reduce both environmental pollution and energy consumption and if properly valorized it could become a material fully complying with ecofriendly construction criteria.

Since a sustainable reuse process of wastes, or by-products, it is starting by an accurate assessment of their production and localization, the main objective of this work was to investigate the yearly availability of sheep wool, as an alternative to common insulation materials (e.g., fiberglass, rock wool, polyurethane foam, polystyrene) or as natural reinforcement fibres for bio composite [11].

In this research study a GIS - based methodology was put forward in order to locate and quantify the waste considered, i.e., sheep wool fibers, its yearly availability and geographical location. In this regard, GIS tool was adopted since it is a suitable tool already used for this kind of application in several studies [12–14].

2 Materials and Methods

2.1 Wool

Sheep wool is an animal fibre; it is composed of 60% keratin protein fibres, 15% moisture, 10% fat, 10% sheep sweat and 5% impurities. In Table 1 is reported a typical chemical composition of wool [15].

Sheep wool is suitable especially as thermal insulating material for roof, external and internal walls in buildings for intensive animal breeding buildings for agroindustry flour mills, olive mills wineries, and anywhere the control of microclimatic conditions is of relevant importance during the agroindustry process [5].

Table 1. Chemical composition of raw sheep wool [15]

Keratin	33%	Carbon	50%
		Hydrogen	12%
		Oxygen	10%
		Nitrogen	25%
		Sulphur	3%
Dirt	26%		
Suint	28%		
Fat	12%		
Mineral water	1%		

By using sheep wool both the insulation indoor air quality (IAQ) and human and/or animal wellness improve. Moreover, through a chemical process (call chemisorption) sheep wool is suitable to neutralize harmful and odorous substances such as Nitrogen Dioxide, Sulphur Dioxide, Toluene, and Formaldehydes. In this study has been considered sheep wool deriving by dairy sheep with a low-quality fleece, unsuitable for textile industry. Rajabinejad et al. [16] compared sheep wool thermal and acoustical behaviors with other traditional insulation materials, that are, glass wool and polystyrene foam. Collected values are similar for all the three materials (Table 2); on the other hand, by considering the embodied energy of sheep wool (the energy used for the production and transportation of a material), it is evident a consistent reduction on this value.

Table 2. Comparation of wool performances and common insulation materials

Insulation material	λ value	R value (100 mm)	Density	Embodied energy	Sound absorption coefficient	Water absorption
	[W/m K]	[m^{2}K/W]	[kg/m^3]	[GJ/m^3]	500–2000 [Hz]	[% wt/wt]
Sheep wool	0,034–0,067	2,5–2,6	18–23	0,11	0,77 (60 mm)	up to 35%
Glass wool	0,032–0,04	2–3	10–30	0,83	0,65 (100 mm)	0,2%
Polystyrene foam	0,033–0,035	2,5–2,8	30–50	3,03	0,35 (50 mm)	0,03 – 0,1%

In this study has been considered sheep wool deriving by dairy sheep with a low-quality fleece, unsuitable for textile industry.

2.2 Methodology

In this study, a GIS - based model to locate and quantify the yearly amount of livestock co-production, i.e., sheep wool, coming from dairy sheep breeding, was put forward and was applied in a study area located in Southern Italy, highly characterized by this kind of breeding.

Study Area

Sicily is a southern region of Italy divided into 9 provinces with the Ionian Sea to the east, the Tyrrhenian Sea to the north and the Mediterranean on the remaining coasts. Sicily is the largest Italian region, and the largest island in Mediterranean with an extension of 25,707 km^2 (Fig. 1). Sicily is also the second region with the highest number of sheep, after Sardinia (Table 2).

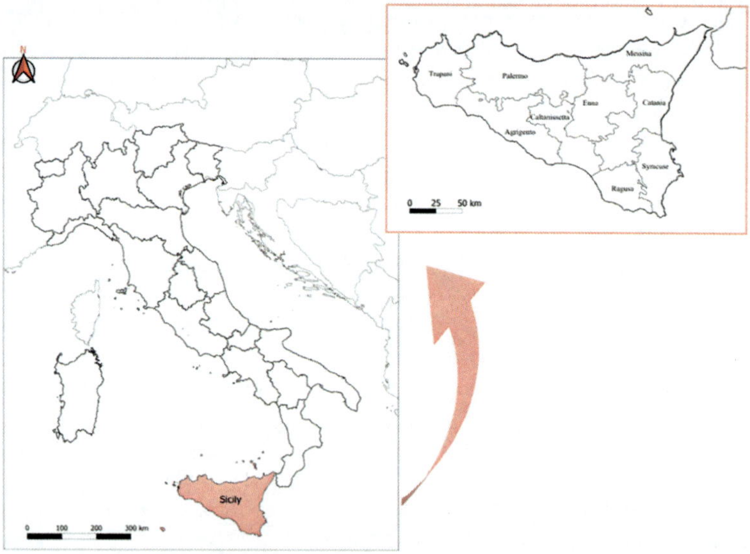

Fig. 1. Study area and geographic position.

This area of southern Italy it is strongly characterized by the breeding of dairy sheep, whose fleece is not suitable for textile industry. In Italy the number of sheep is around 7,4 million with the prevalent number in Sardinia (3,039,160), Sicily (737,819), Lazio (580,322), and Tuscany (321,534) (Table 3).

Table 3. Consistency of the Italian sheep population divided by region. (Source processing IZS-December 2020)

Italian regions	Number of sheep
Sardegna	3,039,160
Sicilia	737,819
Lazio	580,322
Toscana	321,534
Calabria	211,270
Puglia	205,632
Basilicata	183,828
Campania	177,960
Abruzzo	158,582
Marche	119,764
Piemonte	119,357
Lombardia	114,963
Umbria	95,365
Trentino Alto Adige	74,458
Veneto	69,424
Molise	55,977
Emilia Romagna	50,827
Liguria	9,945
Valle d'Aosta	1,904
Total	6,346,310

Data Analysis

With the aim of to evaluate a possible alternative use of agricultural waste, i.e., sheep wool, a methodology was put forward to locate and quantify the availability of sheep's wool in the study area.

In this research, an extensive database was improved by analyzing data supplied by National Zootechnical Registry (December 2020) of the Italian Ministry of Health (IZS), in order to quantify both the sheep breeder surface areas and the related wool production for evaluating their distribution at territorial level by GIS analyses to obtain the geographical areas where the production of this by-product is higher.

The base maps used in the GIS software included the Regional Technical Maps (RTM 2008) as the base map for carrying out thematic maps.

By analysing available data on IZS, the provinces with the highest surface areas dedicated to sheep wool breeding and with the highest amount of sheep wool production were identified and localised on GIS software.

In detail, the QGis software (ver. 3.10.11), an open-source GIS software, was used since it represents a decision support tool suitable to collect, to organize, to analyse, and to localise geographical data. All the GIS analyses were performed by using QGis software; by combining data provided by the base maps, data acquired from the database, and by adopting the Jenks tool available in the QGIS software, several maps were produced.

In detail three step were followed:

1. **Data preparation for GIS-based analyses** (National Zootechnical Registry (December 2020)
2. **GIS-based analyses for carrying out the maps**
3. **Quantification and location of the yearly sheep wool fibres (SWF)**

By using spatial analysis GIS tools, the acquired data were elaborated and applied for providing distribution for both cultivation surface areas and SWF production within the study area. In detail, a territorial distribution of surfaces and production areas were obtained, by classifying the Sicilian provinces through the application of Jenks tool plug-in, with the aim of maximizing the differences among the classes.

The obtained results were reported in GIS and, by using the Jenks tool, were elaborated for evaluating those territorial areas highly characterised by this kind of potential recyclable natural fibres, with the aim of estimating those areas, within each province, characterised by highest density of breeder, to obtain the geographical areas where this by-product is highly concentrated, and therefore the problem of its disposal is potentially relevant. The selected areas could be useful for localizing future collection centres by reducing environmental impact due to the logistic and supply phase.

A tailored methodology was developed to quantify the SWF amount suitable as potential reinforcements for bio-composite or natural insulation materials.

3 Results and Discussions

Data recorded from IZS database were elaborated and deeply analysed in order to consider only data related to sheep wool breeding and the related wool production. Data were organised for showing at territorial level the distribution of this breeder within Sicilian provinces by developing a GIS map.

As first step, the acquired data were used for providing the distribution of areas dedicated to sheep breeding, by showing them at territorial level. Then, by using Jenks tool, available in QGIS software, the Sicilian provinces were grouped in four different classes based on their sheep breeding areas. As shown in Fig. 2, Palermo, followed by Agrigento and Enna are the provinces with the main surface area dedicated to the sheep breeding. The last are the provinces of Syracuse and Ragusa, characterized by an high extension of protected crops and few free surfaces for breeding activities [17].

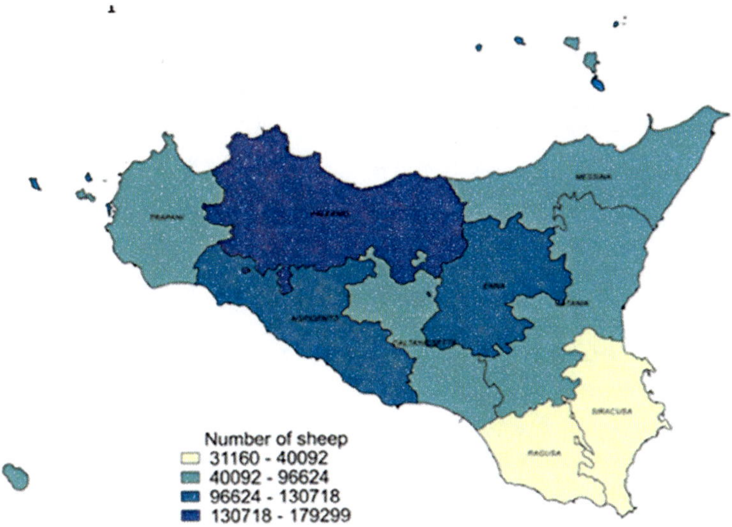

Fig. 2. Distribution within Sicily region of sheep

Since the research paper was focused on estimating the amount and localization of natural fibres that could be reused as potential reinforcements for bio-composite and insulation materials, i.e., sheep wool, it has been also investigated the number of yearly sheared sheep. As illustrated in Fig. 3, the results of the elaborations showed a different distribution within Sicilian provinces and confirmed Palermo, Enna and Agrigento as provinces, with the highest number of yearly sheared sheep.

Data showed in Fig. 4 confirmed that the province of Palermo, with almost 170 t/year, followed by Enna and Agrigento are those ones with the highest availability of potential recyclable natural fibres, i.e., sheep wool.

Then, in view of realizing new eco building materials, the results achieved in this study by developing the GIS-based model could represent a first step to achieve the objective and could be useful for planning tailored collection centres by paying attention to the transport costs during the logistics and supply phase (i.e., from the breeder 'area to the collection centres), with the aim to reduce environmental and social impacts, and costs [18]. Moreover, these could be used as basic information for the analysis of the environmental impact related to the logistics and supply phase of sheep wool fibres to the collection centres.

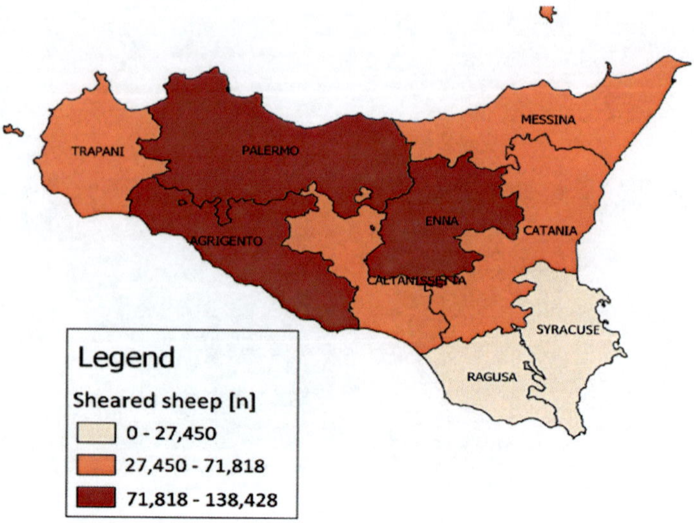

Fig. 3. Distribution within Sicily region of the yearly sheared sheep wool

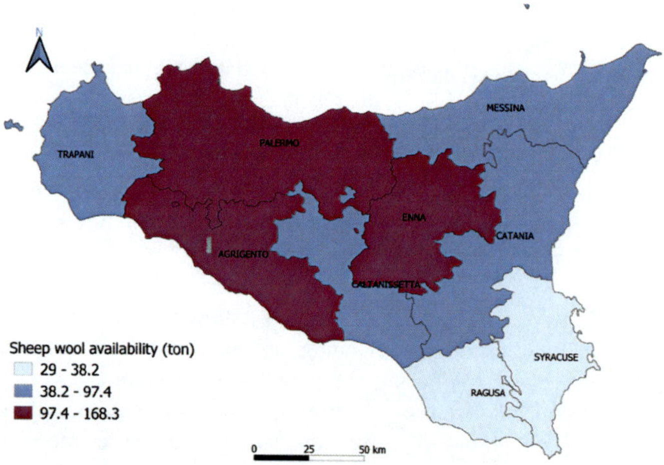

Fig. 4. Distribution within Sicily region of the recyclable sheep wool as potentially resources for eco-friendly materials

In Fig. 4 is reported the distribution within Sicily region of the recyclable sheep wool yearly production, as potentially resources for eco-friendly materials. By considering the number of Sicilian sheep and an average production of 1,00 kg for sheep, the amount of sheep wool was quantified and reported in GIS.

4 Conclusions

Sustainability and energy efficiency in building sector is based on both reductions of CO_2 emissions and energy demand [19].

In this context, the valorization of sheep wool as potential sustainable raw material, suitable to optimize passive indoor conditioning systems, healthy for both production process and application, locally available, with a low manufacture energy consumption, it is a challenge to achieve green building procedure.

This research study was carried out by combining data supplied by National Zootechnical Registry of the Italian Ministry of Health (IZS), and GIS-based maps, fulfilling the proposed aim of the research. In detail, the sheep wool breeding was deeply investigated by elaborating the data related breeding surface areas and wool production for producing GIS maps. The achieved results by the developed GIS-based model contributes to fill the gap in the knowledge of the production and localisation of sheep wool as potential thermal acoustical insulation material and as reinforcement fibres for bio-composites, by supporting green building materials production. Moreover, by defining a new industrial chai as potential source of value in accordance with the statement of circular economy could be produced.

Then, the obtained results could be useful to plan suitable locations of collection centres by reducing the distances from the areas where these by-products are highly produced (i.e., from the breeding fields to the collection centres), with an advantage in terms of costs but also for social impacts and environmental pollution.

In the next future in order to convert this waste in a real resource for building sector it is advisable to improve co-operation between breeders (i.e., for a pre-shearing wash phase), waste collectors, transporters, and recyclers, to ensure the maximum amount of SWF recovery and reuse.

References

1. Toop, T.A., Ward, S., Oldfield, T., Hull, M., Kirby, M.E., Theodorou, M.K.: AgroCycle - Developing a circular economy in agriculture. Energy Procedia **123**, 76–80 (2017). https://doi.org/10.1016/j.egypro.2017.07.269
2. Liuzzi, S., Sanarica, S., Stefanizzi, P.: Use of agro-wastes in building materials in the Mediterranean area: a review. In: Energy Procedia, Elsevier Ltd, pp. 242–249 (2017). https://doi.org/10.1016/j.egypro.2017.08.147
3. Siddi, M.: The European green deal: assessing its current state and future implementation. In: FIIA Work. Pap vol. 14. (2020)
4. Barreca, F., Praticò, P.: Environmental indoor thermal control of extra virgin olive oil storage room with phase change materials. J. Agric. Eng. **947**, 208–214 (2019). https://doi.org/10.4081/jae.2019.947
5. Parlato, M.C.M., Porto, S.M.C.: Organized framework of main possible applications of sheep wool fibers in building components, Sustain **12** (2020). https://doi.org/10.3390/su12030761
6. Zach, J., Hroudova, J., Brozovsky, J.: Study of hydrothermal behavior of thermal insulating materials based on natural fibers. Int. J. Archit. Environ. Eng. **8**, 995–998 (2014)
7. Alexander K.: Building your passivhaus, Archit. J. (2008)

8. Huson, M.G.: 3 - Properties of wool. In: Bunsell, A.R. (ed.), Handb. Prop. Text. Tech. Fibres, 2nd edn, Woodhead Publishing, pp. 59–103 (2018). https://doi.org/10.1016/B978-0-08-101272-7.00003-1
9. del Rey, R., Alba, J., Rodríguez, J.C., Bertó, L.: Characterization of new sustainable acoustic solutions in a reduced sized transmission chamber. Buildings. **9** (2019). https://doi.org/10.3390/buildings9030060
10. Korjenic, A., Klaric, S., Had ic, A., Korjenic, S.: Sheep wool as a construction material for energy efficiency improvement. Energies **8**(6), 5765–5781 (2015). https://doi.org/10.3390/en8065765
11. Rivera-Gómez, C., Galán-Marín, C.: Biodegradable fiber-reinforced polymer composites for construction applications. In: Nat. Fiber-Reinforced Biodegrad. Bioresorbable Polym. Compos. (2017). https://doi.org/10.1016/B978-0-08-100656-6.00004-2
12. Valenti, F., Porto, S.M.C., Dale, B.E., Liao, W.: Spatial analysis of feedstock supply and logistics to establish regional biogas power generation: a case study in the region of Sicily. Renew. Sustain. Energy Rev. **97**, 50–63 (2018). https://doi.org/10.1016/j.rser.2018.08.022
13. Selvaggi, R., Valenti, F., Pecorino, B., Porto, S.M.C.: Assessment of tomato peels suitable for producing biomethane within the context of circular economy: A gis-based model analysis, Sustain **13** (2021). https://doi.org/10.3390/su13105559
14. Akinci, H., Özalp, A.Y., Turgut, B.: Agricultural land use suitability analysis using GIS and AHP technique. Comput. Electron. Agric. **97**, 71–82 (2013). https://doi.org/10.1016/j.compag.2013.07.006
15. Bradbury, J., Chapman, G.: The chemical composition of wool. Aust. J. Biol. Sci. (2016). https://doi.org/10.1071/bi9640960
16. Rajabinejad, H., Bucişcanu, I.I., Maier, S.S.: Practical ways of extracting keratin from keratinous wastes and by-products: a review. Environ. Eng. Manag. J. **15**, 1131–1147 (2016)
17. Parlato, M.C.M., Valenti, F., Porto, S.M.C.: Covering plastic films in greenhouses system: a GIS-based model to improve post use suistainable management. J. Environ. Manage. **263** (2020). https://doi.org/10.1016/j.jenvman.2020.110389
18. Díaz-Palacios-Sisternes, S., Ayuga, F., García, A.I.: A method for detecting and describing land use transformations: an examination of Madrid's southern urban-rural gradient between 1990 and 2006. Cities. (2014). https://doi.org/10.1016/j.cities.2014.03.010
19. Zuo, J., Zhao, Z.Y.: Green building research-current status and future agenda: a review. Renew. Sustain. Energy Rev. **30**, 271–281 (2014). https://doi.org/10.1016/j.rser.2013.10.021

Agriculture 4.0, Automation, Remote Control, Robot and Innovative Vehicle

Remote Controls of Solar Drier Micro-plants for Process Standardization

Tiziana Maria Piera Cattaneo[1,2](✉), Carlo Bisaglia[3], Alessandro Cammerata[4], Annamaria Stellari[3], and Elio Romano[3]

[1] Research Centre for Engineering and Agro-Food Processing, CREA, Via G. Venezian, 26, 20133 Milan, Italy
tiziana.cattaneo@crea.gov.it

[2] DAFNE, Tuscia University, Viterbo, Italy

[3] Research Centre for Engineering and Agro-Food Processing, CREA, Via Milano, 43, 24047 Treviglio (BG), Italy

[4] Research Centre for Engineering and Agro-Food Processing, CREA, Via Manziana, 30, 00189 Roma, Italy

Abstract. This paper aims to study the efficiency of two different transformation processes. The verification of the suitability of the chosen digital methodologies for the monitoring of drying processes and their standardization was also considered.

A drying (with fan active at the top: M1), and a dehydration (with fan always active at the bottom: M2) micro-systems operating with solar energy were implemented with Arduino sensors for the detection of the main physical parameters. Parallel transformation tests on eggplant slices were carried out. The performance of two solar drier micro-plants was compared by monitoring the main involved variables: temperature (T), humidity (RH), time (t), weight (W), fan (F). The monitoring of transformation processes required the optimization of configuration: design, programming, and positioning of sensors.

The temperature reached by the configuration M2 was more dependent on the external temperature; during the hottest hours of the day, it reached higher levels than the configuration M1, in which the fan is programmed to counteract thermal excesses. The constant activity of the fan in M2 allowed the reduction of the night RH% not contrasted by M1, in which, in those hours, the fan is inactive. The trend of weight loss (W), showed the effect of night-time humidity, highlighting an increase in the recorded weight.

The recorded curves relating to temperature and humidity did not show statistically significant differences between the upper and lower areas of the dryer with active fan (M2).

Keywords: Sensors · Digital agriculture · Traceability · Food transformation

1 Introduction

The 2030 Agenda for Sustainable Development, adopted by all United Nations Member States in 2015, provides a shared blueprint for peace and prosperity for people and the planet, now and into the future. At its heart are the 17 Sustainable Development

M. Biocca et al. (Eds.): SHWA 2020, LNCE 252, pp. 167–175, 2022.
https://doi.org/10.1007/978-3-030-98092-4_18

Goals (SDGs), which are an urgent call for action by all countries - developed and developing - in a global partnership. They recognize that ending poverty and other deprivations must go hand-in-hand with strategies that improve health and education, reduce inequality, and spur economic growth – all while tackling climate change and working to preserve our oceans and forests. In particular, inside the Goal 17 it is declared the promotion "*of the development, transfer, dissemination and diffusion of environmentally sound technologies to developing countries on favorable terms, including on concessional and preferential terms, as mutually agreed*" [1].

In these Goals [2] and targets, they are setting out a supremely ambitious and transformational vision. A world where human habitats are safe, resilient and sustainable and where there is universal access to affordable, reliable and sustainable energy. The world incline towards alternative sources of energy. While wind and hydro energy are also making progress, it is solar energy that has become the favored one due to its increased affordability. The preservation of foods by drying is the time-honored and most common method used by humans and the food processing industry. Dehydration of food is one of the most important achievements in human history, making humans less dependent upon a daily food supply even under adverse environmental conditions [3]. Though in earlier times drying was dependent on the sun, nowadays many types of sophisticated equipment and methods are used to dehydrate foods. During the past few decades, considerable efforts have been made to understand some of the chemical and biochemical changes that occur during dehydration and develop methods for preventing undesirable quality losses. Foods can be divided into three broad groups based on the value added through processing by drying. In the case of cereals, legumes, and root crops, very little value is added per kilogram processed. More value per unit mass is added to foods such as vegetables, fruits, and fish; and considerably more to high-value crops such as spices, herbs, medicinal plants, nuts; bioactive materials; and enzymes. Earlier, only sun drying was used for drying. In this process, foods are directly exposed to the sun by placing them on the land or left hanging in the air. The main disadvantages of this type of drying are (i) contaminations from the environment, (ii) product losses and contaminations by insects and birds, (iii) floor space requirements, (iv) difficulty in controlling the process, and (v) bad odor [4]. When the climate is not particularly suitable for air drying or better quality is desired, mechanical air drying is mainly used. However, sun drying is the cheapest method of drying foods. Nowadays, solar, and mechanical air drying is widely used commercially.

Solar drying is an extension of sun drying that uses radiation energy from the sun. Solar drying is a non-polluting process and uses renewable energy. A solar dryer is considered a reliable application of solar energy for food transformation [5].

Furthermore, the digital agriculture revolution must be considered. Historically, agriculture has undergone a series of revolutions that have driven efficiency, yield and profitability to previously unattainable levels. Market forecasts for the next decade suggest a 'digital agricultural revolution' will be the newest shift which could help ensure agriculture meets the needs of the global population into the future. Digitalization will change every part of the agri-food chain. Management of resources throughout the system can become highly optimized, individualized, intelligent and anticipatory. It will function in real time in a hyper-connected way, driven by data. Value chains will become traceable and coordinated at the most detailed level whilst different fields, crops and animals can be accurately managed to their own optimal prescriptions [6].

Integrating multiple sensors into manufacturing systems enables adaptive and flexible automation and increases process adaptivity and quality control. The design of an optimum configuration must be based on a systematic analysis relative to operational and economic considerations. This design should include an evaluation of sensor performance, a decision on how many sensors, of which type they should be, and where to locate them, modes of sensor interaction, and the influence on overall performance based on a cost/benefit analysis. In most real-world applications it is impossible to develop an accurate model to predict sensor performance and therefore, performance must be derived empirically but in a systematic manner [7].

In this context, this paper aims to study the efficiency of two different dehydration processes. Simple digital methodologies were adopted for the monitoring of drying processes. Their suitability for process monitoring and standardization was considered.

2 Materials and Methods

2.1 Micro-drier Systems and Sensors

Two different micro-driers were built at the Research Centre for Engineering and Agro-Food Processing in Rome:

- a drying (with fan active inside the temperature range 35–45 °C at the top: M1), and
- a dehydration (with fan always active at the bottom: M2) micro-system.

Both systems are based on the use of solar energy for food transformation.

The driers are based on different principles: M1 works as a mild oven, blowing hot air to facilitate the evaporation of the product's humidity; conversely M2 acts like a dehydrator forcing humid air outside with the help of fans.

Figure 1 shows a representation of the two drying systems.

The plants were implemented with Arduino sensors (ATmega32U4 Arduino Leonardo AVR® ATmega AVR MCU 8 bit; Arduino UNO Rev3 orig. HIGH QUALITY; Yun Rev 2 Arduino) for the detection of the main physical parameters. In particular, were *in continuous* monitored:

- the in/out RH%, as percentage ratio between the quantity of vapor contained in a mass of air and the maximum quantity (i.e., at saturation) that the volume of air can contain under the same conditions of temperature and pressure;
- the in/out temperature, expressed as degrees Centigrade (°C);
- the light, as the sensor's ability to receive the light signal (millivolts, range between 0 and 1024);
- the fan intensity, as a value within a range between 0 (fan stopped) and 255 (maximum activity = 1800 rpm);
- the weight loss%, as the weight variation detected by 4 load cells placed in correspondence with the four base vertices.

Fig. 1. Representation of the two micro-driers prototypes; M1: drier; M2: dehydrator.

The sensors connection and optimization, and the experiments were carried out at the Research Centre for Engineering and Agro-Food Processing in Treviglio. The drying processes were carried out by applying the simultaneous use of sensors and considering the sensors and the site (position) as independent variables, while fan, weight, temperature, and relative humidity (RH) as dependent variables. The processes were monitored until water activity values below 0.6 and constant weight were reached (50 h). The whole dataset was sent to a remote computer connected via Wi-Fi for data collection and processing. The values of the different variables were standardized to allow reading on the same graph.

The values measured by the sensors and stored in an SD-card in the Arduino board, acquired with a frequency of $3\ 10^{-3}$ Hz, were processed with descriptive statistical analysis to verify their trend over time and tests were applied to variance (ANOVA) [8] and Tukey's [9] post-hoc test for observing the significant difference between the means. The analysis of the time series, of the variance and of the significance of the difference between the means of the values was developed through the software R [10].

2.2 Samples and Determination of Water Activity at the End Point

Fresh eggplant samples were used as model of raw material to be dried. Samples were cut into slices of 3 mm thickness and arranged side by side in parallel into the two plants. Figure 2 schematically shows the samples arrangement.

Water activity (a_w) was determined at the end of the drying with AquaLab 4TE benchtop water activity meter (METER Group, Inc. USA) [11].

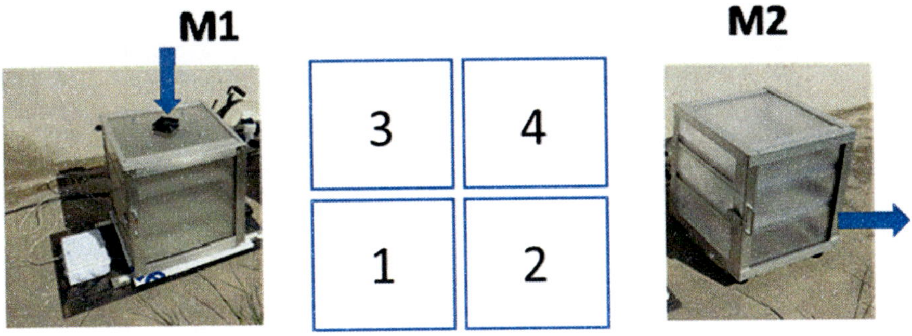

Fig. 2. Samples rearrangement inside the two micro-plants.

3 Results and Discussion

The normalized data, monitored throughout the process, for each variable and relating to the three sites (M1, M2, Ext) are reported in Fig. 3.

The diagrams report the trend of the different sensors monitored in each site. Regarding to M1, fan and relative humidity (RH) showed reverse trends, with an increase of RH during the night, when fan was inactive. Temperature trend was, as expected, in agreement with the fan trend.

In the M2 micro-plant, the temperature reached values higher than in M1 during the day, also if the maximum value was always below 48 °C.

The difference between the temperature and humidity values recorded in the M1 type were statistically significant compared to the M2 type (ANOVA p-value < 0.05).

The temperature reached in M2 with the fan programmed for a constant number of revolutions is more dependent on the external temperature. During the hottest hours of the day, it reaches higher levels than M1 in which the fan is programmed to counteract thermal excesses. The constant activity of the fan in M2 allowed the reduction of the nighttime RH% not contrasted in M1 in which the fan is inactive during the night.

The external monitoring confirmed that the environmental RH was always higher than inside the micro-plants, where the fan could insure forced escape of humid air. The weight trend, measured by an external balance placed at the bottom of micro-plants, allowed to follow the weight loss during the whole experiment.

Figure 4a highlighted the results showed previously, analysing the collected data sensor by sensor, between sites and between positions (Fig. 4a).

The time interval in which the fan was active in M1 is reported top left; the weight loss trend showed the effect of night humidity, highlighting an increase in the weight recorded. The temperature and humidity trends between the two micro-plants, and in comparison, with the external values, agreed with the comments reported before.

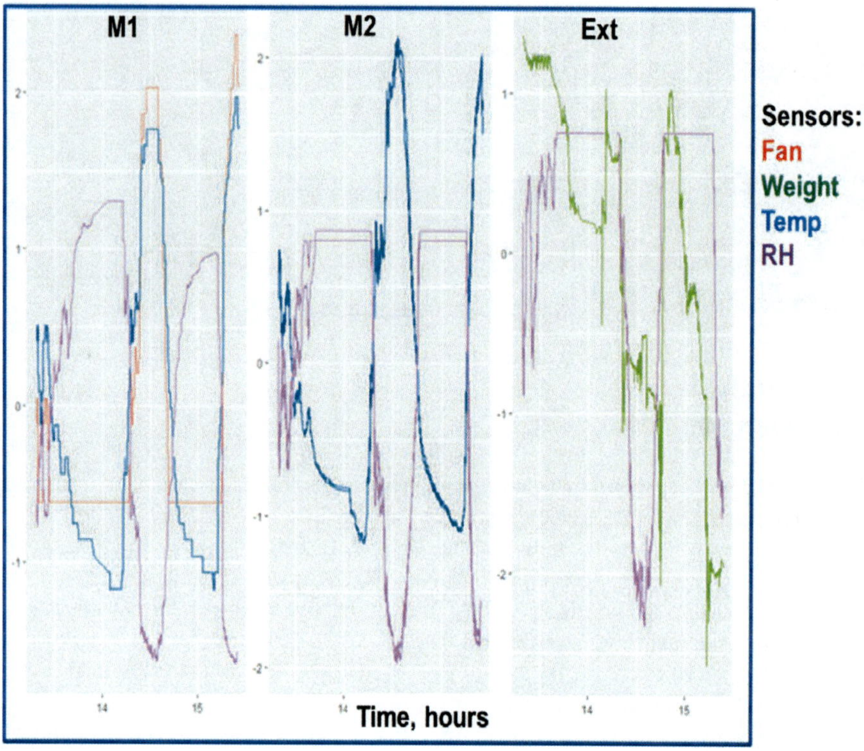

Fig. 3. Variables monitoring between sites. M1: drier; M2: dehydrator; Ext: external environment.

The two last diagrams reported in Fig. 4b, compare the data collected in two different positions within the same plant. The recorded curves relating to temperature and humidity did not show statistically significant differences between the upper and lower area of the dryer when the fan was active or activated (ANOVA p-value = 0.689). The differences between internal and external values of RH were confirmed by the Tukey's post-hoc test at 0.95 confidence level.

Water activity values, reported in Table 1 as single tray and average value inside each micro-drier system, at the end of the drying process did not always be found higher than 0.500, proving the safety of the final dried product. The micro-plant M2 seemed to be able to reach less a_w values, corresponding to limit pollution.

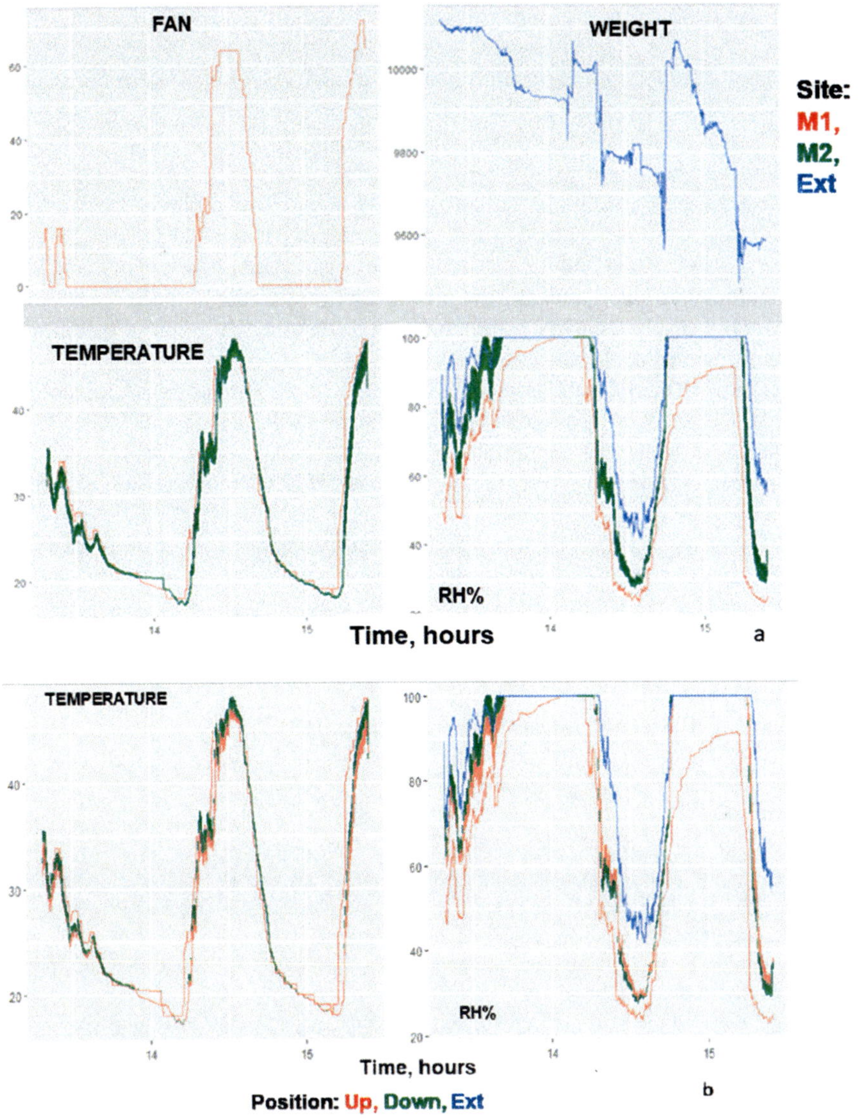

Fig. 4. a) Variables monitoring between sites and between positions. M1: drier; M2: dehydrator; Ext: external environment. b): Up/Down = trays.

Table 1. Water activity (a_w) values at the end of drying processes. T1 (M1); T2 (M2); av = mean values; 1–4: samples position

T1.1	T1.2	T1.3	T1.4	T1av	T2.1	T2.2	T2.3	T2.4	T2av
0.4895	0.4872	0.4791	0.5004	0.4890	0.4679	0.4891	0.4989	0.4866	0.4856

4 Conclusions

The efficiency of two different transformation processes was proven. Both micro-plants worked to reach a good level of dehydration after 50 h.

The chosen digital methodologies were suitable to in continuous monitor the dehydration process.

The process standardization can be easier by using M1, in which fan worked at controlled temperature, and working as a mild oven.

Conversely, M2 suggested higher probability in maintaining the original nutritional value of the dried products and working as «natural dehydration». In this case, in fact, only unbound water is removed from the product, ensuring better preservation of the nutritional principles of the fresh product.

The choice of their use is strictly related to the final aim of transformation. For example, if the need to standardize the process prevails, the use of M1 is recommended; if it is possible to accept limited climatic variations that can affect the duration of the process, M2 is better suited to the production of high quality dehydrated products.

Food safety was ensured by water activity values always lower than 0.500.

Acknowledgements. Technical support: Alex Filisetti, Ivan Carminati, Gianluigi Rozzoni, CREA-IT, Treviglio, Bergamo.

Funder: Italian Ministry of Agriculture, Agridigit project, sub-project Agrofiliere (D.M. 36503/7305/2018, 20/12/2018).

PhD course: DAFNE-EEE, Tuscia University, Viterbo, ITALY.

References

1. Homepage. https://sustainabledevelopment.un.org/; transforming our world: the 2030 agenda for sustainable development, A/RES/70/1. United Nations, Accessed 28 Sept 2021
2. Homepage. https://sdgs.un.org/goals. Accessed 28 Sept 2021
3. Erbersdobler, H. F.: In: Concentration and Drying of Foods, MacCarthy, D., (ed.), pp. 69–87. Elsevier, London (1986)
4. Araya-Farias, M., Ratti, C.: Dehydration of foods: general concepts advances in food dehydration. In: Ratti, C. (ed.) Advances in Food Dehydration, pp. 1–36, CRC Press Taylor & Francis Group, LLC, Boca Raton (2009)
5. Imre, L.L.: Solar drying. In: Handbook of Industrial Drying. Mujumdar, A.S. (ed.) p. 357. Marcel Dekker, New York (1987)
6. Trendov, N.M., Varas S., Zeng, M.: Digital technologies in agriculture and rural areas. In: Briefing Paper, pp. 1–2, Food and Agriculture Organization of the United Nations, Rome (2019)
7. Edan, Y., Nof, S.Y.: Sensor economy principles and selection procedures. IIE Trans. **32**, 195–203 (2000)
8. Tabachnick, B., Fiddell, L.: Experimental Designs Using ANOVA. California State University, Northridge (2020)

9. Haynes, W.: Test di Tukey. In: Dubitzky, W., Wolkenhauer, O., Cho, K.H., Yokota, H. (eds.) Encyclopedia of Systems Biology. Springer, New York, NY (2013). https://doi.org/10.1007/978-1-4419-9863-7_1212
10. Development Core Team, R: A Language and Environment for Statistical Computing; R Foundation for Statistical Computing: Vienna, Austria (2008). http://www.R-project.org. ISBN 3-900051-07-0. Accessed 7 July 2020
11. Decagon Device: Aqualab Water Activity Meter Operator's Manual. Decagon Devices Inc (2008)

Rovitis 4.0: An Autonomous Robot for Spraying in Vineyards

Marcello Biocca[1](✉), Letizia Aiello[2], Cristiano Baldoin[3], Cristian Bolzonella[4], Giuseppe Bugin[4], Pietro Gallo[1], Massimo Gardiman[2], Francesco Meneghetti[5], Federico Pallottino[1], Giorgio Pantano[6], Matteo Pantano[6], Jurij Rakun[7], Peter Lepej[7], Denise Vicino[8], Nicola Vicino[8,9], and Alessandro Zanzotto[2]

[1] CREA – Consiglio per la ricerca in agricoltura e l'analisi dell'economia agraria, Centro di ricerca Ingegneria e Trasformazioni agroalimentari, Monterotondo, Rome, Italy
marcello.biocca@crea.gov.it
[2] CREA – Consiglio per la ricerca in agricoltura e l'analisi dell'economia agraria, Centro di ricerca Viticoltura ed Enologia, Conegliano, TV, Italy
[3] Dipartimento TESAF, Università di Padova, Padua, Italy
[4] CIRVE, Università di Padova, Padua, Italy
[5] Confagricoltura Veneto, Venice, Italy
[6] Azienda Agricola Giorgio Pantano, Candiana, PD, Italy
[7] Faculty of Agriculture, University of Maribor, Maribor, Slovenia
[8] CET Electronics, Zenson di Piave, TV, Italy
[9] Terre Grosse Soc. Agr. s.s, Zenson di Piave, TV, Italy

Abstract. The research project Rovitis 4.0 was aimed to design and develop two prototypes of robot, connected to a Decision Support System (DSS), capable of automatically spraying vineyards. Through these machines a safer application of pesticides is possible, allowing operators to maintain distance from the application zone and during the preparation of the phytoiatric mixture and reducing the use of plant protection products by a variable rate precision spraying system. During the three-year activity of the project, several parameters have been tested, such as: safety and performance of the autonomous driving system; effectiveness of the correct robot-DSS interaction; effectiveness and quality of robotic treatments; economic sustainability and environmental benefits of the innovation. The first prototype of robot, thanks to the development of an autonomous drive system, was able to operate alone through the entire 2020 season. The second prototype was equipped with the precision spraying system based on vegetation recognition. It was able to instantaneously adapt the spraying to the size and position of the target, allowing a 43% reduction of copper-based products in the 2020 season. The overall results showed that a robotic management of the vineyard can lead to environmental and economic advantages even for medium size farms. This paper describes the features of the robots and shows the main findings obtained during the testing of the machines.

Keywords: Precision agriculture · Digital agriculture · Participative projects · DSS · Safety · Autonomous robotics

M. Biocca et al. (Eds.): SHWA 2020, LNCE 252, pp. 176–185, 2022.
https://doi.org/10.1007/978-3-030-98092-4_19

1 Introduction

Over the course of history, mechanization and automation increased crop output numerous orders of magnitude. Nowadays, advancements in the domain of robotics can significantly accelerate this trend (Fountas et al. 2020; Oliveira et al. 2021).

Agricultural robots are autonomous or semi-autonomous systems that can be operated in several stages of the process (Siciliano and Khatib 2016) and the central purposes for computerization of crops are reducing the time and effort required for performing cultivation practices and improving the efficiency in the use of inputs in agriculture. Robotization can be considered the most advanced chapter of digital and precision agriculture (Pallottino et al. 2018).

Spraying pesticides to protect plants from pests and diseases is a common agricultural practice that requires time, readiness of execution and technical competences. In this case, robots can significantly increment the performance and the safety of this operation, avoiding repetitive and demanding activities, reducing the workload, the need for manpower and the contact with substances and contaminated equipment. This allows the repositioning of human labor on other activities (Marinoudi et al. 2019).

Some robots devoted to crop spraying have been realized in the last years, but most of them have remained just in a prototypal stage (Jian-Sheng 2014; Gonzalez-de-Santos et al. 2017; Danton et al. 2020; Chaitanya et al. 2020), and practically no solutions are at present available on the market. Furthermore, even fewer findings are available on spraying vineyards (Oberti et al. 2016; Cantelli et al. 2019).

In this context, the project (Rovitis 4.0 n.d.) aimed to develop two prototypes of robot capable of autonomously spraying the vineyards with precision application of plant protection products. This paper describes the features of the robots and shows the main results obtained during their testing. The project based on a previous activity funded by Regione del Veneto (Dodich project n.d.) and leaded by "Azienda Agricola Giorgio Pantano S.s.".

2 Material and Methods

2.1 The Research Group

At the beginning of the project, an Operational Group (O.G.) funded by Regione del Veneto in the framework of the European Innovation Partnership (EIP-AGRI n.d.) was formed. The O.G. includes two viticultural farms, three research partners, two technological companies, and a union of agricultural producers. The O.G. leader was the farm "Azienda Agricola Giorgio Pantano". The project lasted three years.

2.2 Field Test Sites

The field evaluation of the performance of the two prototypes was carried out at "Azienda Giorgio Pantano" in the province of Padua ($45^{\circ}13'35.18''N11\circ59'2.70''E$), under conventional management and at "Terre Grosse Soc. Agr. S.s." in the province of

Treviso (45∘40′27.72″N12∘29′25.54″E), under organic farming. The vineyard varieties were Glera in the first farm, and Sauvignon and Manzoni bianco in the second farm.

Fig. 1. The first prototype (1), operating at "Azienda Giorgio Pantano", in the configuration of the first year of activity; the second prototype, operating at "Terre Grosse" (2) with the stereo cameras for canopy inspection (3) and the Pulse Width Modulation (PWM) nozzles (4).

2.3 Development of Prototypes

A self-propelled tracked vehicle was prototyped by a partner company as electro-mechanical platform for the robot. The vehicle mounts a diesel engine of 28.2 kW and has a total mass of 1000 kg. It transports all the hardware components, with ad hoc electronics developed for the control of the machine, sensors and the spraying unit (Fig. 1). The decision for this configuration was driven by the initial findings on autonomous driving in vineyards achieved by previous research projects and research activities done in the framework of the "Dodich" project. Such initial findings included but were not limited to: need of a small platform with skid-steer drive and adequate power provided by a propulsion unit (e.g., diesel engine), autonomous navigation based on fusion of sensor data coming from laser scanners, Inertial Measurement Units (IMU) and wheel encoders and a cascade control algorithm (Fig. 2).

An autonomous navigation software and an image analysis software have been developed for the autonomous driving of the prototype, as well as for the recognition of vegetation and the spraying of plant protection products at variable rates. The prototype was equipped with an integrated safety system for stopping or turning off the machine in case of dangerous situations.

The equipment was connected to the cloud, both for remote control and for the interaction with a Decision Support System (DSS), made available by a partner, getting operational suggestions and integrating the information collected by the on board sensors. In this way it is possible to program the specific activities to be carried out by the robot, in an integrated manner with the forecast models and with various agricultural operations. During the initial phase of the prototype development, thanks to the use of the old Dodich platform (Fig. 2) and the ROS system (Robot Operating System), new driving algorithms were implemented and tested. Those included: localization, path planning and a cascade control. For the localization an approach based on sensor

Fig. 2. The initial prototype developed by "Azienda Giorgio Pantano" in previous research projects.

fusion of data coming from IMUs, wheel encoders, LIDAR sensor (i.e., 3D laser scanner) and a single GPS (Global Positioning System) RTK (Real Time Kinematic) receiver was used. For the path planning, an approach based on a global plan and a local plan was employed. Through this strategy, on one hand, the global plan was used to elaborate driving strategies (i.e., trajectories) which included fixed obstacles, on the second hand, the local plan was used to modify the initial trajectories to account also for temporary obstacles as proposed by Lin et al. (2009). For the cascade control, the trajectories elaborated by the planners were then translated to linear and angular velocities, afterwards those were then sent to the Electronic Control Unit (ECU) and electrical commands were elaborated to move the vehicle. Through this architecture the robot was able to proceed autonomously from the shelter station, move to the vineyard, perform the phytosanitary treatment and return to the shelter, correctly recognizing the operating environment.

In the second year of activity, an alternative autonomous driving system was also developed and tested on the second prototype. It entails the use of stereo cameras and image processing software based on artificial intelligence, which, in addition to the three-dimensional reconstruction of the environment, realizes the semantic analysis (therefore with the ability to distinguish people, ditches, walkable paths, obstacles, vegetation, etc.) of the workspace. The large computing capacity required for this type of processing did not allow a complete field test of this system, which will be carried out in the next future after a hardware upgrade process.

In addition to the ability to drive autonomously, the issue of reducing the environmental impact was addressed by developing a variable rate precision spraying system. It is based on the use of special stereo cameras mounted on the front of the robot (Fig. 1), which frame the vegetative wall on both sides and allow, through an image analysis software, to recognize the presence of the canopy and computing its size in terms of volume and projected area, then estimating through an algorithm the total framed leaf area. Based on this information, the flow of the phytoiatric mixture is modulated in real time at the level of a single nozzle, thanks to the use of Pulse Width Modulation (PWM) nozzles (Teejet) (Fig. 1).

The precision spraying unit adopted a dynamic system to continuously regulate the distance of the bars with the nozzles from the vegetation, allowing a more precise targeting of the canopy at an optimal distance.

This robot is also equipped with a PPPs (plant protection products) automatic mixing and injection apparatus (patent pending), providing an on-line mixing of different phytoiatric products (contained as concentrated solution in four tanks) directly on the machine, based on a system of peristaltic pumps connected to a 200 L tank containing clean water.

As an alternative to the on board mixing, a station for automatic refilling the robot with PPPs and water without the presence of the operator was also prototyped. The station was entailed by a robot gantry system which, through the usage of object recognition was able to move to the target location by solving the inverse kinematics through the camera parameters. Moreover, an injection apparatus based on the Bernoulli principle (patent-pending) was integrated for mixing also of power based PPPs (Fig. 3).

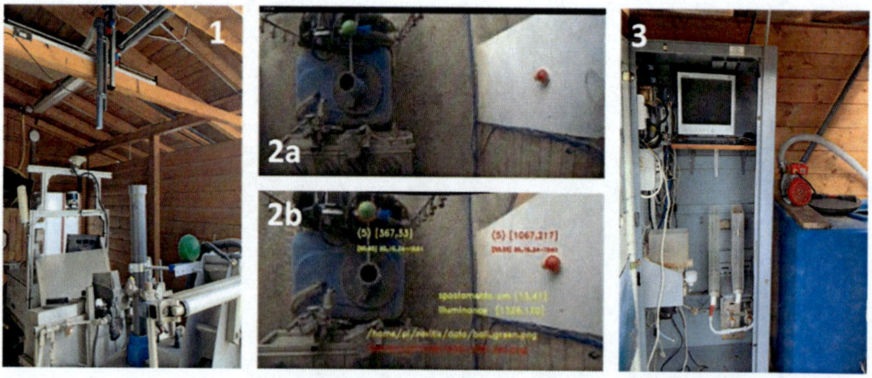

Fig. 3. The prototype of autonomous refilling developed by "Azienda Giorgio Pantano". The system was composed of a gantry system with cameras (1), an intelligent system based on image processing to calculate end-positions for the robot (2a–b), and a mixing solution based on the Bernoulli principle for powders and liquids (3).

2.4 Evaluation of Economic Feasability and Environmental Benefits

Another part of the project regarded the assessments of the operational effectiveness in the field and of the economic convenience in various scenarios characterized by different investment costs, company size and vineyard management systems.

To represent the degree of environmental pressure due to the pesticides use, the PIC indicator (Potential Impact Class, ARPA Toscana) on water, environment and health (Franchi 2017; Bolzonella et al. 2019) was used.

2.5 Evaluation of the Efficacy and Quality of Treatments

In the third year of activity, the experimental design for the evaluation of the efficacy of the phytosanitary protection entailed three experimental plots: the first treated with the robot, the second treated with the conventional air-blast sprayer and the third not treated. This scheme was replicated in the two farms and the observations were conducted during one complete vegetative season. Every treatment was carried out on the same day and with the same PPPs in the two experimental farms, using the conventional sprayers and the robots.

The development of the main fungal diseases (downy mildew and powdery mildew) was assessed in the two partner farms. On leaves and grapes, the following elements were periodically evaluated: 1) first symptoms of the disease; 2) count of new organs affected by downy mildew; 3) evaluation, in some phenological phases, of the disease indices.

Meteorological data were recorded during the vegetative season.

Regarding the quality of distribution of the robots, water sensitive papers (WSP) were used (Otto et al. 2015; Biocca et al. 2021). In the first year of research, tests aimed to evaluate treatments performed at high and medium volume application rate, were carried out in the conventional farm (Pantano). The tests were replicated during the third year, evaluating the first prototype with a modification of the spraying group and the second prototype with the variable rate system operating in the organic farm (Terre Grosse).

3 Results and Discussion

The Rovitis 4.0 Project represents a tangible demonstration of the effectiveness of the EIP-AGRI partnership which assembles partners with various complementary skills and experiences, allowing the development and the implementation of the innovation.

As overall result, in the first farm ("Azienda Giorgio Pantano") the first prototype carried out autonomously the total of 13 scheduled treatments with fungicides during the 2020 season in the experimental plot. Moreover, through the layered autonomous driving control architecture the old Dodich machine was retrofitted for operation, thus demonstrating the interoperability of the autonomous navigation system architecture.

As for the second prototype (operating in the "Azienda Terre Grosse") the use of the variable rate system was remarkable. Considering all the treatments, the innovative system allowed to save 43% of the total quantity of the applied copper-based products, compared to the standard management.

The first evaluations on the execution times by the autonomous vehicles showed longer working times for the robot due to its lower speed compared to the company tractor. The speeds were subsequently increased to 4 km h^{-1} and there is the prospect of equating the speed of the tractor in future versions. The maximum robotic surface with one unit is currently 20 hectares in a single body, due to instrumental limits. These limits can possibly be overcome with the combined and coordinated use of multiple machines (fleets of robots).

Regarding the assessments of the impact of the robot use on the environment, water and human health, the PIC indicator identified a positive reduction of this index (Table 1). Effects are evaluated on a scale with a range from 1 (little impact) to 5 (very impactful). The compartment PIC related to environment, water and heath was calculated by averaging the sum of the PIC of the active principles used on the basis of the farm register of treatments.

The analysis highlights in general an average level of PIC for water and environment and medium-low level for health.

Higher values in the organic vineyards for water and environment compartments are the consequence of a greater use of substances that bioaccumulate. The PIC values for all fields with innovation are lower thanks to the reduction in the dosages of plant protection products obtained with the dynamic atomizer.

Table 1. Compartment PIC Value with and without innovation

	PIC water		PIC environment		PIC health	
	Without innovation	With innovation	Without innovation	With innovation	Without innovation	With innovation
Conventional vineyard	2.15	1.18	1.82	1.00	1.43	0.78
Organic vineyard	3.00	1.65	3.37	1.85	1.00	0.55

Considering the savings of phytosanitary products in conventional and organic experimental vineyards, the economic feasability of the innovation was assessed, considering different levels of investment in the robot purchase between 30′000 and 150′000 euros with a life span of the machine of 10 years.

Table 2 reports the payback time of different level of investment (euros) and surface (hectares) in conventional and organic vineyards. The payback time has been calculated comparing the vineyards management costs of Rovitis 4.0 with subcontractor services.

In the simulations we assume a number of phytosanitary treatments of 13 per year for conventional vineyards and 20 in organic vineyards. A cost of phytosanitary distribution by subcontractor equal to 50 euros per treatment. An annual cost of plant protection products of 750 euros per hectare for conventional vineyards and 600 euros per hectare for organic vineyards.

As for phytosanitary protection, the untreated plots of Sauvignon, Manzoni bianco and Glera were affected by downy mildew, with diffusion index values between 85 and 100% at the end of the season. In contrast, in the theses treated with the robots the leaves and grapes resulted healthy, similarly to those treated with the farm conventional sprayer. The situation of substantial uniformity of the results was also confirmed with the surveys on the production, estimated with samples at harvest, both in quantitative (kg grapes per vine) and qualitative terms (sugar content, pH, total acidity, malic acid, tartaric acid). From the analysis of the data collected, no significant differences were

Table 2. Return of Investment (years) in conventional and organic vineyards. The bold data highlight payback time minor than 5 years considered the maximum acceptable limit.

Vineyards surface (ha)	Conventional vineyards Level of investment (euro)				Organic vineyards Level of investment (euro)			
	30K	50K	100K	150K	30K	50K	100K	150K
1	30.4	50.6	101.3	151.9	23.6	39.4	78.7	118.1
5	6.1	10.1	20.3	30.4	**4.7**	7.9	15.7	23.6
10	**3**	5.1	10.1	15.2	**2.4**	**3.9**	7.9	11.8
15	**2**	**3.4**	6.8	10.1	**1.6**	**2.6**	5.2	7.9
20	**1.5**	**2.5**	5.1	7.6	**1.2**	**2**	**3.9**	5.9
25	**1.2**	**2**	**4.1**	6.1	**0.9**	**1.6**	**3.1**	**4.7**
30	**1**	**1.7**	**3.4**	5.1	**0.8**	**1.3**	**2.6**	**3.9**

highlighted between the theses, except for some slight deviations in a few qualitative parameters of little practical relevance.

Regarding the quality of distribution assessed with the WSP, at the Pantano farm, the new configuration of the machine, with the arrangement of the nozzles vertically and not arched and with an application rate of 750 L ha^{-1} showed the best results. The prototype tested at the Terre Grosse farm, with active variable rate and on line PPPs mixing and injection has allowed on the one hand a consistent reduction in the quantity of product compared to the conventional sprayer thesis, on the other hand it has highlighted some spots with insufficient supply of product, especially in the lower part of the canopy. It is expected to easily overcome these problems with the improvement of the synchronization of the nozzle control with the advancement of the machine and/or with the adoption of nozzles with a higher spray angle, to take advantage of the partial overlap of spray plumes. However, these critical issues had minimal repercussions from perspective of phytosanitary protection, which in any case was comparable to treatments performed in the traditional way.

4 Conclusions

Rovitis 4.0 promises to be an important solution for the near future in order to ensure sustainability for companies in the medium term. This will require increasingly targeted and timely interventions for being able to conduct a more efficient crop protection and at the same time less impacting on the environment. The balance of the first observations was therefore decidedly encouraging and stimulating further experimentation, to confirm, in different years and locations, what was observed in the preliminary tests. The activities carried out and the results achieved by the operating group of the Rovitis 4.0 Project represent an important starting point for future lines of research, where to further develop the design idea of a robot for viticulture, also considering the fast evolution of components and of research.

In conclusion:

- The use of robots increased work safety allowing the operators to maintain distance from both the application zone and during the mixture preparation, thanks to the on board mixing and injection system or, as an alternative, the station for automatically refilling the robot with PPPs and water.
- In the experimental test conditions, the plots treated with the two prototypes showed no significant differences in fungal disease development in comparison with the plots treated with a conventional sprayer.
- Economic evaluation shows that with an investment from 30′000 to 150′000 euros (and a return of investment of at most 5 years), thanks to saving in distribution by subcontractor costs and pesticides costs, Rovitis 4.0 is convenient starting from 10 to 30 hectares of vineyard surface. Furthermore, it is necessary to take in account that the considered level of investment, given considering current costs, will lower in the next future thanks to technological progress and in the view of an industrialization of the production.
- A reduction of the PIC indicator indicates a positive effect of the innovation on the environment and human health.
- The Rovitis 4.0 Project also represents a tangible demonstration of the effectiveness of the Operational Group PEI-AGRI partnership.

Acknowledgements. Research funded by Regione del Veneto, P.S.R. 2014–2020, Operational Group Project "Rovitis 4.0" for the deployment of an autonomous robot connected to a DSS for a sustainable and efficient management of the vineyard.

References

Biocca, M., Cutini, M., Romano, E., Pallottino, F., Gallo, P.: Evaluation of drift-reducing nozzles for pesticide application in hazelnut (*Corylus avellana* L.). AgriEngineering **3**(2), 230–239 (2021)

Bolzonella C., Lucchetta M., Teo G., Boatto V., Zanella A.: Is there a way to rate insecticides that is less detrimental to human and environmental health? Glob. Ecol. Conserv. **20,** e00699 (2019)

Cantelli, L., Bonaccorso, F., Longo, D., Melita, C.D., Schillaci, G., Muscato, G.: A small versatile electrical robot for autonomous spraying in agriculture. AgriEngineering **1**(3), 391–402 (2019)

Chaitanya, P., Kotte, D., Srinath, A., Kalyan, K.B.: Development of smart pesticide spraying robot. Int. J. Recent Technol. Eng. **8**(5), 2193–2202 (2020)

Danton, A., Roux, J.C., Dance, B., Cariou, C., Lenain, R.: Development of a spraying robot for precision agriculture: an edge following approach. In: 2020 IEEE Conference on Control Technology and Applications (CCTA), pp. 267–272 (2020)

Dodich project. http://www.ethics.it/dodich/. Accessed 25 Sept 2021

EIP-AGRI. https://ec.europa.eu/eip/agriculture/en. Accessed 25 Sept 2021

Fountas, S., Mylonas, N., Malounas, I., Rodias, E., Hellmann Santos, C., Pekkeriet, E.: Agricultural robotics for field operations. Sensors **20**(9), 2672 (2020)

Franchi, A.: Pesticides class of potential impact – CIP. An indicator to drive sustainable choices, p. 34. ARPA Toscana, Firenze (2017)

Gonzalez-de-Santos, P., et al.: Fleets of robots for environmentally-safe pest control in agriculture. Precis. Agric. **18**(4), 574–614 (2016). https://doi.org/10.1007/s11119-016-9476-3

Jian-sheng, P.: An intelligent robot system for spraying pesticides. Open Electr. Electron. Eng. J. **8**(1), 435–444 (2014)

Lin, Y.C., Chou, C.C., Lian, F.L.: Indoor robot navigation based on DWA*: velocity space approach with region analysis. In: 2009 ICCAS-SICE, pp. 700–705. IEEE (2009)

Marinoudi, V., Sørensen, C. G., Pearson, S., Bochtis, D.: Robotics and labour in agriculture. A context consideration. Biosyst. Eng. **184**, 111–121 (2019)

Oberti, R., et al.: Selective spraying of grapevines for disease control using a modular agricultural robot. Biosys. Eng. **146**, 203–215 (2016)

Oliveira, L.F.P., Moreira, A.P., Silva, M.F.: Advances in agriculture robotics: a state-of-the-art review and challenges ahead. Robotics **10**(2), 52 (2021)

Otto, S., Loddo, D., Baldoin, C., Zanin, G.: Spray drift reduction techniques for vineyards in fragmented landscapes. J. Environ. Manag. **162**, 290–298 (2015)

Pallottino, F., Biocca, M., Nardi, P., Figorilli, S., Menesatti, P., Costa, C.: Science mapping approach to analyze the research evolution on precision agriculture: world, EU and Italian situation. Precis. Agric. **19**, 1011–1026 (2018)

Rovitis 4.0. http://www.rovitisveneto.it. Accessed 25 Sept 2021

Siciliano, B., Khatib, O.: Handbook of Robotics. Springer, Cham (2016). https://doi.org/10.1007/978-3-319-32552-1

Automatic Image Labelling for Deep-Learning-Based Navigation of Agricultural Robots

Dario Calogero Guastella[1(✉)], Gaetano Rua[1], Domenico Longo[2], and Giovanni Muscato[1]

[1] Dipartimento di Ingegneria Elettrica, Elettronica e Informatica, Università di Catania, Catania, Italy
dario.guastella@dieei.unict.it

[2] Dipartimento di Agricoltura, Alimentazione e Ambiente, Università di Catania, Catania, Italy

Abstract. The aim of this work is to define a procedure for the automatic labeling of images used for the training of a deep neural network which is used to learn a direct mapping from images to steering angles and collision probabilities. A state-of-the-art convolutional neural network for robotic vehicle navigation is used, which has been adapted to work for ground vehicles. Steering angles and collision probabilities are then used to generate respectively the angular and the linear velocity commands to drive a tracked vehicle through rough unstructured terrains, as those typically encountered in agricultural applications.

Keywords: Tracked vehicle · Rough terrains · Perception-driven navigation

1 Introduction

An open issue in robotics for agricultural applications is the fully autonomous navigation of a ground robot in challenging terrains, such as vineyards terraces or hillside crops [1, 2]. Recent approaches to address this problem are based on deep learning methods, capable of effectively combining environment perception (typically coming from an onboard camera) and control [3, 4]. However, to implement a navigation strategy based on the deep learning approach it is necessary to acquire images and data from sensors and to properly label the images in order to create a training dataset. Image labeling is both crucial and tedious for a human operator, if this must be done manually for a very large number of images, as usually required in deep learning applications.

The aim of this work is to define a procedure for the automatic labeling of images used for the training of a deep neural network which is used to learn a direct mapping from images to steering angles and collision probabilities. A state-of-the-art convolutional neural network for robotic vehicle navigation is used [5], which has been adapted to work for ground vehicles. Steering angles and collision probabilities are then used to generate respectively the angular and the linear velocity commands to drive a tracked vehicle through the rough unstructured terrains.

M. Biocca et al. (Eds.): SHWA 2020, LNCE 252, pp. 186–194, 2022.
https://doi.org/10.1007/978-3-030-98092-4_20

2 Related Work

Among the most used learning structures in deep learning, there are certainly the Convolutional Neural Networks (Convnet or CNN) [6]. DroNet [5] is the convolutional neural network which has been considered in our work for the autonomous navigation strategy of an Unmanned Ground Vehicle (UGV). It is designed for the autonomous driving of a quadrotor in urban environments. This CNN has been trained from images collected by cars or bicycles. The control policy resulting from the learning is able to keep driving along the street and to avoid collisions with other vehicles. However, when the drone is in a natural environment, for example in a forest or in a crop, the solution is unable to navigate.

The considered network provides two output values, namely a collision probability and a steering angle, given the input image acquired from the onboard camera. The former is a value ranging from 0 to 1 (the greater the number, the greater the probability of being in front of an obstacle), whereas the latter is the steering that the robot must perform to avoid the obstacle. Then these two values are given as an input to a control algorithm to derive the velocity and steering commands to the vehicle.

The first part of the network structure is shared by the two tasks (collision probability and steering angle), whereas the second part of the structure is divided into two fully connected layers, one for the steering angle prediction, which is posed as a regression problem and one for the collision probability prediction, which is posed as a binary classification problem.

3 Experimental Setup

The UGV employed for this work is the U-Go Robot (Unmanned Ground Outdoor Robot) [7, 8], shown in Fig. 1. It is a tracked vehicle designed for inspection, navigation and transportation in tough environments. It is a 1.2×0.6 m platform with two 650 W DC motors differentially driving the rubber tracks. The on-board sensors are:

- Stereo Camera ZED Camera
- Attitude and Heading Reference System (AHRS);
- Differential Global Positioning System (DGPS) receiver;
- Laser range finder.

Moreover, there are two computing units:

- a Raspberry Pi that executes algorithms for navigation;
- an NVIDIA Jetson TX1 Module used for the deep learning inference.

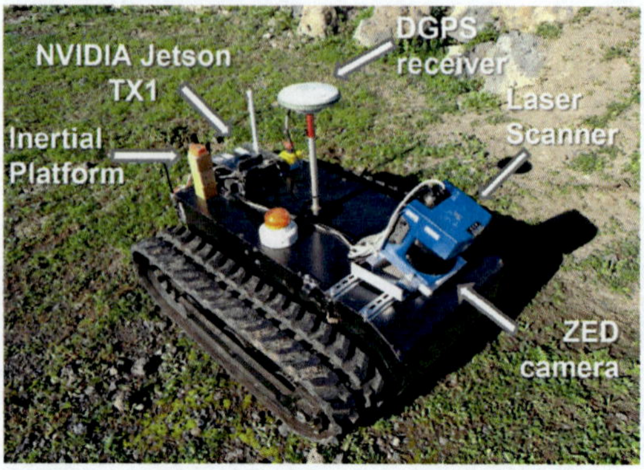

Fig. 1. The mobile platform used for the experimental session and its onboard sensing devices.

4 Proposed Solution

To extend the application spectrum of the network in [5], we aim at retraining it to implement a navigation strategy for a UGV. To this end, a proper training dataset, with a sufficient number of images, must be created.

4.1 Image Acquisition

Images have been acquired during a 2-h manual driving session of the vehicle, along with the distances, measured by an onboard 2D laser scanner, vehicle pitch angles, and the driving commands given by the human operator. The vehicle has been manually teleoperated so that when the robot has no obstacles in front of it, it goes straight, whereas if there is a frontal obstacle (wall, sidewalk, tree etc.) then the operator steers the vehicle away from the obstacle. Both RGB and depth images are recorded.

4.2 Image Labeling

The following step is to label each image. Steering angles are derived by numerical integration over time of the velocity commands given by the operator. Collision probability labels are derived by the distance measurements of the laser scanner.

To calculate the probability of collision for each image starting from the scan measurements (i.e. to label the images with the collision probability). As per definition of probability, this parameter must have a value ranging from 0 to 1. Therefore, the array of scan measurements is first normalized. Since the laser scanner has a 180° frontal view coverage, whereas the ZED camera has a horizontal FOV of 86.5087°, only the scan measurements in the range of −43° to 43° are kept.

Moreover, thresholds related to the measured distance are defined:

- For distances less than minimum (0.4 m), it is assumed that the robot is in front of an obstacle and the collision probability is equal to 1;
- For distances between minimum and maximum, the robot is leaving or approaching an obstacle, so the collision probability must have an intermediate value between 0 and 1;
- For distances greater than maximum (1.4 m), the robot will have no obstacles in front of it, so the probability of collision will be equal to 0;
- In the case of so-called negative obstacles, such as cliffs, we treat these types of obstacles as normal obstacles, so the collision probability will be 1 when the distance is greater than the expected maximum value (in absence of obstacles, the detected average distance is about 1.45 m, because the laser scanner is tilted downwards by about 22°).

Finally, whenever a high pitch in absolute value is detected, the corresponding images must be labeled manually because it is not possible to distinguish from the range finder readings if the robot either in front of an obstacle or at the end or at the beginning of a descent (the same for the climb).

In order to give higher weight to central readings and to partially neglect peripheral ones (consistently with the vehicle size) a combination of two vertically symmetrical sigmoidal functions has been adopted (Fig. 2). The left sigmoid is calculated with the formula:

$$\sigma_{left}(x) = \frac{1}{1 + e^{-x}}$$

Where x represents the orientation values (from −43° to 43°), while the right sigmoid is equal to:

$$\sigma_{right}(x) = 1 - \sigma_{left}$$

By concatenating the two sigmoids along the x-axis, we get our desired weighting profile. By element-wise multiplication of the array of raw measurements with the weighting profile, the central values will be multiplied by numbers close to one, while the lateral values will be attenuated (up to be nullified). This means that if a close obstacle with an angle of 43° is detected, the weight referred to this angle will be close to zero, so will be given less importance. If we have an obstacle around the 0 angle (perfectly frontal), then the corresponding weight will be close to 1. In this way whenever the space in front of the vehicle is traversable, i.e. no obstacles are present, the image is essentially labeled as traversable and the vehicle can go straight, even if a close side obstacle is detected. Finally, the collision probability of the image will be the maximum of *the complement to one* of the array weighted measurements.

Once the labeling process is finished, all the images are mixed together and then split into three groups for training, testing and validation (respectively 70%, 20% and 10% of all the images).

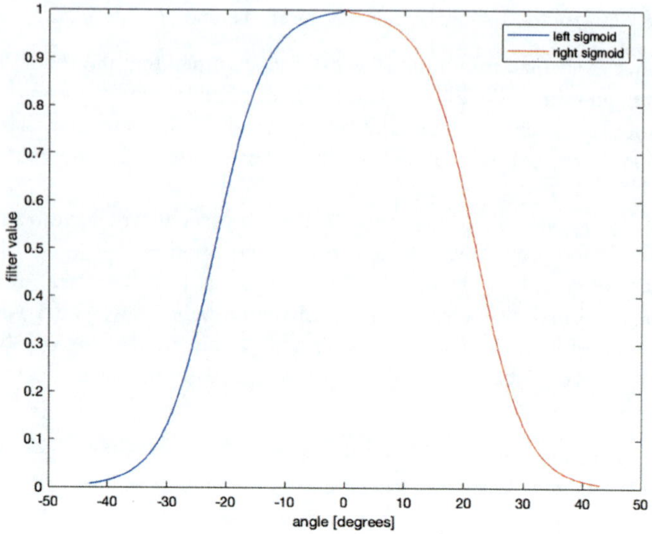

Fig. 2. Weighting profile.

5 Results

The neural network has been retrained with around 4000 images labelled with the approach described above. The trained network proved to be capable of properly regressing both the collision probability and the steering angle. In particular, a fair balance between overestimation and underestimation of the collision probability was observed, which is crucial for the generation of safe linear velocity commands.

In this section we report some graphical examples showing how the images are interpreted in terms of collisions. Four relevant examples among the whole dataset will be illustrated and discussed. The RGB and depth images will be displayed together with two plots: the first plot shows the raw measurements as they are acquired from the laser scanner; the second plot shows the filtered measurement according to the filtering shape.

In Fig. 3 we have a collision example, since we are in front of a sidewalk. In the upper part of the figure the associated steering angle and the collision probability are reported, where the latter is equal to 1 in this case. As we can see from the third subplot representing the real raw measurements, the obstacle is located at a distance smaller than one meter in the frontal direction. This subplot is useful to represent the profile of the detected obstacle. The fourth subplot, for each measurement, draws the weighted measurements.

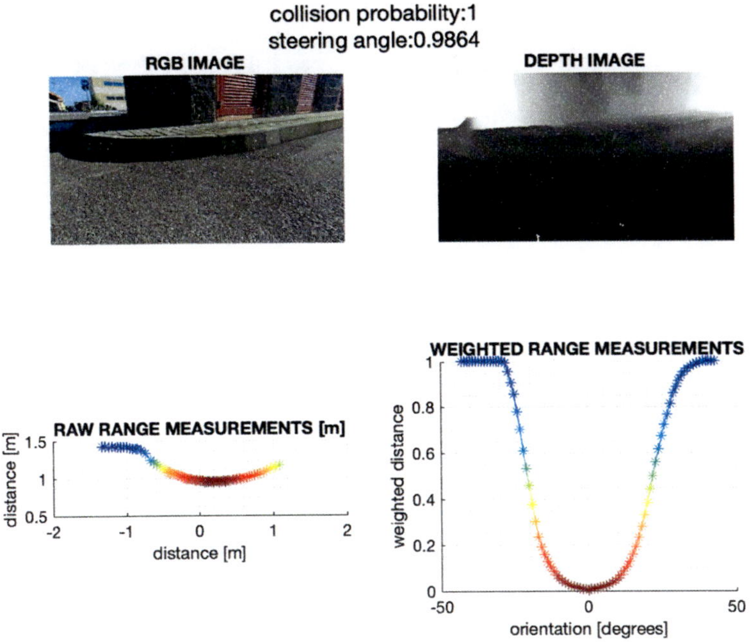

Fig. 3. Frontal obstacle example: RGB and depth images related to their raw and weighted measurements.

In Fig. 4 a different situation is shown in which there is a wall on the left side of the robot path. In this case the collision probability label tends to zero because the wall is not in front of the robot, so it does not represent a dangerous obstacle. The third subplot shows its presence on the left. The fourth subplot, instead, shows how we interpret the situation. Since the wall is located laterally, the associated measurements weigh less, then the result is that we do not see any obstacle.

Figure 5 and Fig. 6 look very similar, although they represent two different situations. In both figures, on the left we have a cliff, i.e. a negative obstacle. The difference is that in Fig. 5 the negative obstacle is located laterally, so it does not actually represent an obstacle while, in Fig. 6, the cliff is seen more frontally. Between the two cases, in fact, we can see an important difference in the fourth subplot, where the weighted measurements show a peak in the second case with respect to the first. The collision probability is consequently higher when the negative obstacle is frontal.

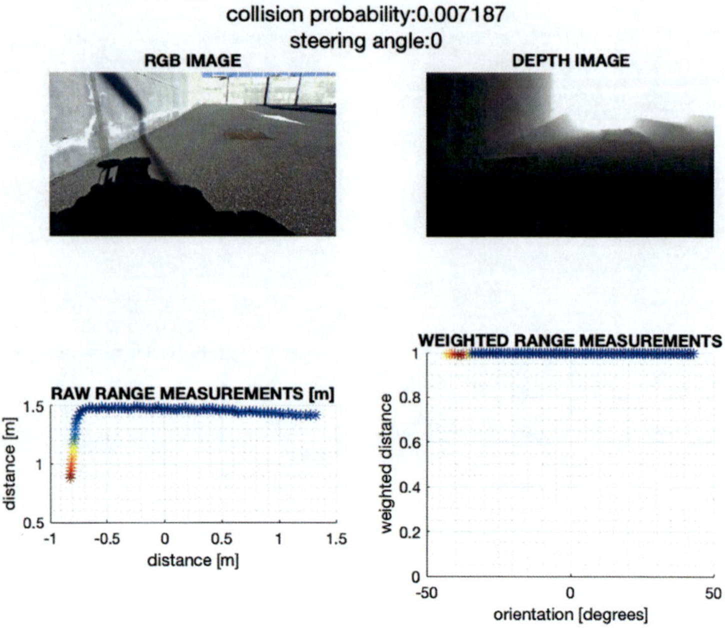

Fig. 4. Lateral obstacle example: RGB and depth images related to their raw and weighted measurements.

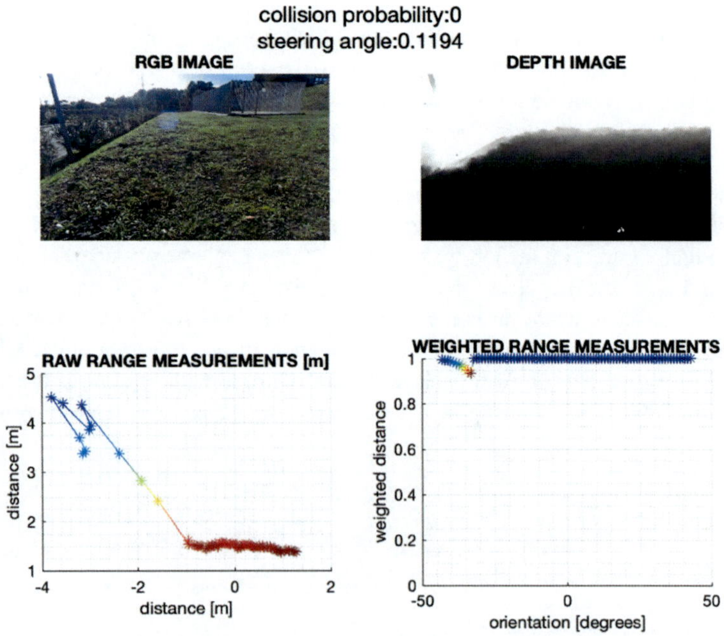

Fig. 5. Lateral negative obstacle example: RGB and depth images related to their raw and weighted measurements.

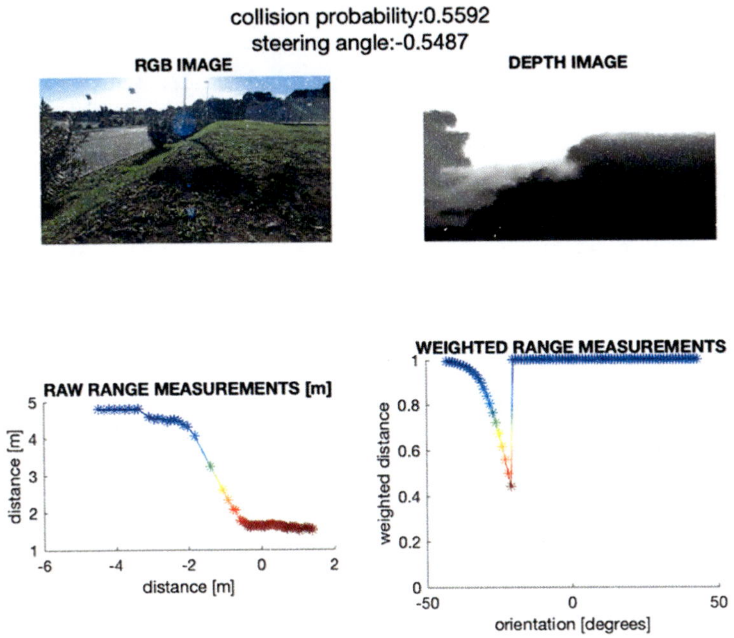

Fig. 6. Frontal negative obstacle example: RGB and depth images related to their raw and weighted measurements.

6 Conclusions

An image labeling algorithm for the creation of a dataset for the training of a convolutional neural network has been presented. Each image is labeled with a steering angle and a probability of collision. Since a large number of images is considered, an algorithm for the automatic labelling is crucial for the sake of efficiency. To implement this algorithm, it was necessary to consider both the distance and the orientation of the obstacles from the vehicle.

As a future development, the possibility to handle the automatic labeling of negative obstacles will be investigated, by integrating over time the attitude of the vehicle and raw laser readings in order to identify either climbs or descents while the vehicle is traversing the environment.

Funding. This work is funded by the University of Catania in the framework of the project "Safe and Smart Farming with Artificial Intelligence and Robotics" - Programma Ricerca di Ateneo UNICT 2020-22 linea 2.

References

1. Vougioukas, S.G.: Agricultural robotics. Ann. Rev. Control, Robot. Auton. Syst. **2**, 365–392 (2019)
2. Roshanianfard, A., Noguchi, N., Okamoto, H., Ishii, K.: A review of autonomous agricultural vehicles (the experience of Hokkaido university). J. Terrramech. **91**, 155–183 (2020)
3. Guastella, D.C., Giovanni, M.: Learning-based methods of perception and navigation for ground vehicles in unstructured environments: a review. Sensors **21**(1), 73 (2021)
4. Grigorescu, S., Trasnea, B., Cocias, T., Macesanu, G.: A survey of deep learning techniques for autonomous driving. J. Field Robot. **37**(3), 362–386 (2020)
5. Antonio, L., Maqueda, A.I., del-Blanco, C.R., Davide, S.: DroNet: learning to fly by driving. In: IEEE Robotics and Automation Letters, Vol. 3 (2018)
6. Liu, W., Wang, Z., Liu, X., Zeng, N., Liu, Y., Alsaadi, F.E.: A survey of deep neural network architectures and their applications. Neurocomputing **234**, 11–26 (2017)
7. Bonaccorso, F., Longo, D., Muscato, G.: The U-Go Robot, a multifunction rough terrain outdoor tracked vehicle. In Proceedings of the 2012 World Automation Congress WAC2012 (2012)
8. Cantelli, B., Longo, M., Schillaci, M.: A small versatile electrical robot for autonomous spraying in agriculture. AgriEngineering **1**(3), 391–402 (2019). https://doi.org/10.3390/agriengineering1030029

Smart Machinery and Devices for Reducing Risks from Human-Machine Interference in Agriculture: A Review

Pierluigi Rossi, Pier Luigi Mangiavacchi, Danilo Monarca(✉), and Massimo Cecchini

Department of Agriculture and Forest Sciences (DAFNE), University of Tuscia, Via San Camillo De Lellis, 01100 Viterbo, Italy
monarca@unitus.it

Abstract. Wireless devices for smart farming can be of great help in remote sensing, machinery automation and precision farming, but they can also provide safety enhancements by reducing risks like exposure to physical agents, chemicals, and tractor overturn. Although technology advances can be beneficial to agricultural safety, it must be said that the integration of wireless devices with farming machinery represents a possible new source of hazards: loss of communication, wrong timing, incorrect or partial information could lead to accidents even without any preexistent or latent failure as a trigger. This review aims therefore to frame the evolution of safety advances in wireless and smart farming applications for agriculture with a focus on safety designs, fail-safe system engineering and wireless network infrastructures such as Bluetooth, GSM, Wi-Fi 802.11 and 802.15 like ZigBee, Lo-Ra technologies and even unmanned aerial vehicles (UAV).

A set of 53 papers starting from 1987 until 2021, involving practical applications, have been deeply analyzed and as a result advantages and disadvantages for every solution have been identified. Further conclusions have been drawn about limits and perspectives of sensing systems for obstacle avoidance like lasers, image processing, Bluetooth low-energy, RFID passive systems and Artificial Intelligence (AI).

Keywords: Agriculture · Remote control · Wireless · Tractor · IoT · Smart farming · Work safety

1 Introduction

The development of new wireless technologies applied in agriculture has led to major improvements in automation and smart farming solutions, including a variety of enhancements for safety systems. The introduction of such innovating systems, however, cannot make hazards disappear and the introduction of autonomous vehicles could even call for a bigger effort at design stage to provide a safer use of such machinery.

This research is part of a project called SMARTGRID, funded by the Italian National Institute for Insurance against Accidents at Work (INAIL), which aims to

M. Biocca et al. (Eds.): SHWA 2020, LNCE 252, pp. 195–204, 2022.
https://doi.org/10.1007/978-3-030-98092-4_21

develop a safety management system that handles hazards related to the presence of remotely controlled vehicles and workers on foot in agricultural and forestry working environments. To do so, state-of-art research has been consequently carried out to frame all the safety improvements since the 80's that, with the introduction of wireless systems or smart technologies, produced interesting innovations for agricultural applications.

2 Methods

The state of art research has been performed on the Scopus Database. Selected criteria focused on wireless communication protocols and on the specific sectors of the project, such as agricultural and forestry activities. On the other hand, since many smart solutions can be applied to a lot of working environments, filtering has been necessary to avoid applications that don't fit the project's purposes: such sectors can be identified as food traceability, transportation, disaster management and other specific areas like medicine, mineral processing, etc. as shown in Table 1.

Table 1. SCOPUS search criteria

Criteria	Content	Search field(s)
Wireless communication sources	Mobile electronic devices, remote, wireless, BLE, Bluetooth, RFID, internet of things, IOT	Article title, abstract or keywords
Objectives	Safety, accidents	Keywords
Selected sectors	Agriculture, tractor, chainsaw	Keywords
Excluded sectors: food traceability and market	Market sector, food, traceability, tracking, supply chain, cold chain, market structure, drinking water, breeding, water quality	Article title, abstract or keywords
Excluded sectors: transportation	Road safety, maritime safety, Aircraft accidents, Truck drivers, automotive, public works, roadkill, traffic control, drone detection	
Excluded sectors: danger and disaster management	Asbestos, disaster management, suicide, terrorist, acid rain, forest fire, radiation	
Other excluded processes or specific sectors	Medicine, dams, mineral processing, outdoor testbeds, static load tests, spectroscopy, health education, disease	

Given the selected search criteria and provided extra filtering on the results which didn't fit the purpose of the review, 53 papers have been found and they cover the time span from 1987 to 2021.

3 State of Art

From the dataset of 53 papers in the previously illustrated SCOPUS search, 49 articles and 4 reviews have been found. Since 1987 sensors and network infrastructures have evolved, and practical applications have changed accordingly. Most of the papers focus on roll-over, obstacle avoidance and other risks that could be easily reduced if vehicles could be controlled remotely or if dangers could be detected immediately. The availability of newer sensors and a better wireless network reliability let management systems rely on new technologies and meet safety requirements.

3.1 Articles

For a better understanding of how new technologies let safety systems evolve in agriculture, the dataset of 49 papers from the SCOPUS search has been subdivided into periods of 5 years, enlightening their specific features (Table 2).

Table 2. Evolution of publications' subjects over time

5-year periods	Subject
1987–1991	Radio-controlled tractor [1], overturn model simulations [2]
1992–1996	–
1997–2001	ROPS design for pre-ROPS tractors [3], remote controls in upset tests of agricultural tractors [4, 5]
2002–2006	Fire protection [6], conceptual framework for autonomous vehicles [7], obstacle detection with lasers [8], autonomous tractor for fruit picking [9], model tractors for safety education [10]
2007–2011	Machines in the data network [11], child safety driver assistant system [12], platooning systems [13], remote plant nursing robots [14], cellular for GNSS assistance [15], remote temperature monitoring based on CAN bus [16], vibration powered wireless system for safety purposes [17]
2012–2016	Remote controls on a tractor for wood harvesting [18], tractor rollover detection mobile application [19]–[21], remote piloted helicopter for agricultural use [22], Bluetooth safety device for electric chainsaws [23, 24], assistive viewing technology for safety in agriculture [25], wireless system for danger detection on equipment [26], four wheel steered autonomous tractor [27], remote control of cherry harvesters [28], lidar system for pesticide drift measurement [29], automatic ROPS deployment [30], energy-harvesting sensors for agricultural safety [31], autopilot based on GPS and steer-by-wire [32], perception of auto-guidance technology [33]
2017–2021	Tablet app to control safety critical functions [34], master-slave remote operations with robots [35], Internet of Things (IoT) monitoring system [36], Blockchain for safety in agriculture [37], GPS collars for goats' behaviour monitoring [38], wireless video monitoring [39], UAVs for boundaries recognition [40], autonomous all-terrain vehicle [41], machine learning to detect abnormal sensors [42], IoT device for farmers' safety [43], LoRaWAN networks [44, 45], Deep Learning in IoT for agriculture [46], IoT for odour pollution prevention [47], 6LoWPAN network for IoT in agriculture [48]

Autonomous Tractors and Remote Controls. According to Scopus Database, the first article ever published to describe the implementation of remote controlling in agriculture is the attempt from Kubota in 1987 where a radio control system managed a tractor remotely [1]. The tractor, powered by a 20 HP diesel engine with hydrostatic transmission, was controlled from a transmitter was that provided the possibility to deal with engine stopping, tractor's speed, mower height, rotation of the mowing blades and two different steering system. A wide range of automatic devices were also added to enhance operational safety, such as automatic brakes and differential locking systems.

Another frontier of smart farming, yet involving automation and wireless networks altogether, concerns the possibility to control many tractors via one single, leading, manned tractor [13]. The platooning system was mounted on a 265 kW – four-wheel drive - Fendt 936 Vario model. An AgGPS-252 receiver was attached to the cab's roof of the first tractor to transmit its position with an accuracy of 2.5 cm. An already present NavController system was used to compensate roll, pitch and yaw movements of the vehicle. The autonomous tractor which followed the leading manned tractor exploited an auto-guide system already installed in the vehicle to identify its position. Both tractors have been equipped with a computer in order to interface GPS receivers and the control units, a system which could also exploit CAN protocol or serial interfaces to communicate with external peripherals.

Obstacle Detection, Collision Avoidance and Anomaly Sensors. Many articles focus on accidents prevention, which are critical for the safe management of autonomous agricultural vehicles. A way to develop a safety system with such capabilities is to create an algorithm to operate an onboard range finder-based obstacle detector [8], that by performing accurate analysis of lasers signals could determine shape and relative motion of the obstacle, within a radius of 8 m. Artificial intelligence, such as machine learning [42], has also been used to detect anomalies in IoT sensors.

Tractor Rollover Prevention and Detection Systems. This is the most critical hazard related to agricultural vehicles and remained critical despite technological improvements over time. Applications have been developed to detect and report tractor rollovers [19, 20, 21], with sensors and models designed to estimate the stability of the machine. When the application detects an accident, it automatically sends an alert with GPS information and time. False alarms can be prevented by action of the operator, who can cancel the emergency message if stability index is 0 but rollover did not happen. Another type of solution involves automatic rollover protection systems [30], called ROPS, since in many accidents rollbars have been removed or lowered in order to reduce the height of the vehicle while performing specific operations.

Smart Safety Applications in Forestry. Forestry is a well-known hazardous sector and indeed operations which involve chainsaws have a significant impact on accident statistics. To enforce safety, a smart solution was developed to prevent accidents if sensors mounted on equipment reveal latent dangers [24, 25]. An electric chainsaw was operated using a Nintendo Wii-mote controller to gather data and transfer it to a computer through Bluetooth connection. The data collected was analyzed in order to distinguishing between normal and dangerous movements, and in case of danger the electric chainsaw would have stopped.

Wireless Network Infrastructures for Safety Purposes. Issues concerning wireless signal strength and reliability could be solved by better infrastructures but also by UAVs [22, 40]. Unmanned Aircraft Systems are an emerging technology for many purposes, including data collection: although shorter ranges improve connectivity, higher altitudes provide better quality signal at greater distances since the angle favors an easier avoidance of obstacles as trees or bushes. Use of such technologies could be of great importance in agriculture and forestry, since the need to establish a reliable connection in large and open spaces where even finding good GSM mobile connection might sometimes represent a struggle for operators.

3.2 Reviews

With the availability of many new wireless network infrastructures which allowed the development of many practical applications which involved them, a consistent number of review articles has been found on Scopus database as shown on Table 3.

Table 3. Review articles

Title	Year	Author
Smart machines, remote sensing, precision farming, processes, mechatronic, materials and policies for safety and health aspects [49]	2018	Colantoni, A.; Monarca, D.; Laurendi, V.; Villarini, M.; Gambella, F.; Cecchini, M
A survey on LoRa for IoT: Integrating edge computing [50]	2019	Sarker, V.K.; Queralta, J.P.; Gia, T.N.; Tenhunen, H.W.; Westerlund, T
A review of practice and implementation of IoT for smallholder agriculture [51]	2020	Antony, A.P.; Leith, K.; Jolley, C.; Lu, J.; Sweeney, D.J
A review on smart IoT based farming [52]	2020	Farooq, H.; Rehman, H.U.R.; Javed, A.; Shoukat, M.; Dudley, S

Smart Machines for Safety in Agriculture [49]. This review covers a wide range of aspects of smart technologies applied to health and safety in agriculture. Smart sensors and web-based applications, like also monitoring systems, greatly improved the safety standards for these machines and the reduction of workers' exposure to physical agents, noise and vibrations were also considered. Precision spraying system, precision livestock farming, big data from the Web or weather stations for decision making, satellite-based tractor rollover risk reduction technologies and geospatial analysis for the identification of the most suitable harvesting areas in zones affected by adverse conditions and harsh terrains have been the focus of the review.

Long Range (LoRa) and Low-Power Wide-Area Networks (LPWAN) [50]. The review showed how these networks can provide better connectivity to IoT devices that can work at lower bitrates and showed a comparison to other connection networks like Bluetooth, Edge, UAVs, Bluetooth Low Energy (BLE), Wi-Fi, ZigBee and GSM connections. The review shows how country-specific regulations over LoRa radio bands can represent limitations. Switching from the most common wireless networks like Wi-Fi and Bluetooth to LoRa networks can also bring issues if low-energy and low data-rate are mission-critical aspects for the selected IoT devices.

Internet of Things in Agriculture [51, 52]. IoT solutions for smart farming have been analyzed from the hardware point of view, since the type of microcontroller units and involved technologies already indicate their possible usages, for example about indoor or outdoor applications or if there's a required level of automation needed for the system. Alongside its rapid growth, IoT seems not to be fully exploited yet to achieve important results in increasing the arable lands and minimizing human intervention. From the technological point of view, some MCUs are common, and this can be of great help since allows developers to rely in an already existing, worldwide, community of coders.

4 Conclusions

A reliable and fault-tolerant safety system must always be considered as a core process of every remote-control system for any agricultural activity. From the described state of art, the following aspects are the most relevant:

- Since safety-related wireless applications are intended to allow remote control or to prevent possible accidents in which a remote-control system is involved, the operator of the remote control must be in a predefined range that allows visual inspection of what the vehicle is doing;
- Mobile phone applications can be of great help in sending automatic alerts in case of accidents, same goes with Bluetooth low energy devices for people nearby farming operations. On the other hand, they often require a mobile internet connection to work properly;
- For platooning systems of autonomous tractors which follow a leading manned tractor, the operator on the leading tractor must supervise the operation of the whole system;
- Use of ZigBee networks increases the feasibility of Wi-Fi networks in harsh conditions, but as goes for LoRa networks, an increased range comes with lower bandwidths;
- Unmanned Aerial Systems can play a key role in improving connectivity among machines working on large areas and in areas with lack of a good GSM mobile connection;
- Internet of Things can be a valuable instrument for easy and low-cost measuring systems, but low-energy devices can get the upper hand in agricultural activities where MCUs can only rely on batteries;
- Machine learning and deep learning AI algorithms could enhance obstacle recognition and avoidance with both laser sensors and image processing, but it must be taken into account that they get "trained" on specific conditions or with data coming from a specific site, which means they might stop working outside of the domain in which they have been trained. That issue can be irrelevant for those companies that make equipment for a small number of local customers, but of course for such

activities AI could represent an expensive solution. For bigger companies AI would not be expensive, but on the contrary the algorithms might provide different results in different areas.

Further developments should consider low-energy modules, energy-harvesting devices, and even passive modules, in association with a reliable wireless infrastructure for longer ranges. Artificial intelligence can also prove to be decisive for safety applications through image processing and signal analysis, but algorithms must be trained over a reasonable amount of data.

References

1. Okuyama, S., Fukuda, M., Ito, K., Tanaka, S.: The radio-controlled tractor from Kubota. SAE Technical Papers, p. 871638 (1987). https://doi.org/10.4271/871638
2. Song, A., Huang, B.K., Bowen, H.D.: Simulating a powered model wheel-tractor on soft ground. Trans. Am. Soc. Agric. Eng. **32**(1), 2–11 (1989)
3. Ayers, P.D.: ROPS design for pre-ROPS tractors. J. Agromed. **4**(3–4), 309–311 (1997). https://doi.org/10.1300/J096v04n03_15
4. Lutz, T.J., Homce, G.T.: Remote controls on an agricultural tractor for performing ASAE/SAE field upset tests. In: 2000 ASAE Annual International Meeting, Technical Papers: Engineering Solutions for a New Century, vol. 2, pp. 5371–5379 (2000)
5. Lutz, T.J., Homce, G.T.: Remote control of an agricultural tractor in SAE field upset tests. Int. J. Veh. Des. **34**(3), 286–296 (2004). https://doi.org/10.1504/IJVD.2004.004013
6. Torvi, D.A.: Fire protection in agricultural facilities: a review of research, resources and practices. J. Fire Prot. Eng. **13**(3), 185–215 (2003). https://doi.org/10.1177/1042391503033368
7. Nielsen, J.D., Bendtsen, J.D., Nielsen, K.M., Sørensen, M.: A conceptual framework for design of embedded systems and data communication for autonomous vehicles. In: IFAC Proceedings Volumes (IFAC-PapersOnline), vol. 37, no. 8, pp. 980–985 (2004)
8. Kise, M., Zhang, Q., Noguchi, N.: An obstacle identification algorithm for a laser range finder-based obstacle detector. Trans. Am. Soc. Agric. Eng. **48**(3), 1269–1278 (2005)
9. Katupitiya, J., Eaton, R., Cole, A., Meyer, C., Rodnay, G.: Automation of an agricultural tractor for fruit picking. In: Proceedings - IEEE International Conference on Robotics and Automation, vol. 2005, pp. 3201–3206 (2005). https://doi.org/10.1109/ROBOT.2005.1570603
10. Hoover, B., Driskill, D.: Design and implementation of model agricultural tractors for the use of safety education. In: 2005 ASAE Annual International Meeting (2005)
11. Autermann, L., Kluge, A.: Agriculture machinery in the data network - Requirements regarding communication | Die Landmaschine im Datennetzwerk - Anforderungen an die Kommunikation. In: Conference: Agricultural Engineering - Land-Technik 2008: Landtechnik regional und International, pp. 387–395 (2008)
12. Quendler, E., Diskus, C., Pohl, A., Buchegger, T., Beranek, E., Boxberger, J.: Child safety driver assistant system and its acceptance. J. Agromed. **14**(2), 82–89 (2009). https://doi.org/10.1080/10599240902779394
13. Zhang, X., Geimer, M., Grandl, L., Kammerbauer, B.: Method for an electronic controlled platooning system of agricultural vehicles. In: ICVES 2009 - 2009 IEEE International Conference on Vehicular Electronics and Safety, pp. 156–161 (2009). https://doi.org/10.1109/ICVES.2009.5400187

14. Larsen, J.J., Nadimi, E.S., Hermansen, P.: Development of a novel safety system for remote plant nursing robots. In: ASABE - 7th World Congress on Computers in Agriculture and Natural Resources 2009, WCCA 2009, pp. 304–309 (2009)
15. Diglys, D.: The use of characteristic features of wireless cellular networks for transmission of GNSS assistance and correction data. In: Proceedings of the International Conference on Information Technology Interfaces, ITI, pp. 141–146 (2010)
16. Zhu, Q., Zhu, D., Su, X.: Distributed remote temperature monitoring and acquisition system based on CAN bus. In: 2010 Prognostics and System Health Management Conference, PHM 2010 (2010). https://doi.org/10.1109/PHM.2010.5413439
17. Scorcioni, S., Bertacchini, A., Dondi, D., Larcher, L., Pavan, P., Mainardi, G.: A vibration-powered wireless system to enhance safety in agricultural machinery. In: IECON Proceedings (Industrial Electronics Conference), pp. 3510–3515 (2011). https://doi.org/10.1109/IECON.2011.6119877
18. Magagnotti, N., Pari, L., Spinelli, R.: Re-engineering firewood extraction in traditional Mediterranean coppice stands. Ecol. Eng. **38**(1), 45–50 (2012). https://doi.org/10.1016/j.ecoleng.2011.10.006
19. Liu, B., Koc, A.B.: SafeDriving: a mobile application for tractor rollover detection and emergency reporting. Comput. Electron. Agric. **98**, 117–120 (2013). https://doi.org/10.1016/j.compag.2013.08.002
20. Liu, B., Koc, A.B.: Mobile app for tractor rollover detection and emergency notification. In: American Society of Agricultural and Biological Engineers Annual International Meeting 2013, ASABE 2013, vol. 6, pp. 4416–4424 (2013)
21. Liu, B., Koc, A.B.: Field tests of a tractor rollover detection and emergency notification system. J. Agric. Saf. Health **21**(2), 113–127 (2015)
22. Trester, R.: Yamaha RMax remote piloted helicopter for agricultural use. In: AUVSI Unmanned Systems 2013, vol. 2, pp. 1768–1776 (2013)
23. Gubiani, R., Pergher, G., Cividino, S.R.S., Lombardo, R., Blanchini, F.: A Wii-controlled safety device for electric chainsaws. J. Agric. Eng. **44**, 690–693 (2013). https://doi.org/10.4081/jae.2013.(s1):e138
24. Cividino, S.R.S., et al.: An improved safety device for electric chainsaws. Contemp. Eng. Sci. **8**(25–28), 1229–1244 (2015). https://doi.org/10.12988/ces.2015.56180
25. Ehlers, S.G., Field, W.E.: Assistive viewing technology to improve agricultural safety. In: American Society of Agricultural and Biological Engineers Annual International Meeting 2014, ASABE 2014, vol. 7, pp. 5361–5385 (2014)
26. Bertacchini, A., Napoletano, G., Dondi, D., Larcher, L., Pavan, P.: Optimized energy-aware wireless system for identification of the relative positioning of articulated systems in the free space. IEEE Sens. J. **14**(5), 1682–1692 (2014). https://doi.org/10.1109/JSEN.2014.2302214
27. Oksanen, T.: Accuracy and performance experiences of four wheel steered autonomous agricultural tractor in sowing operation. In: Mejias, L., Corke, P., Roberts, J. (eds.) Field and Service Robotics. STAR, vol. 105, pp. 425–438. Springer, Cham (2015). https://doi.org/10.1007/978-3-319-07488-7_29
28. Larbi, P.A., Vong, C.N., Karkee, M.: A study of operator performance for a mechanical sweet cherry harvester: comparison between manual and remote-controlled operation. J. Agric. Saf. Health **21**(3), 145–157 (2015)
29. Gregorio, E., Rocadenbosch, F., Sanz, R., Rosell-Polo, J.R.: Eye-Safe lidar system for pesticide spray drift measurement. Sensors **15**(2), 3651–3670 (2015). https://doi.org/10.3390/s150203650

30. Martin-Gorriz, B., et al.: Dynamic simulation modelling of agricultural tractors with automatic deployment of ROPS. In: 2016 American Society of Agricultural and Biological Engineers Annual International Meeting, ASABE 2016 (2016). https://doi.org/10.13031/aim.20162380636
31. Bertacchini, A., Larcher, L.: Wireless sensor systems with energy harvesting capabilities for safety enhancement in agricultural vehicles. In: Proceedings of the IEEE International Conference on Industrial Technology, pp. 1950–1955, May 2016. https://doi.org/10.1109/ICIT.2016.7475065
32. Liu, J., Yuan, J., Cai, J., Tao, C., Wang, L., Cheng, W.: Autopilot system of agricultural vehicles based on GPS/INS and steer-by-wire. Nongye Gongcheng Xuebao/Trans. Chin. Soc. Agric. Eng. **32**(1), 46–53 (2016). https://doi.org/10.11975/j.issn.1002-6819.2016.01.006
33. Bleazard, T., Pate, M.L., Lawver, R.G., Israelsen, C., Hatch, R.: Farmers' perceptions on the utility of auto-guidance technology to improve operator safety. In: 2016 American Society of Agricultural and Biological Engineers Annual International Meeting, ASABE 2016 (2016). https://doi.org/10.13031/aim.20162457078
34. Oetzel, K.: Tablet app to control safety critical functions on farming machines, vol. 2017, no. 2300 (2017)
35. Zhao, Y., Qi, H., Xu, W., Li, J., Zhao, J.: Research on master-slave controlling remote operation of live working robots in agricultural distribution network. In: Proceedings of 2018 IEEE 3rd Advanced Information Technology, Electronic and Automation Control Conference, IAEAC 2018, pp. 1401–1405 (2018). https://doi.org/10.1109/IAEAC.2018.8577683
36. Ehlers, S.G., Tormoehlen, R.L.: Agricultural security monitoring and safety alert system: implementation of wireless video on the farmstead. J. Agric. Saf. Health **25**(4), 155–168 (2019)
37. Shyamala Devi, M., Suguna, R., Joshi, A., Bagate, R.: Design of IoT blockchain based smart agriculture for enlightening safety and security. In: Somani, A.K., Ramakrishna, S., Chaudhary, A., Choudhary, C., Agarwal, B. (eds.) ICETCE 2019. CCIS, vol. 985, pp. 7–19. Springer, Singapore (2019). https://doi.org/10.1007/978-981-13-8300-7_2
38. Muminov, A., Na, D., Lee, C., Kang, H.K., Jeon, H.S.: Modern virtual fencing application: monitoring and controlling behavior of goats using GPS collars and warning signals. Sensors **19**(7) (2019). https://doi.org/10.3390/s19071598
39. Liu, S., Guo, L., Webb, H., Ya, X., Chang, X.: Internet of things monitoring system of modern eco-agriculture based on cloud computing. IEEE Access **7**, 37050–37058 (2019). https://doi.org/10.1109/ACCESS.2019.2903720
40. Wang, L., Wang, Y., Jiangjiang, Z., Wang, X., Wang, S.: Research on boundary recognition and extraction method of field operation area based on UAV remote sensing images. IFAC-PapersOnLine **52**(30), 231–238 (2019). https://doi.org/10.1016/j.ifacol.2019.12.527
41. Chou, H.Y., Khorsandi, F.: Developing and testing an autonomous all-terrain vehicle to experimentally test rollover incidents. In: 2019 ASABE Annual International Meeting (2019). https://doi.org/10.13031/aim.201901456
42. de Souza, P.S.S., et al.: Detecting abnormal sensors via machine learning: an IoT farming WSN-based architecture case study. Meas. J. Int. Meas. Confed. **164** (2020). https://doi.org/10.1016/j.measurement.2020.108042
43. Eeshwaroju, S., Jakkula, P., Ganesan, S.: Smart stick an IoT based product idea for farmers and senior citizens. In: 2020 International Conference on Computing and Information Technology, ICCIT 2020 (2020). https://doi.org/10.1109/ICCIT-144147971.2020.9213751

44. Kloibhofer, R., Kristen, E., Davoli, L.: LoRaWAN with HSM as a security improvement for agriculture applications. In: Casimiro, A., Ortmeier, F., Schoitsch, E., Bitsch, F., Ferreira, P. (eds.) SAFECOMP 2020. LNCS, vol. 12235, pp. 176–188. Springer, Cham (2020). https://doi.org/10.1007/978-3-030-55583-2_13
45. Kloibhofer, R., Kristen, E., Ameri, A.: LoRaWAN with HSM as a security I prove-ment for agriculture applications - Evaluation. LNCS, vol. 12853 (2021)
46. Khalil, R.A., Saeed, N., Masood, M., Fard, Y.M., Alouini, M.-S., Al-Naffouri, T.Y.: Deep learning in the industrial Internet of Things: potentials, challenges, and emerging applications. IEEE Internet Things J. **8**(14), 11016–11040 (2021). https://doi.org/10.1109/JIOT.2021.3051414
47. Zhang, J., Xing, L., Cheng, G., Wang, Y., Han, S.: Diffusion model simulation of odor pollutants in livestock and poultry farms based on environmental monitoring of the Internet of Things. In: E3S Web of Conferences, vol. 261 (2021). https://doi.org/10.1051/e3sconf/202126103018
48. Bhale, P., Biswas, S., Nandi, S.: LIENE: lifetime enhancement for 6LoWPAN network using clustering approach use case: smart agriculture. In: Krieger, U.R., Eichler, G., Erfurth, C., Fahrnberger, G. (eds.) I4CS 2021. CCIS, vol. 1404, pp. 59–75. Springer, Cham (2021). https://doi.org/10.1007/978-3-030-75004-6_5
49. Colantoni, A., Monarca, D., Laurendi, V., Villarini, M., Gambella, F., Cecchini, M.: Smart machines, remote sensing, precision farming, processes, mechatronic, materials and policies for safety and health aspects. Agriculture **8**(4) (2018). MDPI AG. https://doi.org/10.3390/agriculture8040047
50. Sarker, V.K., Queralta, J.P., Gia, T.N., Tenhunen, H., Westerlund, T.: A survey on LoRa for IoT: integrating edge computing. In: 2019 4th International Conference on Fog and Mobile Edge Computing, FMEC 2019, pp. 295–300 (2019). https://doi.org/10.1109/FMEC.2019.8795313
51. Antony, A.P., Leith, K., Jolley, C., Lu, J., Sweeney, D.J.: A review of practice and implementation of the Internet of Things (IoT) for smallholder agriculture. Sustainability **12**(9), 1–20 (2020). https://doi.org/10.3390/su12093750
52. Farooq, H., Rehman, H.U.R., Javed, A., Shoukat, M., Dudley, S.: A review on smart IOT based farming. Ann. Emerg. Technol. Comput. **4**(3), 17–28 (2020). https://doi.org/10.33166/AETiC.2020.03.003

Noise, Vibration, Dust, Endotoxin, Microorganism

Exposure to Heavy Metals in Wood Dust During Dry-Pruning in Vineyard

Pietro Gallo[1], Sabina Failla[2], Marcello Biocca[1(✉)], Enrico Paris[1], Francesco Gallucci[1], Laura Fornaciari[1], and Giampaolo Schillaci[2]

[1] CREA, Consiglio per la ricerca in agricoltura e l'analisi dell'economia agraria, Centro Ingegneria e Trasformazioni Agroalimentari, Monterotondo, Rome, Italy
marcello.biocca@crea.gov.it
[2] Department of Agriculture, Food and Environment (Di3A), University of Catania, Catania, Italy

Abstract. The initial part of winter pruning of the vineyard, also called dry pruning, is carried out mechanically during the period of vegetative rest of the vines. The operation exposes the operators to floating particles of wood, which is classified as a carcinogenic agent. Therefore, a study aimed to investigate the workers exposure was carried out in vineyards located in Sicily (Italy), in five lots situated on the north face of Mt Etna (CT). The wood dust (both total suspended particles and respirable fraction) was sampled by active filter sampling carried out by means of personal air samplers. The collected wood dust underwent to gravimetric and chemical analysis, including heavy metals. Apart from the fact that the obtained values of inhalable dust do not exceed the legal threshold of concentration, the first results suggest that attention should be paid to the exposure of the operators to wood dust generated during mechanical pruning of vineyards. Potential risks for the health from longer times of exposure should be carefully evaluated.

Keywords: Particulate matter · Safety · Airborne particles · Mechanical pruning

1 Introduction

Agricultural and forestry workers are routinely exposed to high levels of airborne dust particles that have been linked to adverse health outcomes. Multiple possible sources, can generate inhalable inorganic or organic dust during different agricultural practices. For example soil particles can be generated during soil tillage (Smith and Lee 2003; Baker et al. 2005), during harvesting (Pagano et al. 2015) or it can be raised from unpavented roads adjacent to fields and houses (Watson et al. 2000). Other sources include the abrasion dust from seeds coated with pesticides which are spread during sowing operations (Biocca et al. 2015), powdery pesticides products (Dowling and Seiber 2002) including sulfur to combat fungal diseases (Griffith et al. 2015), storage of agricultural crops and livestock rearing, especially in broiler and chicken farms (Costa et al. 2012).

M. Biocca et al. (Eds.): SHWA 2020, LNCE 252, pp. 207–215, 2022.
https://doi.org/10.1007/978-3-030-98092-4_22

Wood is also a significant dust source. Coltural such as tree felling, pruning and chipping wood dust can be produced and inhaled by agricultural and forestry workers.

Exposure to wood dust may cause respiratory and dermal symptoms and diseases. The most serious problem araising from wood dust exposure is the risk of developing cancer, mainly nose and sinus adenocancer (Marchi et al. 2017). The International Agency for Research on Cancer (IARC) categorized dust from wood as carcinogenic for humans (IARC 1995).

The inhalation of wood dust depends on particle fraction size, movement of air around the body and respiratory rate. The greatest danger to respiratory organs is the respirable component with a particle size of less than 10 μm (Očkajová et al. 2018).

The European Directive 2004/37 on the "Protection of workers against risks deriving from exposure to carcinogens or mutagens at work", classifies as carcinogenic works involving exposure to hardwood dust and establishes an occupational exposure limit (OEL) of 5 mg m^{-3} (value already indicated in the European Directive 1999/38) measured over a period of 8 h as an inhalable fraction. The Scientific Commission for Occupational Exposure Limits (SCOEL) of the European Union indicates that occupational exposures to wood powders above 0.5 mg m^{-3} induce pulmonary effects and should therefore be avoided. In particular, the value of the SCOEL, coming from an elaboration of the exposures existing in the past, refers to the total dust since only recently has the concept of inhalability been introduced. Using the correlation factor to convert the total powder exposure to inhalable of 2–3 mg m^{-3}, the SCOEL in 2003 adopts an OEL of 1–1.5 mg m^{-3}. The countries of the European community have implemented the 2004 Community Directive and the recommendations of the SCOEL in a different way. In the U.S.A., for example, both the ACGIH (American Conference of Governmental Industrial Hygienists) and the NIOSH (National Institute for Occupational Safety and Health) have adopted a more restrictive limit of 1 mg m^{-3}.

The Directive 2004/37 has been amended by Directive 2017/2398, to be implemented by 17 January 2020, lowering the limit to 2 mg m^{-3} with a transition of five years from 16 January 2018 (entered into force) to 17 January 2023 during which the limit shall be 3 mg m^{-3}.

In recent years there has been increasing effort to understand the nature of the health impacts of different types of particles that originate from different emission sources (Hime et al. 2018).

Particulate matter also contains a great variety of elements, with a very different percentage distribution between particles of both natural (wood - Alagić et al. 2016; volcanos - Calabrese et al. 2016 and sea - Perrino 2010) and anthropic origin.

Among the elements contained in the particulate matters, heavy metals take on particular importance. Exposure limits for arsenic (Ar), lead (Pb), cadmium (Cd) and nickel (Ni) are set by the Italian laws DM 60/2002 and D.Lgs. 152/2007.

The occupational exposure to wood dust is largely investigated during operations carried out indoor, mainly related to the sawing of fresh and stored wood and in the furniture industry. However, exposure can occur also during a number of different agricultural practices generally carried out in open air. Some studies have addressed forest operators' exposure, mainly taking into account the respirable wood dust fraction in chainsaw operation (Marchi et al. 2017; Dimou et al. 2020) or chipping operation (Magagnotti et al. 2013; Fornaciari et al. 2015; Bisaglia and Romano 2018).

The winter pruning of the vineyard, also called dry-pruning or pre-pruning, is performed annually during the period of vegetative rest of the vines. This topic results less investigated, but previous tests carried out during pruning, showed large variability in the extent of potential exposure to inhalable dust wood (Bonsignore et al. 2010).

Pre-pruning has the following purposes: to ensure the maintenance of the shape and size of the individual vines to facilitate all crop operations, adjust the load of gems by stock, choose the best gems in relation to their production capacity, distribute the gems optimally on each vine, obtain the renewal vegetation at the desired points, achieve equilibrium between the productive and vegetative phase, the desired quality and the lignification branches (Baldini and Intrieri 1984; Palliotti et al. 2018). With this work, a large part (80–95%) of the wood produced during the year is removed and it is commonly believed that this operation has a beneficial effect on the vineyard and on its growth.

Winter pruning should be performed in the period between the fall of the leaves and the beginning of the vegetative growth; it is therefore an intervention that can be performed over a long period of time (Baldini and Intrieri 1984; Palliotti et al. 2018).

In the pruned-spur cordon-trained form, which is the most widespread training form in Italian vineyards, it is possible to perform mechanical pruning that requires 3–5 h per hectare, lowering the costs compared to the manual one and improving the operator's weel-being (Romano et al. 2019).

The mechanical pruning is carried out by machines equipped with cutting members that ram the branches with 2–3 gems, while advancing at high speeds of about 1.4–1.8 m s^{-1} (5.0–6.4 km h^{-1}). In general, mechanical pruning is adopted when the mechanization allows to contain the manual work below of the 22–24 h ha^{-1} threshold (Castaldi et al. 2013; Tomasi et al. 2012).

Mechanical pruning is now widespread and has also been adopted by leading companies whose quality level is internationally recognized.

Faced with these advantages, workers engaged in these pruning are exposed to noise and vibration stresses and to the hazardous effects of floating particles of wood dust. During mechanical pre-pruning, a large amount of dust is produced so that some workers reported that the tractor air filter could be occluded by the dust deposits (personal communications).

In this context, the work aims to investigate the workers exposure from wood dust coming from dry-pruning of vineyards situated in Castiglione di Sicilia land (Catania province, Sicily) on the north face of Mt Etna.

2 Material and Methods

2.1 Test Sites

The trials were carried out in five lots located in three farms situated on the north face of Mt. Etna (Fig. 1 and Table 1).

The vineyards (cultivar Nerello mascalese) were trained as pruned spur cordon.

All the tests were carried out in January 2019 in absence of wind.

Table 1. Features of air sampling in the experimental plots

Test N.	Site code	Sampling head	Normal volume sampled (L)	Sampling time (min)	Row lenght (m)
1	TOR_1_A	Dorr-Oliver	111.3	66	1088
2	TOR_2_A	TSP	181.0	60	1088
3	TOR_3_B	TSP	250.7	83	4450
4	TOR_4_B	Dorr-Oliver	137.6	83	4450
5	TOR_5_C	Dorr-Oliver	146.8	87	3730
6	TOR_6_C	TSP	266.4	88	3730
7	FIR_1	Dorr-Oliver	101.0	60	3360
8	FIR_2	TSP	182.8	60	3360
9	RAC_1	Dorr-Oliver	100.0	59	2625
10	RAC_2	TSP	182.6	60	2625
Average			**166.02**	**70.6**	**3050.6**

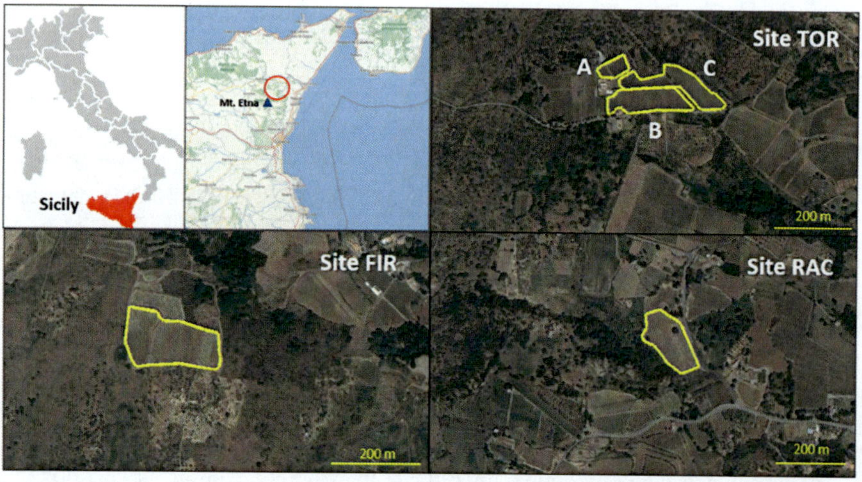

Fig. 1. Localization of the experimental plots.

2.2 Machine

The pre-pruning machine (Fig. 2) was a straddling machine with a double stack of 9 blades (discs) joined to an inverted U-shaped carrying framework. Height and width regulation, as well as the opening of the two columns are controlled by oleodynamic cylinders commanded by electrovalves. The machine was mounted on a 55 kW double traction tractor. The forward speed was maintained constant to 4 km h^{-1}. The velocity of blades was measured with an optical tachometer and it was equal to 14.1 rad s^{-1}.

Fig. 2. The pruning machine (1); the position of the personal air sampler with the Dorr-Oliver sampling head (2) and the personal air sampler for the TSP, placed on the tractor, close to the operator seat (3).

2.3 Dust Sampling

The wood dust total suspended particles (TSP), and respirable fraction (4 µm) were sampled by active filter sampling carried out by means of two air samplers (Zambelli, mod. Ego Plus). The first personal sampler (Fig. 2), equipped with a Dorr-Oliver head for respirable fraction, was fixed on the chest of the operator driving the tractor, and operated at a constant flow of 1.7 L min^{-1}. A second sampler (for TSP) (Fig. 2) was placed on the tractor, near the cab, and it operated at a constant flow of 3.0 L min^{-1}. The samplers were equipped with 0.45 µm PTFE Millipore diskette filters.

The mass concentration of the dust collected on the PTFE filters was determined by gravimetry using an analytical balance (Gibertini Elettronica) (sensibility 1 µg). All filters were conditioned for 48 h at 50% relative humidity before weighing (US EPA 2016).

After the tests, samples lignified shoots were collected in the same vineyards and all samples (filters and shoots) underwent to chemical analysis.

2.4 Chemical Analyses

Chemical analyses were carried out at the LASER-B lab of CREA IT.

The metal content was determined on both biomass and filters used in personal samplers. Both types of samples were pretreated in a microwave mineralization process. To prepare the biomass and filter samples, an aliquot of each sample was transferred to special Teflon containers and subjected to acid attack using a microwave digester (Start D, Milestone, Italy). The determination of the metal content was carried out by ICP-MS (Inductively Coupled Plasma–Mass Spectrometer 7700 Agilent). The etching mixture used was composed of 6 ml of HNO_3 (65% v/v) and 3 ml of H_2O_2

(30% v/v). The calibration of the instrument was carried out using multi-element standards (Standard mix, concentration 10 ppm in metal), prepared in aqueous solution acidified with 1% of HNO_3 (65% v/v).

3 Results and Discussion

The values of both respirable and TSP dust air concentrations during the dry-pruning of the vineyards are shown in Fig. 3. The average concentration value of respirable particles (0.9 mg m^{-3}) and TSP (1.2 mg m^{-3}) were similar; both data showed some variability (respirable dust ranged from a minimum of 0.5 to a maximum of 1.7 mg m^{-3}, and TSP ranged from 0.9 to 2.0 mg m^{-3}).

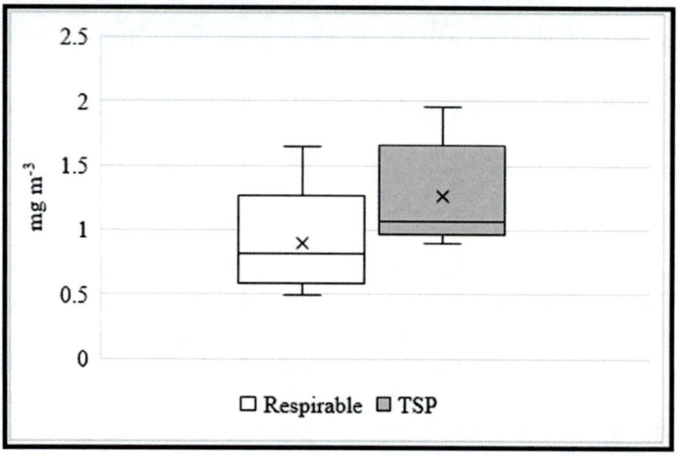

Fig. 3. Respirable and TSP (Total Suspended Matter) (number of samples = 10) (box limits represent: 25 and 75 percentiles; line within box: median; cross within the box: mean; whisker ends: minimum and maximum).

As regard the chemical content of the fresh wood (Fig. 4), the abundance of the analyzed elements in grape shoots decreased in the patterns as follows: K > Mg > Sr > Al > Fe > Mn > Ga > Ba > Cu > Ni > Tl > Li > Pb > Cr > As > Ag > Cd > Bi > Co > Be. In all sites the chemical composition of the wood samples appears similar, with the exception of Nickel, found only in one plot (1.8 mg kg^{-1}), and Aluminum that shows some variability.

The chemical content of the dust is shown in Fig. 5. Data refer to 9 filters because a sample (TOR_6_C) was considered an outlier, showing values of the order of tens or hundreds of times higher than the average, depending on the analyzed elements.

Metal content data appear very variable. In particular, the samplings coded TOR_1 and FIR_2 showed the highest concentration of metals.

The air concentrations of the metals Ar, Pb and Ni (those with a legal exposure limit) are below the limit.

In all dust particle sizes, the descending pattern of analyzed elements is the follow: Al > Mg > K > Ga > Fe > Sr > As > Ni > Cu > Cr > Co > Pb > Mn > Li > e. Therefore, with respect to the wood composition, it is possible to note that an enrichment of some elements (i.e. Al) occurs in the air particles.

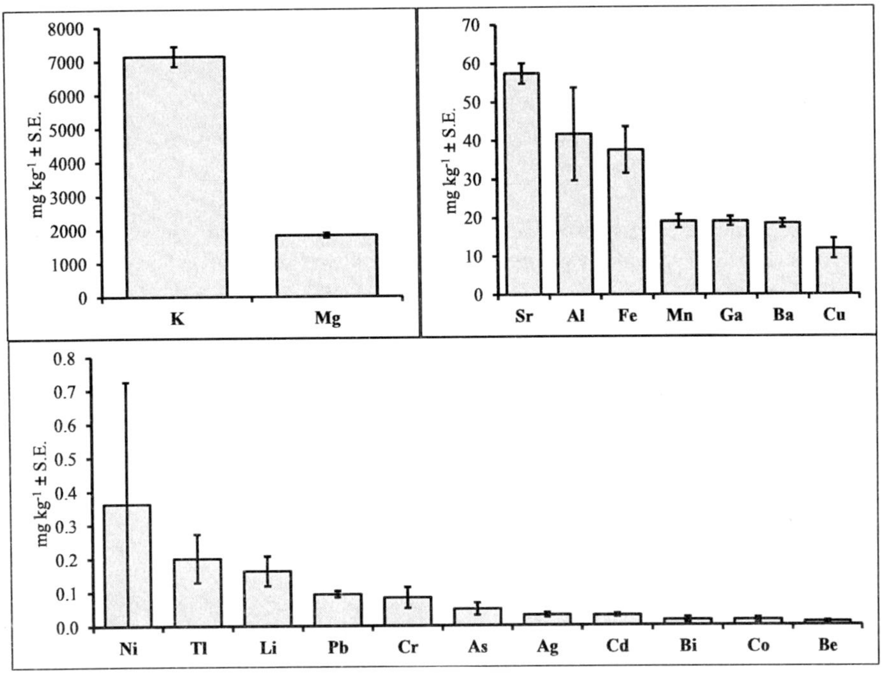

Fig. 4. Grape fresh wood element content (mg kg^{-1} ± standard error). Number of samples = 5.

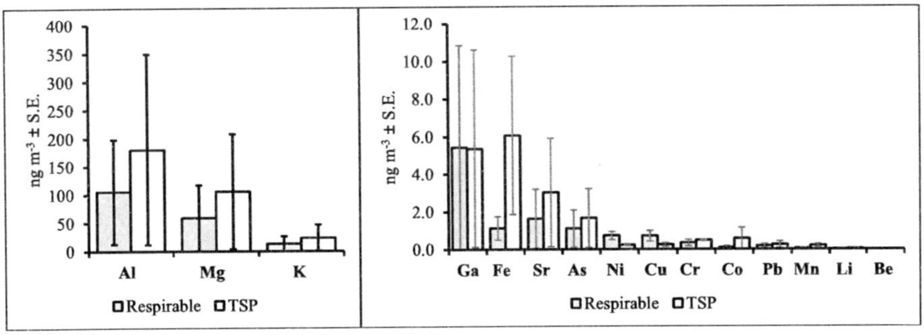

Fig. 5. Elements content in respirable and TSP wood particulate matter (ng m^{-3} ± standard error). Number of samples = 9.

4 Conclusions

This study contributed to evaluate the potential exposure of operators to air-borne particulate matter originated during the mechanical dry-pruning of vineyards. Original findings on the chemical composition of the collected dust were given.

In general, data showed some variablity according to the sampling site. In fact, many variables affect the air concentrations of particulate matter, such as: the number of sources emitting the agents (tractors, operators, wood from pruning), the production speed in relation to the production capacity, the type and position of the source and the dispersal of the agents due to air movements. Furthermore, the variables are connected with individual actions and behaviors like, for example, the closeness of the person to the sources and the duration of the task.

Obtained values of inhalable wood dust do not exceed the European legal threshold of concentration, as well as heavy metals are below the legal limits set by Italian law. Nevertheless, attention should be paid to the exposure to wood dust generated during mechanical pruning of vineyards also considering other concurrent sources of potential inhalable dust (atmosphere). Potential risks due to longer times of exposure should be carefully evaluated.

Chemical content reflects the volcanic soil composition typical rich in metals and shows high variability according to the site of sampling.

Further investigations will clear the contribution to the air-borne particulate matter of the emissions generated by the volcano Etna and of the possible contribution of raised soil particles.

References

Alagić, S.Č, Tošić, S.B., Dimitrijević, M.D., Petrović, J.V., Medić, D.V.: The characterization of heavy metals in the grapevine (*Vitis vinifera*) cultivar Rkatsiteli and wild blackberry (*Rubus fruticosus*) from East Serbia by ICP-OES and BAFs. Commun. Soil Sci. Plant Anal. **47**(17), 2034–2045 (2016)

Baker, J.B., Southard, R.J., Mitchell, J.P.: Agricultural dust production in standard and conservation tillage systems in the San Joaquin Valley. J. Environ. Qual. **34**(4), 1260 (2005)

Baldini, E., Intrieri, C.: Meccanizzazione della vendemmia e della potatura. Clueb, Bologna (1984)

Biocca, M., Fanigliulo, R., Gallo, P., Pulcini, P., Pochi, D.: The assessment of dust drift from pneumatic drills using static tests and in-field validation. Crop Prot. **71**, 109–115 (2015)

Bisaglia, C., Romano, E.: Utilization of vineyard prunings: a new mechanization system from residues harvest to CHIPS production. Biomass Bioenerg. **115**, 136–142 (2018)

Bonsignore, R., Romano, E., Caruso, L., Schillaci, G.: Wood dust production and VOC (volatile organic compound) emissions during mechanical pre-pruning in vineyards. In: Proceedings of "International Conference Ragusa SHWA2010. "Work Safety and Risk Prevention in Agro-food and Forest Systems", Ragusa Ibla Campus, Italy, 16–18 September 2010, pp. 360–367 (2010)

Calabrese, S., et al.: Mount Etna volcano (Italy) as a major "dust" point source in the Mediterranean area. Arab. J. Geosci. **9**(3), 1–12 (2016). https://doi.org/10.1007/s12517-015-2165-0

Castaldi, R., et al.: La potatura meccanica rivoluziona la gestione del vigneto. L'Informatore Agrario **3**, 61–65 (2013)

Costa, A., Ferrari, S., Guarino, M.: Yearly emission factors of ammonia and particulate matter from three laying-hen housing systems. Anim. Prod. Sci. **52**(12), 1089–1098 (2012)

Dimou, V., Malesios, C., Chatzikosti, V.: Assessing chainsaw operators' exposure to wood dust during timber harvesting. SN Appl. Sci. **2**(11), 1–11 (2020). https://doi.org/10.1007/s42452-020-03735-6

Dowling, K.C., Seiber, J.N.: Importance of respiratory exposure to pesticides among agricultural populations. Int. J. Toxicol. **21**(5), 371–381 (2002)

Hime, N.J., Marks, G.B., Cowie, C.T.: A comparison of the health effects of ambient particulate matter air pollution from five emission sources. Int. J. Environ. Res. Public Health **15**(6), 1206 (2018)

IARC: Wood dust and formaldehyde. IARC Monogr. Eval. Carcinog. Risks Hum. **62**, 1–405 (1995)

Magagnotti, N., Nannicini, C., Sciarra, G., Spinelli, R., Volpi, D.: Determining the exposure of chipper operators to inhalable wood dust. Ann. Occup. Hyg. **57**(6), 784–792 (2013)

Marchi, E., et al.: Analysis of dust exposure during chainsaw forest operations. iForest **10**, 341–347 (2017)

Fornaciari, L., et al.: Noise, vibration and dust emissions of a forestry chipper. In: Failla, S. (ed.) Proceedings of International Conference Ragusa SHWA "Safety Health and Welfare in Agriculture Agro-food and Forestry Systems", Lodi, Italy, 8–11 September 2015, pp. 99–106 (2015)

Griffith, C.M., Woodrow, J.E., Seiber, J.N.: Environmental behavior and analysis of agricultural sulfur. Pest Manag. Sci. **71**(11), 1486–1496 (2015)

Očkajová, A., Kučerka, M., Krišťák, L., Igaz, R.: Granulometric analysis of sanding dust from selected wood species. BioResources **13**(4), 7481–7495 (2018)

Pagano, M., et al.: Dust emissions with a new hazelnut mechanical harvesting prototype. In: Failla, S. (ed.) Proceedings of International Conference Ragusa SHWA "Safety Health and Welfare in Agriculture Agro-food and Forestry Systems", Lodi, Italy, 8–11 September 2015, pp. 112–116 (2015)

Palliotti, A., Poni, S., Silvestroni, O.: Manuale di Viticoltura. Edagricole, Bologna (2018)

Perrino, C.: Atmospheric particulate matter. Biophys. Bioeng. Lett. **3**(1), 35–43 (2010)

Romano, E., Caruso, L., Longo, D., Vitale, E., Schillaci, G., Rapisarda, V.: Investigation of hand forces applied to a pruning tool – pilot study. Ann. Agric. Environ. Med. **26**(3), 472–478 (2019)

Smith, L.J., Lee, K.: Soil as a source of dust and implications for human health. Adv. Agron. **80**, 1–32 (2003)

Tomasi, D., et al.: Mechanical pruning, no pruning and manual pruning: effects on grape composition and health status of Pinot g. and Cabernet s. varieties in the "Piave" AOC area of Veneto Region. In: 1st International Workshop on Vineyard Mechanization and Grape and Wine Quality, Piacenza, 27–29 June (2012)

US EPA: Quality Assurance Guidance Document 2.12, Monitoring PM2.5 in Ambient Air Using Designated Reference or Class I Equivalent Methods. U.S. E.P.A., Office of Air Quality, Triangle Park (2016)

Watson, J.G., Chow, J.C., Pace, T.G.: Fugitive dust emissions. In: Davis, W.T. (ed.) Air Pollution Engineering Manual, 2nd edn., pp. 117–135. Wiley, New York (2000)

Dynamic Characteristics of the Seats Equipping Old Agricultural Tractors

Christian Preti[1], Roberto Deboli[1], and Angela Calvo[2](✉)

[1] STEMS (Institute of Sciences and Technologies for Sustainable Energy and Mobility of C.N.R, Italian National Research Council), Strada delle Cacce 73, 10135 Turin, Italy
christian.preti@stems.cnr.it

[2] DISAFA (Department of Agricultural, Forest and Food Sciences and Technologies), Largo P. Braccini 2, 10095 Grugliasco, Turin, Italy
angela.calvo@unito.it

Abstract. The seat is the device that minimize the vibration transmitted to the tractor driver's body by the tractor platform. All seats for agricultural machines require homologation before their installation on the machine, in accordance with the Council Directive 78/764/EEC (Annex XIV), amended with the Delegated Regulation 1322/2014 and with the Delegated Regulation 830/2018. In Italy there is still a large number of underused tractors (less than 200 h per year) that are more than 30 years old. Low cost upgrades are often necessary, even though the seat is an expensive item. The first worn out element of the seat is usually the damper (or shock absorber), but when replaced by the users, only the dimension is considered, not its damping properties. Aim of the work was to perform resonance tests to investigate the dynamic characteristics of seats using different type of dampers. Tests were conducted in laboratory with six different types of dampers and with two different static masses: 40 and 80 kg. The results showed that the seat suspension was nonlinear, with the transmissibility mostly depending on the amount of the inertial mass. The distribution of the vibration energy was also different, in function of the rigidity of the shock absorber.

Keywords: Vibration · Seat · Agricultural machines · Damper

1 Introduction

In Italy many farmers (even though with little farms of few hectares) have their own underused tractor (less than 200 h per year), that is often more than 30 years old [1].

Tractor upgrades are therefore necessary (also due to the new regulations), but many farmers do not have money enough for the extraordinary maintenance of these machines, often uncomfortable, especially for the transmitted whole-body vibration (WBV) to the driver [1].

The vertical vibration mainly arises from the interaction of the tires with the ground irregularities and it is filtered through the platform to the seat.

Agricultural tractors are equipped with suspension seats, that provide a substantial reduction of the vibration enhancing the driver's comfort. The work of the seat in the

M. Biocca et al. (Eds.): SHWA 2020, LNCE 252, pp. 216–222, 2022.
https://doi.org/10.1007/978-3-030-98092-4_23

reduction of the vibration is important, because the human body is very sensitive to the low frequency whole body vibration (WBV) [2].

On the other hand, the seat of an agricultural tractor is a wearing and expensive element of a tractor, and it has a lower duration than the machine (around the half lifetime) [3]. There are cheap seat models in the market, but many farmers prefer to repair the seat on their own and to substitute the worn elements. They are not often able to correctly find the correct spare parts and the result is an apparently fixed seat, that probably does not correctly work (especially in term of vibration attenuation).

One of the most wearable component of a seat is the damper (or shock absorber), that controls and reduces the frequency resonance.

The spring-damper system is not hermetically sealed under the seat, and for this reason it is exposed to dust that compromise its functionality, and for this reason it must be periodically substituted, because the farmer has any system to verify the integrity of the component. The substitution of a damper is not however a simple task. The farmers, when obliged to the substitution of the shock absorber, considers only the physical dimension of the piece to be replaced, and they don't care of its damping characteristics, even though the wrong choice of the damper may compromise the seat functionality.

The goal statement of this work was to perform resonance tests to investigate the dynamic characteristics of a seat using different type of dampers. The vibration transmission of the suspension seat along the vertical direction was examined with different seat configurations and loading conditions (as requested by the current standards, the Council Directive 78/764/EEC, Annex XIV, [4], amended with the Delegated Regulation 1322/2014 [5] and with the Delegated Regulation 830/2018 [6]).

Tests were conducted in laboratory with six different types of dampers and with two different static masses: 40 and 80 kg. The signal was transmitted to the hydraulic vibrating bench (where the seat was placed) and the frequency weighted accelerations were simultaneously measured on the bench platform and on the seat. The Seat Effective Amplitude Transmissibility (SEAT) was calculated and discussed.

2 Material and Methods

2.1 Tested Dampers

In Fig. 1 there is an open damper, and are here visible the piston, the cylinder (where the piston moves) and the little hole where the oil flows through the cylinder from one chamber to another.

The diameters of the holes of the dampers tested in this work were: 0.1 mm; 0.13 mm; 0.15 mm; 0.18 mm; 0.21 mm and 0.25 mm.

The volumes of chambers, the length and the diameter of the pistons were the same for all the dampers.

The adjustment of the mass positioned on the seat (the seat pan must be positioned in the middle of the travel of the suspension system) was obtained turning the knob located in the upper part of the suspension.

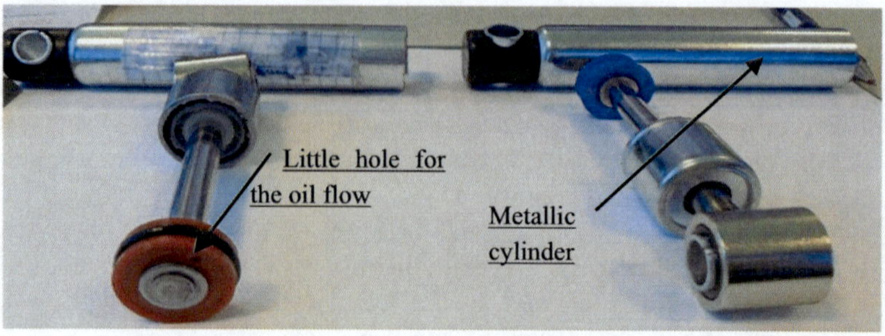

Fig. 1. An open damper showing the little hole for the oil flow.

2.2 Test Conditions

The tests were conducted in accordance with the Council Directive 78/764/EEC (Annex XIV), [4], amended with the Delegated Regulation 1322/2014 [5], and with the Delegated Regulation 830/2018 [6].

The damping test includes two steps: the first is research the resonant frequency of the suspension; the second is the calculation of the transmissibility of the suspension at the resonant frequency.

During each test, the seat was loaded with two ballasts: 40 kg and 80 kg, and they were secured to the seat in order to prevent the movement of the mass.

The seat was always adjusted in function of the weight of the ballast and of the damper type to allow the full working suspension stroke in accordance with the manufacturer's instruction.

Sinusoidal oscillations at different frequencies (±15 mm of amplitude) were generated to the bench to find out the resonance frequency. The applied frequency started from 0.5 Hz up to the highest frequency where the seat did not hit the limit stop pads. The constant frequency interval was 0.1 Hz, both increasing and decreasing the frequency.

After determining the resonant frequency of the suspension, the transmissibility at resonance frequency has been calculated.

2.3 Instrument Characteristics

The bench, located at the STEMS CNR laboratories in Turin, is able to simulate the vertical vibration in the driver's seat fixing point. Vibration is generated by an electro-hydraulic device (MTS System Corp.; Digital Controller Model Flex test 40 and 793 System Software, 2019).

Two ICP accelerometers PCB (356B41 model, 100 mV g^{-1} sensitivity, 10 g mass) were used. An accelerometer was placed on the seat surface and fixed on a flat rubber plate in accordance with the ISO 10326-1:2016 [7]. Another accelerometer was fixed to the bench near the seat to measure the vibration at the seat attachment of the bench.

The measurement chain was calibrated after and before each test.

The acquired signals were stored on a laptop using a National Instruments data acquisition card (NI9234) and later processed using the LabView software.

2.4 Data Acquisition, Transmissibility and Resonance

The seat and the bench acquired acceleration were weighted in accordance with the Delegated Regulation 1322/2014 [5] (Appendix 6).

The r.m.s. value (the square root of the mean square) of the weighted seat vibration acceleration (awS) and the r.m.s. value of the weighted vibration acceleration at the seat attachment on the bench (awB) were simultaneously acquired. The measurement of each frequency was approximately 60 s.

The efficiency of each damper was evaluated by the seat effective amplitude transmissibility (SEAT) value (Eq. 1). The SEAT was calculated at intervals of 0.1 Hz, in the frequency range from 0.5 Hz to the highest frequency where the seat did not hit the limit stop pads (all the tested damper reached 2 Hz, some dampers reached 2.5 Hz, in function of their characteristics).

$$SEAT = \frac{a_{wS}}{a_{wB}} \tag{1}$$

where:

SEAT: seat effective amplitude

a_{wS}: weighted seat acceleration (ms^{-2})

a_{wB}: weighted vibration acceleration at the seat attachment on the bench (ms^{-2})

The seat transmissibility at the resonance frequency was the greatest transmissibility in the measurement interval. The resonance frequency was the frequency at which the transmissibility was the greatest.

3 Results and Discussion

3.1 Data Obtained on the Bench

The observation of the accelerations obtained on the bench is the starting point to evaluate the damper response in function of the vibration input and it simulates what happens on the platform of the agricultural tractor. In our test the accelerations always increased with the augmentation of the frequency (with a growing increase rate), independently by the mass of the ballast.

3.2 Data Obtained on the Seat

Accelerations on the seat were quite different among them, depending on the damper type. High differences were also observed in function of the ballast (Fig. 2).

The acceleration growth is inversely proportional to the diameter of the hole in the damper: this is mostly evident with the ballast of 80 kg (Fig. 2).

The steepness is not always rising: with the ballast of 40 kg, the decrease is mostly evident in the holes of 0.25 mm and 0.18 mm after 1.5 Hz. With the 80 kg ballast, the slope decreases after 1.5 Hz in all the dampers, except the smallest (damper holes of 0.1 and 0.13 mm).

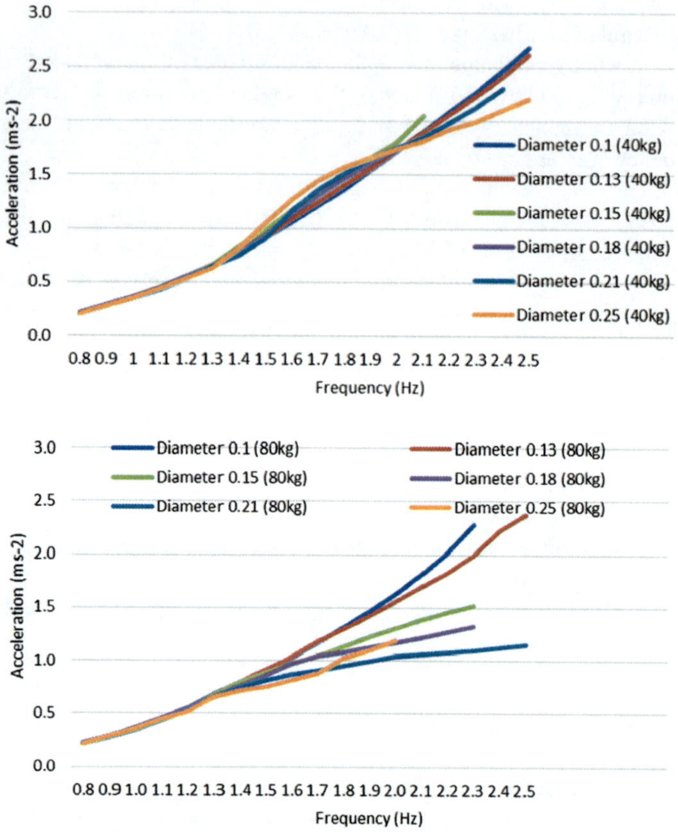

Fig. 2. r.m.s. accelerations obtained on the seat with the ballast of 40 kg (top) and 80 kg (bottom) for different diameters of the damper holes.

The examined dampers do not linearly increase in function of the frequency, as also observed by Qiu [8], and it probably depends on both the properties of the suspension components (as the foam rubber of the padding) and the characteristics of the structure of the suspension.

3.3 Transmissibility and Resonance

The seat effective amplitude transmissibility (SEAT) parameter represents the efficiency of the seat isolation, and therefore the efficacy of the damper. A SEAT value lower than 1 means that the seat reduces the vibration coming from the platform.

Figure 3 shows critical data present with the 40 kg ballast, where the resonance appears at 1.6, 1.7 and 2 Hz (this unique case is produced by the 0.1 mm hole damper).

With the 80 kg ballast, the resonance is at 1.2–1.4 Hz, in line with Qiu (2017), except the 0.1 mm hole damper (Fig. 3). Dampers work better with the heavier ballast, except the damper with the 0.1 mm hole.

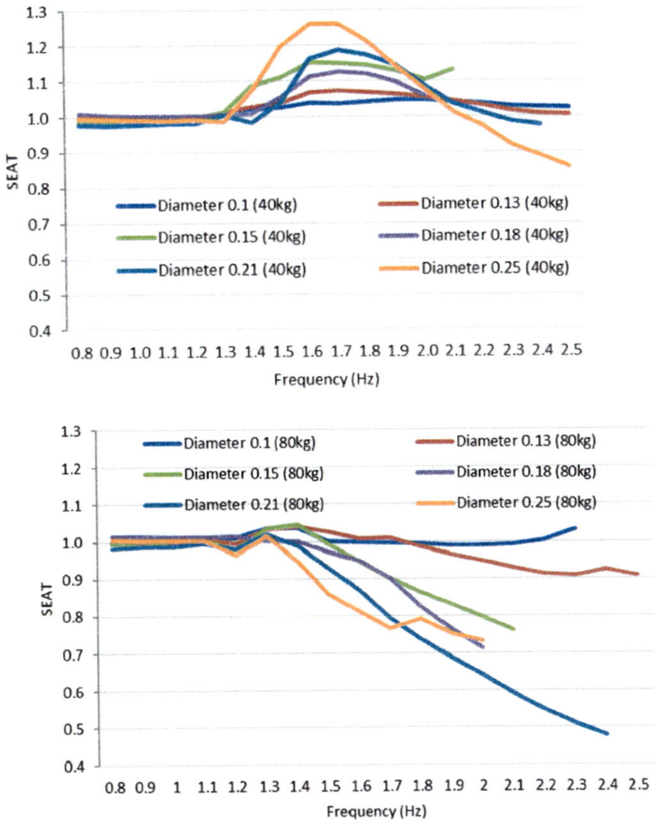

Fig. 3. SEAT values with the 40 kg (top) and the 80 kg ballast (bottom), for different diameters of the damper holes.

Higher is the diameter of the hole damper, lower is the SEAT value at the frequencies higher than 2 Hz.

4 Conclusions

The dampers of the seat of an agricultural tractor have usually a shorter lifetime than the seat, and they need to be replaced when the seat hits the top and the bottom end stop while driving also over a smooth surface.

As many farmers are not able to understand the importance of the correct substitution of the shock absorber, the risk is to replace whatever the damper is. Unfortunately, not all dampers are the same, as found in this study.

The results showed also that the seat suspension had a nonlinear trend in function of the frequency, with the transmissibility depending on the amount of the inertial mass. The distribution of the vibration energy showed different results on the basis of the rigidity of the shock absorber.

As well as it is important to train the agricultural tractor drivers to a correct weight adjustment and damper setting, it is otherwise important to sensitize the farmers for the correct maintenance of the shock absorber of the seat of their tractors.

Acknowledgement. Many thanks to the seat manufacturer COBO, for providing the material for the laboratory tests.

References

1. ENAMA (Ente NAzionale Meccanizzazione Agricola) Homepage. www.enama.it. Accessed 18 June 2021
2. Griffin, M.J.: Handbook of Human Vibration. Academic Press, London (1990)
3. CEMA: Whole-body vibration in agriculture. Practical User's Guide. Homepage. www.cema-agri.org. Accessed 29 June 2021
4. Council Directive 78/764/EEC of 25 July 1978 on the approximation of the laws of the Member States relating to the driver's seat on wheeled agricultural or forestry tractors (1978)
5. Commission Delegated Regulation (EU) No. 1322/2014 of September 19, Supplementing and Amending Regulation (EU) No. 167/2013 of the European Parliament and of the Council with Regard to Vehicle Construction and General Requirements for the Approval of Agricultural and Forestry Vehicles (2014)
6. Commission Delegated Regulation (EU) 2018/830 of 9 March 2018 amending Annex I to Regulation (EU) No 167/2013 of the European Parliament and of the Council and Commission Delegated Regulation (EU) No 1322/2014 as regards the adaptation of the vehicle construction and general requirements for the approval of agricultural and forestry vehicle (2018)
7. ISO 10326-1: Mechanical vibration. Laboratory method for evaluating vehicle seat vibration. Basic requirements, International Standard, Geneva (2016)
8. Qiu, Y.: Dynamic characteristics of a suspension seat determined in laboratory study. J. Ergon. **7**(6), 1–6 (2017)

Hand Arm Vibration: Comparison Between Laboratory and in Field Tests

Emanuele Cerruto, Giuseppe Manetto(✉), and Rita Papa

Department of Agriculture, Food and Environment, University of Catania, Via Santa Sofia, 100, 95123 Catania, Italy
giuseppe.manetto@unict.it

Abstract. Vibration is the most important and known risk connected with the use of modern portable harvesters for olive harvesting. At the same time, acceleration measurement is a quite complex task, being affected by several factors, very difficult to standardize. In this research one commercial machine was used and acceleration values were measured both in laboratory, using a special test bench purposely developed, and in field, under ordinary working conditions. The results showed that laboratory tests produced on average acceleration values 24% higher than field tests, especially near the hand-grip, and wider data dispersion. Both results may depend on the depth with which the harvester head is inserted into the test bench, which in turn affects the resistance encountered by the sticks. So, further tests are necessary to better control these aspects in such a way to develop a device useful to standardize the measurement conditions for these harvester machines.

Keywords: Safety · Olive harvesting · Vibration exposure · Hand-arm system · Test bench

1 Introduction

Vibration exposure of agricultural workers is one of the most important topics concerning safety and comfort. The Italian National Institute for Insurance against Accidents at Work (INAIL) points out that in the period between 2012 and 2017, among the occupational diseases caused by physical agents, the largest increase was recorded for lumbar herniated disc. Among the causes of this pathology, in addition to the manual handling of loads, there is the transmission of vibration to the whole body.

The problem is exacerbated by the increase in mechanization level of many activities recognized in the last decades and may regard both the whole body (Marsili et al. 2002; Servadio et al. 2007; Cerruto et al. 2010; Pessina et al. 2012; Cutini et al. 2017; Deboli et al. 2017) and the hand-arm system (Monarca et al. 2008; Pascuzzi et al. 2009; Dewangan and Tewari 2009; Çakmak et al. 2011; Bernardi et al. 2018; Aiello et al. 2019; Manetto et al. 2020).

The transmission of the vibrations through the hand-arm system is consequent to the use of vibrating tools equipped with handles; it is a rather complex phenomenon because involves several factors (biodynamic characteristics, mechanical impedance of the hand, posture, muscle tension, hand forces, handle size, use of isolators), which interact with

M. Biocca et al. (Eds.): SHWA 2020, LNCE 252, pp. 223–232, 2022.
https://doi.org/10.1007/978-3-030-98092-4_24

the intensity of the vibration and its way of introduction and propagation in the organism (Dong et al. 2005; Aldien et al. 2006; Besa et al. 2007; Concettoni and Griffin 2009; Deboli and Calvo 2009; Tewari and Dewangan 2009; Adewusi et al. 2013).

Hand-Arm Vibration (HAV) is probably the most important risk connected with the use of portable harvesters, whose spread, especially in the olive sector, is due to reduction in harvesting costs (Aiello et al. 2019). Users are mainly interested in productivity and very often underestimate ergonomics and safety aspects. Regular exposure to hand-transmitted vibration can result in symptoms and signs of peripheral vascular, neurological and other disorders collectively known as the hand-arm vibration syndrome (HAVS) (Åstrom et al. 2006; Griffin 2008; Azmir et al. 2015; Calvo et al. 2018).

Acceleration values reported in the literature present great variability due to the many factors influencing the measures (type of machine, worker features and ability, tree characteristics), so it can be of great interest to standardize the measuring conditions, especially under load (Deboli et al. 2014; Lenzuni et al. 2016). In this study, using the same equipment described in Deboli et al. (2014), one electrical machine was tested both in laboratory and in field by two users and acceleration values were compared.

2 Material and Methods

2.1 The Harvester

The portable harvester used for the experimental tests is driven by an electric motor powered by an external 12 V DC battery. Its harvester head carries the motor and 2 groups of 4 carbon fiber sticks, each one 250 mm long and with diameter of 5 mm (Fig. 1). The running frequency, according to the manufacturer specifications, is about 1400 beats per minute under load conditions; an electronic card and a software control the rotation speed of the motor, reducing its value to about 400 rpm in idling condition, so to decrease energy requirements, noise and vibration level. The aluminum-made rod is telescopic, with lengths of 1.55–2.60 m and diameters of 25–28 mm.

2.2 The Test Bench

A suitable test bench was used in laboratory to standardize load conditions for olive portable harvesters. It consists of a square frame 50 × 50 cm with a mesh of 5 × 4 cm (vertical × horizontal) synthetic threads (Fig. 2). The horizontal threads must be tensioned manually and then locked by means of clamps, while the vertical ones are pulled by 1 kg masses. This system ensures an almost constant resistance to the sticks of the harvester head, simulating the canopy action. The frame is applied to a wood structure, made stable by means of suitable weights on its base. The working height is from 140 cm to 190 cm above the ground, so allowing testing portable harvesters with an angle of the rod of about 45°, as during the harvesting of olives from trees.

Fig. 1. The harvester head.

Fig. 2. The test bench.

2.3 The Experimental Activity

In laboratory, to simulate the harvesting action as on an effective olive tree, the sticks of the harvester head were inserted into the frame of the test bench at a depth of about 8 cm (Fig. 3) and, after 5–10 s, they were extracted and then re-inserted in a different position of the frame. Instead, in the field the tests were carried out simulating the harvesting on an olive tree (Fig. 4). In both working conditions, the machine was operated by two workers, tested only with the minimum rod length (1.55 m) and 5 replications, each lasting 1 min, were carried out.

Vibration measurements were carried out by using three mono axial accelerometers DJB Instruments Ltd (United Kingdom), model A/123/S, screwed on to the mutually orthogonal faces of a small cube tied to the rod by a metallic clamp (Fig. 5). The reference axes were selected according to the basicentric coordinate system defined by the ISO 5349-1:2001 regulation: x-axis perpendicular to the palm surface area, y-axis parallel to the longitudinal axis of the rod, and z-axis directed along the third metacarpus bone of the hand.

Fig. 3. The laboratory test.

Fig. 4. The in field test.

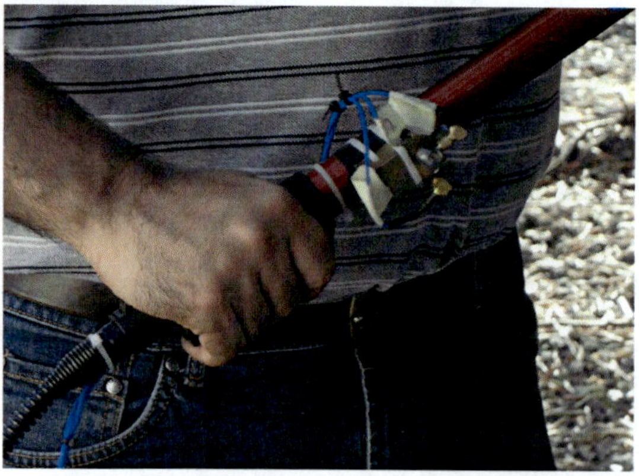

Fig. 5. The accelerometers applied to the rod.

Considering that the operator continuously changes the hand position on the rod during the work, 4 measurement points were chosen in laboratory tests to assess how the vibration level changes along the rod: M1 near the hand-grip, M2 at 50 cm from M1, M3 at 85 cm from M1 and M4 at 100 cm from M1 (Fig. 6). Instead, in field, taking into account the results of the laboratory tests, the measurements were carried out only in M1 and M3.

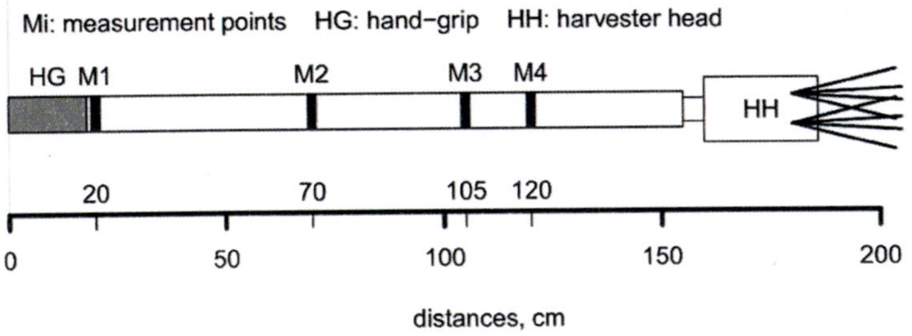

Fig. 6. Schematic view of the portable harvester with the measurement points on the rod.

The accelerometer signals were acquired with a PC by means of a dB4 four channel acquisition unit and recorded in the hard disk by means of the recording module of the dBFA Suite software (01 dB-Metravib, France). Subsequently all the recorded signals were analyzed by using the post-processing module of the same dBFA Suite software, that allows for several post-processing analyses, among which narrow band analysis (FFT), 1/3 octave analysis, and frequency weighting for the hand-arm system according to the ISO 5349-2:2001 regulation.

The signals were analyzed in the range 5.6–1400 Hz (third of octave bands from 6.3 to 1250 Hz) by applying the 1/3 octave analysis. The frequency weighted RMS (Root Mean Square) accelerations were computed for each axis (a_{hwx}, a_{hwy}, and a_{hwz}) and then the global acceleration a_{hw} was calculated as:

$$a_{hw} = \sqrt{a_{hwx}^2 + a_{hwy}^2 + a_{hwz}^2}, \tag{1}$$

from which the daily vibration exposure value, $A(8)$, standardized to an 8-h reference period was obtained:

$$A(8) = a_{hw}\sqrt{\frac{T}{T_o}}, \tag{2}$$

being $T_o = 8$ h and T the total exposure time (hour) associated with a_{hw}.

All acceleration data were statistically analyzed to detect differences related to measurement points and/or to working conditions (laboratory vs in field tests) and/or to operators. All computations and graphical representations were performed by means of the open source software R (R Development Core Team 2019).

3 Results and Discussion

Figure 7 summarizes mean values of global weighted acceleration with respect to measuring point, test condition, and operator. As a first consideration, the results showed no operator influence: on average, acceleration was 13.3 m/s^2 for the operator A and 13.1 m/s^2 for the operator B. Results then showed an increasing trend in acceleration values when the measuring point moved from M1 to M4 (from the hand-grip to the harvester head). The lowest value (about 11.6 m/s^2) was measured in correspondence of the hand-grip, the highest (about 15.1 m/s^2) on the rod, near the harvester head. Considering the test condition, acceleration values measured in laboratory using the test bench were on average significantly higher than those measured in field during harvesting (14.1 vs 11.3 m/s^2). However, considering the interaction between measuring point and test condition, acceleration values measured in laboratory were not significantly different in the measuring points (ranging from 13.3 in M2 to 15.1 m/s^2 in M4), whereas during harvesting they were significantly higher in M3 (13.6 m/s^2) with respect to M1 (9.1 m/s^2). This result, quite common in the literature (Çakmak et al. 2011; Manetto and Cerruto 2013; Calvo et al. 2014), implies that, during harvesting, the two hands are differently exposed and then safety aspects should be assessed for the most exposed hand (left hand for a right-handed operator, positioned near the measuring point M3 in this study).

The daily vibration exposure value $A(8)$, assuming a daily exposure time of 4 h and considering the most exposed hand (13.6 m/s^2), was on average 9.6 m/s^2. This value must be compared with the daily exposure action value of 2.5 m/s^2 and the daily exposure limit value of 5.0 m/s^2 established by the European Directive 2002/44/EC in order to implement proper safety actions (European Commission 2002).

Data variability in laboratory tests was higher than in field tests (Fig. 8). In fact, the coefficient of variation ranged from 5.3 to 6.7% in field tests and from 13.6 to 19.3% in laboratory tests. This result could be due to the thickness of the vegetation that reduces the influence of the different length of the sticks inserted into the canopy. On the contrary, in laboratory tests the sticks encounter a resistance that depends on the depth with which the harvester head is inserted into the test bench: the higher the depth, the lower the resistance and vice versa.

The acceleration component analysis showed that the lowest values were measured along the y-axis in both laboratory and field tests (Fig. 9). This was consistent with the kinematics of the harvester machine: there were no alternating movements along the y-axis. In addition, being the sticks applied eccentrically to the longitudinal axis of the rod, this produced higher acceleration values along the x-axis.

The greater variability of data in laboratory tests also concerned the acceleration components, especially along the x-axis, where in M1 values higher than in field were measured. The coefficient of variation ranged from 5.2 to 12.5% in field and from 14.3 to 37.8% in laboratory.

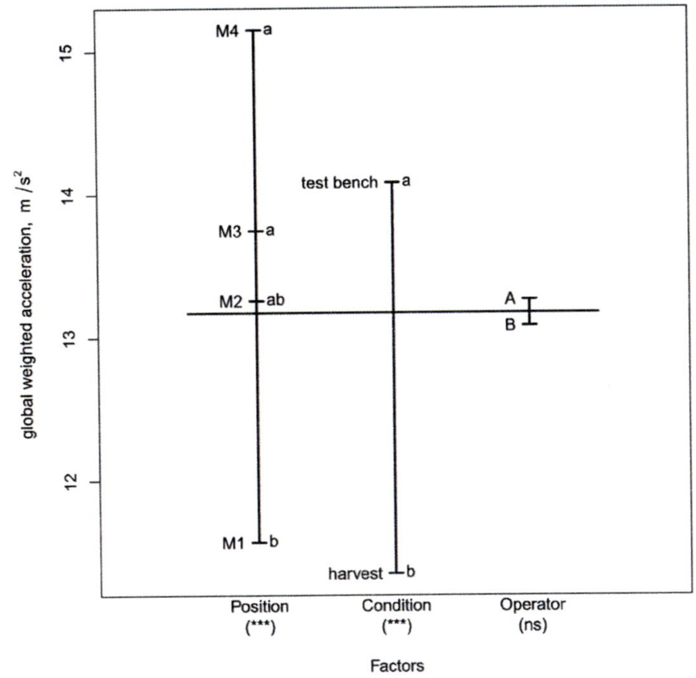

Fig. 7. Mean values of global weighted acceleration (mean separation by Tukey HSD test at $p = 0.05$).

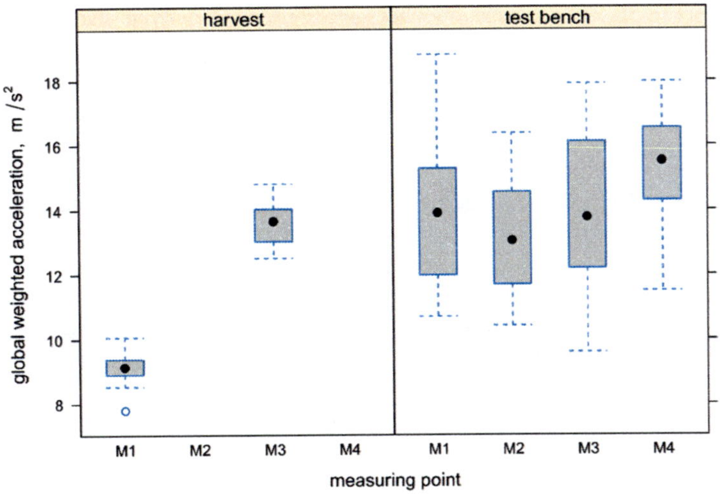

Fig. 8. Box plot of acceleration values in laboratory and in field tests.

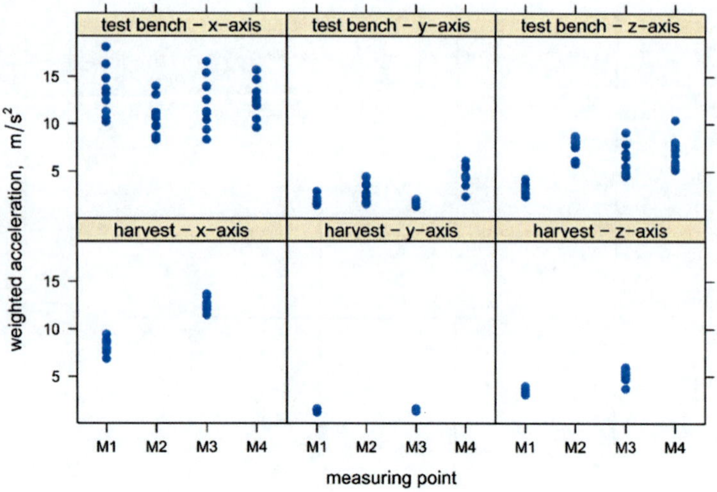

Fig. 9. Acceleration components in laboratory and in field tests.

4 Conclusions and Perspectives

The main results of the present study allow drawing the following conclusions:

- The studied harvester machine exposes the two hands to different acceleration values in field tests, confirming the results obtained in other studies with different types of harvester. The A(8) value, computed for the most exposed hand, exceeds the limit value of 5.0 m/s^2 established by the European Directive 2002/44/EC, so appropriate precautions must be taken during its use to prevent hand-arm vibration disorders.
- On average, acceleration values measured in laboratory were significantly higher than those measured in field (ratio equal to 1.24). In addition, laboratory data presented dispersion wider than field data. Both results may depend on the depth the harvester head is inserted into the test bench, which in turn affects the resistance encountered by the sticks: the higher the depth, the lower the resistance and vice versa.
- Further studies are needed to better simulate the effects of the thickness of the canopy or to ensure, during the laboratory tests, a constant insertion of the machine sticks into the test bench. A different approach, being developed at our Department, aims to calibrate the resistance of the wires of the test bench by means of springs of appropriate stiffness.

Acknowledgements. This research was financed by the University of Catania with the research project "Safe and Smart Farming with Artificial Intelligence and Robotics, WP1: Vibration at the hand-arm system", PIAno di inCEntivi per la Ricerca di Ateneo 2020/2022, Linea di intervento 2: Dotazione ordinaria per attività istituzionali dei dipartimenti.

References

Adewusi, S., Rakheja, S., Marcotte, P., Thomas, M.: Distributed vibration power absorption of the human hand-arm system in different postures coupled with vibrating handle and power tools. Int. J. Ind. Ergon. **43**, 363–374 (2013)

Aiello, G., Vallone, M., Catania, P.: Optimising the efficiency of olive harvesting considering operator safety. Biosys. Eng. **185**, 15–24 (2019)

Aldien, Y., Marcotte, P., Rakheja, S., Boileau, P.-E.: Influence of hand–arm posture on biodynamic response of the human hand–arm exposed to z_h-axis vibration. Int. J. Ind. Ergon. **36**, 45–59 (2006)

Åstrom, C., Rehn, B., Lundström, R., Nilsson, T., Burström, L., Sundelin, G.: Hand-arm vibration syndrome (HAVS) and musculoskeletal symptoms in the neck and the upper limbs in professional drivers of terrain vehicles—a cross sectional study. Appl. Ergon. **37**, 793–799 (2006)

Azmir, N.A., Ghazali, M.I., Yahya, M.N., Ali, M.H., Song, J.I.: Effect of hand arm vibration on the development of vibration induce disorder among grass cutter workers. Procedia Manuf. **2**, 87–91 (2015)

Bernardi, B., Quendler, E., Benalia, S., Mantella, A., Zimbalatti, G.: Occupational risks related to vibrations using a brush cutter for green area management. Ann. Agric. Environ. Med. **25**(2), 255–258 (2018)

Besa, A.J., Valero, F.J., Suñer, J.L., Carballeira, J.: Characterisation of the mechanical impedance of the human hand-arm system: the influence of vibration direction, hand–arm posture and muscle tension. Int. J. Ind. Ergon. **37**, 225–231 (2007)

Çakmak, B., Saraçoğlu, T., Alayunt, F.N., Özarslan, C.: Vibration and noise characteristics of flap type olive harvesters. Appl. Ergon. **42**, 397–402 (2011)

Calvo, A., Deboli, R., Preti, C., De Maria, A.: Daily exposure to hand arm vibration by different electric olive beaters. J. Agric. Eng. XLV, 424, 103–110 (2014)

Calvo, A., Romano, E., Preti, C., Schillaci, G., Deboli, R.: Upper limb disorders and hand-arm vibration risks with hand-held olive beaters. Int. J. Ind. Ergon. **65**, 36–45 (2018)

Cerruto, E., Emma, G., Manetto, G.: Whole body vibrations of tractor driver in citrus orchard. In: International Conference on "Work Safety and Risk Prevention in Agro-food and Forest Systems", Ragusa, Italy, 16–18 September (2010)

Concettoni, E., Griffin, M.: The apparent mass and mechanical impedance of the hand and the transmission of vibration to the fingers, hand, and arm. J. Sound Vib. **325**, 664–678 (2009)

Cutini, M., Brambilla, M., Bisaglia, C.: Whole-body vibration in farming: background document for creating a simplified procedure to determine agricultural tractor vibration comfort. Agriculture **7**, 84 (2017)

Deboli, R., Calvo, A.: The use of a capacitive sensor matrix to determine the grip forces applied to the olive hand held harvesters. Agric. Eng. Int. CIGR J. **11**, 1–10 (2009)

Deboli, R., Calvo, A., Preti, C., Inserillo, M.: Design and test of a device for acceleration reproducibility of hand held olive harvesters. Int. J. Ind. Ergon. **44**, 581–589 (2014)

Deboli, R., Calvo, A., Preti, C.: Whole-body vibration: measurement of horizontal and vertical transmissibility of an agricultural tractor seat. Int. J. Ind. Ergon. **58**, 69–78 (2017)

Dewangan, K.N., Tewari, V.K.: Characteristics of hand-transmitted vibration of a hand tractor used in three operational modes. Int. J. Ind. Ergon. **39**, 239–245 (2009)

Griffin, M.J.: Measurement, evaluation, and assessment of peripheral neurological disorders caused by hand-transmitted vibration. Int. Arch. Occup. Environ. Health **81**, 559–573 (2008)

Dong, R.G., McDowell, T.W., Welcome, D.E., Smutz, W.P.: Correlations between biodynamic characteristics of human hand–arm system and the isolation effectiveness of anti-vibration gloves. Int. J. Ind. Ergon. **35**, 205–216 (2005)

European Commission: European directive of 25 June 2002 on the minimum health and safety requirements regarding the exposure of workers to the risks arising from physical agents (vibration) 2002/44/EC. Off. J., L **177/13**, 13–19 (2002)

ISO (International Organisation for Standardisation): ISO 5349-1:2001. Mechanical vibration - measurement and evaluation of human exposure to hand-transmitted vibration - part 1: general requirements. International Organisation for Standardisation Publications, Geneva, Switzerland (2001)

ISO (International Organisation for Standardisation): ISO 5349-2:2001. Mechanical vibration - measurement and evaluation of human exposure to hand-transmitted vibration - part 2: practical guidance for measurement at the workplace. International Organisation for Standardisation Publications, Geneva, Switzerland (2001)

Lenzuni, P., Deboli, R., Preti, C., Calvo, A.: A round robin test for the hand-transmitted vibration from an olive harvester. Int. J. Ind. Ergon. **53**, 86–92 (2016)

Manetto, G., Cerruto, E.: Vibration risk evaluation in hand-held harvesters for olives. J. Agric. Eng. XLIV(s1), e142, 705–709 (2013)

Manetto, G., Cerruto, E., Papa, R.: Effects of rod and oscillating frequency on the vibrations transmitted to hand-arm system by four olive portable harvesters. In: Coppola, A., Di Renzo, G.C., Altieri, G., D'Antonio, P. (eds.) MID-TERM AIIA 2019. LNCE, vol. 67, pp. 671–679. Springer, Cham (2020). https://doi.org/10.1007/978-3-030-39299-4_73

Marsili, A., Ragni, L., Santoro, G., Servadio, P., Vassalini, G.: Innovative systems to reduce vibrations on agricultural tractors: comparative analysis of acceleration transmitted through the driving seat. Biosys. Eng. **81**(1), 35–47 (2002)

Monarca, D., et al.: Transmission of vibrations from portable agricultural machinery to the Hand-Arm System (HAV): risk assessment and definition of exposure time for daily action and exposure limits. In: International Conference on "Innovation Technology to Empower Safety, Health and Welfare in Agriculture and Agro-food Systems", Ragusa, Italy, 15–17 September (2008)

Pascuzzi, S., Santoro, F., Panaro, V.: Investigation of workers' exposure to vibrations produced by portable shakers. Agric. Eng. Int. CIGR J. **11**, 1–10 (2009)

Pessina, D., Facchinetti, D., Bonalume, V.: Evaluation of vibration levels improves the efficiency of modern tracklaying tractors. J. Agric. Eng. XLIII, e7, 43–47 (2012)

R Core Team: R: a language and environment for statistical computing. R Foundation for Statistical Computing, Vienna (2019). https://www.R-project.org. Accessed 16 May 2021

Servadio, P., Marsili, A., Belfiore, N.P.: Analysis of driving seat vibrations in high forward speed tractors. Biosyst. Eng. **97**, 171–180 (2007)

Tewari, V.K., Dewangan, K.N.: Effect of vibration isolators in reduction of work stress during field operation of hand tractor. Biosys. Eng. **103**, 146–158 (2009)

Fisherman's Exposure to the Noise Emitted by a Prototype of an On-Board Rotating Sorting Machine for Clams (*Chamelea Gallina*)

Laura Fornaciari[1], Daniele Pochi[1], Corrado Costa[1], Giuseppina Mascilongo[2], Marcello Biocca[1], Renato Grilli[1], and Roberto Fanigliulo[1](✉)

[1] Consiglio per la ricerca in agricoltura e l'analisi dell'economia agraria (CREA), Centro di Ricerca Ingegneria e Trasformazioni agroalimentari, Via della Pascolare 16, Monterotondo, 00015 Rome, Italy
roberto.fanigliulo@crea.gov.it
[2] Istituto Zooprofilattico Sperimentale dell'Abruzzo e del Molise (IZSAM) "G. Caporale", Campo Boario, 64100 Teramo, Italy

Abstract. In clam fishing, the efficiency in selecting the commercial size depends on fisherman's ability and on-board equipment functionality. Therefore, it is essential to increase the efficiency of the selection equipment. On this purpose, an innovative sorting machine was realized, less impacting on the clams to be selected, benthic fauna and operators' health. Unlike traditional machines, the new one is not based on vibrations. It consists of two rotating drums provided with holes and spacing of different sizes. A series of spray washes gently cleans the clams removing the sand and the undersized juvenile that are directly thrown back into the sea. Thereby, the stress on the clams is significantly reduced and, at the same time, the conditions of comfort, safety and health protection of the operators are improved. In light of this, in the present work, the emitted levels of noise were measured under normal conditions of use in one day work, as specified by Italian law (Legislative Decree 81/08) and according to the ISO 1999:2013 standard, with both conventional and innovative sorting machine mounted on two boats that fished in parallel on the northern coast of Abruzzo (Adriatic Sea, Italy).

Keywords: Striped clam · Selectivity · Innovative gear · Commercial size · Adriatic sea

1 Introduction

Clams are a high value seafood product in the European Union, and Italy is the main producer (mainly from the Adriatic Sea). In last years, the fishery experienced a drop in catches, and such reduction spurred the Italian government to impose a number of management regulations on equipment characteristics (technical measures), number of fishing days per year (fishing effort or inputs) and daily catches per individual (catches or outputs). In Italy, over 90% of clam catches comes from the central-northern Adriatic Sea, with an annual quantity that currently approaches 50,000 tons. Fishing for

M. Biocca et al. (Eds.): SHWA 2020, LNCE 252, pp. 233–241, 2022.
https://doi.org/10.1007/978-3-030-98092-4_25

striped clams (*Chamelea gallina* L.) by means of hydraulic dredges and the on-board selecting by means of sorting machine are of paramount importance for the local economy in the Adriatic area [1, 2].

The latter operation, due to the high levels of noise emitted by the used equipment, represents a very impacting activity on the health of the fishermen.

Although it has been observed that fishermen have a higher incidence of chronic disease than the general population, their production sector is rarely subjected to health and hygienic surveillance programs. The activity of sea fishing involves a series of unfavourable conditions that can affect the health and safety of workers.

These conditions are linked to various critical factors ranging from the environmental characteristics and effects [3] in which one operates (fishing vessels, weather conditions, etc.) to the various risk factors linked to the particularity of the work cycle (noise, vibrations, ergonomics, fatigue factors, accident risk, etc.), to finish with the need for integration between national and international regulatory sources.

The most important data concerns noise-induced hearing loss. Peretti *et al.* showed that this disease affects almost 2/3 of fishermen, with a significant trend as the length of service increases [4]. Among the categories investigated, it emerges that motorists and captains have a greater risk of developing hearing loss from noise than sailors, due to their proximity to noise sources. It is known, in fact, that the main source of noise inside boats is represented by the engines that produce high noise levels even in fishing boats of less than 30 m in length. Same consideration can be made for agricultural sector, where various research has reported noise-induced hearing loss emitted by tractors engine [5, 6].

Currently, the safety of navigation at national level is regulated by the D.P.R. n. 435/91 [7] "Regulation for the safety of navigation and human life at sea" and by Italian Legislative Decrees 271/99 and 298/99 [8, 9], relating to the protection of the health and safety of seafarers on board merchant and fishing vessels, pending the issuing of provisions that coordinate the regulations laid down by Italian Legislative Decree 81/08 with the shipping sector [10].

In this work, the investigation involved two boats dedicated to the commercial fishing of clams, set up in a different way. The first boat "MP Luigi Padre" was equipped with a traditional turbofan dredger and a traditional three-level vibrating sorting machine for on-board selection of the clams; the second "MP Dora" instead featured an innovative turbo-blower dredger and an innovative sorting machine for the selection of clams.

In order to detect and compare the sound pressure levels emitted by the traditional vibrating sorter and by the innovative rotating sorter, CREA-IT conducted a series of on-board measurements, during normal fishing operations, using specific instrumental chains, in accordance with international standards.

2 Materials and Methods

In striped clam fishing, *Chamelea gallina* (L.), the efficiency in selecting the commercial size depends on the on-board equipment functionality. Therefore, it is essential to increase the efficiency of the sorting equipment. On this purpose, an innovative rotating sorting machine was realized, less impacting on the clams, on the benthic fauna and on the health of the operators (Fig. 1).

Fig. 1. The innovative rotating sorting machine.

Unlike traditional vibrating machines, the new sorter is based on two rotating drums featuring with holes and spacing of different sizes. A series of spray washes gently cleans the clams removing the sand and the undersized juvenile that are directly thrown back into the sea. Thereby, the stress on the clams is significantly reduced and, at the same time, the conditions of comfort, safety and health protection of the operators are improved.

The traditional vibrating sorting machine (Fig. 2) consists of a three-level overlapping sieve featuring decreasing hole diameters, from the top to the bottom, which respectively select the largest clams, the medium ones and those below the allowed size (currently it is not possible to select specimens smaller than 2.2 cm); the latter are directly thrown into the sea. The selection takes place through the grids, described above, whose tilting motion is transmitted by two vibrators on the sides of the machine. The structure of the sorters is entirely made of stainless steel and materials suitable for contact with food products. Both sorters were hydraulically powered.

Fig. 2. The traditional vibrating sorting machine.

A fishing area was repeatedly sampled along the Abruzzo coast (central Adriatic Sea, Italy), around the harbour of Giulianova, on depths ranging from 5 m to 10 m.

Two fishing boats dedicated to the commercial fishing of clams and equipped with hydraulic dredges were used for sampling:

- the MP Luigi Padre (registration 8PC 601, overall length 14.90 m, volume of the boat 60 m^{-3} (gross tonnage: GT = 14) engine power 110.3 kW), mounting the traditional vibrating sorting machine.
- the MP Dora (registration 7PC 366, overall length 15.95 m, volume of the boat 68 m^{-3} (gross tonnage: GT = 16), engine power 110.3 kW), equipped with the innovative rotating sorting machine.

Both vessels were equipped with radar, echosounder and GPS. In order to limit the variability in clam fishing due to differences in the bottom, the operations were always carried out with vessels proceeding in parallel at a distance of at least 50 m from each other.

As usual with hydraulic dredging, the equipment was lowered and operated from the bow of the fishing vessel, which pulled it backwards, keeping the engine at a speed of about 1300 min^{-1}, needed to proceed with a constant speed of 2,96 m s^{-1}. Each haul, except for special needs, lasted 10 min, equal to about 500 m pull. At the end of the haul, the dredge was raised, and all its content unloaded on a "tray", from where a cochlea sent it to the on-board sorter.

The noise measurements were carried out on board, during normal operation of both traditional and innovative sorter, during the selection of the clams. The surveys, on both boats, were carried out during the following phases:

1) during the boat transfer from the port to the fishing spot.
2) during dredge lowering and pulling, with the sorter off, at constant speed of 2,96 m s^{-1}.
3) during dredge lowering and pulling at constant speed of 2,96 m s^{-1}, with the sorter on, during the on-board selection of clams.

In order to characterize the operator's daily exposure to noise risk, three repetitions were carried out for each test thesis, using a sampling time equal to the total duration of the entire operation which includes: the dredge lowering, pulling and subsequent lifting on board; the unloading of the catch (conveyed into the sorter), the washing and sorting of clams.

The noise level was measured according to the methodology reported in the international standard ISO 1999: 2013 [11].

For all measurements performed, the values of the equivalent continuous level of A-weighted sound pressure ($L_{Aeq,\ Ti}$) and the maximum instantaneous C-weighted sound pressure (p_{peak}) were measured and calculated in the frequency range between 20 Hz and 20 kHz, in thirds of an octave. The noise measurements were carried out in the workstation, with the operator positioned at 0.60 m from the clam's sorter. Since the operation requires continuous manual interventions to select the several clam sizes, the presence of the operator in front of the sorter is essential.

The measurement was performed by positioning the microphone at 0.1 m from the left ear of the worker exposed to noise. To simplify the operation, the microphone was fixed to a helmet the operator wore during the tests (Fig. 3).

Once the surveys had been carried out, the daily personal exposure levels ($L_{EX,\ 8h}$) were determined using the formula reported in ISO 1999:2013.

Finally, the maximum usage times were calculated in order not to exceed the threshold values established by Legislative Decree 81/08.

Fig. 3. Clam sorting by means of the innovative rotating machine. The fisherman wears the protective helmet equipped with the microphone.

To measure the noise transmitted to the operator's ear, the following instrumental chain was used:

- Sound level meter (Soundbook n. 06220 standard) - eight-channel signal acquisition/processor with special "Samurai" software (Fig. 4).
- B&K microphone calibrator mod. 4231.
- B&K microphone capsule mod. 4189 with relative extension cable and windscreen.
- helmet with microphone capsule holder.

Before and after the day of measurements, the microphone cap was calibrated using the supplied calibrator verifying that the deviation from the initial calibration level did not exceed 0.3 dB, as required by the standard ISO.

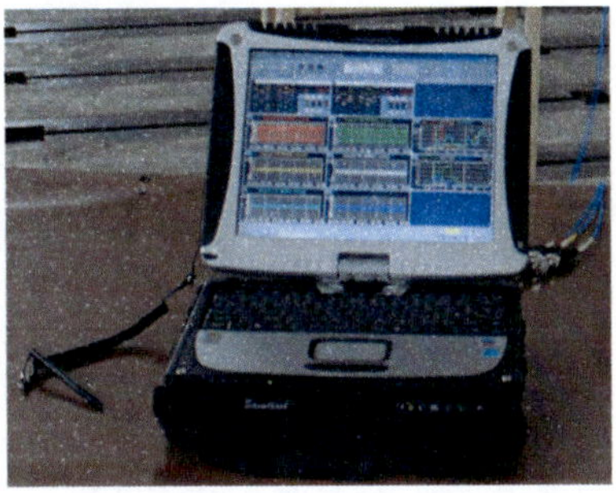

Fig. 4. Soundbook eight-channel signal acquisition/processor with special "Samurai" software.

3 Results

Table 1 shows the values of the equivalent continuous levels of A-weighted sound pressure (L_{Aeq}, $_{Ti}$), and the maximum values of the C-weighted instantaneous sound pressure (p_{peak}), which are always lower than the threshold values established by D. Lgs. 81/08. We can notice that the measured C-weighted maximum sound pressure values (p_{peak}) remain below the threshold of 135 dB (C), while the A-weighted sound pressure levels ($L_{Aeq,Te}$) are higher than the exposure limit value, equal to 87 dB (A), only during the clam sorting phase (sorter ON) with both sorters. Specifically, it should be noted that with the innovative sorter the $L_{Aeq,Te}$ values are significantly lower than with the traditional one.

Table 1. Noise levels at the operator's ear during the three phases of clam fishing.

Sorting machine	Test thesis	L_{Aeq} dB (A)	p_{peak} dB (C)
Traditional	Transfer from the port	80.8	107.3
	Clam fishing (sorter OFF)	84.7	109.5
	Clam fishing (sorter ON)	101.5	118.2
Innovative	Transfer from the port	83.2	114.1
	Clam fishing (sorter OFF)	85.1	111.8
	Clam fishing (sorter ON)	91.5	115.6

As to the other fishing phases, we can observe that during the transfer from the port to the fishing spot the values vary from 80 dB (A) for a fairly new and well maintained boat to 85 dB (A) for a rather old boat. However, under current conditions and assuming professional use equal to 8 h a day, the operator would exceed the exposure limit value. To avoid exceeding the limit value, it is at least necessary to reduce the working time and use the suitable Personal Protective Equipment (PPE). To this end, the threshold values of the times of personal daily exposure to noise were calculated for the different risk bands and reported in Tables 2 and 3.

Table 2. Maximum working daily time for the use of the equipment in order not to exceed the limits in the D. Lgs 81/08. MP Luigi Padre - Traditional sorter.

Test thesis	L_{Aeq}	Limit value 87 dB (A)		$L_{EX,8h}$	Upper action value 85 dB (A)		$L_{EX,8h}$	Lower action value 80 dB (A)		$L_{EX,8h}$
	dB (A)	Hour	Min	dB (A)	Hour	Min	dB (A)	Hour	Min	dB (A)
Transfer from the port	80.8	–	–	–	–	–	–	6	30	79.9
Clam fishing (sorter OFF)	84.7	–	–	–	–	–	–	2	45	80
Clam fishing (sorter ON)	101.5	0	15	86.4	–	–	–	–	–	–

The results show that it is not possible to hypothesize the free use of the traditional sorter, since the short time of use allowed by Legislative Decree 81/08 would be inadequate for the selection of clams, even if a rotation of operators were applied. The use of the PPE and the evaluation of their efficiency prior to purchase is therefore mandatory.

Table 3. Maximum working daily time for the use of the equipment in order not to exceed the limits in the D. Lgs 81/08. MP Dora - Innovative sorter.

Test thesis	L_{Aeq}	Limit value 87 dB (A)		$L_{EX,8h}$	Upper action value 85 dB (A)		$L_{EX,8h}$	Lower action value 80 dB (A)		$L_{EX,8h}$
	dB (A)	Hour	Min	dB (A)	Hour	Min	dB (A)	Hour	Min	dB (A)
Transfer from the port	85.1	–	–	–	8	0	85.0	2	30	80.0
Clam fishing (sorter OFF)	83.2	–	–	–	–	–	–	3	45	79.9
Clam fishing (sorter ON)	91.5	3	0	87.3	1	45	84.9	0	30	79.5

On the other hand, on board the MP Dora, the worker who operates the innovative sorter could work only 30 min remaining below the lower action value of 80 dB (A) and 105 min to remain below the upper action value of 85 dB (A). In this second case, the operator is obliged to undergo annual health surveillance and must use adequate PPE. If he would like to work 3 h a day, the employer, in addition to the previous prescriptions, must subject the worker to training courses and inform him about the risks from noise exposure.

Figures 5 and 6 show the frequency analysis from 20 to 20.000 Hz relating to the noise tests during the on-board selection of clams (sorter ON), and with the contemporary dredge lowering and pulling at constant speed of 2,96 m s^{-1}. The highest noise values are concentrated at low frequencies, especially with the use of the innovative rotating sorting machine. Further tests could be conducted to study why the noise produced by the rotating sorter is concentrated at low frequencies. In Fig. 5, during clam sorting with the innovative sorter, we can also see a significant decay of the noise levels between 1000 and 20.000 Hz, due to the several rotation speed of drums during the several working phases in relation to the worked load of clams.

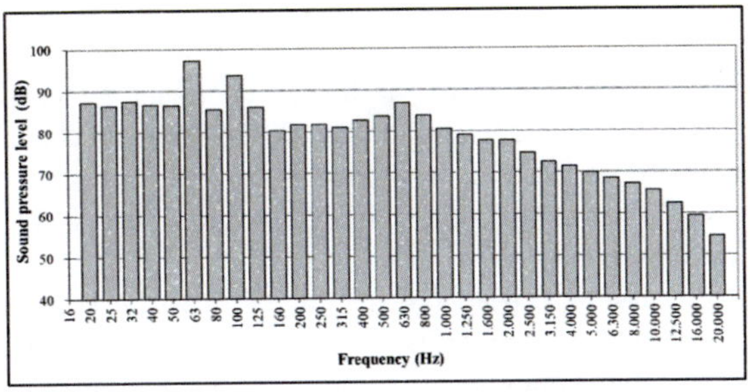

Fig. 5. Frequency analysis during clam sorting with the innovative sorter - MP Dora.

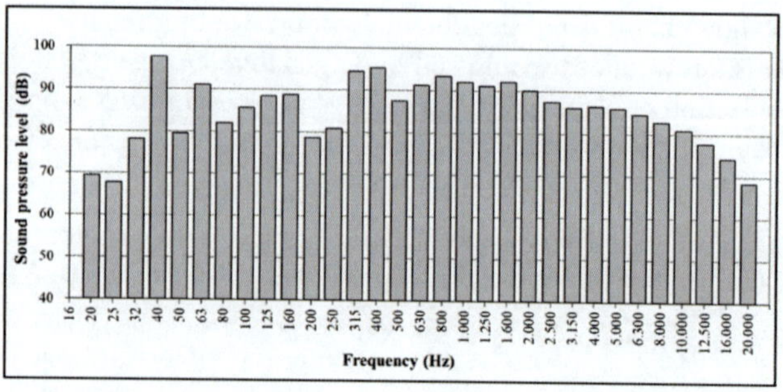

Fig. 6. Frequency analysis during clam sorting with the traditional sorter - MP Luigi Padre.

4 Conclusions

The results of this research indicate the innovative rotating sorting machine as a possible improvement to the clam fishing equipment to make the selection more effective and safer, to promote a more sustainable fishery and to reduce the fisherman's exposure to high noise levels due to the use of traditional vibrating sorting machines.

The measured noise levels are higher than the exposure limit value $L_{EX,\ 8h}$ equal to 87 dB (A) for the clam sorting operator, while they are between the two action values, $L_{EX,\ 8h}$ 80 dB (A) lower action value and $L_{EX,\ 8h}$ 85 dB (A) higher action value, during navigation, as established by Legislative Decree 81/08. It was also found that the noise, to which the operators on board the ship is exposed, includes tonal components in frequency (less than 2000 Hz) in contrast to what is usually measured in industrial activities. Form this point of view, the low frequency components, associated with a long-term, constant exposure to noise, could cause harmful effects on not only on the cochlear and vestibular apparatus of the ear, but also on the cardiac, vascular and gastric apparatus as well as sleep disturbances (extra-auditory effects) [12].

Our work aims to be a contribution to the study of noise risk in a production sector that has numerous employees in the face of equally numerous occupational diseases, including hearing loss from noise. Several investigations show that the L_{pA} dB (A), detected in various on-board environments, is remarkably high and widespread, especially in boats equipped with fixed blade helices, in which the noise increases during the fishing phase, due to higher engine speed [13]. The various operations (transfer from the port to the fishing spot, towing or hauling of the dredge and subsequent recovery, or sorting) can also affect the level of exposure since the exposed subjects operate in different places of the boat. The operating conditions are never identical and in many respects are also linked to environmental conditions. The sea conditions and therefore the ease of navigation, the winds, the sea currents and more generally the meteorological situations, may determine a request for greater or lesser engine power with consequences on noise levels.

The sound level surveys carried out are therefore to be considered preliminary investigations which, however, underline the existence of the problem related to the noise present on board the fishing boats, due to not insulated old engines and vibrating sorting machines. The boat Luigi padre equipped with a relatively new engine, showed low noise levels, especially during the transfer from the port. Despite this, activating the traditional sorting machine, caused an overall noise level higher than in the boat Dora, equipped with an old engine and the innovative rotating sorter.

Funding. This work has been funded by the Italian Ministry for Policies regarding Food, Agriculture, within the framework of the measure 1.39 of the program FEAMP 2014 2020.

References

1. Ferretti, M., Lombardo, F., Romani, G.: Metodi di pesca dei molluschi bivalvi: vongolare tradizionali e turbosoffianti. Quaderni ICRAP Pesca **1**, 157 (1989)
2. Froglia, C.: Clam fisheries with hydraulic dredges in the Adriatic Sea. In: Caddy, J.F. (ed.) Marine Invertebrate Fisheries: Their Assessment and Management, pp. 507–521. Wiley, Hoboken (1989)
3. Dayton, P.K., Thrush, S.F., Agardy, M.T., Hofman, R.J.: Environmental effects of marine fishing. Aquat. Conserv. Marine Freshw. Ecosyst. **5**, 205–232 (1995)
4. Peretti, A., Nataletti, P., Bonfiglio, P., Di Bisceglie, A.P.: Rumore nei battelli da pesca. G. Ital. Med. Lav. Ergon. **35**(4), 211–214 (2013)
5. Pochi, D., et al.: First investigation on the applicability of an active noise control system on a tracked tractor without cab. J. Agric. Eng. **44**, 757–760 (2013)
6. Fanigliulo, R., Del Duca, L., Fornaciari, L., Grilli, R., Tomasone, R., Pochi, D.: Efficiency of an ANC system in the tractor cabin under controlled engine workload. Noise Control Eng. J. **68**(5), 339–357 (2020)
7. Decreto del Presidente della Repubblica 8 novembre 1991, n. 435 – Approvazione del regolamento per la sicurezza della navigazione e della vita umana in mare. – Gazzetta Ufficiale della Repubblica Italiana n. 17 del 22 gennaio 1992 (Suppl. Ordinario n. 12)
8. Decreto Legislativo 27 luglio 1999, n. 271 – Adeguamento della normativa sulla sicurezza e salute dei lavoratori marittimi a bordo delle navi mercantili da pesca nazionali, a norma della legge 31 dicembre 1988, n. 485. – Gazzetta Ufficiale della Repubblica Italiana n. 185 del 9 agosto 1999 (Suppl. Ordinario n. 151)
9. Decreto Legislativo 17 agosto 1999, n. 298 – Attuazione della direttiva 93/103/CE relativa alle prescrizioni minime di sicurezza e di salute per il lavoro a bordo delle navi da pesca. – Gazzetta Ufficiale della Repubblica Italiana n. 201 del 27 agosto 1999
10. Decreto Legislativo 9 aprile 2008, n. 81 – Attuazione dell'articolo 1 della Legge 3 agosto 2007, n. 123 in materia di tutela della salute e della sicurezza nei luoghi di lavoro. - Gazzetta Ufficiale della Repubblica Italiana n. 101 del 30 aprile 2008
11. Norma ISO 1999 Acoustics - Determination of occupational noise exposure and estimation of noise-induced hearing impairment (2013)
12. Caradonna, L., et al.: La valutazione del rischio rumore e vibrazioni nel settore della pesca delle marinerie di Bari, Mola di Bari e Monopoli. In: Atti del 32° Congresso Nazionale di Igiene Industriale e Ambientale. Università degli Studi dell'Insubria 24–26 giugno, Varese, Italy (2015)
13. Rapisarda, V., Valentino, M., Bolognini, S., Fenga, C.: Il rischio rumore a bordo dei pescherecci: alcune considerazioni sulla prevenzione e protezione degli esposti. G. Ital. Med. Lav. Ergon. **26**(3), 191–196 (2004)

Comparison of Whole-Body-Vibration Exposure Between Quarry and Farm Activities

Andrea Colantoni[1], Massimo Cecchini[1(✉)], Danilo Monarca[1], Guido Alfaro Degan[2], and Gianluca Coltrinari[2]

[1] Department of Agriculture and Forest Sciences (DAFNE), University of Tuscia, Via San Camillo De Lellis, 01100 Viterbo, Italy
cecchini@unitus.it

[2] Department of Engineering, University of Roma Tre, Rome, Italy

Abstract. The aim of this paper consists of comparing the whole-body vibration (WBV) exposure for the workers in two different work contexts, quarry activities and farming. In the first one, special vehicles employed, such as dumper, load-hump-dump (LHD) vehicle, diggers, while in farming activities mainly the tractors are used for many different tasks. All vehicles used during the work activities represent a danger source vibration for the drivers. The measurements of whole-body vibration (WBV), according to the ISO 2631 standard, were performed on many different vehicles during experimental campaigns in quarry sites and farms. Field measurements are conducted in real work conditions with the aim of characterize the real exposure of the workers. The signal of fore-and-aft and vertical accelerations was acquired on the driver's seat by a tri-axial accelerometer. The data acquisition system used permits the analysis of the signal measured in terms of acceleration values and frequencies. The acceleration spectrum measured on the vehicles was analyzed in terms of band third octave. Finally, the data were compared in order to highlight the differences of the drivers' exposure by the parameter A(8). The data carried out from the measurements show a variation in WBV exposure levels between drivers that works in quarry sites and the workers of farm. The dose exposure results similar for all workers while the frequency spectrum highlights relevant differences at low frequencies.

Keywords: Tractor · Quarry site · Whole body vibration · Human health · Trucks

1 Introduction

It is recognized that musculoskeletal disorders (MSDs) in the workplace are most prevalent in quarry and farm operations [1, 2]. A negative relationship between whole-body vibration (WBV) exposure and MSDs, is the cause of many human disorders, especially low back pain [3]. Back disorders have been identified as the largest source of early permanent disability among tractor and machinery drivers at quarry sites [4]. A dose-response relationship was established showing that increased duration of WBV exposure was associated with increased risk of injury [5, 6]. Variable postures, anthropometric characteristics of drivers, spinal geometry, and individual spinal

M. Biocca et al. (Eds.): SHWA 2020, LNCE 252, pp. 242–251, 2022.
https://doi.org/10.1007/978-3-030-98092-4_26

tolerance have been identified as important factors codetermining the effects of occupational WBV on human health. Although it has been established that there is an exposure-response relationship between human health outcomes and WBV, the mechanisms are not currently well understood by researchers [7]. In addition, impulsive exposures have also recently been identified as a possible risk factor for low back pain [8]. Occupational drivers, such as bus trucks, dump trucks, and tractors, are a group of workers who have been widely reported to be at increased risk of LBP [9]. ISO 2631-1 (1997) provides a health caution zone without quantitative information to allow an assignment of a certain exposure condition within this zone, considering associated factors such as posture, exposure time, and/or age [10]. The European Union (2002) developed daily action and exposure limits for acceptable vibration exposures based on weighted average vibration (Aw) and vibration dose value (VDV). Italian law, in order to prevent adverse health effects, states that the employer must assess the vibration exposure of each worker for each specific task. It is based on the calculation of daily exposure A(8) expressed as equivalent continuous acceleration over an 8-h period, calculated according to ISO 2631-1. It is known that mainly variations in vibration magnitude within a given class of vehicles can be found under field conditions based on specific operating conditions. Another relevant aspect is the uncertainty in the evaluation of the daily vibration exposure parameter A(8). It is affected by both the uncertainty in the evaluation of the vibration magnitude under field conditions and the uncertainty in the evaluation of the exposure duration [11]. The latter is due to the difference of the many tasks in ordinary daily activities. For operations with heavy equipment, the driver often has to maintain uncomfortable postures (including static sitting) for long periods of time, which can also lead to a number of musculoskeletal disorders. A review of recent Canadian mining accident statistics in Ontario revealed that 16% of all traumatic injuries occurred during the operation of haul trucks [12]. There have been numerous studies exploring the levels of exposure to WBV experienced during the operation of large earth moving equipment and specific vehicles, such as diggers, excavators, and dump trucks. In addition, during off-highway vehicle operations, such as transport trucks, there is the potential for high-magnitude shocks due to rough terrain, heavy loading, and specific soil operations. Driving agricultural tractors has also been associated with dangerously high vibration. In fact, all-terrain vehicles (ATV) are widely used by farmers during their daily work. ATV use creates prolonged WBV exposure and mechanical shock, often in constrained sitting postures with spinal torsion, factors that can be detrimental to the driver's musculoskeletal system. To prevent potential human health risks and reduce WBV, two different strategies can be adopted for vehicle drivers. The first involves the vehicle manufacturer's design. The employer can choose machinery with an appropriate suspension design and cab. Using model-based development, axle and cab suspensions for off-highway mobile machinery have been developed extensively in recent decades. These mechanical elements play a relevant role in the strategy to reduce WBV exposure. In addition, seats are also important elements in reducing vibration transmission to the driver. It is necessary to point out that standard codes for different groups of machines specify which spectral class the seat must meet. The second way of the strategy concerns the skills and behavior of the operator. The operator has a great influence on the dynamic response and thus on WBV exposure. It is emphasized that operator training is important to reduce vibration exposure. The typical WBV exposure of agricultural tractor operators in the range of 0.7–1.0 m/s^2. In addition, the longitudinal and vertical dynamic response plays an important role in assessing whole-body occupational exposure. Further experiments will show that front axle suspension and seat suspension have a significant influence on longitudinal and vertical dynamics [13]. Tractor tire types and vehicle age may also influence drivers' WBV exposure [14].

2 Materials and Methods

The study was conducted by two different experimental campaigns: the first one was realized in the farm of Tuscia University while the second one in a limestone quarry 100 km far from Rome. The farm is managed by the university while the mines is operated by the same company. The study protocol included the whole-body vibration analysis for drivers of tractors and quarry machineries. Before the beginning of the measurements was conducted the analysis of the activities of the tractors in the farm with different mechanical machineries and the same process was take off for the quarry vehicles in order to understand the specific task each vehicle. The study was conducted firstly in the quarry site with the same driver and after with another driver, employee of the same company. Instead in the field tests in the farm, all the measurements were realized on the same driver. All the participants were interviewed before the start of the measurements regarding the ordinary time exposures, speed vehicles and the modalities of operations. In the first measurement campaigns the total of two vehicles were measured while in the farm were tested 2 different tractors (John Deere and New Holland).

Fig. 1. The excavator during loading material and dumper and instrument on driver's seat.

The measurements were realized during the ordinary activities: moving, loading and dumping the rock material (quarry sites) and plowing, sowing, harrowing (farms) (Fig. 2).

Fig. 2. The instrument for the measurement the farm and the tractors during the activities.

The vehicles employed in the experimental campaign can carry out the tasks described above. Operators of the tractor and mining vehicles were instructed to perform their normal duties during the data collection period in order to replace the real conditions. The percentage of time spent in each work phases was defined on the basis of the driver's interview. Accelerometers (Table 1) are the most common transducers used in whole body vibration measurements and produce an electrical output proportional to acceleration. In the field tests, WBV exposure were measured according to the standard procedure of the ISO 2631-1 (1997) guidelines. A seat pad accelerometer was fixed on the seat cushion surface between driver and seat, with regards to direction and anatomical positioning (Fig. 1). The duration of measurements was kept sufficiently long to ensure reasonable signal precision and to ensure that the vibration is typical of the exposures of the drivers for the specific tasks. The acquisition time was of 3 min for each specific task. According to ISO 2631-1 (1997) guidelines the signal was acquired in a three-axis coordinate system to consider the entry points of vibration in worker's body: the x-axis to measure vibration in the anterior posterior direction, the y-axis in the medial lateral direction, and the z-axis in the vertical direction. The guidelines are applicable to the vibration in the frequency range from 0.5 Hz to 80 Hz, transmitted to the body as a whole through the seat pad.

Table 1. Characteristics of instruments.

Model	Sensivity	Position
Traxial accelerometer SEN027-PCB	X axis:10.52 mV/(m/s^2) Yaxis:10.55 mV/(m/s^2) Z axis:10.58 mV/(m/s^2)	Driving seat of the truck in the position which corresponds to the driver's sacroiliac joints (his buttocks)

The frequency-weighted root mean square acceleration (rms) was calculated using the following expression:

$$A(8) = a_w \sqrt{\frac{T_{esp}}{8}}\ \mathrm{m/s^2} \qquad (1)$$

where:

A(8) is the daily dose expressed as [m/s^2];

T_{exp} is the exposure time expressed as [hour];

a_w is the equivalent continuous rms acceleration also expressed as [m/s^2].

As described above, the Italian decree 81/2008 establishes that the evaluation of the daily exposure A(8), expressed as the equivalent continuous rms acceleration over an height hour period, allows the evaluation of the risk due to whole-body vibration exposure. It is calculated as the highest (rms) value of the frequency-weighted accelerations, determined on the three orthogonal axes following the expression:

$$a_w = \max\left(K_x\, a_{wx},\ K_y\, a_{wy},\ K_z\, a_{wz}\right) \tag{2}$$

where $a_{w,max}$ is the maximum value among $1.4 \cdot a_{w,x}$, $1.4\ a_{w,y}$ and $1 \cdot a_{w,z}$ for a seated or standing worker.

3 Results

The following Fig. 3 highlights the trend of solicitation for the measurement realized on the excavator in the quarry site. The green line shows the signal along z – axis, the red line for y axis and finally the blue line is the signal of x-axis. The graph shows the acceleration values (m/s^2) on y-axis and the time expressed in terms of seconds (s) on x-axis.

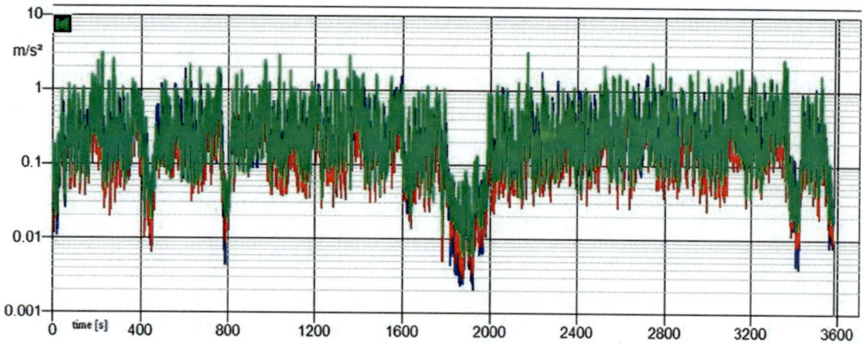

Fig. 3. Time history of accelerations (excavator). (Color figure online)

The frequency spectrum of the measurement is showed in Fig. 4.

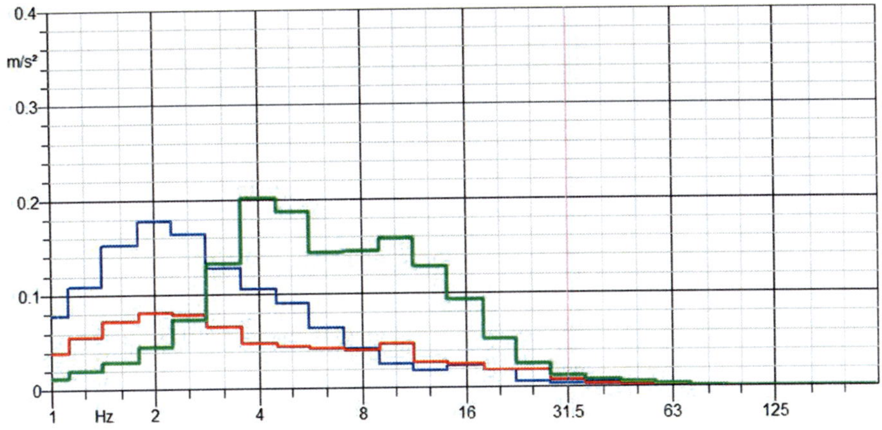

Fig. 4. Frequency spectrum (excavator). (Color figure online)

The same graphs are shown, with reference to the dumper, in Figs. 5 and 6.

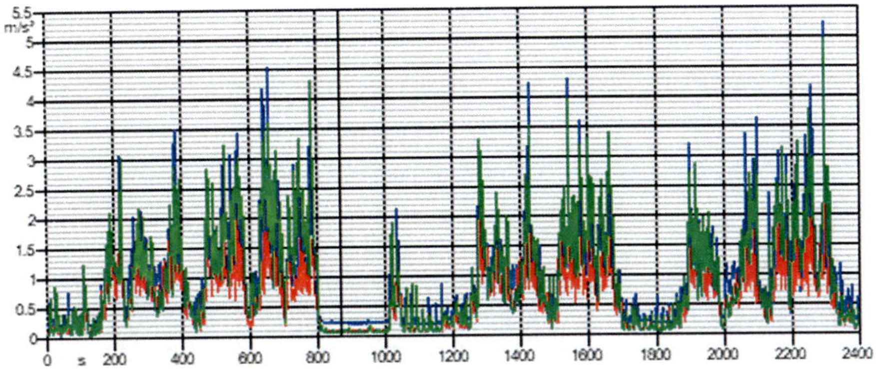

Fig. 5. Time history of accelerations (dumper). (Color figure online)

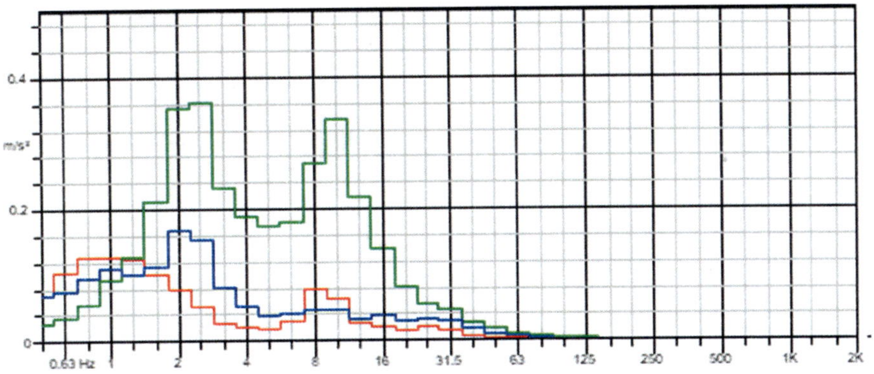

Fig. 6. Frequency spectrum (dumper). (Color figure online)

The colours of the lines are the same used in the previous graph. The figure shows that the vibration signal of each axis is defined in the range from 1 Hz until 31.5 Hz. This aspect has been noted in the measurements for the quarry vehicles. In the Figs. 7, 8, 9 and 10, the signals acquired on the tractors New Holland and John Deere are reported. In order to highlight the different of z-axis respect to the other ones, in these figures the line is coloured by black theme.

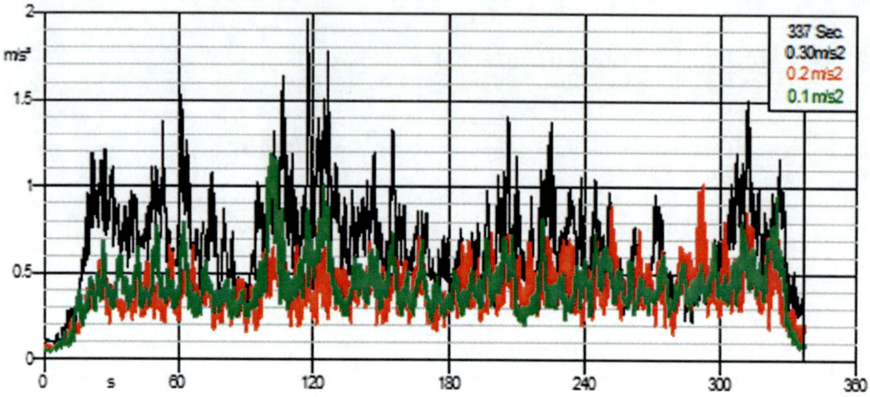

Fig. 7. Time history of accelerations (tractor New Holland). (Color figure online)

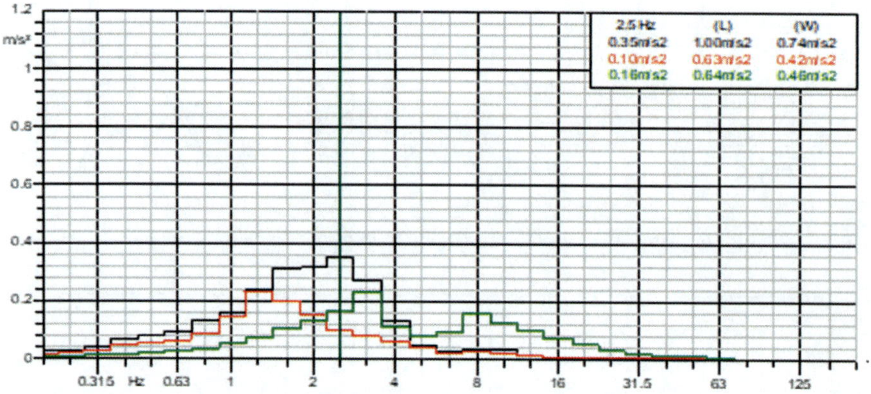

Fig. 8. Frequency spectrum (tractor New Holland). (Color figure online)

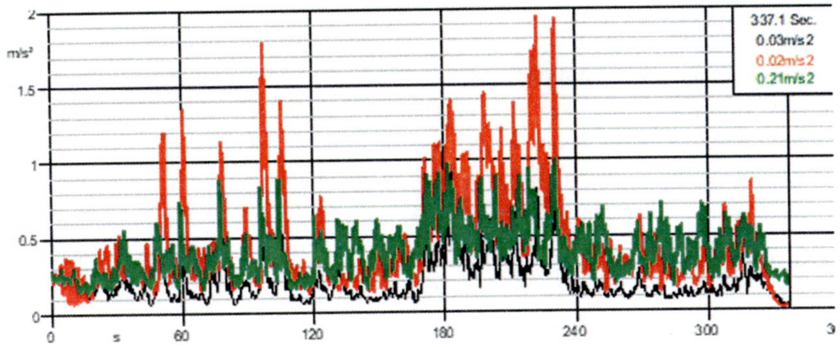

Fig. 9. Time history of accelerations (tractor John Deere). (Color figure online)

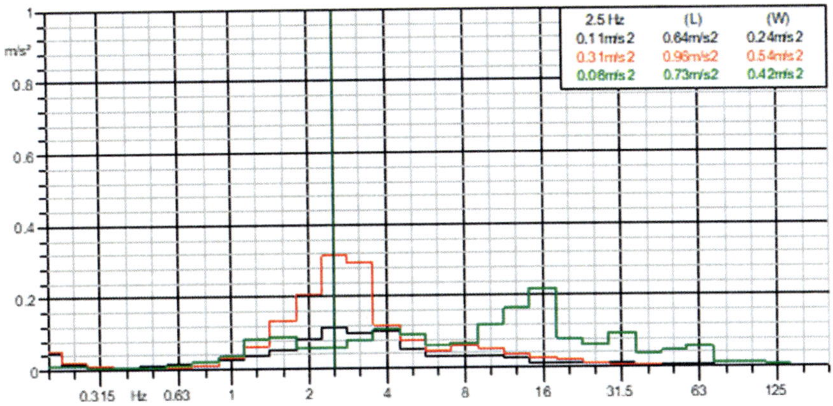

Fig. 10. Frequency spectrum (tractor John Deere). (Color figure online)

Table 2 shows the comparison of the results for each vehicle employed in the quarry (first and second column) and farm (third and fourth column).

Table 2. Comparison of the data obtained from the measurement campaign.

Parameters	Excavator	LHD Vehicle	New Holland	John Deere
Task	Moving ground	Load/Haule/Dump of the ground	Plowing	Sowing
Frequency-weighted rms accelerations (m/s^2)	a_{wx} = 0.351	a_{wx} = 0.229	a_{wx} = 0.169	a_{wx} = 0.189
	a_{wy} = 0.199	a_{wy} = 0.261	a_{wy} = 0.184	a_{wy} = 0.178
	a_{wz} = 0.445	a_{wz} = 0.451	a_{wz} = 0.646	a_{wz} = 0.356
Dominant frequencies (Hz)	DF_x = 2	DF_x = 1.6	DF_x = 2	DF_x = 2
	DF_y = 2	DF_y = 1.3	DF_y = 4	DF_y = 1
	DF_z = 4	DF_z = 2	DF_z = 8	DF_z = 16
Estimated A(8) (m/s^2)	0.49	0.45	0.42	0.37

The results obtained in farm activities shows similar trend regarding the driver that conducted the tractor in the field tests. The maximum stress results on the z axis in the vehicle at the quarry site while in agricultural activities we can observe that for some operations the main stressed axis is Y. The values obtained from the measurements result to be within the thresholds defined by Italian law, generally in the quarry activities are lower respect to the ones of farm operations.

4 Conclusions

The driver's exposure was investigated on different vehicles in two work contexts, quarry sites and farm. The whole-body vibration exposure was assessed based on ISO 2631 (1997). It was determined that the data obtained in both experimental campaigns are similar in terms of acceleration values and frequency spectrum. However, in the farm activities, the use on the tractor of soil tillage implements, has a relevant role on the driver's exposure while the specific task in the quarry site are more regular. This aspect highlights that the assessment of the exposure in the farm activities result to be more complex. The range of frequencies appear similar for all vehicles in quarry sites and farm. This allows a consideration; the human body is solicited by a similar way for each vehicle. In conclusion, the present study demonstrates that all the activities with vehicle can generate serious health hazards for the driver and at the same time the way in order to reduce the exposure can be similar for each activity. It's necessary to point out that the present study had some limitations; the measurement of WBV level experienced through autumn and did not encompass all seasons of the year but this aspect doesn't influence the seasonal variations of WBV exposure. Results of the field tests showed the possibility of further study in order to comprehend better the parameters that influence the exposure for the driver.

References

1. Cecchini, M., Piccioni, F., Ferri, S., Coltrinari, G., Bianchini, L., Colantoni, A.: Preliminary investigation on systems for the preventive diagnosis of faults on agricultural operating machines. Sensors **21**(4), 1547 (2021)
2. Bovenzi, M., Betta, A.: Low-back disorders in agricultural tractor drivers exposed to whole-body vibration and postural stress. Appl. Ergon. **25**(4), 231–241 (1994)
3. Alfaro Degan, G., Antonucci, A., Coltrinari, G., Lippiello, D.: Safety-oriented multi-task approach to manage whole-body vibration exposure among quarry operators. WIT Trans. Built Environ. **189**, 43–53 (2019)
4. Cann, A.P., Salmoni, A.W., Eger, T.R.: Predictors of whole-body vibration exposure experienced by highway transport truck operators. Ergonomics **47**(13), 1432–1453 (2004)
5. Sediel, H., Heide, R.: Long-term effect of whole-body vibration: a critical survey of the literature. Int. Arch Occup. Environ. Health, 58, 1e26 (1986)
6. Griffin, M.J.: Discomfort from feeling vehicle vibration. Veh. Syst. Dyn. **45**(7–8), 679–698 (2007)

7. Johanning, E., Fischer, S., Christ, E., Gores, B., Landsbergis, P.: Whole-body vibration exposure study in us railroad locomotives DAN ergonomic risk assessment. AIHA J. **63**, 439–446 (2002)
8. Rehn, B., Lundström, R., Nilsson, L., Liljelind, I., Järvholm, B.: Variation in exposure to whole-body vibration for operators of forwarder vehicles aspects on measurement strategies and prevention. Int. J. Ind. Ergon. **35**, 831–842 (2005)
9. Lines, J., Stiles, M., Whyte, R.: Whole body vibration during tractor driving. J. Low Freq. Noise Vib. Active Control **14**(2), 87–104 (1995)
10. Kirstein, J.C.: Suspension system optimisation to reduce whole body vibration exposure on an articulated dump truck. Doctoral dissertation, University of Stellenbosch, Stellenbosch (2005)
11. International Organization for Standardization: ISO 2631-1, Mechanical vibration and shocks - evaluation of human exposure to whole-body vibration, Part 1, General requirements (1997)
12. Cvetanovic, B., Zlatkovic, D.: Evaluation of whole-body vibration risk in agricultural tractor drivers. Bulgarian J. Agr. Sci. **19**(5), 1155–1160 (2013)
13. Eger, T., Salmoni, A., Cann, A., Jack, R.: Whole-body vibration exposure experienced by mining equipment operators. Occup. Ergon. **6**(3, 4), 121–127 (2006)
14. Boshuizen, H.C., Bongers, P.M., Hulshof, C.T.: Self-reported back pain in fork-lift truck and freight-container tractor drivers exposed to whole-body vibration. Spine **17**(1), 59–65 (1992)

Operators' Exposure to Hand-Arm Vibration in Small Olive Growing

Angela Calvo[1(✉)], Roberto Deboli[2], and Christian Preti[2]

[1] DISAFA (Department of Agricultural, Forest and Food Sciences and Technologies), Largo P. Braccini 2, 10095 Grugliasco, Turin, Italy
angela.calvo@unito.it

[2] STEMS (Institute of Sciences and Technologies for Sustainable Energy and Mobility of C.N.R - Italian National Research Council), Strada delle Cacce 73, 10135 Turin, Italy

Abstract. Many little sloping olive groves are present in Mediterranean countries, where the olive cultivation is here feasible with brushcutters, motorhoes, chainsaws, backpack mistblowers and hand-held olive harvesters. All the mentioned machines are unfortunately carrier of possible hand-arm vibrations (HAV) risks onset for the operators. Many studies concern the specific vibration risk of the above mentioned machines, but there is a research shortage on the HAV operators' exposure during the entire cultivation season. Aim of this work was to study the operator's exposure to the HAV risk, using both the ISO 5349-1 standard (as requested by the European Directive 2002/44) and the ISO/TR 18570, to correctly evaluate the vascular risk. Vibration transmitted by a motorhoe, a chainsaw, a backpack mistblower and a hand-held olive harvester were measured in an olive grove located in the Liguria Region (Italy, North West). Measurements were performed in field throughout the growing season by one operator (from November 2019 to April 2021, delayed due to the pandemic emergency). The simultaneous acquisition of the vibration along the three directions (X, Y and Z) were performed. The signals were frequency weighted using the weighting curves W_h (ISO 5349-1 standard), and W_p (ISO/TR 18570). The daily operators' exposures *A(8)* and $E_{p,d}$ were calculated and compared with the current regulation. The motorhoe, the chainsaw and the hand-held olive harvester showed the highest values, independently by the weighting curve. Only the backpack mistblower showed lower hand-arm vibration values.

Keywords: Hand-arm vibration risk · Manual olive harvesting

1 Introduction

Due to the small cultivated surfaces and to the terrain conditions, the olive cultivation is performed with portable machines in many sloped olive groves located in the Mediterranean countries [1]. The pruning is carried out using shears or chainsaws, whereas brushcutters, motorhoes hand-tractors are used for the soil aeration and cleaning. Backpack mistblowers are used for the distribution of pesticides and hand-held harvesters (beaters, combs, flaps) for the olive harvesting. Even though these machines aid the manual works, on the other hand all of them are responsible of the

M. Biocca et al. (Eds.): SHWA 2020, LNCE 252, pp. 252–259, 2022.
https://doi.org/10.1007/978-3-030-98092-4_27

possible onset of hand arm vibration (HAV) risks, as observed by many Authors [2–4]. On the other hand, most of the owners of these olive groves are owner occupier people, that means that they are not obliged to the personal risk evaluation for their farm work. They are moreover unaware of the HAV risks that they encounter during the olive cultivation and they never attended any safety training (because it is a personal choice and these courses are not paid by anyone). During the COVID-19 pandemic farmers have been persuaded to improve their little cultivation [5]. The result is an unprotected farm work, often carried out with a light mechanization and small machines, not perceived as dangerous as it could be. The use of portable machines and of hand-tractors, or motorhoes, is not considered risky as an agricultural tractor or other farm operating machines. If injuries may be lower (may be, but it is not always true, especially when a chainsaw or a hand-tractor is used), on the other hand the physical risks and, among them, the hand arm vibration exposure must not be undervalued.

In this preliminary study we considered some common portable machines used during the cultivation of a little olive grove located in a terrace terrain of the Liguria Region (North-West Italy). The machines were a motorhoe, a chainsaw, a mistblower and a beater. These machines were used for the pruning, the soil work, the pesticide distribution and the olive harvesting. In this work the daily exposure for each used tool *A(8)* was calculated, as requested by the European Directive 2002/44 [6] that permits the use of these machines if they don't exceed the daily limit value of 5 ms^{-2}, on the basis of the vibration total values weighted with the W_h weighting curve (ISO 5349-1) [7]. Recent studies, however, have highlighted that for specific frequencies (between 20 and 200 Hz) the use of the W_h weighting curve is not sufficient, because it might miss the vascular risk onset, which may cause the Reynaud disease [8]. For this reason, we also calculated the $E_{p,d}$ daily exposure, using the vibration total values weighted with the W_p weighting curve, as introduced in the ISO/TR 18570 standard [9]. This curve was introduced to fill this gap, because the $E_{p,d}$ exposure takes into account also the frequencies heavily cut by the W_h weighting curve. To make a comparison with the well-known vibration daily exposure *A(8)*, calculated by the vibration total value W_h weighted, also the W_p weighted daily vibration exposure $A_p(8)$ was calculated and commented (as proposed by the ISO/TR 18570). The $A_p(8)$ is the daily vibration exposure normalised to an 8-h day, frequency weighted according to the hand-arm vascular weighting W_p.

2 Material and Methods

2.1 Test Location and Machines

The olive grove was located at Pietra Ligure (44°08′55.49″N 8°16′58.14″E°). It had an extension of 0.8 hectares, with 150 cultivated trees (variety *Olea Europea*). The farmer was 45 years old (mass 78 kg, height 178 cm) and he used a motorhoe, a manual shear, a chainsaw, a mistblower and a beater. His field work was monitored during the month of November 2019 (during the olive harvesting), January 2020 (during the trees pruning), partially in the month of February 2020 (at the beginning of the soil working) and then during April-June 2021 (during the pesticide distribution).

The characteristics of the machines are in Table 1 (the shear was not investigated, being it a manual tool).

Table 1. Characteristics of the machines used during the olive cultivation.

Machine	Engine type	Engine characteristic	Power (kW)	Mass (kg)	Other characteristics	Use (h day^{-1})
Motorhoe	Petrol	450 cm^3	10.76	178	0.8 m tillage width	3
Mistblower	Petrol	77.4 cm^3	3.4	12.5	6200 max speed min^{-1}	2.5
Chainsaw	Petrol	48.7 cm^3	2.2	6.4	14000 max speed min^{-1}	2
Beater	Electric	12 V	0.36	2.8	1300 strokes min^{-1}	4

2.2 Measurement Chain

The acceleration along the directions X, Y and Z (using the standards ISO 22867:2021 [10], ISO 5349-2:2001 [9], and EN ISO 20643 [11]) were acquired using a tri-axes accelerometer (PCB, model SEN020, mass 10 g, sensitivity 1 mV/g, previously calibrated), mounted on the handle (or handles) of the machine.

Three acceleration were acquired on each handle of each machine during the field work and the acquisition time was large enough to reach a stabilized signal (around 3 min). The vibration data (not filtered, filtered using the *Wh* weighting filter of the ISO 5349 and filtered by the W_p weighting filter of the ISO/TR 18570) were stored on a PC and therefore processed by the LabView program.

The *Wh* filter applied to the original one-third octave band signals led to the root mean squares (r.m.s.) a_{hwx}, a_{hwy} and a_{hwz}, and therefore to the vibration total value a_{hv} (Eq. 1).

$$a_{hv} = \sqrt{a_{hwx}^2 + a_{hwy}^2 + a_{hwz}^2}\left(\mathrm{ms}^{-2}\right) \tag{1}$$

On the other hand, the W_p filter applied to the original one-third octave band signals led to the root mean squares (r.m.s.) a_{px}, a_{py} and a_{pz} and therefore to the vibration total value a_{pv} (Eq. 2).

$$a_{pv} = \sqrt{a_{px}^2 + a_{py}^2 + a_{pz}^2} \quad \left(\mathrm{ms}^{-2}\right) \tag{2}$$

The highest r.m.s. was used when two handles were present (European Standard 2002/44/CE).

2.3 Calculation of A(8), $E_{p,d}$ and $A_p(8)$

The *A(8)* (ms^{-2}) vibration daily exposure was calculated for each machine, considering both the vibration total value a_{hv} and the daily use *T*, and therefore compared first with the daily exposure action value (2.5 ms^{-2}) and then with the daily limit value (5 ms^{-2}), accordingly to the requests of the European Directive 2002/44/EC (Eq. 3).

$$A(8) = a_{hv}\sqrt{\frac{T}{8}} \qquad (ms^{-2}) \tag{3}$$

where:

a_{hv}: vibration total value, W_h weighted
T: daily use of the vibrating machine (h)

To also consider the vascular risk of the hand-arm system, the daily vibration exposure value $E_{p,d}$ was calculated (Eq. 4). This parameter was compared with the daily threshold for the risk of the white finger onset.

$$E_{p,d} = \sqrt{a_{pv}^2 T} \qquad (ms^{-1.5}) \tag{4}$$

where:

a_{pv}: vibration total value, W_p weighted
T: daily use of the vibrating machine (s)

To make a comparison with the more known vibration daily exposure *A(8)*, calculated by the a_{hv} vibration total value, the W_p weighted daily vibration exposure $A_p(8)$ was also calculated (Eq. 5).

$$A_p(8) = \frac{E_{p,d}}{\sqrt{28800}} \qquad (ms^{-2}) \tag{5}$$

where 28800 is the reference time in seconds of 8 h (ISO 18570).

3 Results and Discussion

The average vibration total values (W_h and W_p weighted) are visible in Fig. 1. The machine with the lowest hand-arm vibration values (both a_{hv} and a_{pv}) is the mist-blower, as also observed by other Authors [12]. The vibration produced by this machine is surely not dangerous for the operator's health by the ISO 5349-1, but the a_{pv} value is not so low, and some comments will be necessary when the $A_p(8)$ will be later discussed. Other than the obvious higher a_{pv} acceleration data for all the investigated machines, particularly noteworthy is the 'low' average a_{hv} vibration total value of the chainsaw (5.8 ms^{-2}) in face of a very high a_{pv} vibration total value (59.6 ms^{-2}). It is also interesting to notice that the chainsaw shows a vibration total value W_h weighted lower than the same data of the motorhoe (13.5 ms^{-2}), whereas its vibration total value W_p weighted is much higher (59.6 ms^{-2} against 35.0 ms^{-2}, Fig. 1). This fact means that the chainsaw shows high acceleration data in the frequency interval cut by the W_h

weighting curve, as is clearly visible in Figs. 2 and 3. In Fig. 3, in fact, it is visible the fundamental of the chainsaw, that reaches more than 39 ms^{-2} at the frequency of 160 Hz, heavily cut by the W_h weighting curve (Fig. 2). The same machine has other high W_p weighted acceleration values also at the frequency of 125, 200 and 400 Hz (where the W_p weighting curve keeps unchanged the linear acceleration values initially acquired by the accelerometers, Fig. 3).

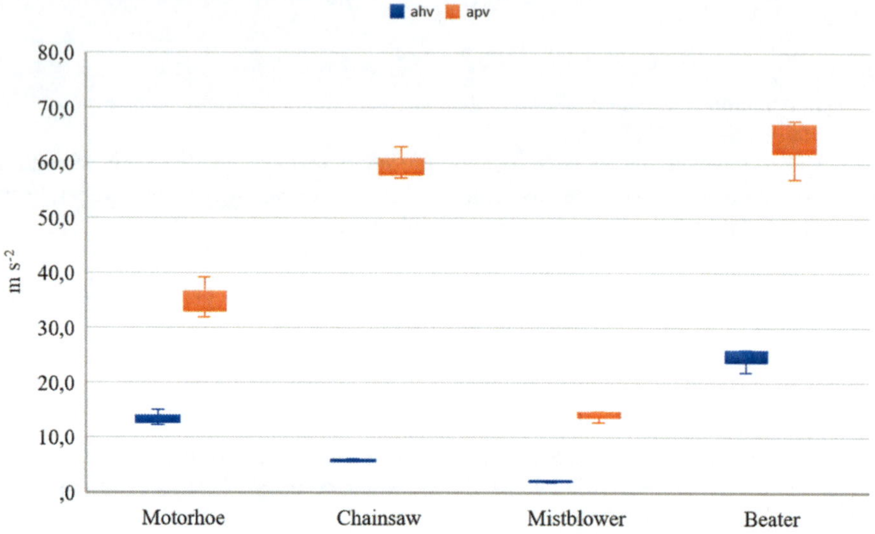

Fig. 1. a_{hv} and a_{pv} average vibration total values of the vibrating machines with the error bars.

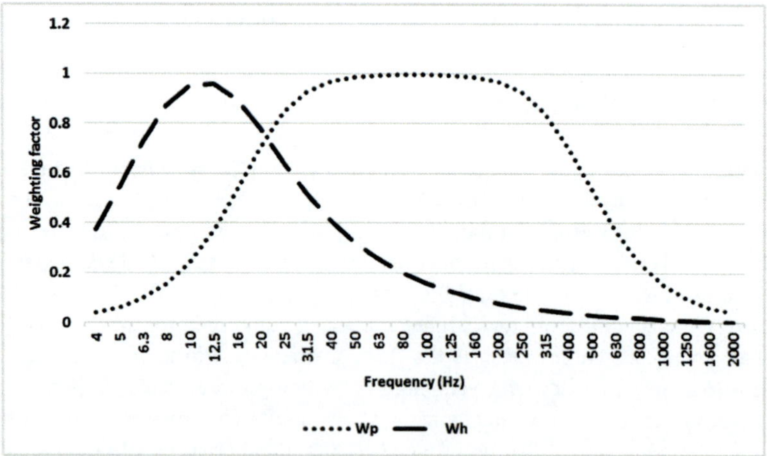

Fig. 2. W_h and W_p weighting curves.

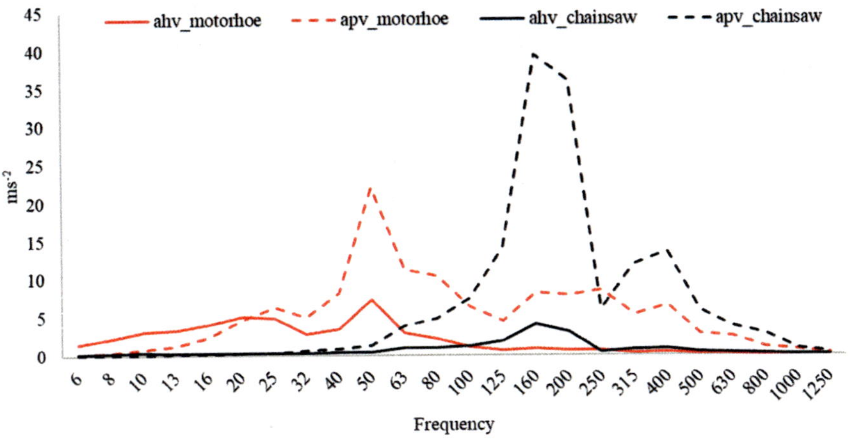

Fig. 3. Frequency weighted analysis of the chainsaw and the motorhoe.

The motorhoe shows a daily exposure *A(8)* equal to 8.3 ms^{-2} against 2.9 ms^{-2} for the chainsaw (used for only 2 h per day, while the motorhoe is used 3 h per day). On the contrary, the $A_p(8)$ of the motorhoe is 'only' 21.5 ms^{-2}, against the 29.8 ms^{-2} of the chainsaw (Table 2). This is not a mistake or an erroneous interpretation, but the concern of the use of a weighting curve (W_p) that does not cut the acquired acceleration data of the chainsaw as the W_h weighting curve does (as previously discussed).

From the Directive 2002/44/EC perspective, the *A(8)* of the chainsaw is slightly higher than the daily exposure action value (2.5 ms^{-2}), against the *A(8)* of the motorhoe that heavily exceeds the daily limit value (5 ms^{-2}), and that could not be used for the 3 h per day, but only for 1 h per day (not to exceed this daily limit value).

Even though there are not daily limit values provided by any Directive for the $A_p(8)$, it is evident that the chainsaw has an $A_p(8)$ higher than the motorhoe (29.8 ms^{-2} versus 21.5 ms^{-2}).

Table 2. Daily exposure to the hand-arm vibration (*A(8)*, $E_{p,d}$ and A_p *(8)*) of the operator when using the observed machines in the olive grove.

Machine	Use	ahv		A(8)		apv		Ep,d		Ap(8)	
		Mean	St. dev.	Mean	St. dev.	Mean	St. dev.	Mean	St. dev.	Mean	St. dev.
	h day^{-1}	ms^{-2}		ms^{-2}		ms^{-2}		$ms^{-1.5}$		ms^{-2}	
Motorhoe	3	13.5	1.43	8.3	0.85	35.0	3.71	3640.4	374.34	21.5	2.21
Chainsaw	2	5.8	0.29	2.9	0.15	59.6	3.01	5059.8	260.55	29.8	1.54
Mistblower	2.5	2.0	0.16	1.1	0.11	14.1	1.14	1337.6	126.63	7.9	0.75
Beater	4	24.6	2.25	17.4	1.07	63.7	5.85	7647.6	470.68	45.1	2.77

Brammer et al. [13] defined the daily exposure threshold for the onset of the vibration white finger (VWF) in the range 1150–1750 $ms^{-1.5}$. Table 2 shows only the mistblower with an $E_{p,d}$ in these range limits. This fact means that in theory all the others machines are not safe for the operator for the VWF onset.

Another aspect regards the very high values of the $E_{p,d}$ of the beater and the chainsaw, more than five times higher than the mistblower the first, and slightly less than four times the second.

4 Conclusions

The motorhoe, the chainsaw and the hand-held olive harvester showed values above and beyond the admitted law limits as *A(8)*. Only the backpack mistblower showed lower hand-arm vibration values. On the other hand, however, the $A_p(8)$ (the corresponding *A(8)* calculated using the W_p weighted acceleration data) was not low.

Without contesting the European Directive 2002/44, that introduced for the first time the daily exposure limits for the hand-arm vibration exposure, the proposal is to put side by side both the *A(8)* and the $A_p(8)$ daily exposure values, to better estimate the hand-arm vibration risk.

References

1. Aiello, G., Vallone, M., Catania, P.: Optimising the efficiency of olive harvesting considering operator safety. Biosys. Eng. **185**, 15–24 (2019)
2. Manetto, G., Cerruto, E., Papa, R.: Effects of rod and oscillating frequency on the vibrations transmitted to hand-arm system by four olive portable harvesters. In: Coppola, A., Di Renzo, G.C., Altieri, G., D'Antonio, P. (eds.) MID-TERM AIIA 2019. LNCE, vol. 67, pp. 671–679. Springer, Cham (2020). https://doi.org/10.1007/978-3-030-39299-4_73
3. Deboli, R., Calvo, A., Preti, C.: Vibration and impulsivity analysis of hand held olive beaters. Appl. Ergon. **55**, 258–267 (2016)
4. ENAMA (Ente NAzionale Meccanizzazione Agricola) Homepage. www.enama.it. Accessed 18 June 2021
5. Meuwissen, M.P.V., et al.: Impact of Covid-19 on farming systems in Europe through the lens of resilience thinking. Agric. Syst. **191**, 1–12 (2021)
6. Directive 2002/44/EC of the European Parliament and of the Council of 25 June 2002 on the minimum health and safety requirements regarding the exposure of workers to the risks arising from physical agents (vibration) (sixteenth individual Directive within the meaning of Article 16(1) of Directive 89/391/EEC), Luxembourg (2002)
7. ISO 5349-1: Mechanical vibration. Measurement and evaluation of human exposure to hand-transmitted vibration. International Standard, Geneva (2001)
8. Bovenzi, M., Pinto, I., Picciolo, F.: Risk assessment of vascular disorders by a supplementary hand–arm vascular weighting of hand-transmitted vibration. Int. Arch. Occup. Environ. Health **92**(1), 129–139 (2018). https://doi.org/10.1007/s00420-018-1363-y
9. ISO/TR 18570: Mechanical vibration. Measurement and evaluation of human exposure to hand-transmitted vibration. Supplementary method for assessing risk of vascular disorder. International Standard, Geneva (2017)

10. ISO 22867: Forestry and gardening machinery. Vibration test code for portable hand-held machines with internal combustion engine. Vibration at the handles. International Standard, Geneva (2021)
11. EN ISO 20643: Mechanical vibration. Hand-held and hand-guided machinery. Principles for evaluation of vibration emission. International Standard, Geneva (2008)
12. Calvo, A., Preti, C., Caria, M., Deboli, R.: Vibration and noise transmitted by agricultural backpack powered machines critically examined using the current standards. Int. J. Environ. Res. Public Health **16**, 1–20 (2019)
13. Brammer, A.J., Pitts, P.M.: Frequency weighting for vibration-induced white finger compatible with exposure-response models. Ind. Health **50**, 397–411 (2012)

Analysis of Mechanical Vibrations Generated by Tractors: A First Approach to the Absorption of the Human Body

Rino Gubiani[1(✉)], Nicola Zucchiatti[1], Ugo Da Broi[2], and Carlo Moreschi[2]

[1] Department of Agricultural, Food, Environmental and Animal Sciences (D4IA), University of Udine, Via delle Scienze 208, 33100 Udine, Italy
rino.gubiani@uniud.it
[2] Department of Medicine, Forensic Medicine, University of Udine, P.le S.Maria della Misericordia 15, 33100 Udine, Italy

Abstract. The aim of this work was to evaluate the level of vibrations produced by different types of tractors - also used with different operating machines connected – and the magnitude of vibrations impacting the whole human body.

Experimental tests concerning vibrations were carried out on several tractors of different age and size, with different connected operating machines. All the parameters of the tractors and operating machines tested were recorded (speed, tires, type of soil, type of operation, etc.). Measures were also carried out measuring to the weight and height of the driver. Vibrations were measured and analyzed with an instrument aimed at measuring acceleration on the three axles every second. Noise was also measured.

Results show that the amount of vibrations increases with increasing speed and vibrations during work in vineyard rise in hill vineyard due to uneven ground caused by runoff.

The test made with an old tractor with an undamped seat and only the safety frame with a healthy weight range person show that the vibrations on the z axis are very high given the high speed for the type of terrain, uncultivated field and dirt road.

In other test with a tractor equipped with a cushioned cab and a mechanical seat with parallelogram suspension, we note the reduction of the component on the vertical z axis, while the y and x components increase due to rolling, pitch and serpentine movements. Also in this case, the component on the x axis increases due to the contribution of the arms on the steering wheel.

However, our study suggested that further and exhaustive studies, both from the point of view of new tractors equipped with shock absorbers and from the point of view of the effect on drivers, should be encouraged.

Keywords: Vibrations · Tractors safety · Workers health · Vineyard workers

M. Biocca et al. (Eds.): SHWA 2020, LNCE 252, pp. 260–271, 2022.
https://doi.org/10.1007/978-3-030-98092-4_28

1 Introduction

The vibrations transmitted by agricultural machinery are generally high, but the extent of this risk factor and the prevalence of related diseases are still far from being fully known. The estimated postural overload in tractor drivers, however, is greater than in many other categories of workers such as excavators, wheel loaders and tram drivers (Gores et al. 2010; Peretti et al. 2019).

This is probably due to the characteristics of the sector, such as (Peretti et al. 2015): a large number of family-run farms or farms with few workers, work activities that differ considerably depending on the season and the crops grown, discontinuous use of machinery, the presence of vehicles with different characteristics, highly variable levels of exposure and little health surveillance. Unfortunately, these aspects also have negative repercussions on the information-training of workers and on risk awareness. Moreover, in our country, machines are particularly obsolete because about 75% of the tractors in circulation are over 25 years old, partly because it has always been particularly difficult to renew these vehicles. In recent years, the registration of tractors has been increasingly reduced, reaching an all-time low (less than 18,000 units in 2020) due to the pandemic that has affected the purchase of new vehicles.

Vibrations, on the other hand, are a very complex phenomenon and are correlated to a number of factors such as: the operation carried out and the operating machine used, the forward speed of the machine, the profile and type of soil, the engine revs, the characteristics of the seat, etc.

Much research has been carried out to measure the levels of whole-body vibration to which tractor drivers are subjected during agricultural activity (Scarlett et al. 2007; Pessina and Bonalume 2009). These authors found high levels of exposure during transfers and in the use of old tractors, where the seats are often deteriorated or without an efficient suspension system, and this can pose risks to the safety and health of workers if all the technical prevention and protection measures and all the organisational measures that can be concretely implemented in the workplace are not correctly assessed and implemented by the employer.

Vibrations transmitted to the whole body entail risks, in particular lumbago and traumatism to the spinal column. In fact, vibration spectra measured on tractor seats show maximum accelerations between 2.5 and 6 Hz (Bovenzi and Betta 1994; Scarlett et al. 2007; Bonomini et al. 2015). Biodynamic studies have shown that the lumbar spine of a seated subject subjected to vibrations has a resonance frequency precisely between 2 and 6 Hz and the vibrations transmitted by tractors therefore excite the lumbar spine at its resonance frequency and the vertebral area most involved is L4-L-5 where the maximum compressive force develops during all tractor driving operations (including use of the pedals, controls and rear implement control).

One of the factors that has the greatest influence on the increase in the vibration phenomenon is positively correlated with the increase in forward speed (Pessina 1993) and this phenomenon is most evident on dirt roads. Other factors are influenced by the hours of driving, the irregularity of the ground, the non-uniformity and inflation of the tyre and the inclination, the radius of curvature, the base and wear of the seat and suspension.

If the ground is easily mouldable, i.e. wet or soft, it affects the tyre-ground impact negatively (Laib 1995) because the particles that make up the ground affect the behaviour of the tyre. On soil, other studies have been carried out and it has been observed that a dry soil conducts less vibration to the operator in comparison to dry soil and that the driver with the higher protein/body fat mass reduces these vibrational phenomena even more (Hamid 2016). Another interaction is between implements and vibrations, in fact, the response to daily static compression dose on the spine increases slightly with increasing tilling depth (Singh et al. 2019) up to 0.12 m. Thereafter, showing a sudden decrease with increasing milling depth to 0.14 m. This confirms that the vibration transmitted to the driver's body tends to decrease as the milling depth increases. This corroborates the study showing that the interaction between the soil particles and the tilling tool could delay the flow of vibrations due to increased cohesion and friction, according to analyses of the propagation and attenuation characteristics of vibrations to different soils, as shown by other studies (Kim and Lee 2000).

Another important variable is the seat as it can dampen vibrations and thus have greater comfort and in a study (Mehta and Tewari 2010) comparing different materials, thicknesses and densities, it was possible to state that high intensity polyurethane padding is the best material for dampening the vibrations transmitted by the tractor in dynamic conditions.

In addition to this aspect, (Tewari and Prasad 2000) an experiment was conducted to measure the pressure distribution on the seat and backrest influenced by: the seat plane, the curvatures of the backrest and its angle of inclination. It is concluded that a seat with a radius of curvature of 75 cm, a backrest with a radius of curvature of 30 cm and a backrest angle of 10° are the most suitable parameters for operators to improve comfort. It should also be remembered that there are also disorders affecting the cervical rachis (base of the neck, arm and shoulder) and the right side of the body, which is used more because it is the side where the controls are usually located in the cab and is the side of torsion for rear control of the vehicle, and also influencing the disorders of the thigh is the incorrect adjustment of the seat. Other tests were carried out by comparing a wheeled tractor and a tracked one during the execution of some work carried out in a vineyard in Sicily with varying slopes of the terrain (Catania and Vallone 2013), where vibrations are generally not affected by slope variations. The operating machines coupled to the wheeled tractor always record higher vibration values on the seat than the same operating machines coupled to the crawler tractor. This may be attributed to the fact that the wheeled tractor has both a smaller contact surface with the ground and a lower overall mass of the vehicle compared to the crawler tractor.

Through other tests (Vallone and Bono 2016) carried out on the basis of the age of the tractor and the number of working hours, it was possible to identify three classes of vibration risk: up to 3000 working hours, the risk is low, between 3000 and 6000 h, the risk is medium, and above 6000 h, the risk becomes high.

Another variable taken into account is the operators' weight, 63, 69 and 78 kg respectively, it was observed that this also affects the vibration load the operator may experience, so that a person with a higher body mass suffered less vibration disturbance (AL-Mafrachi 2016).

Great emphasis is placed on a study conducted in Italy (Bonalume 2012) aimed to improve the database of vibrational phenomena of vineyard tractors including the major operations that must be carried out in a vineyard, which will be classified and explained below.

On the basis of the state of the art, therefore, the University of Udine has undertaken a study aimed at investigating the vibrational phenomenon related to vineyard operations and the relationship to evaluate the level of vibrations produced by different types of tractors - also utilised with different operating machines connected - and the magnitude of vibrations impacting the whole human body in case of drivers with different Body Mass Index (Kg/m^2).

This kind of study is important because the demand for new tractors in viticulture were:

- higher forward speeds for decreased transport time and displacements;
- more working time for reduce fixed costs;
- possibility to operating with different machines front or rear attached;
- very large offer of models (narrow tractors, tool carriers, half crawler, over-the-row tractors, self-propelled as harvesters).

2 Methods

The tests followed two methodologies:

1) one survey sheet where were harvest of all parameters about tractors when operate with equipment in vineyard and so the data concerning:

   - vineyard tractors as a brand, power, age, hours, model, gearbox, type, structure, suspension, cab, seat, tyres, warn, speed;
   - cultivation: variety, dimension, soil conditions, track conditions, slope;
   - operation: type, machinery, model, attach (mounted, trailer);
   - worker: gender, tall, weight, BMI (Body Mass Index). If BMI is less than 18.5, it falls within the underweight range; if is 18.5 to < 25, it falls within the healthy weight range and if is 25.0 to < 30 or more, it falls within the overweight or obesity range;

2) second test with two tractors (one obsolete) and one new one and with one driver normal BMI: 2 different speed (6–10 km/h); 3 different position to measuring (seat, lumbo-sacral, cervico-dorsal junction) (Table 1).

Table 1. An example of sheet built about tractors tested.

n. test	Tractor Front axle suspension	Power kW	Year	hours	seat	Tires type front	radius inc	type rear	radius inc	speed km/h	Operation	VIBRATIONS a_{RMS} m/s^2 x	y	z	a_{PEAK} m/s^2 x	y	z
1	yes	63	2015	4101	pneu	340/65	18	380/70	28	5,5	topping	0,174	0,381	0,325	1,150	1,706	1,521
2	yes	57	2010	10310	mecc	320/65	24	380/70	24	7,1	spraying	0,247	0,334	0,413	1,941	1,769	1,731
3	yes	77	2019	801	pneu	340/65	18	440/65	28	3,5	soil working	0,302	0,312	0,290	2,516	1,320	1,239
4	not	59	2009	12800	mecc	340/65	18	420/70	28	4,2	shredding	0,316	0,278	0,603	3,101	1,324	5,896
5	not	73	2019	1500	pneu	320/65	20	320/65	20	4,2	travelling	0,375	0,456	0,263	2,153	1,844	1,222
6	not	73	2019	1500	pneu	320/65	20	320/65	20	4,2	spraying	0,461	0,376	0,470	2,499	2,149	2,692
7	yes	71	2018	2039	pneu	280/70	16	360/70	24	4,2	spraying	0,411	0,536	0,400	3,481	2,465	1,802
8	yes	59	2009	12800	mecc	340/65	18	420/70	28	4,2	flamer	0,256	0,238	0,322	2,371	1,475	1,921
9	yes	59	2009	12800	mecc	340/65	18	420/70	28	4,2	flamer	0,198	0,199	0,314	1,573	1,016	1,426
10	yes	59	2009	12800	mecc	340/65	18	420/70	28	4,2	flamer	0,186	0,189	0,345	1,629	1,041	1,632
11	yes	59	2009	12800	mecc	340/65	18	420/70	28	4,2	flamer	0,184	0,185	0,300	1,365	0,834	1,395
12	yes	59	2009	12800	mecc	340/65	18	420/70	28	4,2	flamer	0,191	0,184	0,327	1,120	0,873	1,436
13	yes	59	2009	12800	mecc	340/65	18	420/70	28	4,2	flamer	0,218	0,185	0,269	1,995	0,955	1,289
14	yes	59	2009	12800	mecc	340/65	18	420/70	28	4,2	flamer	0,186	0,189	0,382	1,145	0,859	1,816
15	yes	59	2009	12800	mecc	340/65	18	420/70	28	4,2	flamer	0,374	0,289	0,600	3,098	3,023	5,025
16	not	68	2015	4150	mecc	340/65	18	420/70	28	5,5	fertilizazion	0,320	0,231	0,303	0,950	0,560	0,834
17	not	69	2020	920	mecc	340/65	18	420/70	28	4,2	shredding	0,424	0,467	0,638	4,063	3,841	6,794
18	not	75	2007	3800	mecc	280/70	18	420/70	28	10,2	spraying	0,294	0,519	0,747	2,164	3,654	15,409
19	not	64	2018	2889	pneu	280/70	18	420/70	28	2,2	defoliation	0,151	0,175	0,243	1,637	0,936	1,330
20	not	75	2017	2550	pneu	280/70	16	420/70	24	1,9	defoliation	0,215	0,204	0,379	1,537	1,092	8,693
21	not	75	2017	2218	pneu	280/70	18	420/70	28	3,2	defoliation	0,242	0,236	0,508	4,938	3,742	13,850
22	not	67	2015	3110	pneu	320/65	24	420/85	30	7,2	spraying	0,378	0,598	0,521	3,160	4,784	20,288
23	not	67	2019	1405	pneu	280/70	16	440/65	24	7,5	spraying	0,449	0,473	0,571	3,102	2,921	9,495
24	not	69	2007	10258	pneu	280/70	20	420/70	28	7,5	spraying	0,291	0,481	0,614	2,242	3,104	5,027
25	not	67	2006	6959	pneu	280/70	18	420/70	24	6,2	spraying	0,217	0,333	0,357	1,975	1,517	8,901

As far as the measurements on the driver are concerned, the Whole Body Vibration (WBV) was measured. The driver was a person with a BMI of around 22.

The survey considered two agricultural tractors registered respectively in 1978 (Fiat 680 2W tractor) and 2015 (New Holland T4 FAWD). Both machines are characterised by medium-low power, but they clearly differ in terms of the level of technical progress, especially with regard to equipment capable of improving the operator's driving conditions.

The two tractors under analysis, although technically very different, have similar power ratings: 50 kW for the Fiat 680, 44 kW for the New Holland T4. Both the Fiat 680 and the New Holland T4 have a cab with a platform suspended on silent-blocks and a seat with mechanical suspension. The measurements were carried out at two different speeds and tyre pressures while driving on a dirt road.

The speeds used for the field tests, which were similar for both tractors, were as follows:

- for the Fiat 680 tractor the V1 (low) speed was 10.9 km/h obtained using 1st gear at 1500 rpm engine speed, while the V2 (high) speed was 15.9 km/h with the gearbox in 4th gear and engine speed of 1600 rpm;
- for the New Holland T4 tractor, the V1 speed was 10.6 km/h in 1st gear and 1500 rpm engine speed, while the V2 speed was 16.1 km/h in 4th gear and 1600 rpm engine speed.

Vibrational levels were measured with a triaxial accelerometer, Larson & Davis HVM 200, with a cushion support (Fig. 1), fixed respectively to the seat for lumbosacral measurements, to the back (at L10-L12 level) and to the neck of the operator for the verification of the vibrational impact at the level of the lumbosacral spine points (LS) about the mid-back (LL) cervical-dorsal junction (CD). In this case not exist a specific normative for positioning accelerometers and the choice was selected with medical team.

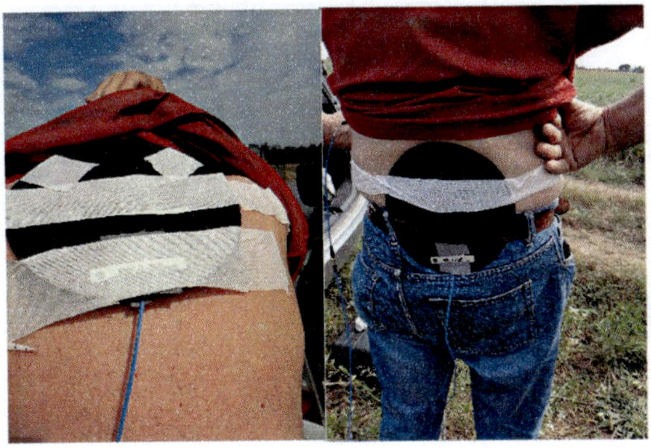

Fig. 1. Triaxial accelerometer positioning over the cervical-dorsal junction and to the back (lumbar area).

All measurements were carried out according to ISO 2631. The filter for x, y and z axis respectively was 1,1,1.4, for CD measurements the cushion was rotate (x axis become z axis) and change filter to1.4 from 1 values.

The time exposure for calculate A(8) was depending of work condition and was long in function to be representative to working day (reported in Run Time in figure in results).

The tests were carried out at the "Antonio Servadei" didactic-experimental farm of the University of Udine.

The field test was carried out by having the agricultural machines operate on a dirt road of approximately 500 m: the aim of this procedure was to have sufficient bearing capacity, so that the soil would have characteristics of stability and not deformability, following repeated stresses transmitted by the tractors in transit.

The experimental scheme therefore envisaged the use of two different tractors at two speeds, with the carrying out of body measurements on the rachis at three points - lumbar spine, lumbosacral spine and at the cervical-dorsal junction - for a total of twelve measurements according to the scheme 2 speeds × 2 tractors × 3 body measurements = 12 measurements.

3 Results

We have measured the various operations in viticulture and are now measuring harvesting operations with the grape harvester and transport trailers (Fig. 2).

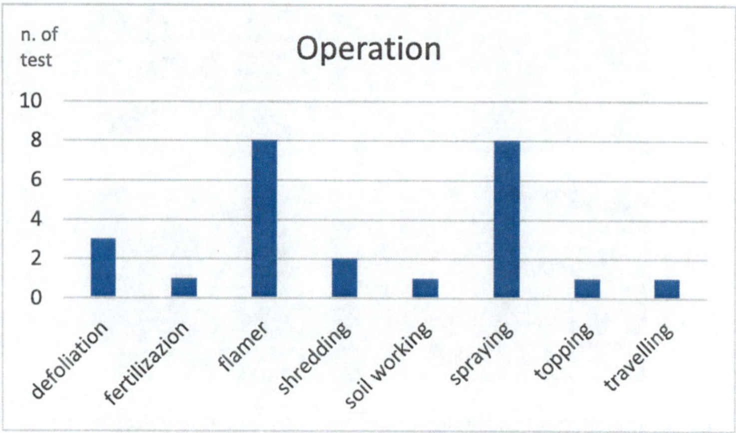

Fig. 2. Number of test in different vineyard operations.

The average of tractors is 8.15 old and with 6948.4 h. This means that in the vineyard the tractors are relatively recent and in fact the average working hours/year is 855.7, which is very high compared to other farms in this area.

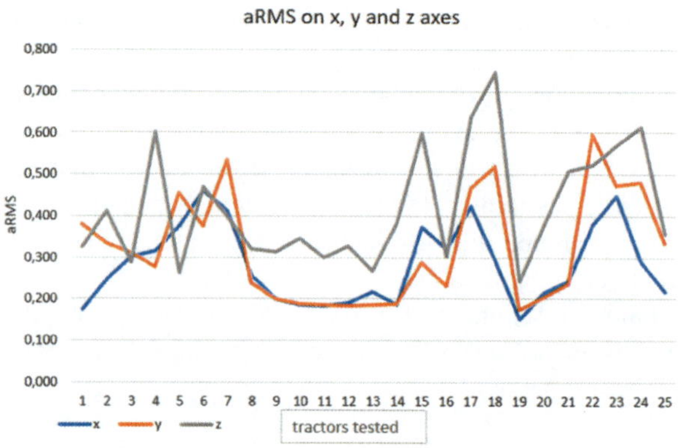

Fig. 3. Sum of Arms for all tractors tested.

The aRMS for the three axles x, y and z (Fig. 3), we can see that few tractors overcome the threshold also half of them were with pneumatic seats and front suspension, are above the warning limit of 0.5 m/s^2 of acceleration stability to be able to work 8 h.

Table 2. Summary sheet for each test with vibration tools and sound meter, first case.

Spraying, vineyard, lowland

Overall Data

Start Date and Tim 2021-lug-28 10:19:55

Run Time (hh:mm: 00:16:17

	x	y	z	Sum	Units	RUMORE dB	
a_{RMS}	0,2168	0,3334	0,3574	0,6616	m/s²	L_Aeq	85,60
MTVV	0,7798	0,8547	1,6243	1,6549	m/s²	L_{cpeak}	122,50
a_{PEAK}	1,9747	1,5175	8,9008	8,9331	m/s²		
a_{MIN}	0,0010	0,0013	0,0011	0,0029	m/s²		
A(8)	0,0559	0,0860	0,0658	0,0860	m/s²		
A(8) Act.	21,72	9,18	15,65	9,18	Hours		
A(8) Exp.	>24	>24	>24	>24	Hours		
VDV	2,4868	3,5591	3,6813	3,6813	m/s^1.75		
Exposure Points				3	Points		

Spraying, vineyard, hill

Overall Data

Start Date and Tim 2021-lug-29 09:29:54

Run Time (hh:mm: 00:10:47

	x	y	z	Sum	Units	RUMORE dB	
a_{RMS}	0,3781	0,5976	0,5214	1,1190	m/s²	L_Aeq	73,90
MTVV	1,3554	2,2934	3,5403	3,5621	m/s²	L_{cpeak}	117,40
a_{PEAK}	3,1602	4,7838	20,2876	20,3304	m/s²		
a_{MIN}	0,0076	0,0088	0,0222	0,0283	m/s²		
A(8)	0,0793	0,1254	0,0782	0,1254	m/s²		
A(8) Act.	7,14	2,86	7,36	2,86	Hours		
A(8) Exp.	>24	11,43	>24	11,43	Hours		
VDV	4,4286	7,1353	7,3663	7,3663	m/s^1.75		
Exposure Points				6	Points		

In Table 2 are shows two examples of the summary sheets for each type of test. We see that in the first case on the left the vibrations are low even though it is a 15 year old tractor, but for example the noise is very high. In the second sheet, the tractor is new with a pneumatic seat, but we are working on hills with mounted mistblower and the vibrations exceed the attention limit of acceleration. These vibrations are due to the typical irregular terrain caused by runoff by the strong and intense rains to which the Friulian hill is subject.

Table 3. Summary sheet for each test with vibration tools and sound meter, second case.

Defoliation, wineyard, lowland

Overall Data

Start Date and Tim 2021-giu-22 14:41:11

Run Time (hh:mm: 00:13:58

	x	y	z	Sum	Units	RUMORE dB	
a_{RMS}	0,1515	0,1753	0,2427	0,4051	m/s²	L_Aeq	80,90
MTVV	0,6322	0,3982	0,4559	1,0104	m/s²	L_{cpeak}	114,90
a_{PEAK}	1,6373	0,9361	1,3301	2,4791	m/s²		
a_{MIN}	0,0123	0,0173	0,0234	0,0394	m/s²		
A(8)	0,0362	0,0419	0,0414	0,0419	m/s²		
A(8) Act.	>24	>24	>24	>24	Hours		
A(8) Exp.	>24	>24	>24	>24	Hours		
VDV	1,9003	1,8114	1,7046	1,9003	m/s^1.75		
Exposure Points				1	Points		

Operation shredding, vineyard lowland

Overall Data

Start Date 2021-giu-22 16:41:05

Run Time 00:13:39

	x	y	z	Sum	Units	RUMORE dB	
a_{RMS}	0,2422	0,2363	0,5076	0,6943	m/s²	L_Aeq	76,40
MTVV	1,8589	1,7629	2,5574	3,7095	m/s²	L_{cpeak}	-
a_{PEAK}	4,9383	3,7422	13,8497	13,9907	m/s²		
a_{MIN}	0,0149	0,0156	0,0584	0,0682	m/s²		
A(8)	0,0571	0,0558	0,0856	0,0856	m/s²		
A(8) Act.	17,40	18,27	7,76	7,76	Hours		
A(8) Exp.	>24	>24	>24	>24	Hours		
VDV	4,6604	3,9213	5,9523	5,9523	m/s^1.75		
Exposure Points				3	Points		

Two more sheets on lowland vineyards with two different operations are show in Table 3, defoliation and chopping. The limits are also lower because the speed is low and the soil is well tilled. In the case of the sheet on the left with chopping, however, the z-value is higher than the limit, which suggests working less than 8 h per day.

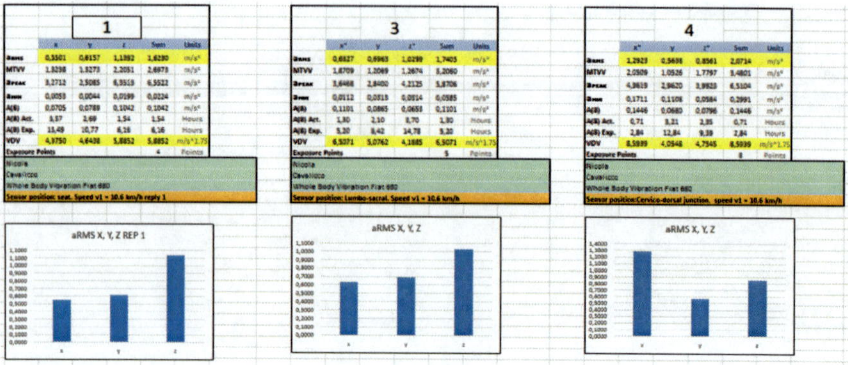

Fig. 4. Different vibration on x, y and z axis (blue bar) for 3 different position (A, B, C) on the body with tractor FIAT 680 at V1 of speed. (Color figure online)

The test made with an old tractor with an undamped seat and only the safety frame with a healthy weight range person (Fig. 4) show that the vibrations on the z axis are very high given the high speed for the type of terrain, uncultivated field and dirt road. In the test with lower speed they were lower but still above the limit of 0.5 m/s^2. In the lumbar-sacral area they do not decrease, while in the cervical area (number 3 on the right) they increase in the x axis due to the contribution of the arms caused by the vibration of the steering wheel.

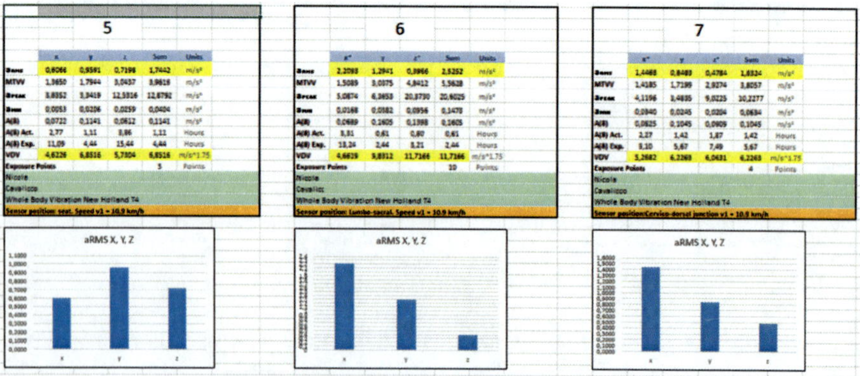

Fig. 5. Different vibration on x, y and z axis (blue bar) for 3 different position (A, B, C) on the body with tractor NH T4 at V1 of speed. (Color figure online)

In this other test with a tractor (Fig. 5) equipped with a cushioned cab and a mechanical seat with parallelogram suspension, we note the reduction of the component on the vertical z axis, while the y and x components increase due to rolling, pitch and serpentine movements. Also in this case, the component on the x axis increases due to the contribution of the arms on the steering wheel (also it is a modern steering wheel equipped with a hydraulic piston). It is obvious, on the other hand, that these are the highest speed measurements. Under these conditions we generally keep a similar speed to the 6 km/h test where vibrations are more contained. It was not possible to compare two different BMIs because one of the drivers had a problem and the test was deleted.

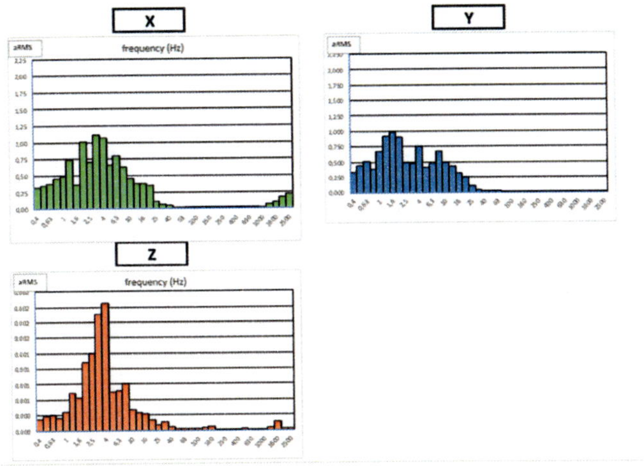

Fig. 6. Frequency analysis over 3 axes, x, y, z.

A frequency analysis of the vibrations shows that vibrations between 2 and 4 Hz are the most common (Fig. 6).

Biodynamic studies have shown that the lumbar spine of a seated person subjected to vibration has a resonance frequency between 2 and 6 Hz. The vibrations transmitted by tractors therefore excite the lumbar spine at its resonance frequency. The result is a mechanical overload caused by vertical displacement of the adjacent vertebral bodies.

If a driver is seated and exposed to vibrations in a vertical direction, he may have a disturbance or problem with the cervical vertebrae at 3–4 Hz, the lumbar vertebrae at 4 Hz, and the scapular girdle at 5 Hz.

4 Conclusions

This is a work in progress and with the collaboration of the physicians we are trying to get the relationship about machinery type an BMI of the driver.

A preliminary analysis was carried out on a suitable number of tractors with different operating machines in the vineyard and it was noted that where the soil is damaged the vibrations are beyond the legal limit.

During tests with two tractors, the high vibration of the seat at high speeds showed, in the three different positions, the increase in aRMS values in the cervico-dorsal area, probably due to the contribution of the steering vibrations.

We are planning future tests in order to obtain a model able, on the basis of the BMI of each person, to see the vibrational limits an operator may undergo in function of the different machines. So in this case the study like overcomes the simple limit 0.5 m/s^2 ruled by the law by for a machine. The real vibrational limit not is due only to a vibrational mode of the tractor but also to the transmission of vibrations on to the human body.

References

Bonomini, F., et al.: Analisi delle vibrazioni su trattori agricoli in normali condizioni di lavoro e in condizioni predefinite. Rivista Italiana di Acustica **39**(3) (2015)

Bonalume, V.: Il comfort del posto di lavoro sulle macchine agricole: esame delle criticità e formulazione di soluzioni. Milano: Università degli studi di Milano. Università degli Studi di Milano, 2012 June 12. (24. Ciclo), Anno Accademico 2010/2011 (2012)

Bovenzi, M., Betta, A.: Low-back disorders in agricultural tractor drivers exposed to whole-body vibration and postural stress. Appl. Ergon. **25**(4), 231–241 (1994)

Catania, P., Vallone, M., Alleri, M., Morello, G., Spartà, G., Febo, P.: Whole body vibrations during field operations in the vineyard. J. Agricult. Eng. **44**(2s), e143 (2013)

Gores, B., Ellegast, R., Hermanns, I., Raffler, N., Rissler, J.: Assessing combined exposures of whole body vibration and awkward posture – further results from application of a simultaneous field measurement methodology. Ind. Heath **48**, 638–644 (2010)

AL-Mafrachi, A.A.A.H.: Whole body, arm-hand vibration and performance drivers tractors during conservation tillage under different velocity and soils. Int. J. Adv. Eng. Manag. Sci. (IJAEMS) **2**(1), 239362 (2016)

Kim, D.-S., Lee, J.-S.: Propagation and attenuation characteristics of various ground vibrations. Soil Dyn. Earthq. Eng. **19**(2), 115–126 (2000)

Laib, L.: Analysis of the vibration-excitation effect caused by deformable soil surfaces. J. Terramechanics **32**(3), 151–163 (1995)

Mehta, C.R., Tewari, V.K.: Damping characteristics of seat cushion materials for tractor ride comfort. J. Terramechanics. **47**(6), 401–406 (2010)

Peretti, A., Bonomini, F., Pasqua di Bisceglie, A.: Analisi delle vibrazioni su trattori agricoli in normali condizioni di lavoro e in condizioni predefinite. Rivista Italiana di Acustica **39**(3), 90–103 (2015)

Peretti, A., Carletti, E., Pedrielli, F., Jacopo, G.: Vibration tests on an agricultural tractor equipped with advanced vibration control systems for different configurations. Atti di convegno (in stampa). Società italiana di Acustica (2019)

Pessina, D., Bonalume, V.: Esposizione a vibrazioni nella filiera viticola. Convegno AIIA, Ischia, 12–16 settembre (2009)

Pessina, D.: Valutazione delle vibrazioni sul sedile di guida di un trattore convenzionale equipaggiato con differenti pneumatici radiali. Atti del convegno nazionale AIGR. Il ruolo dell'ingegneria per l'agricoltura 2000, pp. 429–440 (1993)

Scarlett, A.J., Price, J.S., Stayner, R.M.: Whole-body vibration: evaluation of emission and exposure levels arising from agricultural tractors. J. Terramechanics **44**(1), 65–73 (2007)

Singh, A., Nawayseh, N., Singh, L.P., Singh, S., Singh, H.: Investigation of compressive stress on lumbar spine due to whole body vibration exposure in rotary tillage operation. Int. J. Automot. Mech. Eng. **16**(2), 6684–6696 (2019)

Tewari, V.K., Prasad, N.: Optimum seat pan and back-rest parameters for a comfortable tractor seat. Ergonomics **43**(2), 167–186 (2000)

Vallone, M., Bono, F., Quendler, E., Febo, P., Catania, P.: Risk exposure to vibration and noise in the use of agricultural track-laying tractors. Ann. Agric. Environ. Med. **23**(4), 591–597 (2016)

Abatement Systems Efficiency: Comparison of PM Emitted by Biomass Combustion

Andrea Colantoni[1], Enrico Paris[2], Monica Carnevale[2], Beatrice Vincenti[2], Adriano Palma[1,2], Leonardo Bianchini[1], and Francesco Gallucci[2(✉)]

[1] Department of Agriculture and Forestry Science (UNITUS-DAFNE), Tuscia University, Via San Camillo de Lellis snc, 01100 Viterbo, Italy

[2] Consiglio per la ricerca in agricoltura e l'analisi dell'economia agraria (CREA) - Centro di ricerca Ingegneria e Trasformazioni agroalimentari, Via della Pascolare 16, 00015 Monterotondo (Rome), Italy

francesco.gallucci@crea.gov.it

Abstract. In recent years, the use of renewable energy is an increasingly developed and studied topic. It is very important to evaluate the environmental impact due to the exploitation of such innovative energies. Among the different renewable sources, the energy conversion of biomass makes it possible to obtain energy from naturally and cyclically bioavailable waste products. Biomass burning is the main thermochemical process used to produce heat and energy which provides a useful system for the management and disposal of vegetal residues such as pruning and agroforestry waste that could risk being left on the ground and be eliminated through the practice of open burning. Combustion is a phenomenon that inevitably generates PM and others atmospheric pollutants therefore this also occurs in the case of biomass burning. In this work, the differences in terms of environmental impact between controlled and uncontrolled biomass combustion systems are demonstrated.

Keywords: Biomass burning · Abatement systems · Combustion · PM emission

Agroener Project "Energy from agriculture: sustainable innovations for the bioeconomy" of CREA (Council for Agricultural Research and Analysis of Agricultural Economics) Project Financing MiPAAF Executive Decree no. 26329 of 1 April 2016.

1 Introduction

The use of renewable sources is constantly increasing in the European countries, in fact, the global policy aimed to reduce greenhouse gases emissions. Biomass play an important role in the energy conversion field for heat and power production and to declining fossil fuels, they are available in various forms and can be obtained from different sources (Baxter et al. 1998; Frandsen 2005). Their advantages from which the recent great development and demand is derived, is represented by the capacity to reduce

M. Biocca et al. (Eds.): SHWA 2020, LNCE 252, pp. 272–277, 2022.
https://doi.org/10.1007/978-3-030-98092-4_29

the carbon dioxide emissions and global warming, with benefit on management and disposal solutions (Wang et al. 2012; Werther et al. 2000). Despite such peculiar characteristics, there is a need for the use of dedicated abatement systems to control emissions. Indeed, although biomass are considered CO_2 neutral and the new biomass conversion systems have achieved low carbon monoxide and nitrogen oxides emission levels (Arvelakis et al. 2002), the great disadvantage is represented by the high level of fly ash and particulate matter emissions into the air. PM production from combustion systems depends on furnace type, combustion parameters and physical-chemical characteristics of the biomass. Several abatement system are now available, such as cyclones, electrostatic precipitators, wet scrubbers, bag filter and flue gas condensers (Hamilton et al. 2010). The uncontrolled combustion of biomass (forest fires, use of non-certified domestic fireplaces, combustion of pruning on the sidelines) is called *open burning* and leads to the release into the atmosphere of a higher number of contaminants and pollutants. *Open burning* process has a significant atmospheric impact in terms of PM emitted, and so it can be considered an unacceptable practice for the atmospheric preservation. The use of abatement systems, on the other hand, allows biomass to be used in energy conversion processes by direct combustion. The high level of atmospheric pollutants produced by biomass combustion (Boman et al. 2011; Roden et al. 2009) represents one of the main limit of the use of biomass heating in many European countries and in some Italian regions, because of various environmental issues. It has been studied (Caserini et al. 2005) that PM concentration is strictly associated to the air quality and may be negatively affected by an intense use of biomass. The optimization of the thermochemical conversion process and the use of abatement efficiency systems (Bologa et al. 2010; Migliavacca et al. 2014) are among the most relevant factors in the study of conditions to reduce the pollutant emission. Since pollutants from biomass burning are considered toxic and carcinogenic potential (Dockery et al. 1993; Donaldson et al. 2002), new technologies and systems are being studied to reduce emissions (Nussbaumer 2005). In this work it was evaluated the difference, in terms of TSP (total solid particle) emitted, between open burning (OB) and controlled combustion with only multicyclones filter (C) and with multicyclones + bag filters (C+F).

2 Material and Methods

2.1 Biomass Characterization

The chemical and physical characterization of biomass was carried out at the LASER-B (Laboratory for Experimental Activities on Renewable Energy from Biomass) of CREA-IT. In this study, several fruit tree prunings were used for the combustion tests: samples of plum, citrus, olive, poplar and elm tree were characterized in order to determine the moisture content, the heating value, the ash content, and the elemental composition (Carnevale et al. 2019, 2021; Gallucci et al. 2018; Proto et al. 2021). The moisture content was determinated using a Memmert UFP800 oven at 105 ± 2 °C for 24 h according to the UNI EN ISO 18134-2:2017. The elemental analysis was carried out by the Costech ECS 4010 CHNS-O, to determine the content of carbon (C), hydrogen (H) and nitrogen (N) according to the UNI EN ISO 16948:2015. The higher heating value (HHV) was determined using an Anton Paar 6400 isoperibol calorimeter according to the UNI EN ISO 18125:2018. The lower heating value (LHV) was calculated considering the HHV and the hydrogen content. The ash content was measured using a Lenton EF11/8B muffle furnace according to the UNI EN ISO 18122:2016.

2.2 Experimental Campaign and Sampling System

The biomass were burned in plants with different abatement systems: an 80 kW_{th} boiler (GSA-80, D'Alessandro Termomeccanica Miglianico, CH, Italy) without any abatement system to simulate the open burning process, a 30 kW_{th} boiler (CSA 30-100 GM, D'Alessandro Termomeccanica Miglianico, CH, Italy) equipped with only the multi-cyclonic filter and a 350 kW_{th} demonstrative cogeneration plant based on a moving-grate furnace, with a steam generator (500 kg/h at 1.2 MPa) and equipped with multicyclonic and bag filters. During combustion, an isokinetic sampling system was used to collect the PM subsequently determined gravimetrically. The sampling system is equipped by the HP5 Dadolab probe and the ST5 Dadolab isokinetic sampler respecting the EN13284-1:2017 (Stationary source emissions - Determination of low range mass concentration of dust - Part 1: Manual gravimetric method). PM was sampled by glass microfiber filters (Ø 47 mm) and quantified with gravimetric analysis using a Mettler Toledo AL104 Analytical Balance placed in a conditioned room at 20 °C and 50% humidity. It was assumed that the combustion process could influence the amount of solid particles emission during biomass burning, and the measurements are carried out only during the phase in which the plants are in full load operation in order to avoid the influence of the part load conditions and incomplete burning phenomena (Fig. 1).

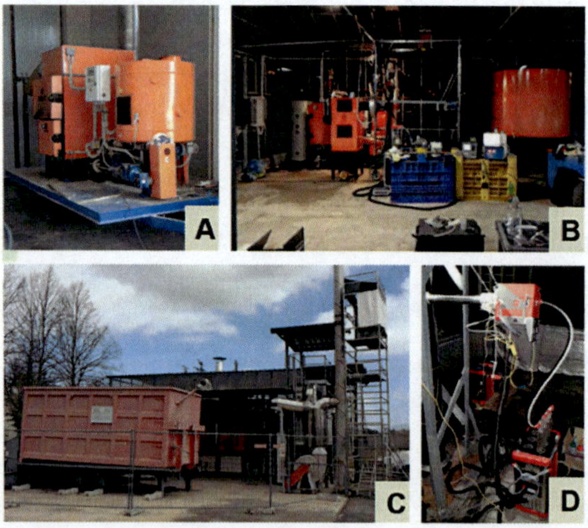

Fig. 1. A) The 80 kW_{th} boiler (GSA-80, D'Alessandro Termomeccanica Miglianico, CH, Italy) without any abatement system; B) 30 kW_{th} boiler (CSA 30-100 GM, D'Alessandro Termomeccanica Miglianico, CH, Italy) equipped with only the multi-cyclonic filter; C) 350 kW_{th} boiler equipped with multicyclonic and bag filters; D) The HP5 Dadolab probe and the ST5 Dadolab isokinetic sampler

3 Results and Discussions

The data used, normalized to be able to compare different plants, were obtained both from direct surveys and from results obtained from previous works (Proto et al. 2021). In order to compare the results obtained in different operative conditions, it was decided to normalize the results about emission from Poplar, Elm combustion and from emission of agricultural biomass burning in the mentioned work. The data are normalized considering the emission factor, and the comparison is shown in Fig. 2.

3.1 Physical and Chemical Characterization

The emissions mainly depend upon the plant and combustion conditions, closely related to the quality of combustion that can be influenced mainly from the operative condition, boiler structure and characteristics but also from the chemical and physical biomass composition. The results of biomass characterization are shown in the following table (Table 1).

Table 1. Physical and chemical characterization of biomass. ([1] Proto et al. 2021; [2] Gallucci et al.)

Biomass	Moisture (%)	LHV (MJ/kg)	C (%)	H (%)	N (%)	Ash (%)
Plum	23.88	18.96	51.09	3.35	2.90	1.03
Citrus [1]	12.31	17.95	48.86	5.75	0.35	1.80
Olive [1]	9.30	18.68	49.58	5.00	0.21	0.90
Poplar [2]	43.3	17.94	47.36	5.58	0.31	3.12
Elm [2]	27.7	19.40	52.11	6.12	0.30	5.09

As regards the sulphur content, it was found for each biomass below the limit of quantification (LOQ).

3.2 PM Emissions

The different tree biomass has been tested in different combustion condition: Open burning (OB), with multi-cyclonic filter (C), with multi-cyclonic + bag filter (C+F). The results obtained from PM monitoring were reported at the oxygen content of 11% as required by Italian Legislative Decree 183/17, and are shown in the Fig. 2.

By standardizing and normalizing the results, it was possible to compare the quantity of PM released with and without abatement systems. It is undoubt that open burning is a phenomenon that generates large quantities of PM, as to constitute a dangerous result for the areas and the neighbouring population to the phenomenon of combustion. Multi-cyclonic filters are able to trap relatively large particles and so they are often used for pre-cleaning processes of the flue gas before the flue gas condensation unit (Carlsson 2008; Strand et al. 2002). The exploitation of two abatement systems together, multi-cyclone and bag filter, allows to significantly reduce the PM emitted. The red line represents the limit of 75 mg/Nm^3 imposed by the legislation in force in Italy for emissions from small boilers.

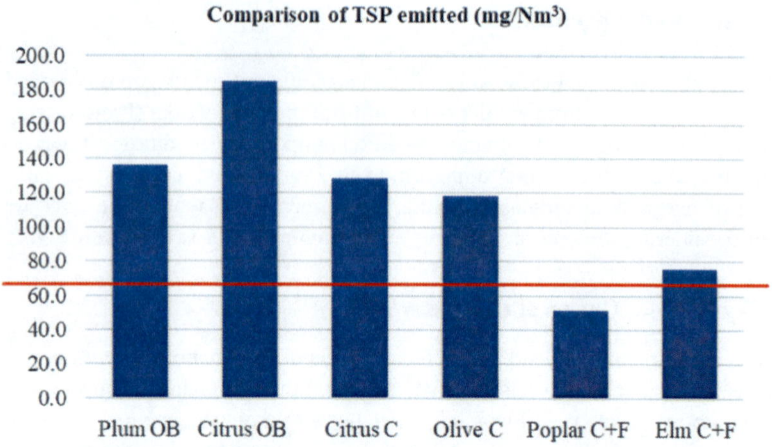

Fig. 2. Comparison of normalized TSP emitted from different essences during combustion in boilers equipped with different abatement system. The red line represents the limit of 75 mg/Nm^3 (Color figure online)

4 Conclusions

PM emissions vary depending on the biomass physical and chemical composition, combustion conditions and abatement systems used. The high level of pollutants and the formation of fly ash in the flue gas, represent the main difference between natural gas combustion and biomass combustion. In particular, PM concentration is a topic that has the attention of the global environmental policies. The use of biomass plants and related abatement systems allow to obtain an efficient energy conversion from agro-forestry waste products. Moreover, the biomass boilers with adequate abatement systems represent the effective method of disposal of residual biomass that otherwise risk being burned in the open field or being involved in fires, processes that have a high atmospheric impact. Finally, the combustion process is not the only critical factor for the PM emission considering optimal combustion and not relevant differences in the composition biomass (lignocellulosic biomass) as the wood chip samples from the different tree used for the study. In order to decrease the PM emitted through the results obtained from the study, the installation of appropriate equipment and abatement system have a significant role.

References

Arvelakis, S., Gehrmann, H., Beckmann, M., Koukios, E.G.: Effect of leaching on the ash behavior of olive residue during fluidized bed gasification. Biomass Bioenergy **22**(1), 55–69 (2002)

Baxter, L.L., et al.: The behavior of inorganic material in biomass-fired power boilers: field and laboratory experiences. Fuel Process. Technol. **54**(1–3), 47–78 (1998)

Bologa, A., Paur, H.R., Ulbricht, T., Woletz, K.: Particle emissions from small scale wood combustion devices and their control by electrostatic precipitation. Chem. Eng. Trans. **22**, 119–124 (2010)

Boman, C., Pettersson, E., Westerholm, R., Bostrom, D., Nordin, A.: Stove performance and emission characteristics in residential wood log and pellet combustion, part 1: pellet stoves. Energy Fuels **25**(1), 307–314 (2011)

Carlsson, K.: Gas Cleaning in Flue Gas from Combustion of Biomass. Thermal Net (2008)

Carnevale, M., et al.: Characterization of emissions from combustion of olive wood chips. In: European Biomass Conference and Exhibition Proceedings (2021)

Carnevale, M., et al.: Characterization of agricultural residues: physical and chemical analysis for energy transformations. In: European Biomass Conference and Exhibition Proceedings (2019)

Caserini, S., et al.: New Insight into the Role of Wood Combustion as Key PM Source in Italy and in Lombardy Region. In: 16th Annual International Emissions Inventory Conference "Emission Inventories: Integration, Analysis, and Communications" Raleigh, North Carolina, 14–17 May 2005 (2005)

Dockery, D.W., et al.: An association between air pollution and mortality in six U.S. cities. N. Engl. J. Med. **329**(24), 1753–1759 (1993)

Donaldson, K., et al.: The pulmonary toxicology of ultrafine particles. J. Aerosol Med. **15**(2), 213–220 (2002)

Frandsen, F.J.: Utilizing biomass and waste for power production - a decade of contributing to the understanding, interpretation and analysis of deposits and corrosion products. Fuel **84**(10), 1277–1294 (2005)

Gallucci, F., et al.: Assessment of organic micropollutants (PCDD/Fs and PCBs) from biomass combustion in a small CHP facility. In: European Biomass Conference and Exhibition Proceedings (2018)

Migliavacca, G., Morreale, C., Hugony, F., Tombolato, I., Pession, G.: Reduction of PM emissions from biomass combustion appliances: evaluation of efficiency of electrostatic precipitators. Chem. Eng. Trans. (2014)

Nussbaumer, T.: Dieselruss und Holzfeinstaub grundverschieden. Holz-Zentralblatt **131**(70), 932–933 (2005)

Proto, A.R., et al.: Assessment of wood chip combustion and emission behavior of different agricultural biomasses. Fuel **289**, 119758 (2021)

Roden, C.A., Bond, T.C., Conway, S., Pinel, A.B.O., MacCarty, N., Still, D.: Laboratory and field investigations of particulate and carbon monoxide emissions from traditional and improved cookstoves. Atmos. Environ. **43**(6), 1170–1181 (2009)

Hamilton, S., Fleming, S., Stewart, R.: The Assessment of Flue Gas Particulate Abatement in Wood Burning Boilers (2010)

Strand, M., et al.: Fly ash penetration through electrostatic precipitator and flue gas condenser in a 6 MW biomass fired boiler. Energy Fuels **16**(6), 1499–1506 (2002)

Wang, L., Hustad, J.E., Skreiberg, Ø., Skjevrak, G., Grønli, M.: A critical review on additives to reduce ash related operation problems in biomass combustion applications. Energy Procedia **20**, 20–29 (2012)

Werther, J., Saenger, M., Hartge, E.-U., Ogada, T., Siagi, Z.: Combustion of agricultural residues. Prog. Energy Combust. Sci. **26**(1), 1–27 (2000)

Occupational Health

Critical Issues in the Application of the INAIL Methodology for the Assessment of Work-Related Stress in the Agricultural Sector

Massimo Cecchini(✉), Leonardo Bianchini, Filippo Cossio, and Simone Riccioni

Department of Agriculture and Forest Sciences (DAFNE), University of Tuscia, Via San Camillo De Lellis, 01100 Viterbo, Italy
cecchini@unitus.it

Abstract. Employers are obliged to evaluate all the risks present within their company. Among the risks, that of work-related stress must also be assessed. In order to comply with this obligation, in Italy the Permanent Consultative Commission, on the provision of Legislative Decree 106/09, published the indications for the evaluation of work-related stress.

Studies carried out previously and confirmed by the National Institute for Occupational Safety and Health (NIOSH), had already shown that this pathology can have harmful effects on workers' health and on their safety, as stress is potentially able to increase the probability of accident events.

INAIL (National Institute for Insurance Against Industrial Injuries) has drawn up a questionnaire tool indicator which turns out to be, currently in the national territory, a practical reference tool used by the employers for carrying out the preliminary evaluation of work-related stress.

The applicability of this methodology to all production sectors, however, could make the risk incorrectly assessed, generating gaps, which could have harmful effects on the health and safety of workers. Based on these considerations, a study was carried out to assess the reliability of the application of the INAIL checklist on farms, in order to propose an adaptation of the same to the analyzed sector.

Thirty-two farms belonging to the agro-zootechnical and agricultural sector located in the Lazio region (Italy) were observed. The questionnaires proposed by the INAIL guidelines were administered in the farms. The questionnaires were submitted to workers and/or their representative for safety and health. Subsequently, the data were processed to identify any critical issues.

The study brought to light some critical issues in the application of the assessment based on the INAIL checklists applied to the agricultural sector. The most significant discrepancy is that relating to the sub-group "Work environment and Work equipment".

Keywords: INAIL guidelines · Health at work · Work safety · Work-related stress · Checklist

M. Biocca et al. (Eds.): SHWA 2020, LNCE 252, pp. 281–294, 2022.
https://doi.org/10.1007/978-3-030-98092-4_30

1 Introduction

The term stress as an effect on human health was coined in 1939 by the Austrian physician Hans Selye [1] who first studied the interactions of certain factors on human biological, physical and mental functions.

The concept was indirectly recognised in 1940 by the World Health Organisation (WHO) when it was established. The first recognised link between stress as a human pathology and work-related causes was made in 1999 by the National Institute for Occupational Safety and Health (NIOSH), which defined the concept of work-related stress and associated it with negative health effects related to work behaviour and actions and with work-related accidents.

Studies [2, 3] have shown that prolonged periods of stress [4], including work-related stress, can affect the health status of an individual, increasing physical and mental disorders [5, 6] that can alter behaviour, making it detrimental to individual health and well-being. This effect, which is also evident in the quality of the final work product, can lead to reductions in work effectiveness and efficiency. Work-related stress can lead to negative consequences for workers and companies.

It was in 2004, with the implementation of the European Agreement on work-related stress concluded in Brussels on 8 October 2004 between UNICE/UEAPME, CEEP and ETUC, that the importance of the topic was officially recognised. This agreement identified work-related stress as an international concern for both employers and employees, and led to the need for joint action to raise awareness and contain this issue.

At the Italian level the need to regulate this risk, defined as ‘psychosocial’, was already understood, especially in rescue workers after catastrophic events. The European Agreement of 2004 was recognised in Italy by trade associations and trade unions on 9 June 2008, through the drafting of the Interconfederal Agreement. Work-related stress became officially recognised at legislative level in 2008, with the issuing of the legislative decree 81/2008 as transposition of Directives 89/391/EEC, 89/654/EEC, 89/655/EEC, 89/656/EEC, 90/269/EEC, 90/270/EEC, 90/394/EEC, 90/679/EEC, 93/88/EEC, 95/63/EC, 97/42/EC, 98/24/EC, 99/38/EC, 99/92/EC, 2001/45/EC, 2003/10/EC, 2003/18/EC, 2004/40/EC, 92/58/EEC, 92/57/EEC, 2004/40/EC, 2002/44/EC, 2006/25/EC, 2004/40/EC.

A national law recognises, for the first time in Italy, the employer’s obligation to assess the risk of work-related stress to which its employees might be subjected; however, the law does not set out the methodologies on how to objectively and practically assess this risk, prompting research groups [7–11] to develop judgment techniques with the intention of identifying, through assessments [12–19], systems to reduce the impact of stress in the workplace.

In Italy, the enactment of legislative decree 106/2009 correcting and supplementing the consolidated act on health and safety in the workplace, entrusts the Permanent Advisory Commission with the obligation to draw up the methodology for assessing risk from work-related stress and the obligation for employers to draw up the specific risk assessment even in the absence of the above methodology.

The assessment methodological path established by the Permanent Advisory Commission is expressed on 18 November 2010 with the publication of a Circular of the Ministry of Labour and Social Policies entitled "Indications of the Permanent Advisory Commission for Health and Safety at Work for the assessment of work-related stress (articles 6, paragraph 8, letter m-quater, and 28, paragraph 1-bis of legislative decree no. 81/2008, as amended)". This circular specifies that the step-by-step methodological approach, proposed by the Permanent Advisory Commission, represents the minimum level of assessment that all employers should follow, without prejudice to the possibility for employers to follow a more articulated path based on specific business needs and complexities [20].

On the basis of the above, a study was carried out to assess the reliability of the application of the INAIL checklist in farms, in order to propose an adaptation of the checklist to the analysed sector.

2 Materials and Methods

The assessment of work-related stress risk, according to the indications of the Italian legislation, is carried out by the employer in collaboration with the Prevention and Protection Service Manager (PPSM) and with the competent doctor after consulting the Workers' Representative for Safety and Health (WHSR) or, in his absence, the Territorial Workers' Representative (TWHSR), considering homogeneous groups of workers exposed to risks of the same type, including managers and supervisors.

The methodological pathway identified by the Permanent Advisory Commission presupposes a preliminary assessment during which the objective, detectable and numerically appreciable indicators are analysed. The indicators to be analysed must, at the very least, belong to three distinct families: "Sentinel Events", "Work Content Factors" and "Work Context Factors", which can also be assessed by means of appropriate checklists. To carry out this phase correctly, it is necessary, especially for the last two indicators, to involve workers, the WHSR/TWHSR or, depending on the size of the company, a group of workers in the assessment. If, at the end of this phase, no risk factors are found, the result can be included in the company's risk assessment document and only a monitoring plan is needed. If, at the end of the preliminary phase, the presence of risk factors is highlighted, it will be necessary to carry out an in-depth assessment that takes into account the subjective perception by means of questionnaires, focus groups or semi-structured interviews. The indications provided by the Permanent Advisory Commission concerning the assessment of risk from work-related stress, leave the employer ample freedom on how to obtain data to reach an assessment that is quantifiable, objective and measurable. In order to standardise the assessment criteria and to provide valid tools to the employers, the Italian National institute for insurance against industrial injuries (INAIL) started to study the applicability to the Italian territory of assessment models and methodologies already used in other EU Countries [21].

The methodology provided by INAIL of the indicator tool questionnaire, translated, contextualised and adapted after years of study of the British HSE methodology, integrated with the proposals of the "National Network for the Prevention of

Psychosocial Distress in the Workplace", is currently a company reference within the national territory, to allow employers to comply with the obligation to assess work-related stress [22, 23].

Preliminary to the assessment are the steps of: setting up an assessment management team; developing a communication strategy and worker involvement; developing a risk assessment plan. These steps are necessary because the assessment of work-related stress risk is a dynamic, step-by-step process, with the need for follow-up steps.

The checklist used is composed of 76 questions, divided into three main macro areas: "Company Indicators Area", "Work Content Area" and "Work Context Area"; each macro area is in turn composed of a series of company indicators that can be grouped into sub-areas.

The "Company Indicators Area", also known as "Sentinel Events", is composed of 10 questions objectively detectable by the company personnel office, with the possibility of providing for each question, based on the observation of the last three years of each parameter indicated, three different answers: decreased, unchanged or increased. Each answer corresponds to a different score that will be added to obtain the total result of the macro area and thus identify the risk level.

The second area related to "Work Content" is composed of four sub-areas: "Work environment and Work equipment", structured in thirteen questions; "Task Planning", articulated in six questions; "Workload", composed of nine questions; "Working Time", structured in eight questions. For each question, in this case, it is possible to provide an answer by choosing between only two options: yes or no.

The third area, concerning the "Work Context", is divided into six sub-areas: "Function and Organisational Culture", characterised by eleven questions; "Role in the Organisation", divided into four questions; "Career Development", structured into three questions; "Decision-making Autonomy", structured into five questions; "Interpersonal Relations at Work", ordered into three questions and finally "Home-Work Interface", divided into four questions. As in the previous area, each question can be answered by choosing between two options only: yes or no. The scores obtained for each sub-area will contribute to the generation of the risk assessment for that area and, overall, to the final identification of the company risk condition.

Each indicator has a response that can increase or maintain the accumulated score constant. At the end of each sub-area there is a table that indicates, according to the score obtained that is compared with numerical intervals, if the criticality in that context is "not relevant", "medium" or "high". For each area, the scores of the analysed sub-areas will be summed up in order to understand if the risk for that context is "not relevant", "medium" or "high". In order to provide the objective assessment, the scores of the areas will be summed up and it will be necessary to compare the value obtained, with the numerical ranges provided to see if the assessment can be considered exhausted with the preliminary assessment with risk level "not relevant", or if it is necessary to proceed to corrective actions with in-depth assessments in case of risk level "medium" or "high" (Table 1).

Table 1. Summary table of the risk level according to the score obtained [21]

Colour code	From	To	Minimum requirements based on the Consultative Committee's indications
	0	58	Analysis of the indicators does not reveal any particular organisational conditions that can lead to work-related stress. If the preliminary assessment reveals a 'non relevant risk', this result is recorded in the Risks assessment report and a 'monitoring plan' must be established, which by way of example can consist in the periodic checking of Sentinel Events.
	59	90	Analysis of the indicators reveals organisational conditions that can lead to work-related stress; corrective actions must be implemented and subsequently verified; in the event that they are ineffective, in-depth analysis must be undertaken. For every condition identified with a MEDIUM score, suitable corrective actions must be adopted (e.g., organisational, technical, procedural, communication or training interventions) that refer specifically to the Work Content and/or Work Context indicators with the highest risk values. The effectiveness of the corrective actions must subsequently be verified, even by means of monitoring carried out using the same checklists; if the actions prove to be ineffective, the assessment moves on to in-depth analysis.
	91	216	Analysis of the indicators reveals a HIGH level of work-related risk, such as to require immediate corrective actions. Corrective actions must be adopted for the criticalities identified and their effectiveness must be then verified; it hey are found to be ineffective, the assessment moves on to in-depth analysis. For every condition identified as having a HIGH score in a single Area, adequate corrective actions must be adopted (e.g., organisational, technical, procedural, communication or training interventions) that refer specifically to the Work Content and/or Work Context indicators with the highest risk values.

INAIL's method for assessing work-related stress is the main national reference used by employers to assess this risk. It is used in most Italian production sectors, including agriculture.

While this versatility facilitates assessment, it may also alter the accuracy of the assessment and lead to incorrect risk estimates.

To this end, research was carried out to verify whether the INAIL checklist is an effective tool for the preliminary assessment of work-related stress in agriculture.

Thirty-two farms belonging to the agro-zootechnical and agricultural sector located in the Lazio region (central Italy) were observed. The survey was carried out during the first half of the year 2018.

The technical moderators who took part in the evaluation were all experienced individuals with qualifications as Prevention and Protection Service Managers in the agricultural sector and beyond.

In order to obtain an objective study, it was necessary to reduce the boundary variables by considering the main ones, also according to the indications provided by the Permanent Advisory Commission.

For each surveyed farm, the following data were recorded: number of workers employed; whether or not the workers had received training in accordance with articles 36 and 37 of legislative decree 81/2008; whether the role of PPSM was carried out by farm staff, by an external professional or by the employer; and the company role of the person interviewed.

In addition to these objective data, the comprehension of some indicator questions by the respondents was observed and, above all, the relevance of the questions to the agricultural macro sector was examined.

3 Results

The processing of work-related stress data, concerning the 32 analysed farms, concerned the determination of the influence on the total risk of the individual parameters belonging to the macro areas: "Work Content" and "Work Context". The area "Company Indicators" was not considered in the analysis because the value associated with its parameters was always null for all the farms.

Graph 1 and Table 2 show respectively the influence of the three macro areas on the determination of the total risk and the percentage weight of the sub-areas on the determination of the risk.

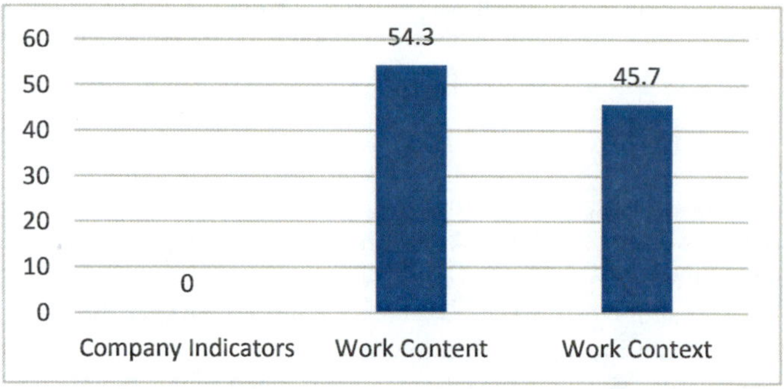

Graph 1. Influence of macro areas on the determination of total risk

Table 2. Percentage weight of sub-areas on total risk

Parameter	% of total risk
Working environment and equipment	38.42
Task planning	2.15
Workload - work rate	10.88
Working hours	2.83
Function and organisational culture	18.99
Role within the organisation	0.26
Career development	18.72
Autonomy in decision-making - control of work	6.63
Interpersonal relations at work	1.12
Home-work interface - work-life balance	0.00

The results show that the parameter relating to the sub-area "Working environment and Working equipment" has a weight of almost 40% of the overall risk value; the other influential sub-areas are "Organisational Function and Culture" and "Career Development" with almost 20% each, while "Workload - Work rate" comes to 10%. Altogether these four sub-areas contribute almost 90% (87.01%) in determining the risk value.

A study of the data recorded in relation to the number of workers (Table 3) showed that most farms, almost 70%, employ one or two workers.

Table 3. % weight of sub-areas according to the number of farm workers

Number of workers	1	2	3	4	5	8	13	25	30
Number of companies with the same number of employees	11	11	1	3	2	1	1	1	1
Working environment and equipment	41	37	33	40	38	29	36	43	42
Task planning	1	4	0	0	0	7	0	7	8
Workload - work rate	11	13	20	8	8	14	0	0	8
Working hours	1	2	0	4	0	0	7	36	0
Function and organisational culture	20	20	27	17	15	14	21	7	17
Role within the organisation	0	0	0	0	4	0	0	0	0
Career development	17	18	20	23	23	21	21	7	25
Autonomy in decision-making - control of work	7	7	0	8	8	14	7	0	0
Interpersonal relations at work	2	0	0	0	4	0	7	0	0
Home-work interface - work-life balance	0	0	0	0	0	0	0	0	0
Total (of each column)	100%								

The differences in weight in percentage of the total risk between the ten sub-areas were also analysed by comparing the cases where the external PPSM was present in the farm or the PPSM role was carried out by the employer.

From the elaboration (Table 4) the greatest variation occurs in relation to working time and role within the organisation. In other words, if the PPSM is not the employer, these two parameters have a greater influence in determining the risk. However, it should be noted that, overall, the two parameters have a marginal weight on the value of the risk, affecting less than 5%.

Table 4. Comparison of the percentage weight of sub-areas in companies with external PPSM or PPSM employer

Parameter	PPSM (employer) %	PPSM (ext) %	Δ
Working environment and equipment	39.06	38.03	0.10
Task planning	2.89	1.71	0.26
Workload - work rate	10.24	11.26	−0.05
Working hours	1.09	3.88	−0.56
Function and organisational culture	18.70	19.17	−0.01
Role within the organisation	0.00	0.42	−1.00
Career development	19.33	18.36	0.03
Autonomy in decision-making - control of work	7.52	6.10	0.10
Interpersonal relations at work	1.19	1.07	0.05
Home-work interface - work-life balance	0.00	0.00	0.00
Total	100%	100%	

Table 5 shows the weights of the various parameters according to whether or not training is present. In the case of the presence of training in the staff, the working environment parameter has a greater influence on the total risk.

Table 5. Comparison of sub-areas in companies with and without training

Parameter	No training	Yes training
Working environment and equipment	36.62	44.83
Task planning	2.14	2.21
Workload - work rate	12.47	5.17
Working hours	1.66	7.02
Function and organisational culture	19.27	17.98
Role within the organisation	0.33	0.00
Career development	18.96	17.87
Autonomy in decision-making - control of work	7.39	3.91
Interpersonal relations at work	1.14	1.02
Home-work interface - work-life balance	0.00	0.00
Total	100%	100%

It should be noted that training was present in only 20% of the farms.

Table 6 shows the weights of the various parameters according to whether the check list was compiled by the employer or by the worker. The variation in the percentage influence of the various sub-areas on the total risk was also observed, comparing the farms in which the preliminary assessment was compiled by one or more workers or in which it was compiled by the employer.

Table 6. Comparison of weights in % of sub-areas between preliminary assessments completed by workers or employers

Parameter	Worker compilation	Compilation employer
Working environment and equipment	34.84	39.42
Task planning	2.68	2.00
Workload - work rate	14.69	9.81
Working hours	1.79	3.13
Function and organisational culture	20.44	18.59
Role within the organisation	0.00	0.33
Career development	20.64	18.19
Autonomy in decision-making - control of work	2.88	7.68
Interpersonal relations at work	2.04	0.86
Home-work interface - work-life balance	0.00	0.00
Total	100%	100%

As can be seen from Table 6, the greatest difference in weighting is for the parameter "Workload - Work rate", followed by "Decision-making Autonomy - Control of Work" and "Working environment and Work equipment". In the case of the document being completed by the employer, the influence of the above parameters on the total risk is about 5% greater in the subgroups "Work environment and Work equipment" and "Decision-making Autonomy - Control of Work"; for the subgroup "Workload - Work rate", however, the trend is reversed in favour of workers. It should be noted that the document was completed by workers in only 20% of cases.

In addition to the work carried out so far, the following is an illustration of the influence on the assessment of the risk of work-related stress of the questions considered by the interviewees to be inconsistent with agricultural activities, due to a lack of understanding of the question itself or due to the regulatory inapplicability of the parameter to the sector under study.

In particular, the following questions were found to be irrelevant:

1) For the macro area "Sentinel Events":

- % absence from work;
- % holidays not taken;
- % internal transfers requested by staff;
- % staff turnover (outgoing-incoming).

2) For the macro area "Work Content", concerning the parameter "Working Time":

- overtime is routinely worked;
- working night shifts is usual;
- there is a fixed or rotating night shift.

3) For the macro-area "Work Context", concerning the parameter "Organisational Function and Culture":

- presence of a company safety management system;
- presence of code of ethics and/or behaviour (excluding disciplinary codes);
- identification of a contact person for listening to and dealing with cases of occupational distress (stress/mobbing…).

4) For the macro area "Work Context", with regard to the parameter "Career Development":

- criteria for career advancement are defined;
- reward systems are in place in relation to good personnel management by managers/leaders;
- reward systems exist in relation to the achievement of safety objectives.

5) For the macro-area "Work Context", concerning the parameter "Home-work interface":

- possibility of reaching the workplace by public transport/company shuttle bus;
- possibility of vertical/horizontal part-time work.

Concerning the parameters of the sub-areas "Company Indicators" and "Home-Work interface", the test results showed that they do not contribute to the increase of the risk since, for each analysed farm, they have a zero value.

About the parameter "Working time", the questions analysed all have a zero value, so they do not contribute to the increase in the risk value.

The analysis of the parameters of the subgroup "Function and Organisational Culture" produced the results shown in Table 7.

Table 7. Score related to the analysis of the parameters of the subgroup "Function and Organisational Culture"

Question	Score	%	Total parameter
Presence of a company safety management system	10	12.05	83
Presence of a code of ethics and/or conduct	3	3.61	
Identification of a contact person for listening to and managing cases of work-related distress	23	27.71	
Other	47	56.63	

The analysis of the parameters of the subgroup "Career development" produced the results shown in Table 8.

Table 8. Score for the analysis of the parameters of the subgroup "Career development"

Question	Score	%	Total parameter
Criteria for career advancement are defined	30	37.04	81
Reward systems exist in relation to good personnel management by managers/leaders	27	33.33	
Reward systems exist in relation to the achievement of safety objectives	24	29.63	

In order to assess the influence of the analysed parameters considered not to be congruent with agricultural activities, a percentage comparison was carried out, the results of which are highlighted in Graph 2.

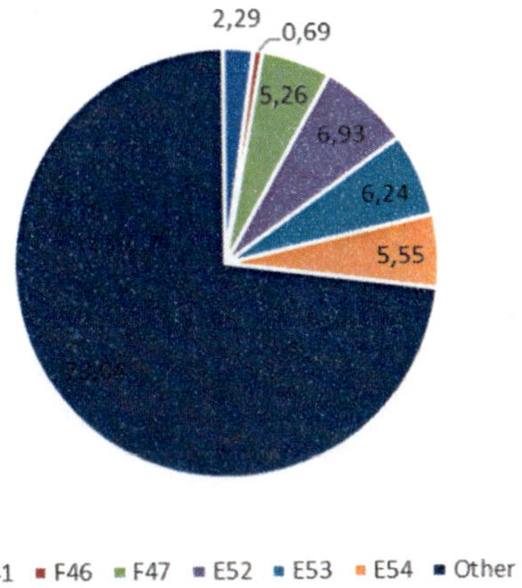

Graph 2. Comparison in % of questions considered to be inconsistent with agricultural activities (F41: Presence of a company safety management system; F46: Presence of a code of ethics and/or conduct (excluding disciplinary codes); F47: Identification of a contact person for listening and managing cases of work discomfort (stress/mobbing….); E52: The criteria for career advancement are defined; E53: There are reward systems in relation to the correct management of personnel by managers/bosses; E54: There are reward systems in relation to the achievement of safety objectives)

The questions analysed, shown in Graph 2, which are considered to have the greatest influence (greater than 5%) on the final risk value are:

- identification of a contact person for listening to and dealing with cases of occupational distress (stress/mobbing…);

- criteria for career advancement are defined;
- reward systems are in place in relation to good personnel management by managers/leaders;
- there are reward systems in place in relation to the achievement of safety objectives. Competence.

As shown in Graph 2, the weight of the analysed questions alone, which were considered to be irrelevant, on the total risk is 26.96%, which means that their elimination from the questionnaire on work-related stress would lead to a reduction in the risk value of about 27%.

4 Conclusions

The results of this study, although the observations are limited to a sample of 32 farms, would seem to highlight some critical issues in the application of the INAIL checklist for the work-related stress assessment in agriculture. The methodology of the preliminary assessment is optimal when applied to medium-large enterprises, mainly in the industrial sector, but is of doubtful effectiveness when applied to small enterprises. The structure of most Italian farms follows a family rather than sectoral type of management. The analysis of the questions that were not considered suitable for the agricultural sector, for reasons related to their comprehension and inapplicability, showed a considerable influence on the result of about 27%: this can lead to an erroneous risk assessment, with negative and dangerous repercussions on the health of the workers and on the farm finances.

Another aspect observed during the investigations was the lack of training of the personnel being assessed: in fact, about 80% of the farms were still not in compliance with the requirements of articles 36 and 37 of legislative decree 81/2008 and more specifically with the requirements of the Agreements between the State and the Regions on training of 2012.

The compilation of the questions by the various company figures also highlighted some differences in the perception of factors that contribute to the quantification of risks, especially between workers and employers. The answers to the questions in the subgroups "Decision-making autonomy - Work control", "Work environment and Work equipment" and "Workload - Work rate" were evident, with almost 5% difference. A difference deemed significant is that relating to the subgroup "Work environment and Work equipment". The indicators that make up this subgroup analyse the presence or exceeding of the legal limits of various risks (noise, vibrations, chemical, carcinogenic, biological, manual handling of loads, etc.) and, in part, the conformity of structures and machines. The results underlined that there is a greater awareness of the risks on the part of the employers than on the part of the workers; this could be a consequence of the lack of training of the workers who show a lower perception of the risks than the employer.

Another important aspect is the fact that some questions, found to be 'not congruent' with the agricultural sector, lead to the undervaluing and discrediting of the entire checklist for the preliminary assessment of work-related stress by those concerned, making it perceived as a document of little use.

In conclusion, the preliminary work-related stress risk assessment checklist is a useful tool, excellent in the analysis of medium-large, well-structured and organised farms, but with limitations in the assessment of smaller agricultural realities, which are also the most widespread in Italy. A valid solution could be to adapt this instrument to the various realities trying to replace the indicator questions that are not relevant or poorly understandable with others of similar calibre in order to provide more precise assessment tools. This step, however, presupposes a cohesive action that requires the professional contribution of specialists in the field such as sociologists and psychologists in collaboration with the various figures of corporate prevention.

References

1. Selye, H.: The stress concept in 1955. J. Chronic Dis. **2**, 583–592 (1955)
2. Monnot, M.J., Beehr, T.A.: Subjective well-being at work: disentangling source effects of stress and support on enthusiasm, contentment, and meaningfulness. J. Vocat. Behav. **85**(2), 204–218 (2014)
3. Niedhammer, I., Sultan-Taïeb, H., Chastang, J.F., Vermeylen, G., Parent-Thirion, A.: Fractions des maladies cardiovasculaires et des troubles de la santé mentale attribuables aux facteurs psychosociaux au travail dans 31 pays en Europe. Archives des Maladies Professionnelles et de l'Environnement **74**(6), 678 (2013)
4. Seňová, A., Antošováb, M.: Work stress as a worldwide problem in present time. Procedia. Soc. Behav. Sci. **109**, 312–316 (2014)
5. Honda, A., Date, Y., Abe, Y., Aoyagi, K., Honda, S.: Work-related stress, caregiver role, and depressive symptoms among Japanese workers. Saf. Health Work **5**(1), 7–12 (2014)
6. Tennant, C.: Work-related stress and depressive disorders. J. Psychosom. Res. **51**(5), 697–704 (2001)
7. Mucci, N., et al.: Work-related stress assessment in a population of Italian workers. The stress questionnaire. Sci. Total Environ. **502**, 673–679 (2015)
8. Rowden, P., Matthews, G., Watson, B., Biggs, H.: The relative impact of work-related stress, life stress and driving environment stress on driving outcomes. Accid. Anal. Prev. **43**(4), 1332–1340 (2011)
9. Mellor, N., et al.: Management standards' and work-related stress in Great Britain: progress on their implementation. Saf. Sci. **49**(7), 1040–1046 (2011)
10. Zoni, S., Lucchini, R.G.: European approaches to work-related stress: a critical review on risk evaluation. Saf. Health Work **3**(1), 43–49 (2012)
11. Atkins, R.C., Harris, P.: Using reflexology to manage stress in the workplace: a preliminary study. Complement. Ther. Clin. Pract. **14**(4), 280–287 (2008)
12. Persechino, B., et al.: Work-related stress risk assessment in Italy: a methodological proposal adapted to regulatory guidelines. Saf. Health Work **4**(2), 95–99 (2013)
13. Ipsen, C., Jensen, P.L.: Organizational options for preventing work-related stress in knowledge work. Int. J. Ind. Ergon. **42**(4), 325–334 (2012)
14. Greiner, A.: An economic model of work-related stress. J. Econ. Behav. Organ. **66**(2), 335–346 (2008)
15. Guthrie, R., Ciccarelli, M., Babic, A.: Work-related stress in Australia: the effects of legislative interventions and the cost of treatment. Int. J. Law Psychiatry **33**(2), 101–115 (2010)

16. Leka, S., Jain, A., Iavicoli, S., Vartia, M., Ertel, M.: The role of policy for the management of psychosocial risks at the workplace in the European Union. Saf. Sci. **49**(4), 558–564 (2011)
17. Reavley, N.J., Ross, A., Martin, A., LaMontagne, A.D., Jorm, A.F.: Development of guidelines for workplace prevention of mental health problems: a Delphi consensus study with Australian professionals and employees. Ment. Health Prev. **2**(1–2), 26–34 (2014)
18. Dollard, M.F., Neser, D.Y.: Worker health is good for the economy: union density and psychosocial safety climate as determinants of country differences in worker health and productivity in 31 European countries. Soc. Sci. Med. **92**, 114–123 (2013)
19. Cooper, C.L., Cartwright, S.: An intervention strategy for workplace stress. J. Psychosom. Res. **43**(1), 7–16 (1997)
20. Ministry of Labour and Social Policies: Circular Letter MLPS n. 23692 of 18/11/2010 Approval of the necessary indications for the assessment of work-related stress risk as per Article 28, paragraph 1-bis, of Legislative Decree n. 81 of 9 April 2008 (2010)
21. ISPESL: The assessment and management of work-related stress. Integrated approach according to the Management Standard HSE model contextualised in the light of Legislative Decree 81/08 (2010)
22. INAIL: The methodology for the assessment and management of work-related stress risk (2011). ISBN 978-88-7484-197-4
23. INAIL: The methodology for the assessment and management of work-related stress risk (2018). ISBN 978-88-7484-119-6

The Figure of the Workers' Representative for Safety and Health (WHSR) in Agriculture: A Survey in Some Italian Regions

Rino Gubiani[1(✉)], Massimo Cecchini[2], Luciano Caruso[3], Simone Riccioni[2], and Nicola Zucchiatti[1]

[1] Department of Agricultural, Food, Environmental and Animal Sciences, University of Udine, Via delle Scienze 208, 33100 Udine, Italy
rino.gubiani@uniud.it

[2] Department of Agriculture and Forest Sciences, University of Tuscia, Via S. Camillo De Lellis, 01100 Viterbo, Italy

[3] Department of Agriculture, Food and Environment, University of Catania, Via Santa Sofia, 100, 95123 Catania, Italy

Abstract. The Workers' Representative for Safety and Health (WHSR) plays a very important role: his impact on safety in the agricultural sector has not been assessed. The aim of this work is therefore to learn more about the figure of the agricultural WHSR.

The research was developed on different techniques aimed at establishing contact with the WHSR and the creation and distribution of a questionnaire to understand the preparation and type of WHSR consisting of 47 questions.

According to the data obtained from the questionnaires, 92% of the respondents were male and 85% were of Italian nationality, with 41% having a high school diploma, 37% a middle school diploma, and 15% a university degree. As regards the finding of the information necessary for the activity of WHSR, a very varied distribution was observed, with a significant proportion of subjects (16%) turning mainly to patronages and trade associations, but also WHSR obtaining the specific information directly from information sheets or legislation (11%). Analysing the single groups of items in the first group, "training in occupational safety and health", we note that as many as 83% of the WHSR have attended safety courses of various titles and often many more than just one, but also others such as "first aid course", "fire-fighting course", etc. In fact, most of them know the "safety procedures" (93%) and the operational figures at work (responsible for the prevention and protection service; competent physician; employer) with a percentage of 95%.

Keywords: Skills in safety · Agricultural accidents · Risk assessment

M. Biocca et al. (Eds.): SHWA 2020, LNCE 252, pp. 295–302, 2022.
https://doi.org/10.1007/978-3-030-98092-4_31

1 Introduction

European directive 89/391 defines "workers' representative with specific responsibility for the safety and health of workers" "any person elected, chosen or designated in accordance with national laws and/or practices to represent workers where problems arise relating to the safety and health protection of workers at work".

In Italy, this figure is regulated by Articles 47 and 50 of Legislative Decree 81/2008, the Italian transposition of Directive 89/391/EEC. There are also different forms of WHSR, such as the territorial TWHSR, the sector SWHSR and the production site PSWHSR, but not exist a lot of literature or research about this figure.

The provision states that workers, through their representatives, 'have the right to monitor the application of the rules on the prevention of accidents and occupational diseases and to promote the research, development and implementation of all appropriate measures to protect health and physical integrity'. The WHSR must be present in all companies or production units, but it only has advisory and propositional functions in preventive matters, powers of control, reporting and appeal, but not decision-making powers (art. 50), which are the responsibility of the employer. This is a very important role, therefore, and its effect on safety in the agricultural sector has never been investigated, and in any case the purpose of this role is to help reduce accidents, which in agriculture are still quite numerous. The most recent INAIL data on accidents in agriculture in Italy show a decreasing trend, with 33,063 accident reports in 2019 of which 165 were fatal.

Looking at the data for the five-year period 2015–2019, the highest number of accident reports with 10,218 cases is found in the North-East; the ranking is followed by the South with 7,137 cases, the Centre with 6,310 cases, the North-West with 5,634 cases and the Islands with 3,764 cases.

More than 30% of the total number of accidents reported in 2019 were in Emilia Romagna (12.8%), Tuscany (8.8%) and Lombardy (8.6%).

The male gender is the one that has shown a higher number of accidents; in 2019, in fact, 27,031 accidents involving men and 6,032 for the female gender were recorded; the percentage of fatal accidents for the male gender stood at 97.6% of the total.

In terms of age, the statistics show that the age group most affected was between 50 and 54 for men and between 55 and 59 for women.

The most common causes of accidents in the five-year period 2015–2019 were: slipping (26.9%), which mainly affected females, and total or partial loss of control of a machine (21.9%), which affected more males.

An analysis of the data for the five-year period 2015–2019 relating to fatal accidents recorded on the Infor.Mo. INAIL system showed that: 90% of fatal accidents involved Italian citizens.

As many as 84% of the fatalities had several years' experience in their job and 93% of the fatalities occurred in farms with less than 10 workers.

The age groups in which more than 60% of the cases were recorded were those aged 51 and over, 33.8% of whom were over 65.

The statistics show that 73.5% of fatal accidents occurred during agricultural work with tractors (more than 75% of cases) such as milling, mowing, sowing and handling materials, and 26.5% during transport to and from work.

The causes of the accidents were, in 49% of cases, overturning and running over of operators on the ground due to the use of equipment, followed by a worker falling from a height of 15% (barns, silos, tanks, etc.), contact with moving working parts 12% and unexpected/inappropriate start-up of machinery and equipment in 10% of cases.

Among the riskiest tasks are those related to land cultivation, with 1/3 of accidents due to falls (especially from structures) and 1/4 due to loss of control of vehicles, land vehicles and portable equipment.

The trend in occupational diseases is different, with an increase compared to previous years. Certainly, the downward trend in accidents can also be attributed to the application of recent legislation on safety at work and the increased action of the WHSR.

In order to better know this figure, INAIL carried out a project (2013–14) entitled "The role of WHSR in the process of detecting the perception of risk in the workplace preventive field" (Buresti et al. 2014) on a sample of 600 WHSR where the sample interviewed was mainly represented by males (79.0%), aged between 45 and 54 years (45.9%), with an average age of 49.08, in possession of a high school diploma (54.2%). WHSR enrolled in the survey operate, for the most part (43.9%), in large companies (250 or more employees) and are mainly in the manufacturing sector (39.9%). In agriculture, farms of this size do not exist, or are very limited in number. 41.7% of the WHSR interviewed considered the level of occupational safety and health (OSH) training for workers to be quite adequate. 94.1% of the sample had received the specific training required by the OSH legislation; 43.6% considered it fairly effective and 34.6% very effective.

From research of Cecchini et al. (2013) it results that the enforcement of the regulation concerning health and safety at workplaces has given strong benefits in prevention activities, in agricultural sector of the province of Viterbo (Central Italy). Nevertheless, it has been noted that some farms comply the laws only from a theoretical point of view, e.g. providing documentation without implementing them in practice (Cecchini et al. 2013).

To date, it is not possible to make a precise estimate of the number of WHSR operating in agriculture: in fact, a specific question submitted to INAIL on 12 November 2015 was answered that such data is not available. However, it is possible to estimate a number of about 10,000 farms with employees and therefore subject to the obligation to appoint the WHSR (the data of the 6th General Census of Agriculture of ISTAT - 2010 - show in fact that about 99% of farms employ exclusively family workers, therefore the figure corresponds to 1% of the total of about 1.5 million active farms). The aim of this work is therefore to learn more about the figure of the agricultural WHSR.

2 Materials and Methods

The survey to learn more about figure of WHSR in terms of skills, preparation, problems, etc. took place in 2019 in the regions of Sicily, Latium, Veneto and Friuli Venezia Giulia with three teams of researchers from the Universities of Tuscia, Catania and Udine.

The research was developed on different techniques to establish contact with WHSR based on:

- listening to the WHSR, according to a plan of interviews and through special "focus groups";
- listening to workers' representative organisations, through interviews with all the representative unions;
- listening to representative farm organisations, through interviews with all representative organisations.
- implementation and distribution of a questionnaire to understand the preparation and typology of the WHSR.

In particular, the questionnaire (Fig. 1) consists of 47 questions, the possible answers to which are:

- multiple choice;
- binary answer yes/no;
- response with score 1–5 (Likert scale type);
- open-ended.

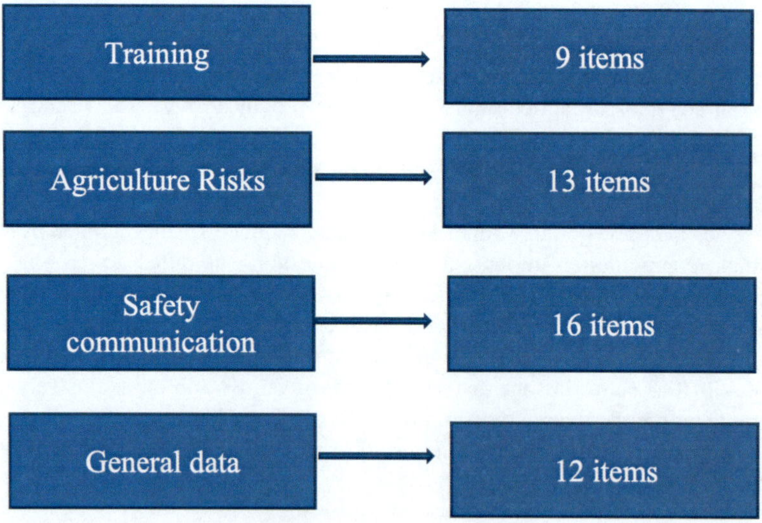

Fig. 1. The structure of questionnaire

Seventy-eight questionnaires were collected and processed using the Excel and SPSS package.

3 Results

From the data obtained from the questionnaires administered to agricultural WHSRs, it was observed that 92% of those interviewed were male and, as it concerns nationality, the majority (85%) were Italians, with a minority from Romania, Egypt, Sri Lanka and Albania. About the educational qualification possessed (Fig. 2), the majority of the interviewees had a high school diploma (41%) or an intermediate school leaving certificate (37%), while graduates (including both first level and specialist degrees) amounted to 15%. With regard to origin, however, the sample of respondents was rather diversified, with 40% from Latium, 11% from Sicily, 14% from Veneto and 10% from Friuli-Venezia Giulia (Fig. 3).

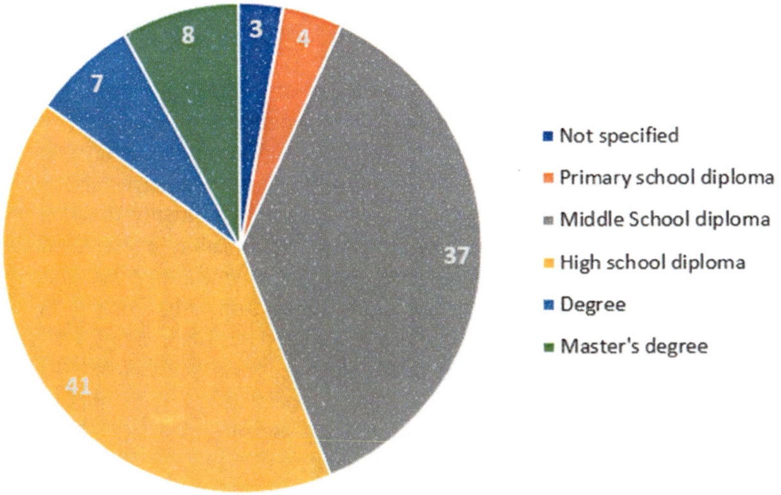

Fig. 2. Educational qualifications of the interviewees.

The majority of the WHSR surveyed (66%) have received specific training for their role in the farm, while 16% say they have no training at all. A fair fraction of respondents claims to have had previous work experience prior to their current role, including both sector-specific tasks (worker, forester) and non-sector-specific tasks (such as clerk, freelancer, teacher). The response to the question on safety communication was also very diverse: while 33% of those interviewed stated that it would be preferable to reduce the written text on each sheet to a minimum, 34% also preferred a reduction in the written text but with references to detailed fact sheets and 23% felt that it would be more useful to have written information, as detailed as possible.

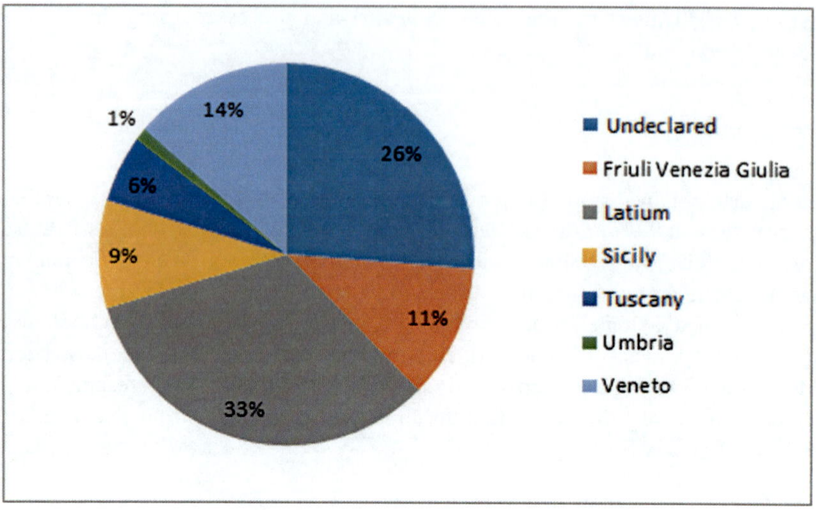

Fig. 3. Regions of origin of the interviewees

Respondents showed a fairly good knowledge of the meaning of specific safety-related terminology (such as the term ROPS). Regarding the retrieval of information necessary for the activity of the WHSR, a very varied distribution was observed, with a significant proportion of subjects (16%) turning mainly to patronages and trade associations, but also WHSR obtaining the specific information directly from information sheets or legislation (11%). However, work-related information is also obtained from specific courses, from company Safety, Prevention and Protection Managers (SPPM), from the employer, from the Internet, from consultants, from colleagues, from personal research. Finally, with regard to the way in which information on safety in the workplace can be accessed, the majority of respondents (32%) would prefer to select the crop operations of interest and display the associated risks accordingly, while a significant proportion (27%) would prefer the direct selection of specific risks and 23% of respondents would opt for the selection of crops of interest.

If we now analyse the individual groups of items in the first group, "occupational health and safety training", we see that as many as 83% of the WHSR have attended safety courses under various titles and often many more than just one, but also others such as "first aid course", "fire-fighting course", etc. In fact, most know the "safety procedures" (93%) and the operational figures (SPPM, competent doctor, employer) (95%).

Linked to these questions is the one concerning the application of the notions in the world of work, where the answers of the respondents stand at a median of 3 (out of 5). However, it is noticeable that often these lessons appeared to be too technical (median of 3 on a scale of 5) with too much material, as is evident from the next question, where the answer is also on the median of 3. In the second series of items "the risks in agriculture and the relative measures of prevention and protection" it can be seen that the WHSR believe they know the risks to which the workers are exposed (median 4) and consider the machines sold in the new market to be safe (4) while the machines used by farms are relatively less safe (3). The introduction of compulsory training in the use of tractors is considered very important (5), while the risks due to noise and vibrations are also considered to be adequately addressed (4). Similarly to the training in the use of tractors, the training in the use of sanitary equipment is also considered important (4), while the risks caused by repetitive movements of the upper limbs (pruning, fruit selection) and the risk due to dust are known (both median of 4).

Top marks are given for compliance with the labels of plant protection products and the use of personal protective equipment (PPE) when handling these products (median of 5 for both). The climatic impact of heat and cold on workers' health is also considered important (4), while maximum attention is given to the provision of a first aid pack (5) and the provision of a telephone set for working alone.

For the third group of items, "communicating safety in the company", everyone considers the role of the WHSR important (median of 4) within the farm and in 75% prevention and protection measures were developed within the farm and to do this in 60% of the cases the suggestions of the local WHSR were taken into account. Communication in terms of safety of both the WHSR and the employer is considered to be quite important (median of 3 for both). Communication by both the SPPM and the competent doctor, also within the farm, is considered more important (median of 4).

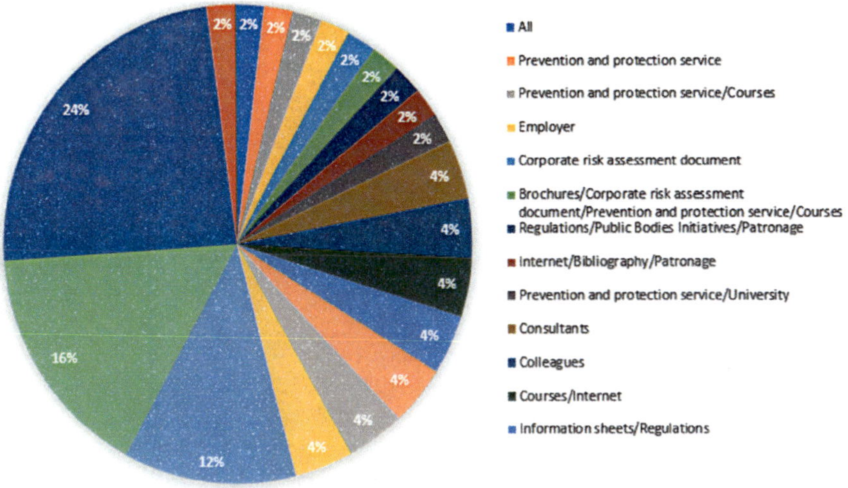

Fig. 4. How information is retrieved

As far as finding information is concerned (Fig. 4), a quarter of it is obtained from patronages and trade associations, while 16% use the Internet and 12% courses. The information is then passed on through a myriad of minor channels, such as work colleagues, consultants, etc., which can be a source of information.

4 Conclusions

In conclusion, the work carried out has made it possible to build an up-to-date database on the knowledge of WHSR on safety in the agricultural sector in various Italian regions and may serve as a database for the drafting of scientific publications in the sector, given the considerable amount of data available, as well as providing a starting point for a general improvement in knowledge in the field of safety in the agricultural sector, aiming for greater uniformity at national level. The long-term goal is to reduce the incidence of accidents by adopting all the best safety practices, increasing training and information for staff at all levels.

It is clear from the work that the WHSR are well informed and trained in their work and are aware that they contribute to the improvement of farm safety. The problems encountered is that at present it is a purely voluntary assignment with no monetary or career return but only personal satisfaction. In the Italian agricultural sector, since most of the farms are small, this makes it difficult to find WHSRs: in fact, their number, especially the active ones, is much lower than it should be. This shortcoming could be replaced, and in part is, by (territorial) TWHSRs operating on several farms. The problem with TWHSRs, however, is that companies do not like the presence of operators from outside the farm in such a sensitive area as safety. The fact of opening farms doors, showing warehouses and machines to staff who then visit other farms is not really perceived favourably. Finally, another problem is that all the courses that WHSRs have to attend (and which, as we have seen, are actually attended by WHSRs who are well prepared) are paid for by the employer and the courses have to be held during normal working hours. This makes companies reluctant to encourage WHSRs to attend such courses, and often only on strictly trusted personnel: in fact, if a WHSR changes job or farm, the result is a zero investment for the latter.

Research has shown, however, that WHSRs can indeed be a point of reference for the safety of a farm, but even today they are still a very unrecognised figure and should be valued and encouraged in order to prevent them from remaining a minor role in the safety chain in the agricultural sector.

References

Cecchini, M., et al.: Survey on the status of enforcement of European directives on health and safety at work in some Italian farms. J. Food Agric. Environ. **11**(3&4), 595–600 (2013)

Buresti, G., Persechino, B., Fontana, L., Rondinone, B.M., Boccuni, F., Iavicoli, S.: Progetto insula. Edizioni, Inail (2014)

European Union, Council Directive No (EC) 89/391 of 12 June 1989 on the introduction of measures to encourage improvements in the safety and health of workers at work (1989)

Impacts of Crops and Livestock Productions

Ammonia Emission from a BAT Farrowing Room with Fully Slatted Flooring and Individual Manure Pan. Results of a Yearly Monitoring Survey

Annamaria Costa[1](✉), Cecilia Salvagnini[2], Eleonora Buoio[1], Fabio Palmeri[3], Andrea Salvagnini[3], and Francesco M. Tangorra[1]

[1] Department of Veterinary Science for Health, Animal Production and Food Safety, Università degli Studi di Milano, Via dell' Università, 6, 26900 Lodi, Italy
annamaria.costa@unimi.it
[2] Department of Biotechnology, University of Verona, Strada Le Grazie 15, 37134 Verona, Italy
[3] Verona, Italy

Abstract. The large amount of pollutants emitted by intensive livestock farming, via the ventilation system, has leaded to the adoption of Best Available Techniques (BAT; ILF BREF) to limit pollutant release into atmosphere, from swine and poultry rearing confinements. In the present study the efficiency at limiting ammonia emission was measured in a farrowing room (BAT 4.7.3.3 "Crates with fully slatted flooring and manure pan solution", ILF BREF) vs. a traditional room with slatted floor on deep pit, in North East Italy.

To evaluate the efficiency level in reducing ammonia emission in a North Italian context, three farrowing rooms equipped with BAT number 4.7.3.3, and three conventional rooms with slatted floor on deep pit, each lodging 14 sows, were studied in three different seasons of the year. In each room, identical for design and management, ventilation was managed by the FRCA control unit system (FANCOM EasyFlow, Panningen, The Netherlands). Waste levels and ammonia concentration were measured (Gas Badge Pro analyser) in 7 different points every 3 days. Ammonia emission was calculated multiplying the ammonia concentration values (ppm, converted to mg m^{-3}) and the ventilation rate (m^3 h^{-1}). Data were analysed using the ANOVA procedure of SAS (Cary, NC), keeping the room (BAT vs traditional) as experimental unit.

The ammonia emitted by the BAT room was lower than that emitted by the reference room. Considering the value of 4.680 kg/ap/yr of ammonia emitted by the reference farrowing room of the study, lower than the one indicated by the ILF BREF for the traditional system (8.3 kg NH_3/ap/yr), the value of 3.234 kg NH_3/ap/yr of ammonia corresponds to reduction of 31%, in line with the ammonia reduction efficiency of the solution "4.7.3.3 Crates with fully or partly-slatted flooring and manure pan" indicated in the ILF BREF manual for data from Netherlands and Belgium (Flanders).

Keywords: Farrowing room · Lift crates · Ammonia concentration · Emission rate

M. Biocca et al. (Eds.): SHWA 2020, LNCE 252, pp. 305–313, 2022.
https://doi.org/10.1007/978-3-030-98092-4_32

1 Introduction

In 1996, the European Union enacted the Directive EU96/61/EC [1], which is also known as "Integrated Prevention and Pollution Control" (IPPC). The purpose of the IPPC was to reduce ammonia (NH_3) emissions into the atmosphere by defining the obligations of industrial and agricultural activities with high pollution potential.

New or existing industrial and agricultural activities, with a high pollution potential are defined in Annex I of this directive and include energy activities, production and processing of metals, mineral industry, chemical industry, management waste and animal husbandry. The IPPC Directive established a procedure for authorization and fixed the minimum requirements with regard to pollutant emissions to air, water and soil, to achieve a higher level of environmental protection.

For farming, this directive compels the application of an Environmental Integrated Permit that covers all forms of emission into the environment, that must be followed by large farms with more than 40.000 poultry, or with 2.000 finishing pigs heavier than 30 kg, or with 750 sows. Specifically, this directive prevents or limits ammonia emissions using sustainable and economic technologies.

In 2003, a panel of specialists of the European Commission published the "Integrated Pollution Prevention and Control (IPPC) Reference Document on Best Available Techniques (BAT) for Intensive Rearing of Poultry and Pigs" [2] to describe the best available techniques for enclosed farms, specifically for pig and poultry production, to address ammonia emissions into the atmosphere.

The term "Techniques" indicates all the technologies, as structural solution and plant, used in the production process to prevent and reduce pollution, and the management choices adopted in the farm (as density of animals, frequency of manure removal, storage).

These techniques are called "Available" when they have passed the experimentation phase and can thus be classified as reliable and are economically sustainable.

The Available technique can be defined as "Best" when it ensures the highest level of protection of the environment, intended as a whole.

In the ILF-BREF manual (2003), the ability of the enclosed facilities for sows, growing and finishing pigs to reduce ammonia emission are compared with the traditional ones [3].

The main reasons to adopt techniques able to reduce the environmental impact of swine farming come from the wide variety of pollutants such as ammonia emitted from animal facilities [5].

Gaseous NH_3 is a predominant pollutant in animal systems. It is originated by the enzymatic decomposition of uric acid or urea in mammals. Ammonia is responsible of negative responses in animal performance and welfare and in workers' health, and it is of great environmental concern because it contributes to soil acidification and increased nitrogen deposition in ecosystems [6].

The first experimental studies demonstrated that pigs reared in enclosed facilities with underfloor waste pits have depressed rate of gain [7] and that the incidence and severity of pneumonic lesions in pigs have been related to the air pollutant levels [8].

Drummond et al. (1980) [9] found that aerial ammonia can reduce body weight gain of piglets, with percentage reductions from controls at 12%, 30% and 29% for 50, 100 and 150 ppm of exposition.

The mechanism inducing growth depression in pigs triggered by ammonia could be related to a consequent reduced amount of feed intake or to a less efficiency in nutrients utilization due to a state of general discomfort or sickness.

Ammonium ion, which leads to the formation of ammonium salts, can be found in piggeries and it can combine itself with dust or other gasses, carrying bacteria and active endotoxins, that are potentially hazardous agents [10].

Ammonia originated in livestock farms is emitted in the surrounding via the ventilation system, so the concentration of the pollutant and the volumes of air extracted from the buildings are the key variables for their emission levels into the atmosphere.

A wide number of structural-plant solutions are presented in the ILF BREF manual [2], with different efficiency in lowering ammonia emission into the atmosphere. One of the most reliable solutions to reduce ammonia emission for growing-finishing pigs and sows facilities, is the adoption of the vacuum system, a technique in which pigs are housed on fully slatted floor and where the manure collected in the underneath pit is promptly removed to the storage tank. This technique can induce an ammonia reduction by 25% and it is widely used in the confinement swine facilities of Northern Italy [11].

Each BAT is presented in the ILF BREF manual together with their efficiency in limiting ammonia emission, on the basis of the performed experiments, that in most of cases, show scenarios of North Europe, as the Netherland or Belgium, Denmark or Spain etc. The need to evaluate the efficiency of a single BAT to reduce ammonia emission is fundamental in different countries [12], since different climatic conditions can vary ventilation rate, affecting considerably the amount of ammonia emitted outside the livestock buildings.

For the above exposed reasons, the present study was aimed at measuring ammonia emission rate from a farrowing room (BAT 4.7.3.3 "Crates with fully slatted flooring and manure pan solution", ILF BREF) vs. a traditional room with slatted floor on deep pit, in North East Italy.

2 Material and Methods

2.1 Location, Animals and Farrowing Rooms

The experimental study was conducted in a pig farm located in Santa Maria di Zevio (VR), in Northern Italy, from November 2018 to August 2019.

The cycles have covered three seasons of the year (November–December; February–March and July–August) to consider different climatic conditions.

Up to 14 sows (Landrace × Large White × Duroc, a typical cross for Parma ham production) were housed in the 6 rooms in the monitoring trial. The sows, with parity room 2 to 9, were randomly distributed in the experimental and reference rooms. Each farrowing cycle lasted 5 weeks (1 week pre-birth and 4 weeks of housing, until weaning), a vacuum period of 3 days was applied for washing and disinfection.

The two types of room were identical for design, ventilation system, feeding system and feed formulation and management. Both solutions were equipped with a heating system, each crate had a heated nest, the wet feed was delivered through the Spot mix feeding system. The experimental and reference rooms differed for the manure pit and manure removal.

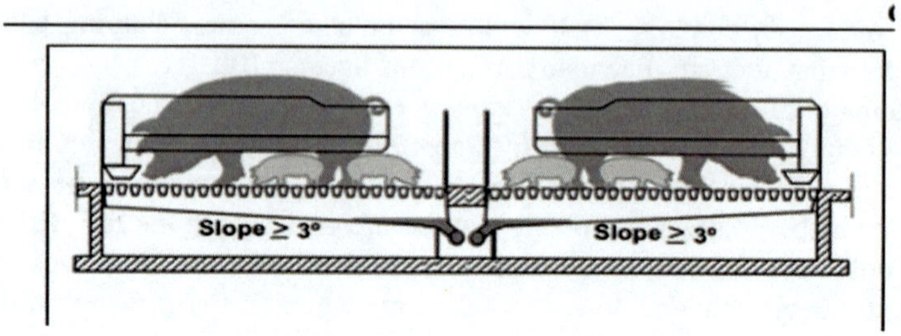

Fig. 1. The experimental room of the trial, BAT classified as 4.7.3.3 "Crates with fully or partly-slatted flooring and manure pan solution"

The experimental rooms present a fully slatted floor with farrowing stalls equipped with individual tanks for the collection of manure driven in a channel with vacuum system and a pan under the stalls, that is a BAT classified as 4.7.3.3 "Crates with fully or partly-slatted flooring and manure pan solution" by the ILF BREF manual. The reference room had a fully slatted floor and a deep manure pit (see Fig. 1).

2.2 Environmental Conditions, Ammonia Concentrations Monitoring

Temperature and ventilation rate were monitored by the FANCOM ventilation system with a 1 h frequency. Relative humidity was monitored by EasyLog EL-USB-2 (LASCAR electronics, Whiteparish, UK) devices, with 15 min frequency rate.

Air exchange was managed by FRCA (Fancom) control unit system, based on a free running impeller (FANCOM EasyFlow, Panningen, The Netherlands), for each room, for continuous, real-time monitoring of the ventilation rate, to maintain the desired temperatures, according to the physiological requirements of the housed animals during the farrowing cycle.

Ammonia concentration was measured five times in the day using Gas Badge Pro Ammonia, Industrial Scientific, Pittsburgh, PA, USA, accuracy ±5%, in each room at 7 locations shown in the Fig. 2 (A, B, C, D, E, F, G), according to Costa (2017). Sampling was carried out at 80 cm of height, at animal level respiratory apparatus.

The Fig. 2 shows the 7 sampling points for ammonia in the room.

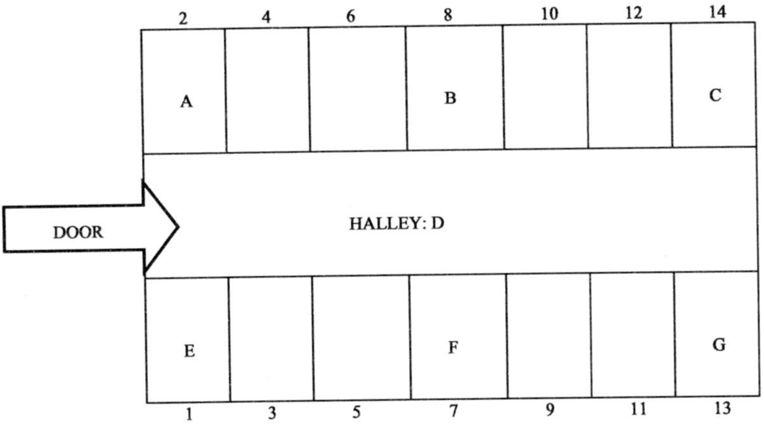

Fig. 2. The seven sampling points for ammonia in the room (A, B, C, D, E, F, G: Ammonia measuring points)

2.3 Calculation of Ammonia Emission

Emission rate was calculated as the product of ammonia concentration for the ventilation rate, as reported in Eq. (1).

$$Ei = Ci \times Vi \tag{1}$$

where Ei = pollutant (ammonia, dust) emission at time i,
Ci = pollutant concentration at time i,
Vi = ventilation rate at time i,
i = time in minute of monitored parameter.

The error of the pollutant emission factor (δE) is limited by the sum of the errors of the pollutant concentration measurement (δC) and the ventilation rate measurement (δV), and it is indicated in Eq. (2).

$$\delta E = \delta C + \delta V \tag{2}$$

2.4 Statistical Analysis

The data collected were submitted to variance analysis (Proc GLM of the SAS 9.4 statistical package, 2019) to assess the effect of the solution type (BAT vs Reference) on ammonia emission during the monitoring year survey.

3 Results and Discussion

3.1 Environmental Parameters in the Farrowing Rooms

Temperature and ventilation rate were similar in the two types of farrowing room, relative humidity was in the range 60%–80% as recommended for the compartment (Table 1).

Table 1. Mean Temperature, ventilation rate, and relative humidity in the two types of farrowing room (temperature and ventilation rate were monitored with a 1 h frequency, relative humidity was monitored with 15 min frequency rate during each production cycle).

Monitoring phase	Mean temperature, °C Mean (Min-Max)		Mean relative humidity, % Mean (Min-Max)		Mean ventilation rate, $m^3 h^{-1}$ Mean ± St. Dev	
	BAT	REF	BAT	REF	BAT	REF
1	22.93 (20.80–29.20)	23.87 (21.30–28.90)	47 (33–66)	48 (35–64)	41 ± 17	66 ± 19
2	23.99 (21–28.7)	23.74 (24.4–29.3)	48.71 (34.1–63.7)	51 (37–68)	52 ± 24	59 ± 8
3	26.19 (25.00; 29.17)	26.44 (22.60–29.60)	48.51% (45%–73%)	51% (42%–68%)	303 ± 128	322 ± 42

Ventilation rate was in the range for the considered housing solutions in the three monitoring seasons and it was proportional to the increase in temperature and the weight gain of the piglets.

3.2 Ammonia Concentrations and Emissions

In Table 2, mean ammonia concentrations ad particulate matter concentrations in the two types of farrowing room measured in the three measuring seasons are reported.

Ammonia was converted from ppm to mg/m^3 in order to compare the results obtained with the standards reported in the ILF BREF (for traditional halls the reference is 8.3 kg of NH_3/ap/yr in the farrowing room).

A wide variability in ammonia concentration in livestock houses is reported in literature. Our data are similar to those reported by other researchers. Ammonia concentrations ranging from 3 to 12 mg m^{-3} [15], 12 to 30 mg m^{-3} [16], 6.26 to 10.43 mg m^{-3}, [17] have been reported in different studies.

Table 2. Mean ammonia concentrations and emissions from the two types of farrowing room

Monitoring phase	Mean Ammonia concentration, mg m^{-3} ± St. Dev		Mean ammonia emission Kg/y/ animal place	
	BAT	Reference	BAT	Reference
1	5.37 ± 3.18	5.71 ± 1.51	3.234	4.680
2	4.05 ± 1.44	4.69 ± 1.18		
3	2.44 ± 0.49	2.14 ± 0.08		

As expected, the ammonia concentration values decreased markedly with increasing ventilation rate, values are in line with the available literature [13, 14].

Ammonia presented limited concentrations, as it usually occurs in farrowing rooms.

This variation can be affected by ventilation rate [18], relative humidity, animal density, the degree of manure and urine on the floor, and the type of floor and pit underneath [19–23].

Measuring time can affect ammonia concentration and emission, the measurements of ammonia concentrations performed during the day when animal activity and ventilation rate are greatest may inflate the measured values, authors [2] reported that overnight ammonia emission values were lower by 30% in comparison with daily values. Other researchers [20] measured 7% higher daily ammonia emissions from 8 AM to 6 PM in finishing pigs (5.69 compared 5.87 g/d/pig). This day-night variation in ammonia emission is linked to the greatest animal activity [24] during the day, and to the excretory behaviour of the pigs, which also shows a diurnal variation, with peaks in the daytime [20, 25]. Despite of these considerations, the ammonia values emitted by the two types of room of the study were considerably lower than the standard reported by the ILF BREF in 8.3 kg NH_3/ap/yr. Considering the ammonia value of 4.680 kg NH_3/ap/yr emitted by the reference room of this study, the value of 3.234 kg NH_3/ap/yr emitted by the BAT room corresponds to the 31% reduction, in line with the efficiency indicated for the solution 4.7.3.3 Crates with fully or partly-slatted flooring and manure pan.

4 Conclusions

The ammonia emission from the farrowing rooms of the swine farm investigated in this study presented limited values, showing a lower ammonia emission potential in comparison with the value indicated by the ILF BREF manual for this solution. The ammonia emission from the BAT experimental room reached 62% of ammonia reduction in comparison with a standard reference room emitting 8.3 kg NH_3/ap/yr. In conclusion, further studies are needed to investigate the environmental impact of swine husbandry in different climatic zones.

References

1. Directive EU96/ 61/EC. http://eur-lex.europa.eu/LexUriServ/LexUriServ.do?uri=CELEX:31996L0061:en:HTML
2. European Commission (EC): ILF BREF: Integrated prevention pollution control (IPPC) reference document on best available techniques for intensive rearing of poultry and pigs. Seville (2003). http://eippcb.jrc.ec.europa.eu/reference/BREF/irpp_bref_0703.pdf
3. Loyon Pain, B.F., Weerden, T.J., Chambers, B.J., Phillips, V.R., Jarvis, S.C.: A new inventory for ammonia emissions from U.K. agriculture (Ammonia Special Issue). Atmos. Environ. **32**, 309–313 (1998)
4. Radon, K., Weber, C., Iversen, M., Danuser, B., Pedersen, S., Nowak, D.: Exposure assessment and lung function in pig and poultry farmers. Occup. Environ. Med. **58**(6), 405–410 (2001)

5. Pain, B.F., Weerden, T.J., Chambers, B.J., Phillips, V.R., Jarvis, S.C.: A new inventory for ammonia emissions from U.K. agriculture (Ammonia Special Issue). Atmos. Environ. **32**, 309–313 (1998)
6. Day, D.L., Hasen, E.L., Anderson, S.: Gases and odors in swine confinement buildings. Trans. ASAE **8**, 118–121 (1965). https://doi.org/10.13031/2013.40446
7. Consigliere, R., Costa, A., Meloni, D.: Effects of vermiculite-based additives on macroscopic lung lesions, carcass traits and meat quality in finishing pigs. Large Anim. Rev. **24**, 195–199 (2018)
8. Kovacs, F., Nagy, A., Sallai, J.-N.: The effect of certain environmental factors on the health and production of pigs 2: data on dust and living germ content as well as on the chemical contamination of the air in pig houses of closed system. Hungarian Vet. J. **22**, 496–505 (1967)
9. Drummond, J.G., Curtis, S.E., Simon, J., Norton, H.W.: Effects of aerial ammonia on growth and health of young pigs. J. Anim. Sci. **50**, 1085–1091 (1980). https://doi.org/10.2527/jas1980.5061085x
10. Hartung, J., Seedorf, J.: Characterization of airborne dust in livestock housing and its effects on animal and environment. In: Pedersen, S. (ed.) Dust Control in Animal Production Facilities, pp. 140–154. Danish Institute of Agricultural Science, Aarhus, Denmark (1999)
11. Costa, A., Guarino, M.: Definition of yearly emission factor of dust and greenhouse gases through continuous measurements in swine husbandry. Atmos. Environ. **43**, 1548–1556 (2009). https://doi.org/10.1016/j.atmosenv.2008.11.009
12. Costa, A.: Ammonia concentrations and emissions from finishing pigs reared in different growing rooms. J. Environ. Qual. **46**, 255–260 (2017)
13. Ni, J.Q., et al.: NH3 emission from a large mechanically-ventilated swine building during warm weather. J. Env. Qual **29**, 751–758 (2000)
14. Costa, A., Domeneghini, C.: Pollutants in livestock buildings: Ammonia and dust interplay with the respiratory tract. In: (a cura di): Banhazi, T., Aland, A., Hartung, J. (eds.) Air Quality and Livestock Farming, pp. 49–58. CRC Press, London (2018). ISBN: 9781138027039
15. Koerkamp, P.W.G., et al.: Concentration and emission of ammonia in livestock buildings in Northern Europe. J. Agric. Eng. Res. **70**, 79–95 (1998). https://doi.org/10.1006/jaer.1998.0275
16. Demmers, T.G.M., Burgess, L.R., Short, J.L., Phillips, V.R., Clark, J.A., Wathes, C.M.: Ammonia emissions from two mechanically ventilated UK livestock buildings. Atmos. Environ. **33**, 217–227 (1999). https://doi.org/10.1016/S1352-2310(98)00150-2
17. Seedorf, J., Hartung, J.: Survey of ammonia concentrations in livestock buildings. J. Agric. Sci. **133**, 433–437 (1999). https://doi.org/10.1017/S0021859699007170
18. Gustaffsson, G.: Investigations of factors affecting air pollutants in animal houses. Ann. Agric. Environ. Med. **4**, 203–215 (1997)
19. Fabbri, C., Valli, L., Guarino, M., Costa, A., Mazzotta, V.: Ammonia, methane, nitrous oxide, carbon dioxide and particulate matter emissions in two different buildings for laying hens. Biosyst. Eng. **97**, 441–455 (2007). https://doi.org/10.1016/j.biosystemseng.2007.03.036
20. Aarnink, A.J.A., Keen, A., Metz, J.H.M., Speelman, L., Verstegen, M.W.A.: Ammonia emission patterns during the growing periods of pigs housed on partially slatted floors. J. Agric. Eng. Res. **62**, 105–116 (1995). https://doi.org/10.1006/jaer.1995.1069
21. Aarnink, A.J.A., Van den Berg, A.J., Keen, A., Hoeksma, P., Verstegen, M.W.A.: Effect of slatted floor area on ammonia emission and on the excretory and lying behaviour of growing pigs. J. Agric. Eng. Res. **64**, 299–310 (1996). https://doi.org/10.1006/jaer.1996.0071

22. Blanes-Vidal, V., Hansen, M.N., Pedersen, S., Rom, H.B.: Emissions of ammonia, methane and nitrous oxide from pig houses and slurry: effects of rooting material, animal activity and ventilation flow. Agric. Ecosyst. Environ. **124**, 237–244 (2008)
23. Arogo, J., Westerman, P.W., Heber, A.J.: A review of ammonia emissions from confined swine feeding operations. Trans. ASAE **46**, 805–817 (2003)
24. Costa, A., Borgonovo, F., Leroy, T., Berckmans, D., Guarino, M.: Dust concentration variation in a pig barn in relation to on-line monitored animal activity. Biosyst. Eng. **104**, 118–124 (2009)
25. Jeppsson, K.H.: Diurnal variation in ammonia, carbon dioxide and water vapour emission from an un-insulated, deep litter building for growing/finishing pigs. Biosyst. Eng. **81**, 213–223 (2002). https://doi.org/10.1006/bioe.2001.0025

Safety in Hazelnut Mechanical Harvesting

Pietro Catania(✉), Maria Alleri, Eliseo Roma, and Mariangela Vallone

Department of Agricultural, Food and Forest Sciences, University of Palermo, Viale delle Scienze ed. 4, 90128 Palermo, Italy
pietro.catania@unipa.it

Abstract. Hazelnut cultivation occupies an area of about 13,000 ha in Sicily, 10,000 of which are concentrated in the province of Messina, specifically in the pedo-mountain municipalities of the Nebrodi. These are heterogeneous plants by cultivar and planting time. To date, the hazelnut represents a scarcely mechanized marginal crop, with a characteristic training system as the multi stemmed bush. The harvest takes place as soon as the hazelnuts reach maturity and detach from the herbaceous dome that surrounds them falling to the ground. Before harvesting, swathing is carried out by hand using rakes, or facilitated by backpack blowers, or mechanically by means of a swather or with a tractor-mounted blowing tubes. Harvesting is done by hand directly from the ground, with implications for the health of the operators; it requires a high need for manpower, difficult to find. Labor productivity remains very low (10 kg/h per worker). The aim of the research was to study the mechanization of hazelnut harvesting in the Sicilian environment.

Keywords: Health risk · Working capacity · Working productivity

1 Introduction

Hazelnut (*Corylus avellana*, L.) belongs to the family of Betulaceae and is one of the world's major nut crops. It is a deciduous plant with multi stemmed shrub habit. The stems born from numerous adventitious buds at the level of the collar and reach heights from 4 to 7 m; the number of stems varies and in optimal conditions it can exceed the number of 30/40 per bush. This continuous emission of suckers allows the plant to renew the epigeal part; in case of damage by abiotic or biotic factors, a new sucker quickly replaces the deteriorated or damaged one. The germination, on the other hand, is subsequent to anthesis and takes place in March as soon as the temperatures begin to rise.

The initially taproot root system is soon replaced by a dense bundled system that develops mostly in the first 50 cm of soil up to a maximum of 2 m [1]; the density of this type of root system allows the plant to retain the soil, in fact hazelnut is considered suitable for containing erosive phenomena and hydrogeological instability [2].

The fruit is a modified achene; it can be solitary or grouped in infructescences up to a number of 4 hazelnuts. It is dry indehiscent with a single loggia (rarely 2) wrapped in a herbaceous dome that covers it more or less entirely depending on the cultivar, which dries up when ripe, letting the fruit fall to the ground. The ripening and fall of the fruit

M. Biocca et al. (Eds.): SHWA 2020, LNCE 252, pp. 314–323, 2022.
https://doi.org/10.1007/978-3-030-98092-4_33

takes place between late August and early October depending on the cultivar and the growing environment. The outer pericarp changes from green to brown with maturation and its consistency increases until it becomes woody when ripe, while the seed represents the edible part of the fruit.

The fruits generally consist of 40–60% of shell which is a waste product of the hazelnut processing industry but is used as a fuel; the remaining part, the seed, is the edible part.

The primary destination of hazelnuts is the confectionery industry, 90% of hazelnuts are in fact processed and only 10% of world hazelnut production is destined for fresh consumption.

The hazelnut seed has an oil content is between 62 and 72% of the dry weight. The oil is 95% composed of fatty acids such as oleic, linoleic, palmitic and stearic acids. About 70% of oleic acid is present. The quantity and quality of the fatty acids also determines the shelf-life of the product, in fact a greater quantity of linoleic acid determines a greater self-oxidation of the product, therefore a lower shelf life [3, 4].

Antioxidant compounds such as tocopherols (vitamin E) counteract the oxidation of fats, ensuring a better shelf life of hazelnuts. The perisperm, which is ingested only with the hazelnuts consumed as they are (the toasted ones are peeled, then it is removed), contains most of these antioxidant compounds (polyphenols) together with the tannins. The presence of these unsaturated fats together with compounds with antioxidant properties such as vitamin E (α-tocopherol) and vitamin B6 give hazelnuts a very important role in the human diet as amply demonstrated by numerous researches [5].

These healthy properties allowed the consumption of dried fruit and therefore also of hazelnuts constantly increase, involving numerous segments of the population such as sportspeople, diabetics, people with cardiovascular diseases.

Hazelnut worldwide occupies an area of almost 1 million hectares (FAO 2018), with a production of 663,888 t (FAO 2018), or a yield per ha equal to 0.9 t.

Turkey is the first producer in the world (376,000 t), followed by Italy with 100,000 t, then Azerbaijan 23,000 t, United States 21,000 t, Spain and Georgia with 19,000 t.

The hazelnut cultivation occupies an area of about 13,000 ha in Sicily, 10,000 of which are concentrated in the province of Messina, specifically in the pedo-mountain municipalities of the Nebrodi. These are heterogeneous plants by cultivar and planting time. To date, the hazelnut represents a scarcely mechanized marginal crop, with a characteristic training system as the multi stemmed bush.

The harvest takes place as soon as the hazelnuts reach maturity and detach from the herbaceous dome that surrounds them falling to the ground. Before harvesting, swathing is carried out by hand using rakes, or facilitated by backpack blowers, or mechanically by means of a swather or with a tractor-mounted blowing tubes.

Harvesting is done by hand directly from the ground, with implications for the health of the operators; it requires a high need for manpower, difficult to find. Labor productivity remains very low (10 kg/h per worker).

The aim of the research was to study the mechanization of hazelnut harvesting in the Sicilian environment.

2 Materials and Methods

2.1 Harvesting Machine Used During the Tests

The machine used in the tests (T1) is a semi-mounted harvester (Fig. 1), model EU2000 by Chianchia (Cherasco, Italy). It is equipped with a steel frame with a triangle connecting the tractor's three-point hitch. The power is taken from the tractor's PTO, 540 rpm, by means of a regular cardan shaft. It is one of the most widespread machine in mechanical harvest of shell fruits [6].

Fig. 1. Machine used in harvesting tests (T1) connected by a three-point hitch to the tracked tractor.

The harvesting machine weighs 320 kg; its dimensions are: height 1.65 m, width 1.70 m and length 1.10 m. It requires tractors with 22 kW minimum power with a three-point hitch equipped with hydraulic lifting with a capacity of 800 kg.

It is equipped with two suction openings with a diameter of 100 mm and 120 mm and a cage cleaning system with interchangeable rollers of different diameters, depending on the species of nuts to be harvested. In fact, it can be used to harvest other nuts such as chestnuts, almonds, walnuts, pistachios, etc. For hazelnuts, cages with holes having a diameter of 26 mm are always used.

The machine includes the following operating groups:

- Harvesting group: it realizes the suction of the product from the ground with the impurities sending it to the sedimentation chamber. A fan equipped with a 10 propellers impeller and external diameter of 400 mm, creates a depression inside the sedimentation chamber in the external part of which the suction modules are placed. Each module is connected to a flexible hose with a diameter of 100 mm which is managed manually by the operator. The suction air current sends the collected product into an internal hopper, expelling the dust to the outside by centrifugal force through a rubber propeller valve located at the base of the sedimentation chamber. The passage of the product (hazelnuts and impurities) from the sedimentation chamber to the loading hopper is facilitated by a rubber impeller. In this phase, a first cleaning of the product takes place by means of an air current generated by a second fan having 12 propellers and a diameter of 320 mm. This allows the removal of light foreign bodies such as leaves, mowing or shredding residues, lumps of earth, etc.

- Cleaning group: the product is conveyed into the cleaning group consisting of a double rotating cylindrical cage, by means of a screw conveyor. The internal cylindrical grid eliminates the larger bodies, while the external grid separates the hazelnuts with lower granulometry. The internal roller is equipped with a rotating auger expelling the waste material (twigs, stones, etc.) to the outside. The external grid, on the other hand, conveys the hazelnuts into a second loading hopper.
- Sorting and packing group: the hazelnuts are pushed by a current of air from the loading hopper towards three outlets, placed at different heights, where the bags are placed. In the first two, at a lower altitude, the unbroken hazelnuts arrive due to their higher specific weight; in the third exit the hazelnuts with lower specific weight are conveyed: they are considered poor quality hazelnuts (fruits of the previous year, hazelnuts with aborted seeds, holes, etc.) A correct regulation of the air flow allows an efficient selection of the fruits (Fig. 2).

Fig. 2. Packing group: note the two bags connected to the outlets, collecting the unbroken and waste hazelnuts separately.

2.2 Experimental Test Procedure

The tests were divided into two theses, named T1 and T2.

T1 consisted in the use of the machine in plots of 8 m^2 in size, in order to evaluate the capacity and productivity of the machine's work as well as the quality of the collected product in two working conditions: suction machine with only one suction pipe (T1M1) and suction machine with two suction pipes (T1M2). The two T1 harvesting methods were compared with a manual harvesting site (T1H).

T2 was based on the use of the machine, for several days, in three different experimental farms in Messina province in order to assess the real working capacity and productivity of the mechanized site with a machine equipped with two suction tubes (T2M2 thesis).

2.2.1 Experimental Farm for T1 Thesis

T1 tests were carried out at the farm located in the municipality of Ficarra, 640 m asl. T1M1, T1M2 and T1H tests were carried out in three replicates on 8 m^2 plots each, according to a randomized blocks scheme.

The plots were flat, the hazelnut trees over 50 years old and cultivated as multi stemmed bush, with an average number of 6 stems per plant. The plots were made homogeneous in terms impurities in the soil. Before the tests, in fact, the mechanical blower BR 700 by Stihl, was used on all the plots in order to remove leaves and impurities from the soil surface. Subsequently, the T1M1, T1M2 and T1H sites were applied on different plots (Fig. 3).

Fig. 3. Experimental plots used in T1 for tests T1M1, T1M2 and T1H.

2.2.2 Experimental Farms for T2 Thesis

T1 tests were carried out at the farm located in the municipality of Ficarra, 640 m asl.

T2 thesis was performed at three different Sicilian farms named A, B and C.

The farm named A falls within the municipality of S. Angelo di Brolo; it is located at 510 m a.s.l. and is conducted under the conventional regime. The plot under test has a northern exposure and a flat position; hazelnut plants are over 50 years old and cultivated in multi stem bush with an average number of 4 stems per plant. The layout of planting is 6 m × 5 m. The harvested area is 1.00.00 ha wide.

The farm named B falls within the municipality of Gioiosa Marea; it is located at 850 m above sea level and is conducted in organic regime. The plot under test is facing north with flat location, the hazelnut trees are over 50 years old and cultivated in multi stem bush, with an average number of 8 stems per plant. The layout of planting is 6 m × 5 m. The harvested area is equal to 5.00.00 ha.

The farm named C falls within the municipality of S. Angelo di Brolo and is located at 800 m above sea level. The plot under test is exposed to the North-West with flat location, the hazelnut trees are over 50 years old and cultivated in multi stem bush with an average number of 6 stems per plant. The layout of planting is 5 m × 4 m. The harvested area is equal to 1.20.00 ha.

Before harvest, the soil was prepared mowing and shredding the grass and then removing the crop residues by means of a mechanical blower (model BR 700 by Sthill). Then, T2M2 thesis was applied in the entire area of each farm above described.

2.3 Harvesting Site Organization

The mechanized harvesting site consisted of the suction machine connected to the tracked tractor model 5830 F by Landini, 40 kW powered. The number of workers was 3 in the M2 site, one of which was in charge of driving the machine and two in harvesting with the suction hoses, and equal to 2 in the M1 site, one of which employed the machine and one dealing with the suction hose. The manual harvesting site consisted of 2 workers.

2.4 Working Time Analysis

In all the sites used for T1 and T2 theses, the working times were measured according to the CIOSTA (Comité International d'Organisation Scientifique du Travail en Agricolture) methodology [7].

A CASIO chronometer was used for the following time measurements: effective time (TE), auxiliary times (TAS and TAV), idle times (TM), rest time (TR) and finally the utilization time (TU). TAS (*tempus adiuvandi supplendo*) concerned the time for unloading and loading of the bags with the harvested hazelnuts. TAV (*tempus adiuvandi vertendo*) concerned all the maneuvers for positioning the machine near the plot to be harvested.

2.5 Statistical Analysis

The data were subjected to ANOVA and then compared by Tukey's test at the 5% significance level.

3 Results

Figure 4 shows the results obtained in thesis T1.

The manual working site registered a working capacity of 13.28 kg/h. The mechanized harvesting site with a single suction hose (T1M1) provided a working capacity of 62.2 kg/h, while the mechanized harvesting yard with two suction tubes (T1M2) provided a working capacity of 108.5 kg/h. It is noted that the mechanized sites recorded a greater working capacity than the manual one. The T1M1 site, in fact, expressed a working capacity approximately 5 times greater than the manual one and the T1M2 site approximately 8 times greater. The working capacity obtained in all the mechanized sites was higher than that obtained in [8]. As regards the working productivity of the mechanized sites, T1M2 site recorded a value of about 5 kg/h*worker higher than T1M1, a modest increase due to the morphology of the plots.

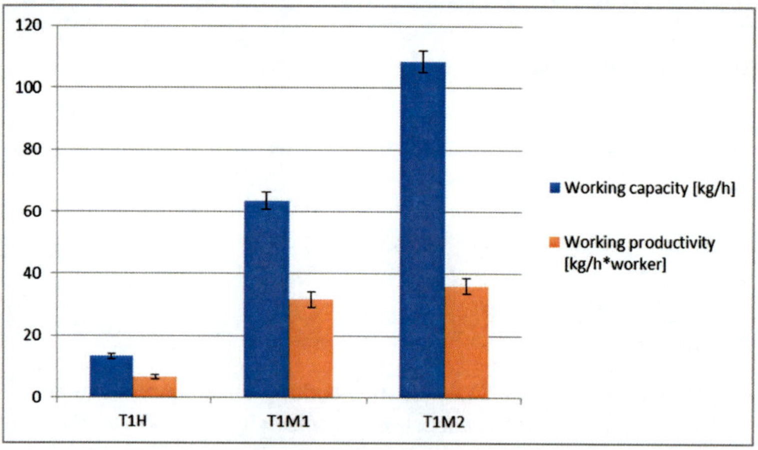

Fig. 4. Working capacity and productivity in thesis T1 for the tests T1H, T1M1 and T1M2.

Table 1 shows the values of total production and yield per hectare of hazelnuts obtained in the three different farms in which the mechanized site with 2 suction pipes (T2M2) was used.

Table 1. Total production and yield per hectare of hazelnuts obtained in the three different farms.

Farm	Area [ha]	Harvested product [kg]	Product yield [kg/ha]
A	1.0	3,384	3,384.00
B	5.0	6,806	1,361.20
C	1,2	1,380	1,150.00

The highest production yield was obtained in farm A with values about 2.5 greater than B and about 3 greater than C. The differences in the production values obtained in the three plots are attributable to the different field conditions, such as the quantity of fruit on the ground and the presence of impurities.

Figure 5 shows the values of working capacity and productivity obtained in the three farms. The highest working capacity was obtained in farm A, equal to 313 kg/ha, then in B with 180 kg/ha, and finally in C with 138 kg/ha. A similar trend was obtained with regard to working productivity.

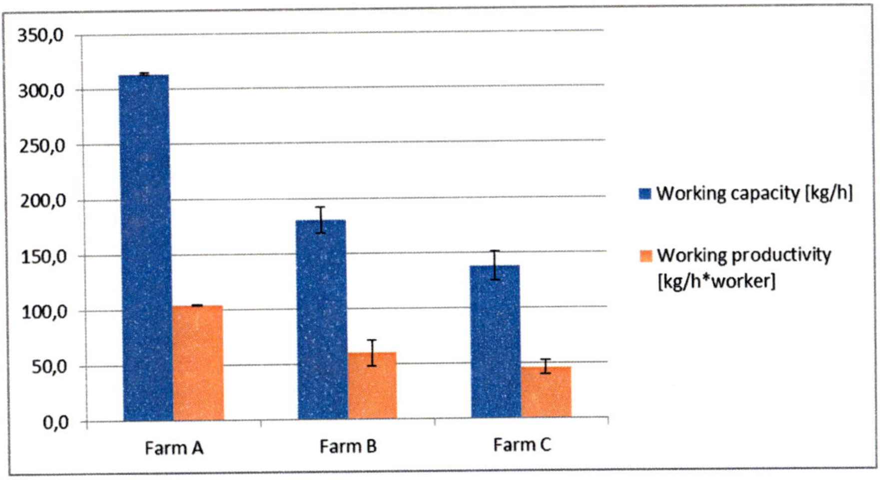

Fig. 5. Working capacity and productivity in thesis T2 applied in farms A, B and C.

The differences in the working capacity values of the mechanized site employed in the three farms are mainly attributable to the different yields per hectare of product recorded. The working site employed in farm A, in fact, recorded higher auxiliary times (TAS and TAV) compared to sites B and C, due to the greater quantity of bags of hazelnuts filled, as can be deduced from the analysis of the working times in the three farms (Fig. 6).

The highest effective time values (TE) were obtained in farm C; this is due to the lower production yield respect to farms A and B. The lower presence of fruit allowed the site to proceed rapidly, reducing turnaround (TAV) and unloading times (TAS).

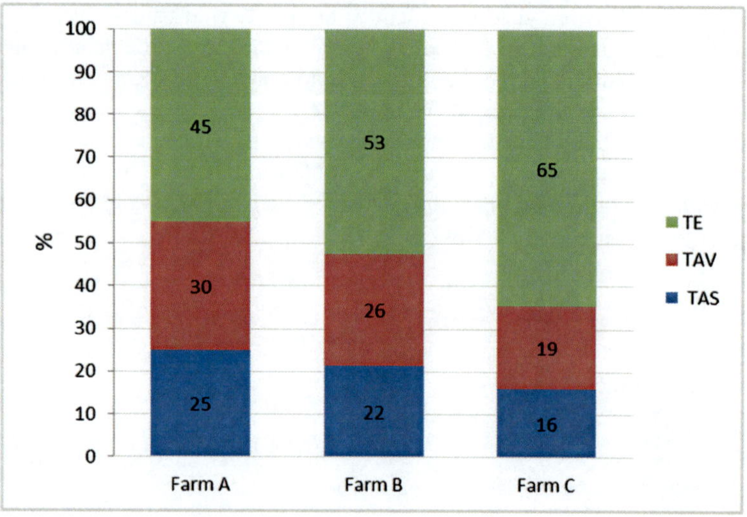

Fig. 6. Working times analysis in T2M2 thesis.

4 Conclusions

The results obtained from the study indicate that the mechanization of hazelnuts in terraced plots with plant age over 50 years can be carried out respecting the operators safety. The use of the mechanized working site, as the first year of application, provided excellent values of working capacity and productivity. It is therefore possible a future diffusion of hazelnut harvesting machines in Sicily as the plots can be managed with cultivation interventions aimed at their use.

This bodes well for a future spread of hazelnut harvesting machines in Sicily as the plots can be cultivated with crop interventions aimed at their use. The application of mechanized working sites for hazelnuts harvest in Sicilian farms would allow to overcome the big obstacle of finding manpower, thus avoiding the economic collapse of the farms.

Mechanization would make it possible to maintain a strong environmental balance represented by the wonderful rural landscape offered by hazelnut trees in the Nebrodi territory of the province of Messina.

References

1. Tombesi, A.: Il nocciolo. Ed. Reda, p. 121 (1985)
2. Bernardi, B., et al.: The assessment of hazelnut mechanical harvesting productivity. Agron. Res. **4**, 1491–1497 (2017)
3. Ozdemir, M., Açkurt, F., Yildiz, M., Biringen, G., Gurcan, T., Loker, M.: Effect of roasting on some nutrients of hazelnut (*Corylus avellana* L.). Food Chem. **73**, 185–190 (2001)

4. Bignami, C., Bertazza, G., Cristofori, V., Troso, D., Kernel quality and composition of hazelnut (*Corylus avellana* L.) cultivars. In: Sixth International Congress on Hazelnut. Tarragona-Reus, Spain, 14th–18th June 2005. Acta Horticulturae, vol. 686, pp. 477–484 (2005)
5. Richardson, D.G.: The health benefits of heating hazelnuts: implications for blood lipid profile, coronary heart disease and cancer risks. Acta Horticolturae **445**, 295–300 (1997)
6. Maccioni, L., Bietresato, M., Borgianni, Y.: From the extraction of currently fulfilled requirements to value curves: a case study in the field of harvesting machines for shell fruits and lessons learnt in engineering design. Appl. Sci. **10**, 3809 (2020)
7. Bolli, P., Scotton, M.: Lineamenti di tecnica della meccanizzazione agricola 1st edn., p. 221. Edagricole, Bologna (1987)
8. Yildiz, T.: Labor requirements and work efficiencies of hazelnut harvesting using traditional and mechanical pick-up methods. Turk. J. Agric. For. **40**, 301–310 (2016)

Precision Farming and Traceability

Purpose-Intended Modification of a Conventional Sprayer Machine for Application in Precision Agriculture

Elio Romano[1(✉)], Giampaolo Schillaci[2], Domenico Longo[2], Carlo Bisaglia[1], and Sabina Failla[2]

[1] CREA – Consiglio per la ricerca in agricoltura e l'analisi dell'economia agraria, Centro Ingegneria e Trasformazioni Agroalimentari, Treviglio, (BG), Italy
elio.romano@crea.gov.it
[2] Department of Agriculture, Food and Environment (Di3A), University of Catania, Catania, Italy

Abstract. The application of precision agriculture (PA) in the management of agricultural practices is increasingly widespread, especially in crops with higher income such as vineyards. The organization of agricultural operations conducted with precision approaches, provides that the machines may have the information collected from previous years (yield and quality of preceding productions), from surveys prior to agricultural practice (aerial or terrestrial drones, satellite maps) or from on-the-go surveys. The possibility of improving this information depends on the availability of machines capable of distributing inputs to the crop in variable rate (VR) mode. Moreover, in crop protection treatments there is the need to greatly improve reaching the target to avoid dispersion of chemicals in the environment. In this paper, a study was carried out to design and apply a machine able to distribute plant protection products, modifying some of its mechanical parts to obtain a modular effect capable of supporting the variability of a prescription map for variable-rate distributions. A conventional sprayer machine has been equipped with a prototype device able to varying the air outlet section and the tilt of the fan blades, to have modular airflows during the distribution of the product in the vineyard. The results from different configurations were statistically elaborated to verify their possible correspondence with the clusters of prescription maps based on the canopy index or spectral indices.

Keywords: Crop protection · Environmental impact · Variable-rate · Axial fan

1 Introduction

Precision practices in field operations could contribute to the reduction of GHG (greenhouses gas) emission thanks to the reduction of inputs for the agricultural field operation. In viticulture, GHG emission is caused by the production and distribution of fertilizers and pesticides, and pesticide application energy usage. Some studies show that machinery usage in viticulture accounted for more than 60% of the total warming potential of the wine production process (Sarri et al. 2020).

M. Biocca et al. (Eds.): SHWA 2020, LNCE 252, pp. 327–336, 2022.
https://doi.org/10.1007/978-3-030-98092-4_34

The most usual machines for the distribution of pesticides are sprayers equipped with axial fan of a 700–800 mm diameter producing air currents characterized by high volume and low velocity, in comparison to the centrifugal fan. Currently, the layout of these kind of axial fans can be set manually in order to select blade pitch, air-outlet width, and rotating speed. This adjustment plays a key role in the efficacy and the environmental sustainability of pesticide applications (Balsari et al. 2008; Doruchowski et al. 2014; Grella et al. 2017; Salcedo et al. 2021).

Experimental results have shown that the spray deposition is directly linked to the airflow pattern obtained from the sprayers and air-volume and direction considerably influence both spray deposit uniformity and off-target losses (Cross et al. 2003; Grella et al. 2020; Hołownicki et al. 2017).

In this context, precision viticulture and smart farming could improve the environmental behavior of the viticulture system (Gil et al. 2011; Sarri et al. 2020).

The aim of the work is to study the behavior of an axial fan different layout to design and to define combinations between blade pitch, air-outlet width, and rotating speed, to obtain three or four different gradations required in precision agriculture in cases of use of prescription maps for VRT (variable rate) distribution.

2 Material and Methods

2.1 Test Sites

The trials were carried out indoor at the inspection centre for sprayers and sprayer machine manufacturers "Antonino Turrisi", (Sferro, Catania, Italy, 37°50′15.8″N 14° 79′64.1″W) to minimize interferences among sprayer fan airflow and environmental conditions. The warehouse was 12 m width and 23 m long with a gable roof with 7.0 m high in the top and 6.0 m high in the bottom. No noticeable obstacles were present inside the building.

The tests were conducted in June and July 2019 with a mean temperature of 32.5 °C in June and 35.7 °C in July and a mean humidity of 26.4% in June and 26.0% in July. Inside the warehouse there was not any significant movement of air caused by wind currents.

2.2 Sprayer Machine

The sprayer machine (hydraulic pulverization) was equipped with an axial helical fan (700 mm diameter), with 8 polyethylene blades (130 mm long); these have an adjustable pitch that assumed five different angles: 20, 26, 33, 40, and 44°, as reported in Failla et al. (2020a; 2020b).

The conventional trailed sprayer was connected to a tractor WD narrow tractor model 3050DT (Goldoni, Modena, Italy) with 35 kW nominal power. For experimental reasons, the tank and the nozzle bar have been eliminated. A purposed metal sheet placed behind the fan increase or decrease the air-outlet width that can be adjusted from 40 mm to 110 mm (Figs. 1 and 2).

Fig. 1. The sprayer machine.

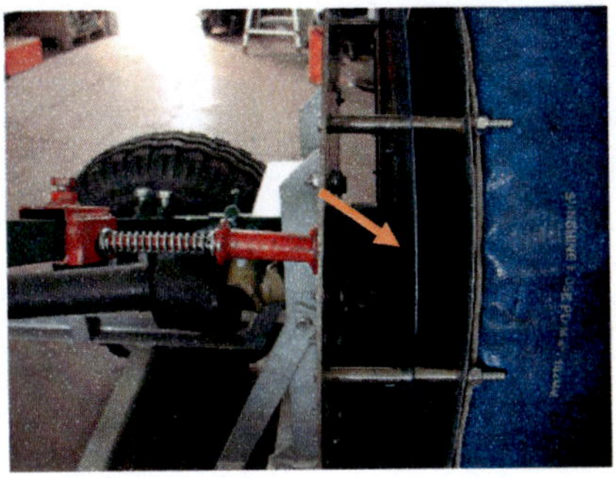

Fig. 2. The purposed metal sheet.

2.3 Test Layout

Thirty tests were performed by combining blade pitch (20, 26, 33, 40, and 44°), air-outlet width (40, 60 and 110 mm), and rotating speed ($\tau_1 = 3.8$ and $\tau_2 = 4.53$), as shown in Failla et al. (2020b).

A three-cup anemometer, mounted on a purposely designed frame, was used for measuring speed of the outgoing air from the fan at 4 distances, 0.50–1.00–1.50–1.80 m from the centre of the fan (horizontal adjustment), both on the right and on the left side.

To estimate power requirements, during fan operation at different layouts, a torque-meter (HBM mod. T30FN, Darmstadt, Germany) with 3 kNm nominal torque and 50 Hz (3000 rpm) nominal rotational speed was used to measure torque and rotational speed for the tractor PTO.

2.4 Data Analysis

All the data were statistically analyzed with the open-source software R (R Development Core Team 2012) to verify the contribution of the measured variables (speed of the outgoing air to the right and left side and the absorbed power) in the characterization of the observed theses and in the possible clustering of the results obtained. Multivariate PCA (Principal Component Analysis) was used for the contribution of each variable (Abdi and Williams 2010) developing the *factoMineR* and *factoshiny* packages (Kassambara 2017) after standardizing the values. The clustering of values was developed by applying the *NbClust* package (Charrad et al. 2014).

3 Results and Discussion

The average values and the respective standard deviations of the experimental tests results are shown in Tables 1 and 2.

Table 1. Mean values and standard deviation of results in rotating speed with τ_1

Blade angle degree	Section mm	Power kW	Air speed dx m s^{-1}	Air speed sx m s^{-1}
		5.78 ± 3.72	**13.73 ± 9.76**	**15.59 ± 7.73**
20		**0.83 ± 0.07**	**11.52 ± 8.15**	**10.79 ± 4.67**
	40	0.92 ± 0.00	12.90 ± 8.51	12.45 ± 7.61
	60	0.81 ± 0.00	12.95 ± 8.09	11.08 ± 2.28
	110	0.77 ± 0.00	8.70 ± 9.46	8.85 ± 2.81
26		**2.00 ± 0.12**	**14.25 ± 11.21**	**12.33 ± 7.38**
	40	1.90 ± 0.00	14.70 ± 10.75	15.83 ± 9.47
	60	2.17 ± 0.00	17.65 ± 10.38	15.23 ± 3.94
	110	1.95 ± 0.00	10.40 ± 14.21	5.93 ± 3.51
34		**7.36 ± 0.42**	**15.53 ± 12.21**	**17.18 ± 6.22**
	40	6.87 ± 0.00	15.65 ± 12.60	20.58 ± 9.50
	60	7.36 ± 0.00	19.78 ± 13.70	16.70 ± 3.37
	110	7.85 ± 0.00	11.15 ± 12.26	14.28 ± 3.63
40		**8.68 ± 0.33**	**14.46 ± 10.02**	**19.23 ± 7.30**
	40	8.27 ± 0.00	12.85 ± 6.97	19.25 ± 8.64
	60	9.05 ± 0.00	18.40 ± 13.12	23.05 ± 7.72
	110	8.73 ± 0.00	12.13 ± 10.77	15.38 ± 4.67
44		**10.03 ± 0.19**	**12.74 ± 7.14**	**18.41 ± 9.46**
	40	9.81 ± 0.00	9.77 ± 4.07	21.75 ± 8.91
	60	10.03 ± 0.00	13.40 ± 4.71	21.25 ± 5.72
	110	10.25 ± 0.00	14.48 ± 10.67	12.23 ± 11.81

The analyzed dataset, represented by the values collected in the 30 tests, did not show values considered to be outliers. The multivariate analysis of the principal components (Principal Component Analysis PCA) (Fig. 3) showed that the first two dimensions of analysis express 88.09% of the total variance of the dataset; this means that 88.09% of the total variability of the distribution of values is explained. This value is greater than the reference value which is equal to 82.92%, the variability explained by this plan is therefore significant (the reference value is the 0.95-quantile of the distribution of the inertia percentages obtained by simulating 2636 tables of data of equivalent size based on a normal distribution). From these observations, it is most likely not necessary to interpret the subsequent dimensions.

The estimate of the most correct number of axes to interpret suggests limiting the analysis to the description of the first 2 axes. These axes exhibit a greater amount of explained variance than those obtained from the 0.95 quantile of the random distributions (88.09% versus 82.92%). This observation suggests that only these axes significantly interpret the dataset. Consequently, the distribution will be studied on these first two axes (Figs. 4 and 5). In the biplot (Fig. 5) the observed conditions are shown

Table 2. Mean values and standard deviation of results in rotating speed with τ_2

Blade angle degree	Section mm	Power kW	Air speed dx m s^{-1}	Air speed sx m s^{-1}
		7.95 ± 4.42	**14.28 ± 9.65**	**15.46 ± 7.52**
20		**2.09 ± 0.14**	**12.35 ± 9.19**	**12.67 ± 5.68**
	40	2.26 ± 0.00	13.83 ± 9.13	15.75 ± 7.03
	60	2.06 ± 0.00	14.10 ± 9.62	14.63 ± 3.43
	110	1.94 ± 0.00	9.13 ± 10.63	7.63 ± 2.28
26		**3.46 ± 0.16**	**16.11 ± 11.91**	**12.81 ± 4.57**
	40	3.55 ± 0.00	16.65 ± 10.27	15.95 ± 1.69
	60	3.59 ± 0.00	20.03 ± 11.37	14.93 ± 1.48
	110	3.25 ± 0.00	11.65 ± 15.44	7.55 ± 3.92
34		**10.09 ± 1.37**	**16.83 ± 12.62**	**15.40 ± 8.28**
	40	9.26 ± 0.00	19.23 ± 13.68	20.35 ± 7.39
	60	9.06 ± 0.00	18.13 ± 10.23	18.87 ± 2.57
	110	11.94 ± 0.00	13.15 ± 16.27	5.33 ± 1.00
40		**11.31 ± 0.10**	**12.46 ± 5.99**	**16.30 ± 4.68**
	40	11.17 ± 0.00	14.90 ± 5.56	19.37 ± 5.25
	60	11.36 ± 0.00	15.37 ± 5.35	18.17 ± 1.90
	110	11.39 ± 0.00	7.10 ± 4.37	11.37 ± 0.76
44		**12.82 ± 0.48**	**12.81 ± 4.97**	**20.32 ± 10.48**
	40	12.33 ± 0.00	11.17 ± 5.00	23.33 ± 11.59
	60	13.43 ± 0.00	15.27 ± 5.05	22.93 ± 7.00
	110	12.70 ± 0.00	12.00 ± 5.86	14.70 ± 12.51

with a denomination that indicates in order: the rotating speed, then the blade angle then the section, so that for example 1_26_11 represents the condition of rotating speed 1, of 26° of blade angle and 110 mm of section.

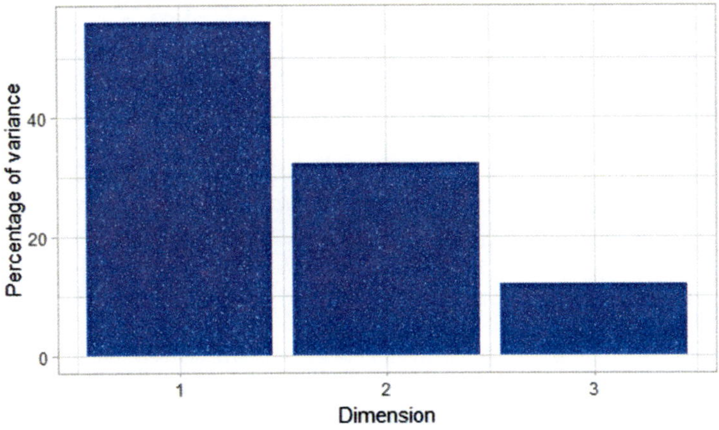

Fig. 3. Breakdown of the overall variance into the three components processed by the PCA.

Dimension 1 opposes values such as 2_44_6, 1_40_6, 2_34_4, 2_34_6 and 2_44_4 (to the right of the graph, characterized by a strongly positive coordinate on the axis) to values such as 2_20_11, 1_26_11, 1_20_6, 2_26_11 and 1_20_11 (to the left of the graph, characterized from a strongly negative coordinate on the axis).

The group containing the values 2_44_6, 1_40_6, 2_34_4, 2_34_6 and 2_44_4 (characterized by a positive coordinate on the axis) is distinguished by high values for the sx_corr and kW variables (the variables are ordered by the most significant).

The group containing the values 2_20_11, 1_26_11, 1_20_6, 2_26_11 and 1_20_11 (characterized by a negative coordinate on the axis) is distinguished by low values for the variables sx_corr, kW and dx_corr (the variables are ordered by the weakest).

Dimension 2 opposes values such as 1_26_6, 2_26_4 and 2_26_6 (at the top of the graph, characterized by a strongly positive coordinate on the axis) to the value of 2_40_11 (at the bottom of the graph, characterized by a strong negative dimension on the axis).

The group containing the values 1_26_6, 2_26_4 and 2_26_6 (characterized by a positive coordinate on the axis) is distinguished by high values for the variable dx_corr.

The group in which the single 2_40_11 is located (characterized by a negative coordinate on the axis) is characterized by low values of dx_corr and high values of kW.

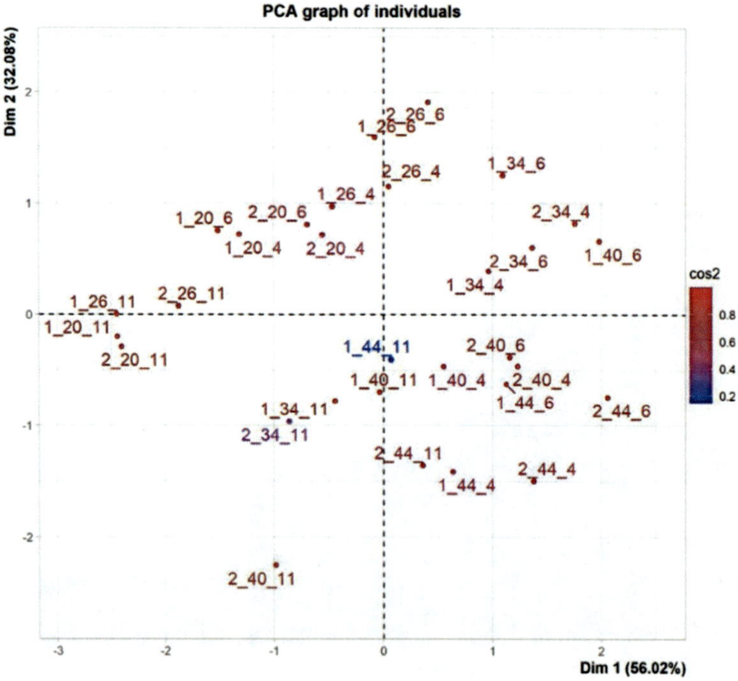

Fig. 4. Distribution of values based on the first two main components of the PCA. The significance of the effect of the components is indicated by the value of cos2.

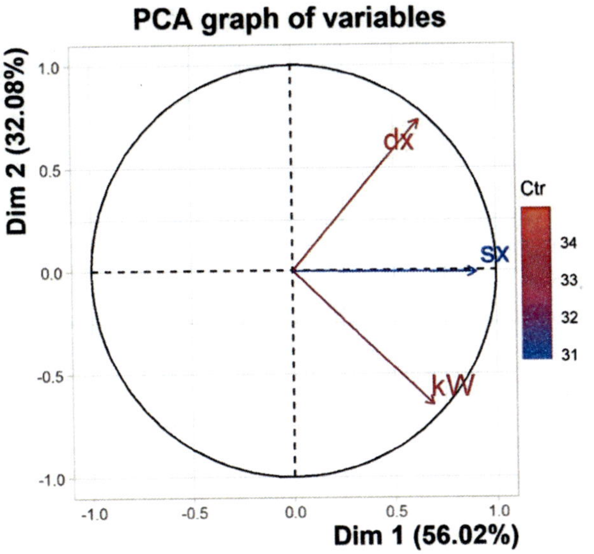

Fig. 5. Variables and their control effect on the distribution of values. Ctr indicates the extent of the control effect of the variables on the observed values.

The Silhouette analysis (Fig. 6), conducted with the most common clustering method of k-medoids and the PAM (Partitioning Around Medoids) algorithm (Kaufman and Rousseeuw 1990), showed clustering in 3 clusters as the most suitable.

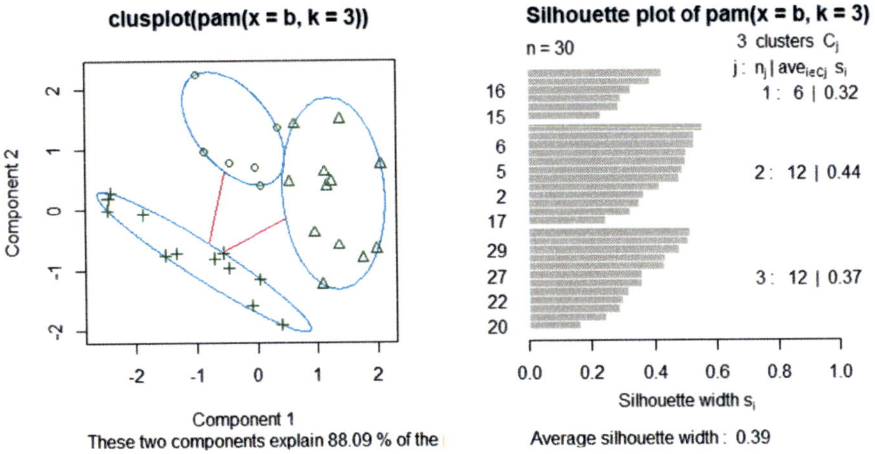

Fig. 6. Clusplot and Silhouette plot for the evaluation of the distribution of values in 3 clusters.

It is therefore possible to identify three clusters (Fig. 7).

Cluster 1 is made up of individuals such as 2_26_11, 1_26_11, 1_20_6, 1_20_11 and 2_20_11. This group is characterized by low values for the variables kW, sx_corr and dx_corr (the variables are ordered by the weakest).

Cluster 2 is made up of individuals such as 2_40_11, 2_44_4, 2_44_6, 2_44_11 and 1_44_4. This group is characterized by high values for the variable kW.

Cluster 3 is made up of individuals such as 2_34_6, 2_34_4, 1_40_6, 1_34_6, 2_26_6, 2_26_4 and 1_26_6. This group is characterized by high values for the dx_corr variable.

All three clusters of the outgoing air speed show how it is possible to reach the same speed with different combinations between engaged gear, blade inclination and opening of the outlet section. But with the same output speed, some combinations can have an effect of increasing the power required and therefore the overall consumption of the spraying operation.

This analysis therefore indicates the possibility of identifying conditions for realizing different flows, with combinations that can affect the power required and therefore consumption in different ways.

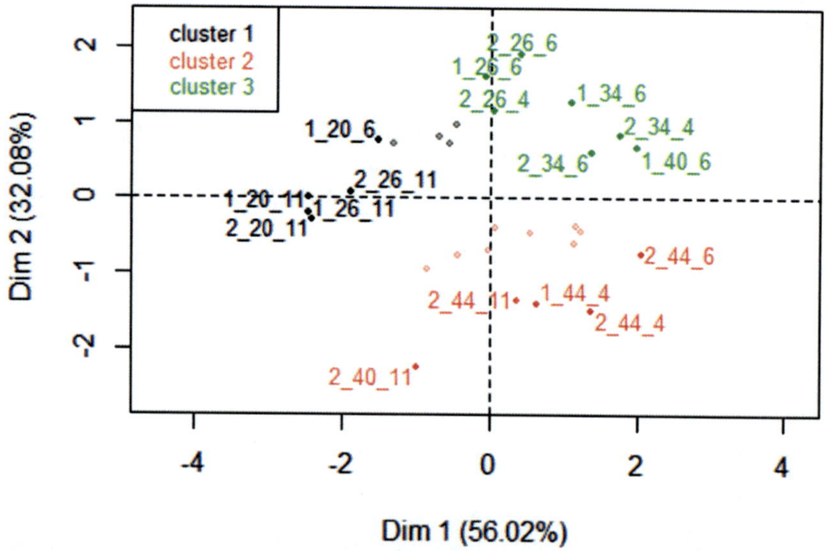

Fig. 7. Ascending hierarchical classification of values. The classification identified 3 clusters.

4 Conclusions

Through this analysis, the study of the observed conditions, helps in the definition of combinations between blade pitch, air-outlet width, and rotating speed, to obtain three or four different gradations required in precision agriculture in the use of prescription maps for VRT distribution (variable rate). Furthermore, in the preparation of the combinations, the manufacturer can opt to favor combinations with reduced consumption and therefore greater sustainability.

Acknowledgments. The authors wish to acknowledge the "Antonino Turrisi" firm in the Catania province (Sicily, Italy) for hosting the experimental tests. The activity presented in the paper was part of the research project titled "Safe and Smart Farming with Artificial Intelligence and Robotics. WP2 - Valutazione prestazioni di ventilatori per macchine irroratrici e loro automazione." Piano di Incentivi per la Ricerca di Ateneo 2020/2022. University of Catania (Italy).

References

Abdi, H., Williams, L.J.: Principal component analysis. WIREs Comp. Stat. **2**(4), 433–459 (2010). https://doi.org/10.1002/wics.101

Balsari, P., Marucco, P., Oggero, G., Tamagnone, M.: Study of optimal air velocities for pesticide application in vineyard. Aspects Appl. Biol. **84**, 417–424 (2008)

Charrad, M., Ghazzali, N., Boiteau, V., Niknafs, A.: NbClust: An R package for determining the relevant number of clusters in a data set. J. Stat. Softw. **61**(6), 1–36 (2014). http://www.jstatsoft.org/v61/i06/

Cross, J.V., Walklate, P.J., Murray, R.A., Richardson, G.M.: Spray deposits and losses in different sized apple trees from an axial fan orchard sprayer: 3. Effects of air volumetric flow rate. Crop Prot. **22**(2), 381–394 (2003)

Doruchowski, G., Balsari, P., Gil, E., Marucco, P., Roettele, M., Wehmann, H.-J.: Environmentally optimised sprayer (EOS)-a software application for comprehensive assessment of environmental safety features of sprayers. Sci. Total Environ. **482–483**, 201207 (2014)

Failla, S., Romano, E., Longo, D., Bisaglia, C., Schillaci, G.: Effect of different axial fans configurations on airflow rate. In: Coppola, A., Di Renzo, G., Altieri, G., D'Antonio, P. (eds.) Innovative Biosystems Engineering for Sustainable Agriculture, Forestry and Food Production. MID-TERM AIIA 2019. LNCE, vol. 67, pp. 691–699. Springer, Cham (2020a). https://doi.org/10.1007/978-3-030-39299-4_75

Failla, S., Bisaglia, C., Schillaci, G., Longo, D., Romano, E.: Sprayer axial fan layout affecting energy consumption and carbon emissions. Resources **9**, 136 (2020b)

Gil, E., Llorens, J., Landers, A., Llop, J., Giralt, L.: Field validation of dosaviña, a decision support system to determine the optimal volume rate for pesticide application in vineyards. Eur. J. Agron. **35**(1), 33–46 (2011)

Grella, M., Gallart, M., Marucco, P., Balsari, P., Gil, E.: Ground deposition and airborne spray drift assessment in vineyard and orchard: the influence of environmental variables and sprayer settings. Sustainability **9**(5), 728 (2017)

Grella, M., Miranda-Fuentes, A., Marucco, P., Balsari, P.: Field assessment of a newly-designed pneumatic spout to contain spray drift in vineyards: evaluation of canopy distribution and off-target losses. Pest Manag. Sci. **76**(12), 4173–4191 (2020)

Hołownicki, R., Doruchowski, G., Swiechowski, W., Godyn, A., Konopacki, P.J.: Variable air assistance system for orchard sprayers; concept, design and preliminary testing. Biosys. Eng. **163**, 134–149 (2017)

Kassambara, A.: Practical Guide to Principal Component Methods in R. STHDA (2017)

Kaufman, L., Rousseeuw, P.: Finding Groups in Data: An Introduction to Cluster Analysis. Wiley, New York (1990)

Marucco, P., et al.: OPTIMA project: development of a smart sprayer for vineyards. Aspect. Appl. Biol. **144**, 77–83 (2020)

R Development Core Team: A Language and Environment for Statistical Computing. R Foundation for Statistical Computing, Vienna (2012). http://www.R-project.org. Accessed 26 Apr 2019

Salcedo, R., Fonte, A., Grella, M., Garcerà, C., Chueca, P.: Blade pitch and air-outlet width effects on the airflow generated by an airblast sprayer with wireless remote-controlled axial fan. Comput. Electron. Agric. **190**, 106428 (2021)

Sarri, D., et al.: Smart farming introduction in wine farms: a systematic review and a new proposal. Sustainability **12**, 7191 (2020)

Farm Management Information Systems: Digital Register of Farm Management in Southern Italy

Costanza Fiorentino[1], Marica Crimaldi[2], Pasquale Libergoli[2], Paola D'Antonio[1](✉), and Vincenzo Scalcione[1]

[1] Department of European and Mediterranean Cultures – DICEM, University of Basilicata, via Lanera 20, Matera, Italy
paola.dantonio@unibas.it

[2] ITS Foundation - New Technologies for Made in Italy Food System, Agri-food Production Sector, S.C. 138 C. da Marangi n. 26, 70010 Locorotondo, Italy

Abstract. The continuous evolution of the agricultural sector pushes farms to consider the use of new technologies. In the urban environment, the issue of environmental sustainability is becoming increasingly important, and the main objective for the future is to guarantee all possible technological innovations, trying to make them perfectly in tune with the environment, real sustainable technologies towards which we are already making many steps forward such as electric cars, photovoltaic panels, stored wind energy and Smart Grid Technology. Precision agriculture is also configured as a management strategy that uses information technologies to collect data from multiple sources, with a view to their subsequent use in the decision-making process. The shared goal of all technologies that fall within the scope of precise farming is the containment of the indeterminacy caused by the infinite variability that is typical of any agricultural land. In this perspective, a software to compile the treatment register offers multiple advantages. It allows you to obtain the result of a complete traceability of agricultural products.

Keywords: New technologies · Smart Grid Technology · Precision farming · Digital register of farm management

1 Introduction

The continuous evolution of the agricultural sector, mainly due to the advent of new economic and social scenarios and the growing importance of environmental protection, encourages farmers to adopt new technologies (Wigboldus et al. 2016). It is possible both to develop low cost and low environmental impact cultivation systems, and to have safe and objective documentation of the productions to meet the traceability needs (Verdouw et al. 2019) expressed by the market and the public administration.

Among the new methodology of planning and "rational" representation of the landscape we find satellite technologies that provide solutions able to make the most of

M. Biocca et al. (Eds.): SHWA 2020, LNCE 252, pp. 337–354, 2022.
https://doi.org/10.1007/978-3-030-98092-4_35

its value, operating in many fields of application, ranging from environmental and territorial monitoring to solutions for defense and security, up to scientific missions and planetary exploration through space data infrastructures, drones and airplanes.

The geomatics is based on two impulse factors:

- The evolution of technologies dedicated to detection, which has made it possible to acquire point and area observations of high precision on a global scale (GPS, Remote Sensing, SAR).
- Hardware and software development for information management (GIS).

In the urban environment, the issue of environmental sustainability is becoming increasingly important, and the main objective for the future is to guarantee all possible technological innovations, trying to make them perfectly in tune with the environment, real sustainable technologies towards which we are already making many steps forward such as electric cars, photovoltaic panels, stored wind energy and Smart Grid Technology.

Many software supporting the farmer in the management of the farm are present on the market: The Farm Management Information Systems (FMIS) are developed to compile past, collect present and predict future information related to certain operations for enhanced decision making (Tummers et al. 2019).

FMISs of the last generation include the Geographical Information System (Zhang and Taylor 2001) and Decision Support Tools (DST) (Wolfert et al. 2014). The latter is often based on Cropping System Simulators (CSS), born for learning purposes and later used for irrigation scheduling (FAO 2016), pest (Siad et al. 2019) and disease prediction (Pan et al. 2010). The Internet allowed FMISs to become outsourced services (e.g., for irrigation service managed by farmers associations) that profited from the growing amount of networked information (e.g., weather networks) (Richardson 1997).

Although most people can see the benefits of using a more precise approach to manage crops with additional information, the tools provided by precision farming and other information technologies have not yet moved into mainstream agricultural management. The increased complexity of the systems inhibits easy adoption and makes calculations as to the financial benefits uncertain. These issues can be resolved by improving the decision-making process through better management information systems, improved data interchange standards and clear management methods (Saiz Rubio et al. 2020).

At the moment, the utilization of scientific models together with the large amounts of data in different formats produced by modern farm machinery, sensors located within the farm, remote sensing, etc. is still an open area of research and new methods are developed continuously (Gondchawar et al. 2016),

The seamless incorporation of new functionality and assisting features into an existing FMIS is of paramount importance (Raja and Vyas 2019).

Access to large amounts of data, placed in databases that are constantly supplied with further elements by local and remote sensor systems, and their processing with mathematical-statistical techniques developed in the field of artificial intelligence studies (in synergy with agronomic research), are the basis of the "decision support systems" (DSS).

The shared goal of all technologies that fall within the scope of precise farming is the containment of the indeterminacy caused by the infinite variability that is typical of any agricultural land. To really do the right thing, the farmer should know moment by moment the state of the soil, of all the communities of living beings that inhabit it, its precise supply of water, air and nutritional elements, its texture and structure, then the state of the environment outside the ground, and finally the condition of the plant organisms residing in that land: all repeated indefinitely on the surface considered (Ferrández-Pastor et al. 2018). The difficulty in managing all these variables has so far hindered the full affirmation of science on experience.

Precision farming is very data intensive and historically linked with site specific activities and management on the field. It has become very clear in recent years that Precision Farming is not limited to site-specific farming (Xu et al. 2018). The use of techniques and methods that form precision farming can provide a wealth of information and tools to handle and apply information properly for any type of farm in any region. This information-driven approach can be used to help improve crop management strategies and proof of compliance through documentation.

With a view to rationalizing agricultural work, to contain its consumption and ecological footprint, and to support its profitability and compliance with regulations, the preparation of a more comprehensive FMIS, based on solutions able to automate the identification and registration of all mechanized activities carried out in the company. It is an IT framework that aims to embrace every area of the farm, in order to provide the farmer with constant assistance in his business.

The specifications of a portal of a farm, from the external stakeholders point of view, revealed that the history of the farm, information about the producers in the form of curriculum vitae, farm location, climatic and soil conditions and, last but not least, farming practices, is the information that the consumers would like to see in it. The consortium believes that further information and communication (ICT) developments in agriculture will include the development of agricultural robotics (Cruz et al. 2020; Davis and Dukes 2015) in collaboration with advanced FMIS systems.

Developing codes of good farming practice, diversifying markets and production systems as well as European standards of sustainable agricultural production systems require implementation of more elaborate management strategies. These have to respect specific ecological conditions, demands from the rural regions and those from the value-added chains. On top of that, these strategies have to be simple, but flexible enough to be adapted easily to changing economic or environmental conditions and they need proof of their compliance. Beyond that, the demand for information about the production processes is growing, both from the perspective of the value-added chains (traceability, Carli et al. 2014) as well as from regional stakeholders in order to fulfill multifunctional objectives by farming. An important prerequisite for farmers to comply with all these different demands is to easily have sufficient and timely information available for decision making or providing documentary evidence. The rapid development of technologies for information and communication, new sensors as well as the vast potentials for providing geo-referenced data (remote-sensing, on-line sensors, public databases etc.) also allows farmers to access new and high quality data and use them as specific information in decision making or process documentation (Navarro et al. 2020, Poonia et al. 2019). With automated data acquisition and handling in an on-farm management information system the farmers can be seen to comply with a rapidly growing demand of standards in the management of the production processes.

Precision agriculture application today involves planning targeted interventions with operating machines able to accurately transfer the prescription to the field vegetation (D'Antonio et al. 2020). Automatic and geo-referenced driving of tractors with satellite correction (RTK) is fundamental in this context. The automatic driving technologies make it possible to read and transmit geo-referenced maps in isobuses that refer to in field crop treatment.

To obtain the result of an automatic transfer of data from the places where the productions take place (fields, stables, etc.) to the company center, and as the activities are often very dispersed throughout the territory, it is essential to use wireless technologies (Jat et al. 2016) or technologies such as new 5G network, which also allows to overcome the limits related to the coverage ranges of wireless networks. In this way, in addition to allowing communication between more or less distant entities, it is possible to create real farm networks, which can be defined as the set of components of the FMIS through which the identification systems and sensors send in time real the data recorded to a server which, after processing, makes them usable to the company management through specific management programs.

An important role in this IT structure is played by the farm management register ('Quaderno di campagna') where all information, from seed and crop treatment to yield, are registered on digital record sheets. It allows - in addition to data recording - the collection and storage of information and the possibility of viewing, through graphic animations on digitized maps, the dynamic execution of the works in the fields. In this 'review' an analysis of the digital register of farm management available in FMISs, both free and for a fee was conducted. Five products were selected which, on the basis of interviews conducted in southern Italy with farmers and operators in the sector, seemed to arouse greater interest.

2 The Farm Management Register 'Quaderno di Campagna' in Italian Farms

All elements of the farm are united in the farm network (Source: dreamstime.com).

Established with Presidential Decree 290 of 2001, the Register of farm management, received a full regulatory definition with Legislative Decree 150 of 2012.

Keeping the Register of farm management has become an obligation for all those who sell or give agricultural products to third parties, while those who use plant protection products in vegetable gardens or family gardens whose product is intended for use are exempt self-consumption.

According to article 16 of decree 150: "Treatment Register means a company form that chronologically reports the list of treatments performed on the different crops, or, alternatively, a series of separate modules each relating to a single agricultural crop." All phytosanitary treatments performed must be registered by the time of collection, and in any case no later than 30 days from the time they are performed (Source: all4officesrl.it).

The Register must be kept for at least 3 years after the one to which the interventions refer and must be kept together with the product purchase invoices and the copy of the purchase forms with the relative danger classification.

The register must be completed by the owner of the company or by agricultural assistance centers subject to the authorization of the territorially competent ASL. Furthermore:

- in the event that the user of the plant protection products is different from the company owner, the latter must in any case sign the Register at the end of the calendar year;
- if the user is neither the owner nor the purchaser of the products, the Register can also be filled in by a different person as long as he has the authorization of the owner (to be kept together with the Register);
- when the treatments are carried out on behalf of third parties, the compilation of the Register is the responsibility of the owner of the company or the treatments are noted by the contractors and validated by the owner;
- In the case of cooperatives of producers that purchase plant protection products, the Register must be kept at the registered office of the association and it is the duty of the legal representative, after delegation from the members, to fill it in.

Again art. 16 establishes that the Farm management register must necessarily contain some data:

- the company's personal data;
- the denomination of the treated crop and its extension expressed in hectares;
- the date of the treatment, the product and the relative quantity used, expressed in kilo-grams or liters, as well as the adversity that made the treatment necessary.

In the Register, it is therefore required to indicate the parcel of the company where the crop is located, the cultivated species, and to specify in which phase of the biological cycle the plants are found at the time the treatment is carried out. In this way it is possible to create a real identity card of the product that arrives on the tables of consumers or in the subsequent links of the supply chain.

The Farm management register must also be completed when the treatments are carried out for the defense of stored foodstuffs or for treatments in an extra-agricultural area (eg. public green areas).

The penalties resulting from the failure to keep the Register of farm management are listed in Article 24, paragraph 13, of Legislative Decree 150/2012:

> "Unless the fact constitutes a crime, the purchaser and the user who does not fulfill the obligations of keeping the Register of crop treatments established by article 16, paragraph 3, are punished with the pecuniary administrative sanction of the payment of a sum to be 500 euros to 1500 euros. In case of reiteration of the violation, suspension from one to six months or the revocation of the authorization is ordered."

3 The Digital Farm Management Register

Southern Italy is often characterized by small and medium-sized farms and has a very fragmented landscape that makes precision agriculture applications as well as the management of machines in the field more complex.

In this context, company management that keeps track of all the agronomic actions carried out, the products used and the related costs used is the starting point for any precision agriculture application.

Farmers in southern Italy have recognized the importance of the digital farm management register, but it remains difficult for them to choose the right product for their needs. The choice of each farmer would fall on free products, but they mainly require software: easy to use, interoperable with other systems and able to cover all farm management needs. The table in the appendix lists the FMIS and DSS software widely used in Italy; they often include an application dedicated to the farm management register. A software to compile the register offers multiple advantages. First of all, it allows the farmer to streamline operations and therefore save time, above all because it allows access to complete and constantly updated databases (pesticides, nutrition products, weather): it is essential for the producer to always know whether the chemicals he is using are allowed by the legislator for the crops he has in place (it may happen to have in stock packages of revoked plant protection products that expose you to the risk of being fined).

Another advantage that comes from the use of software is the ability to print all the documents required by law or by the protocols to which the company adheres (Treatment register, Global Gap, etc.).

With the digital farm management register it is possible to obtain the result of a complete traceability of agricultural products. The legislator asks that each treatment has to be associated with the phenological phase of the plant, the exact day must be indicated, the type and quantity of product used and the cause that made the treatment necessary must be specified. In this way, upon collection, all the data required by the certification systems will be available: a very useful quantity of data, as a statistical resource, to refine production procedures in the years to come. In Fig. 1 is shown the web page of the Italian Health Ministry of the authorized plant protection product.

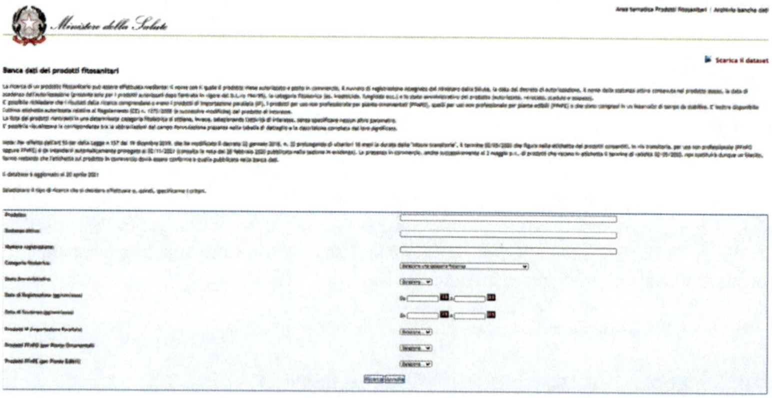

Fig. 1. Ministry of health database (Source: http://www.fitosanitari.salute.gov.it/fitosanitariwsWeb_new/FitosanitariServlet)

Digital registers available for the Italian user can be downloaded and installed locally or reside on an Internet web-site to which you must subscribe. In the first case, however, it is necessary to maintain a connection with the manufacturer's site, as the digital register would not make sense without constant updating of the databases. For this reason, and thanks to the enormous development of resource allocation technologies in the clouds, the market is definitely preferring the online version. The user can access the program from any device and wherever there is an internet connection.

In this regard, the issue of privacy is not negligible: manufacturers ensure the protection of their customers data through security protocols such as periodic back-ups and multi-level passwords, but they have to enter all their sensitive data on a cloud that belongs to a private company. And this need can intimidate the suspicious farmer.

Italian farmers increasingly express the need to have an easy-to-use and interoperable management system to keep warehouses in order, manage staff work, control energy consumption and simplify the maintenance of the treatment register. The huge supply of software on the market often confuses farm operators. There are many software, and often different versions of each are provided, with different packages and tiered prices. There are also free tools that may not satisfy the user's needs.

In this 'review' an analysis of the available free and paid products was conducted. Five products were selected which, on the basis of the interviews conducted with farmers, seemed to arouse greater interest:

1. Wapi
2. GAzie
3. QdC 'Quaderno di Campagna'
4. QdC TRAX
5. DSS HORTA

3.1 WAPI

Wapi (Fig. 2) is a new program available only for mobile devices, it is an app that can be downloaded for free from Google Play, very simple and intuitive. It guides the user through a series of steps that require to establish the production season, define the field (even with a polyline on Google Maps), enter the crop, the calendar of operations, treatments and fertilizations. For all processes, drop-down menus are opened that list the different options very accurately. The pesticide database is updated weekly. Only the products admitted in the most recent database of the Ministry of Health are displayed for each crop: you can consult the label of each product and, you can also choose the most suitable one (Fig. 3). Once the product has been chosen, it is inserted into the virtual warehouse and if a certain quantity remains at the end of the treatment, this is stored for subsequent use. Finally, a register of treatments in pdf is produced, perfectly in line with the legislation.

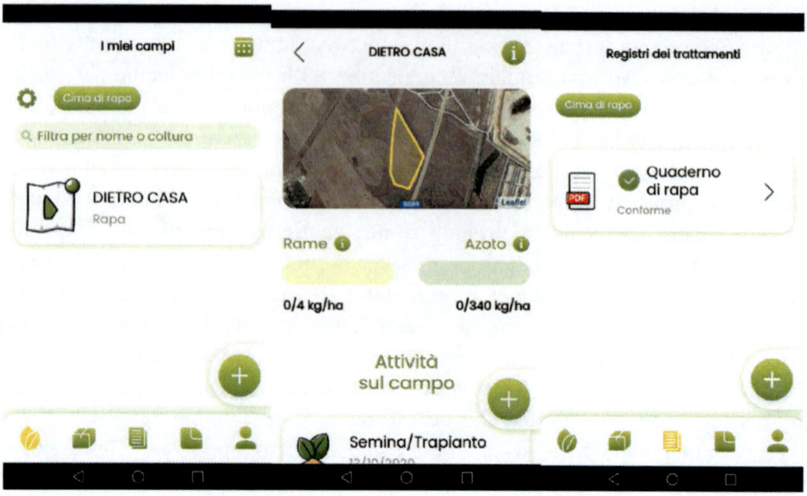

Fig. 2. Three program screens. From the initial screen, which lists all the company fields, we move on to the details of one of these. In the third image from the left, the pdf of the Treatment register.

The application is very simple, convenient and complete. Being connected to the Internet, it does not need to download heavy databases and does not require constant updates. However, it has the limit of being "only" a Register of treatments, it does not offer all the tools of a more complex information system: it is in any case an essential management system, which allows to keep the various fields under control, the warehouse of pesticides, fertilizers and propagation material. It is great for small farms, some limitation should be pointed out:

- if unfortunately, you have to close the session in the middle of the growing season the entered data cannot be recovered, but this problem may affect all non-downloadable software;
- important controls are lacking, for example the distance of the fields from the streams, the interval between treatments, the different doses for different weeds/pests, the maximum fertilization rates, the periods of prohibition, the disposal of manure with different efficiency and many other things.

These are the limits of an application which, however, has decided to prefer lightness and easy to use to absolute completeness.

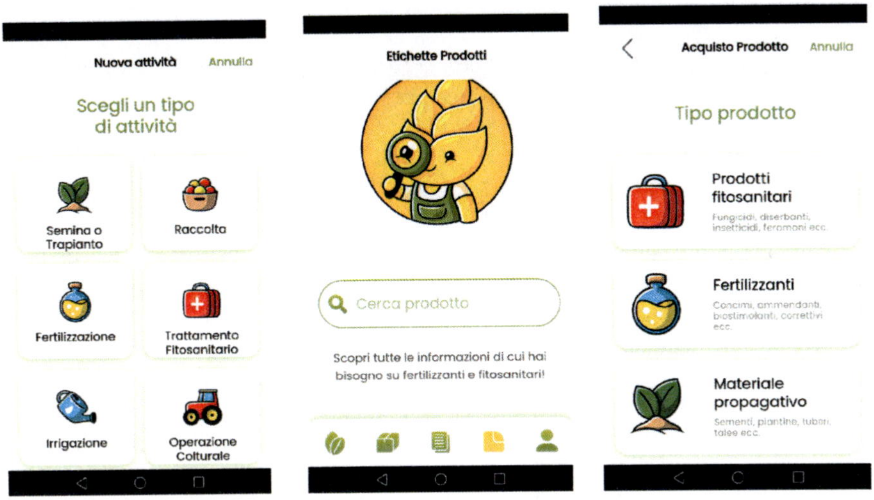

Fig. 3. Three Wapi screenshots. In the center, the search field on the database.

3.2 GAZIE

The farm management register is just a module introduced in 2018 in the GAzie company management system, accessible free of charge from the SourceForge platform. The activation of the module is not automatic (it is therefore necessary to search for it in the preferences). The software can work locally (via a virtual server such as XAMPP) or on the Internet, with real servers: this option allows users to use it also from a smartphone. GAzie implements a lot of business management tools, beyond the agricultural module, and can also be used online, from any mobile or fixed device. It is possible to manage the warehouse, issue any type of sales document (DdT, invoices, etc.) and invoices to the public administration also in electronic format (XML). However, the interface is not very accurate, it requires a certain computer familiarity - at least in the start-up phase - and there are many programming limits that hinder its use (for example on the Homepage screen, the menu box on the left doesn't have a scroll bar, and without a scroll wheel on the mouse, it is not possible to scroll down items). The creator of the Farm management register module states that one of its strengths would be complete warehouse management connected to suppliers and with specific indication of payments; the control of the minimum stock, with the possibility of reordering stocks that are running out. For each plant protection product, the safety and technical data sheets can be inserted, interfacing with the database of the Ministry of Health. In addition, the cultivated fields are identified on Google Maps: there is an indication of the shortage time, the management of metallic copper, limited by law, and nitrogen, with particular reference to the legislation on nitrates. Is very interesting the application for the calculation of the cost of each agricultural production, from sowing to harvest, including operations with agricultural machinery and workers.

3.3 QdC® – 'Quaderno di Campagna'

QdC® - "Quaderno di Campagna" is produced by Image Line, a chain of web sites linked to various agricultural sectors, which has now become an essential reference for its databases. The "Quaderno di Campagna" is an online application, it naturally relies on proprietary databases, therefore it is constantly accessible to an updated register of registrations and to the most recent legislation. The structure is intuitive (the user is permanently assisted) and very complete. Three versions are available: Compulsory Integrated Defense, Voluntary Integrated Defense and Organic Agriculture. The platform is particularly suitable for professional operators, technicians and producer organizations. The weather section is very rich: it is possible to have forecasts and historical weather data limited to the indicated farm, thanks to the processing of a large amount of data from weather satellites and from the national network for meteorological observations of the Air Force. The producer declares that in Italy 50,000 hectares are already cultivated by companies that make use of this register. The menus available in the main screen are the following: Farm, Fields, Warehouses, Monitoring, Operations, Traceability, Prints, Processing.

The Farm menu includes all the personal information, the geolocation (with Google Maps), the list of operators, contractors and agricultural machinery of the farms.

Fields include all the crops residing on the various fields of the farm. After having geolocated the field, the program gives access to customized weather data (relating to quadrants of 1 km $\times$ 1 km) updated daily and historical weather data relating to the same plot (but only for the last 30 days). The two more expensive versions (Voluntary Integrated Agriculture and Organic Agriculture) allow access to the Automatic Rainfall Registration, the Fertilization Plans and Nitrate Directive restrictions and the Irrigation Plans.

With the Warehouses menu it is possible to keep all stocks under control; with the connection to the internal Fitogest database, the technical information on the products that are loaded in the warehouse is always available: stocks, automatically updated; the regulatory situation; labels and safety data sheets; the calculation of the quantities of product used per year and the active substances used for each production unit. The two top versions give access to additional tools: fertilizer warehouse connected to the internal Fertilgest database, the seed and seedlings warehouse and the auxiliary warehouse, always connected to the Fitogest database. The Monitoring menu responds to the increasingly need to reduce the consumption of pesticides and use alternative tools: with this software it is possible to set the intervention thresholds (with 'alert' in case of exceeding), record installation and removal of traps, record catches, results of visual surveys and field inspections, and once the threshold is exceeded, this data can be used to justify the phytosanitary intervention. From the Operations menu the Phytosanitary Treatments submenu opens, in accordance with the current rules. The farmer chooses the field, the product, the dose.

Data on authorizations, safety intervals, intervals between treatments, maximum number of treatments, dose compliance, control of the maximum dose per treatment and per year, buffer zones, availability in warehouse, and the number of treatments per active substance can be recorded. From the Operations menu the Fertilizations submenu opens, it allows to enter the fertilization plans developed and record the fertilizations carried out, under the monitoring of Fertilgest. The program checks the maximum contributions of nitrogen, phosphorus and potassium, signaling any exceeding and preparing the printouts required by current legislation. It is possible to insert the reference to the Nitrates Standard. Finally, always from the Operations menu, it can be recorded all other operations carried out in the farm: from sowing, to irrigation, to the launch of auxiliaries, to transplanting, and even to green manuring, tanning seeds and harvesting.

A separate item is dedicated to the Traceability of production, it responds to the requests of cooperatives or to show the history of their products to the final consumer. For each batch we can create a "unique traceability code" to be printed on the labels and applied to the product packages: it is a simple QR code. Finally, it is possible to print the treatment register and other files so that they are in line with the regulations.

The last utility is the calculation of costs related to each crop: the operating cost of the machines, in connection with the internal database "agricultural machines"; then the remuneration of the operators; and finally, the purchase price of the technical resources.

QdC is a very complete register, accessible from any type of device and constantly updated, but - as for all remote tools, all its own information is provided to a private web-site, with the risk (probably remote) of data loss.

3.4 QDCTrax

The Farm management register is managed online, on a dedicated page of the qdc-trax.com website. It is produced by the historic software house LeonardoWeb and therefore has almost years of evolutionary history.

QDCTrax is an expert system for the management of phytosanitary defense and for the control of adversities, for fertilization, field operations, warehouse management, transfers, traceability. Production cost analysis and statistical data are available (Fig. 4).

(a) (b)

Fig. 4. (a) Main menu screen. (b) The Print and Report submenu gives the possibility to print a large number of documents, including all the documentation for the Global Gap.

The program allows users to simulate the treatments to verify their compliance and correctness. As for access security, this is managed with a multi-level identification system, with personal codes, username and password.

The application is accessible from all devices, portable and fixed. It is very interesting the connection between client farms and certification bodies and customers. CUR traceability codes are assigned which can be issued to the markets or to the transfer companies for each batch produced (batch number).

On the web-site of the farm management register it is not possible to trace the asking price, but there is the possibility of evaluating the software, which is very intuitive and easy to access. Ultimately, however, it does not offer many possibilities more than the simple and free Wapi, and it shares the problem of all applications that rely on the cloud, namely the limitation of privacy.

3.5 DSS Horta

An interesting tool for the farm management register is included in the DSS developed by HORTA, a spin-off company of the Catholic University of Piacenza, which has become a leader with its increasingly numerous DSSs on the market. The information available for each crop relates to the weather conditions, with 7-day forecasts; the periods of infectious risk for the main infections of the crop; to the presence and evolution of insects. The program also provides data on the phenological development of plants, the current level of protection and, if necessary, an updated database on plant protection products, with the optimal dose of product to be distributed. Finally, the program provides a document in compliance with the law.

The register is not an advanced tool like the previous ones, but it is certainly sufficient to comply with the legal constraints.

4 Conclusions

Precision agriculture, not only allows to operate in the field in a timely manner, respecting the multifaceted variability of agricultural ecosystems, but it is also configured as a strategy management that uses information technologies to collect data from multiple sources, with a view to their subsequent use in the decision-making process (Bacco et al. 2019).

Precision farming in Europe uses new technologies in information handling and management as well as in managing the spatial and temporal variability found on all farms. Such explicit information use improves economic returns and reduces environmental impact (Sørensen et al. 2010; Coppola et al. 2020). The introduction of advanced ICT technologies into agriculture will also be a significant progress in all efforts for measurements oriented payments within agro-environmental programs and related efforts to enforce environmentally sound systems in land use within the EU (Kernecker et al. 2021).

The DSS try to fill, as far as possible, this knowledge deficit and to provide the farmer with a description of the state of the field that contains the greatest amount of information (Debaucheet al. 2020; Rose et al. 2016).

The Farm management register is an unavoidable tool as the registration of the treatments carried out, in the manner indicated, has been made mandatory by law. It allows the storage of large amounts of data available to carry out statistical studies on production, on consumption and on the validity of the interventions.

The Farm management register is an unavoidable tool as the registration of the treatments carried out, in the manner indicated, has been made mandatory by law. Alongside the recording of the treatments and products used, and therefore the fulfillment of a legal obligation, the digital farm management register has the possibility of providing additional services to the user: it allows to have large amounts of data available to carry out statistical studies on production, on consumption and on the validity of the interventions. It is possible to associate crop information, labor data, and manage inventory flows.

The Digital Farm management register is a real farm management system, capable of providing important help to the farmers. Among the analyzed software Wapi is a very valid free application, especially for the management of small farms.

QdC, although it is not free, is a complete application, but for farms of significant size and varied activities, certainly the choice of a more complex management software is necessary.

Appendix

FMIS and DSS software widely used in Italy.

Choice	Services	Geographic area	Website
ABACO spa	Sensors and solutions for field monitoring, IoT monitoring platforms, DSS	Emilia Romagna, Lazio, Lombardia, Puglia, Sardegna, Toscana, Veneto	www.abacogroup.eu
ADLER HTE srl	Irrigation management, GIS mapping, other farm management software	Campania, Emilia Romagna, Lombardia, Sardegna	www.adlerhte.com
AGABUNA srl	Sensors and solutions for field monitoring, IoT monitoring platforms		www.agabuna.com
AGRICOLUS srl	Sensors and solutions for field monitoring, Pest Monitoring, Precision Applications		www.agricolus.com
AGRILOOM	Sensors and solutions for field monitoring, Soil analysis and sensors, Monitoring platforms		www.agriloom.com
AGRONICA GROUP srl	Sensors and solutions for field monitoring, Precision Applications, Monitoring platforms		www.agronica.it
AGROROBOTICA srl	Sensors and solutions for field monitoring, Soil analysis and sensors, Monitoring	Toscana	www.agrorobotica.it
ARCHETIPO srl	Sensors and solutions for field monitoring, Precision Applications, Irrigation Management, DSS		www.archetipo-srl.com
BDF srl	Databases, Data management, other farm management software		www.bdfsrl.it
BLUETENATCLES srl	Precision Applications, IoT Monitoring Platforms, Irrigation Management	Trentino Alto Adige	www.bluetentacles.com
CASSANDRA TECH srl	Sensors and solutions for field monitoring, Precision Applications, DSS	Calabria, Emilia Romagna, Lazio, Lombardia, Marche, Piemonte, Puglia, Sardegna, Toscana, Trentino Alto Adige	www.cassandratech.it

(*continued*)

(*continued*)

Choice	Services	Geographic area	Website
CDM TECNOCONSULTING (LUTECH GROUP)	Sensors and solutions for field monitoring, Soil analysis and sensors, Precision Applications	Italia, Abruzzo, Basilicata, Calabria, Campania, Emilia Romagna, Friuli Venezia Giulia, Lazio, Liguria	www.gruppocdm.it
CET ELECTRONICS snc	Sensors and solutions for field monitoring, Pest monitoring, Monitoring platforms	Veneto	www.cet-electronics.com
CLIMATE corp	Precision Applications, IoT Monitoring Platforms, Irrigation Management, DSS		www.climatefieldview.it
CYNOMYS srl	Sensors and solutions for field monitoring, Soil analysis and sensors, Precision Applications,		www.cynomys.it
DELAJO - DEFINE LAND AND JOB	Treatment register		www.delajo.com
DELAVAL spa	Barn/Farm Management, Automation and Robotics		www.delaval.com
DINAMICA GENERALE SPA	Sensors and solutions for field monitoring		www.dinamicagenerale.com
DNAPHONE srl	Quality management		www.dnaphone.it
DRONEBEE, AGRICOLTURA INTELLIGENTE	Sensors and Solutions for Field Monitoring, Pest Monitoring, Precision Applications,		www.dronebee.it
ECOSEARCH srl	Sensors and solutions for field monitoring, Soil analysis and sensors, Precision Applications,		www.ecosearch.it
ELAISIAN	Sensors and Solutions for Field Monitoring, Pest Monitoring, Precision Applications,		www.elaisian.com
ENOGIS srl	Precision Applications, IoT Monitoring Platforms, DSS		www.enogis.it
EVJA	Sensors and solutions for field monitoring, Soil analysis and sensors, Pest monitoring,	Italia, Abruzzo, Basilicata, Calabria, Campania, Emilia Romagna, Friuli Venezia Giulia, Lazio, Liguria	www.evja.eu
EZ LAB srl	Blockchain		www.ezlab.it
FEELERA srl-sb	Supply chain, Blockchain		www.feelera.eu
FINAPP srl	Sensors and solutions for field monitoring		www.finapptech.com
FOS spa	Sensors and solutions for field monitoring, Pest monitoring, Monitoring platforms		www.gruppofos.it
GRAINIT srl	Sensors and solutions for field monitoring		www.grainit.it
GREENTOP	Sensors and solutions for field monitoring, Precision Applications, DSS		www.greentop.it
HK HORTICULTURAL KNOWLEDGE srl	Sensors and solutions for field monitoring, Precision Applications, Monitoring platforms		www.hkconsulting.it
HORTA srl	Irrigation management, DSS (decision support systems), Agronomic consultancy, Treatment register		www.horta-srl.it
IDROPLAN, FARM TECHNOLOGIES srl	Sensors and solutions for field monitoring, IoT monitoring platforms, Irrigation management		www.idroplan.org
IFARMING srl	Sensors and solutions for field monitoring, Soil analysis and sensors, Precision Applications,	Emilia Romagna, Lombardia, Piemonte, Puglia, Toscana, Veneto	www.ifarming.it

(*continued*)

(*continued*)

Choice	Services	Geographic area	Website
IMAGE LINE srl unipersonale	Treatment register, Other farm management software, Food and agri-food product safety		www.imagelinenetwork.com
INFORMATICA AMBIENTALE srl	Soil Analysis and Sensors, Precision Applications, Irrigation Management, DSS		www.iambientale.it
IOAGRI srl	ERP (business resource planning)		www.ioagri.it
IRRIWORKS srl	Precision Applications, DSS (decision support systems), Other farm management software		www.irriworks.com
IRVO	Precision Applications, Irrigation Management, Agronomic Consultancy, treatment register		www.irvos.it
ITPHOTONICS srl	Sensors and solutions for field monitoring, Soil analysis and sensors, Precision Applications,		www.itphotonics.com
LIMES srl	DSS (decision support systems), Download and interpretation of satellite data, GIS mappings		www.limes-tech.it
MECHANIKA.PARTS	Digital market		www.mechanika.parts
METACORTEX srl	Soil analysis and sensors, Precision Applications, DSS (decision support systems), Drones		www.metacortex.it
MYVIGNETO	Pest monitoring, Treatment Register		www.myvigneto.com
NEOMA srl	IoT, Automation and robotics monitoring platforms		www.neoma.tech
NEOS srl	Sensors and solutions for field monitoring, Precision Applications, Irrigation Management, Dow	Italia, Abruzzo, Basilicata, Calabria, Campania, Emilia Romagna, Friuli Venezia Giulia, Lazio, Liguria	www.neos-spinoff.com
NEPTUNE srl	Precision Applications, Irrigation Management, Satellite Data Download and Interpretation, Map		www.neptunewaterleaks.com
NETSENS srl	Sensors and solutions for field monitoring, IoT monitoring platforms, Irrigation management		www.netsens.it
ONO EXPONENTIAL FARMING srl	Automation and robotics		www.onoexponentialfarming.com
PASTELLA FACTORY srls	Sensors and solutions for field monitoring	Italia, Abruzzo, Basilicata, Calabria, Campania, Emilia Romagna, Friuli Venezia Giulia, Lazio, Liguria	www.pastellafactory.com
PESSL INSTRUMENTS gmbh	Sensors and solutions for field monitoring, Soil analysis and sensors, Pest monitoring,		www.metos.at
PHEROMED sas	Sensors and solutions for field monitoring, pest monitoring, DSS		www.pheromedfly.com
PRIMO PRINCIPIO	Sensors and solutions for field monitoring, Pest monitoring, Monitoring platforms		www.wiforagri.com
QdC 'Quaderno di Campagna' Image Line	Register of treatment		https://quadernodicampagna.imagelinenetwork.com/
TRAX QdC Leonardo	Register of treatment		http://www.qdctrax.com/
RURALSET srl	Sensors and Solutions for Field Monitoring, Pest Monitoring, Precision Applications,		www.ruralset.it

(*continued*)

(*continued*)

Choice	Services	Geographic area	Website
SIRbit srl	Sensors and solutions for field monitoring, Soil analysis and sensors, Precision Applications	Italia, Abruzzo, Basilicata, Calabria, Campania, Emilia Romagna, Friuli Venezia Giulia, Lazio, Liguria	www.sirbit.it
SPEKTRA AGRI srl	Precision Application		www.vantage-italia.com
STAM srl	Sensors and solutions for field monitoring, Precision Applications, Monitoring platforms		www.stamtech.com
Studio Agronomico Silvio Fritegotto	Agronomic consultancy, Information	Toscana	www.fritegotto.it
Studio Agronomico Zampolli	Soil analysis and sensors, Precision Applications, DSS (decision support systems),	Lombardia	
SURGE srl	Blockchain		www.surgetech.eu
SYSMAN PROGETTI & SERVIZI srl	Sensors and solutions for field monitoring, Soil analysis and sensors, Monitoring platforms	Puglia	www.sys-man.it
TALPALABS srl	Sensors and solutions for field monitoring, Soil analysis and sensors, Precision Applications	Piemonte	www.talpalabs.it
TECSERVICE srls	Satellite navigation	Basilicata, Campania, Puglia	www.tecservicesrls.it
TERRASYSTEM srl	Sensors and solutions for field monitoring, Soil analysis and sensors, Pest monitoring, Appl	Lazio, Toscana, Umbria	www.terrasystem.it
TESSA AGRITECH SRL	Sensors and solutions for field monitoring, IoT monitoring platforms, Irrigation management	Emilia Romagna, Trentino Alto Adige	www.tessa-agritech.com
TRIMBLE ITALIA srl	Precision Applications, Satellite Navigation, Other Farm Management Software, Market	Italia, Abruzzo, Basilicata, Calabria, Campania, Emilia Romagna, Friuli Venezia Giulia, Lazio, Liguria	www.agriculture.trimble.com
TRK srl	Sensors and solutions for field monitoring, IoT monitoring platforms, DSS	Lombardia	www.trkitalia.it
UPTOEARTH gmbh	Sensors and solutions for field monitoring, Soil analysis and sensors, Precision Applications	Campania, Lazio, Umbria, Veneto	www.uptoearth.eu
VISUAL DRONE ag	Sensors and solutions for field monitoring, Download and interpretation of satellite data, Mappings	Puglia	www.visual-drone.it
WINET srl	Sensors and solutions for field monitoring, Soil analysis and sensors, Precision Applications	Italia, Abruzzo, Basilicata, Calabria, Campania, Emilia Romagna, Friuli Venezia Giulia, Lazio, Liguria	www.winetsrl.com
XFARM srl	Sensors and solutions for field monitoring, Soil analysis and sensors, Pest monitoring,	Italia, Abruzzo, Basilicata, Calabria, Campania, Emilia Romagna, Friuli Venezia Giulia, Lazio, Liguria	www.xfarm.ag
YNNOVA srl	Sensors and solutions for field monitoring, Precision Applications, Monitoring platforms	Italia, Abruzzo, Basilicata, Calabria, Campania, Emilia Romagna, Friuli Venezia Giulia, Lazio, Liguria	www.ynnovahq.com

References:

Bacco, M., Barsocchi, P., Ferro, E., Gotta, A., Ruggeri, M.: The digitisation of agriculture: a survey of research activities on smart farming. Array **3–4**, 100009 (2019)

Carli, G., Canavari, M., Grandi, A.: Introducing activity-based costing in farm management: the design of the FarmBO system. Int. J. Agric. Environ. Inform. Syst. **5**, 69–84 (2014)

Coppola, A., Di Renzo, G.C., Altieri, G., D'Antonio, P.: Lecture Notes in Civil Engineering, Preface (Editorial), vol. 67, pp. 5–7 (2020)

Cruz Ulloa, C., Krus, A., Barrientos, A., Del Cerro, J., Valero, C.: Robotic fertilisation using localisation systems based on point clouds in strip-cropping fields. Agronomy **11**, 11 (2020)

D'Antonio, P., Scalcione, V.N.: Software and satellite technologies for precision agriculture: the potential with, EPH. Int. J. Agric. Environ. Res. **6**(3) (2020)

Davis, S.L., Dukes, M.D.: Methodologies for successful implementation of smart irrigation controllers. J. Irrigat. Drain. Eng. **141**, 04014055 (2015)

Debauche, O., Mahmoudi, S., Elmoulat, M., Mahmoudi, S.A., Manneback, P., Lebeau, F.: Edge AI-IoT pivot irrigation, plant diseases and pests identification. Procedia Comput. Sci. **177**, 40–48 (2020)

FAO. AquaCrop; Food and Agriculture Organization of the United Nations, Rome, Italy (2016)

Ferrández-Pastor, F.J., García-Chamizo, J.M., Nieto-Hidalgo, M., Mora-Martínez, J.: Precision agriculture design method using a distributed computing architecture on internet of things context. Sensors **18**, 1731 (2018)

Gondchawar, N., Kawitkar, R.S.: IJARCCE IoT based smart agriculture. Int. J. Adv. Res. Comput. Commun. Eng. **5** (2016)

Kernecker, M., Busse, M., Knierim, A.: Exploring actors, their constellations, and roles in digital agricultural innovations. Agric. Syst. **186**, 102952 (2021)

Jat, D.S., Madamombe, C.G.: Wireless sensor networks technologies and applications for smart farming. In: Ramesh, C.P., (ed.) Smart Farming Technologies for Sustainable Agricultural Development. IGI Global, Hershey, PA, USA, pp. 25–39 (2018)

Navarro, E., Costa, N., Pereira, A.: A systematic review of IoT solutions for smart farming. Sensors **20**, 4231 (2020)

Pan, Z., Li, X., Yang, X.B., Andrade, D., Xue, L., McKinney, N.: Prediction of plant diseases through modeling and monitoring airborne pathogen dispersal. CAB Rev. **5**, 1–11 (2010)

Poonia, R.C., Gao, X.Z., Raja, L., Sharma, S., Vyas, S.: Smart Farming Technologies for Sustainable Agricultural Development; Advances in Environmental Engineering and Green Technologies. IGI Global, Hershey (2019)

Raja, L., Vyas, S.: The study of technological development in the field of smart farming. In: Smart Farming Technologies for Sustainable Agricultural Development. IGI Global, Hershey, PA, USA, p. 24 (2019)

Richardson, D.: The Internet and rural and agricultural development; Technical report. Food and Agriculture Organization of the United Nations, Rome, Italy (1997)

Rose, D.C., et al.: Decision support tools for agriculture: towards effective design and delivery. Agric. Syst. **149**, 165–174 (2016)

Saiz-Rubio, V., Rovira-Más, F.: From smart farming towards agriculture 5.0: a review on crop data management. Agronomy **10**, 207 (2020)

Siad, S.M., Iacobellis, V., Zdruli, P., Gioia, A., Stavi, I., Hoogenboom, G.: A review of coupled hydrologic and crop growth models. Agric. Water Manag. **224**, 105476 (2019)

Sørensen, C.G., et al.: Conceptual model of a future farm management information system. Comput. Electron. Agric. **72**, 37–47 (2010)

Tummers, J., Kassahun, A., Tekinerdogan, B.: Obstacles and features of Farm Management Information Systems: a systematic literature review. Comput. Electron. Agric. **157**, 189–204 (2019)

Xu, M., David, J.M., Kim, S.H.: The fourth industrial revolution: opportunities and challenges. Int. J. Financ. Res. **9**, 90–95 (2018)

Verdouw, C., Sundmaeker, H., Tekinerdogan, B., Conzon, D., Montanaro, T.: Architecture framework of IoT-based food and farm systems: a multiple case study. Comput. Electron. Agric. **165**, 104939 (2019). [CrossRef]

Wigboldus, S., Klerkx, L., Leeuwis, C., Schut, M., Muilerman, S., Jochemsen, H.: Systemic perspectives on scaling agricultural innovations. A review. Agron. Sustain. Dev. **36**, 46 (2016)

Wolfert, S., Goense, D., Sorensen, C.A.G.: A future internet collaboration platform for safe and healthy food from farm to fork. In: Proceedings of the Annual SRII Global Conference, SRII. IEEE Computer Society, San Jose, CA, USA, 23–25 April 2014, pp. 266–273 (2014)

Zhang, N., Taylor, R.K.: Applications of a field level geographic information system (FIS) in precision agriculture. Appl. Eng. Agric. **17** (2001)

Environment Safety, People Health Protection and Welfare

A Survey on Safety Among Tree-Climber Professional Arborists

Marcello Biocca[1(✉)], Leonardo Bianchini[2], Pietro Gallo[1], Federica Caffaro[3], Eugenio Cavallo[4], and Massimo Cecchini[2]

[1] CREA – Consiglio per la Ricerca in Agricoltura e l'Analisi dell'Economia Agraria, Centro di ricerca Ingegneria e Trasformazioni agroalimentari, 00015 Monterotondo, Roma, Italy
marcello.biocca@crea.gov.it

[2] Department of Agriculture and Forest Sciences, University of Tuscia, 01100 Viterbo, Italy

[3] Department of Education, Roma Tre University, 00185 Rome, Italy

[4] Institute of Sciences and Technologies for Sustainable Energy and Mobility, National Research Council of Italy, 10135 Turin, Italy

Abstract. Rope-based access methods (tree-climbing) are a bunch of techniques to ascend and operate on tree canopies. In urban forestry, climbing trees by ropes is often the only solution to care for and maintain them. Since the tree-climbing is a work at height often associated to the use of dangerous tools, and tiresome gathering information on safety and health issues and risks which may undermine tree-climbing workers is particularly important to develop targeted interventions to enhance their safety. This paper reports the results of a study about worker's risks associated with this technique and on the knowledge and awareness of safety aspects. A questionnaire including 48 questions was submitted to a group of 86 Italian professional climbing arborists. The results of the survey allowed to consider several aspects of the safety of tree-climbing activity pointing out the critical areas where safety interventions are most required. Among others, lack of training and near-misses appear to be significant factors associated with serious accidents occurred during the work activity of tree-climbers.

Keywords: Risk assessment · Urban forestry · Tree rope access · Risk perception

1 Introduction

Rope-based access methods, also named "tree-climbing", are a variety of techniques to ascend and operate in tree canopies (Anderson et al. 2015). In Italy, the emerging profession of arboriculture has contributed to the diffusion of this method among professional arborists. These techniques are especially widespread for the care and maintenance of ornamental trees located in urban areas (Biocca et al. 2021). Several cultural operations can be performed using tree-climbing as ascending technique: pruning, felling of trees, cabling, phytosanitary inspections and others. Tree-climbing allows targeted and selective interventions on tree branches, reaching the canopy's

M. Biocca et al. (Eds.): SHWA 2020, LNCE 252, pp. 357–364, 2022.
https://doi.org/10.1007/978-3-030-98092-4_36

internal portions and accessing trees located in confined areas where machinery cannot enter (Biocca and Bortolini 2019). The work safety and associated risks are particularly important because the tree-climbing is a work at height, is tiresome, and it is often associated with the use of dangerous tools (i.e. saws, chainsaws, etc.).

The sector is still emerging and some authors have studied technical and economic aspects of the technique (Anderson et al. 2015; Biocca et al. 2020). However, scientific literature about risk-taking and accidents frequency among arborists is relatively poor in terms of quantitative and qualitative analysis (Mazzocchi et al. 2015; Bortolini et al. 2016). Particularly underinvestigated is the role played by near misses, i.e. those incidents in which a worker might have been hurt if the circumstances had been slightly different. Wright and Schaaf (2004) showed that near misses and accidents substantially share the same determinants, confirming the idea that near-misses may be considered a proxy of being exposed to the risk of suffering a more severe accident. Near misses have been investigated in different sectors. Still, less is known in the literature about near misses and their determinants in the agricultural and forestry sector, with few exceptions (Lilley et al. 2002; Lundqvist and Gustafsson 1992; Caffaro et al. 2017).

This paper focuses both on the worker's risks associated with tree-climbing and on the knowledge and awareness of safety aspects among professional arborists. In addition, it explores the antecedents of work-related accidents, considering the role played by the operator's sociodemographic characteristics, training, organization of work, near misses and risk perception. Finally, the study concludes a previous part of the survey which was mainly related to the technical aspects of tree-climbing (Bianchini et al. 2021).

2 Material and Methods

Data was collected using an ad-hoc questionnaire. The questionnaire was based on the results obtained in a previous focus group including five professional arborists with different work experience and the authors of the paper. In the focus group safety, technical and professional items were discussed. A draft version of the questionnaire was then submitted to the same participants to the focus group and to additional 10 selected arborists. The questionnaire was reviewed and finalized according to their comments. The final version of the questionnaire included 48 items, divided into five sections (Fig. 1): 1) sociodemographic characteristics of the participants and experience in tree-climbing, 2) education and training, 3) occupational health and safety aspects (history of accidents and near misses and musculoskeletal disorders), 4) organization of work and 5) perceptions and attitude (risky tasks and operations, safety behaviours). Responses were collected through five-point rating scales and multiple-choice questions. The final version of the questionnaire was then disseminated through specialized groups on social media (Facebook) and by e-mail.

Data collected were explored using descriptive statistics and then Exploratory Factor Analysis (EFA) with an Oblimin rotation (extraction method: Principal Component Analysis, PCA) was performed to identify the main dimensions representing the participants' risk perception. Cronbach's Alphas were then computed to measure the

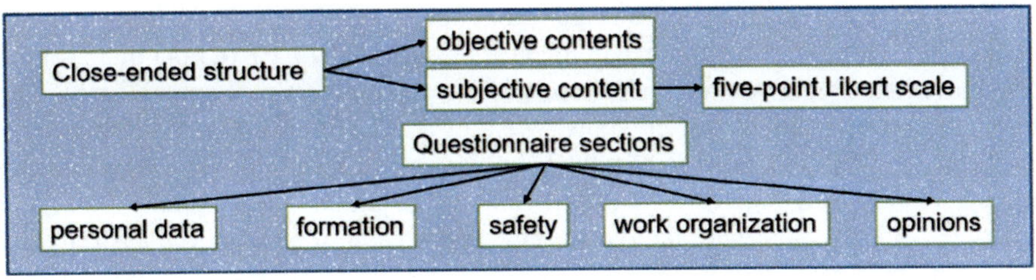

Fig. 1. Structure and contents of the questionnaire employed in the survey.

reliability of the extracted factors. Based on the factor loadings and reliability analysis, two variables ("Low risk activity" and "High risk activity", see the Results section) were then calculated as the average of the scores of the items belonging to Factor 1 and Factor 2, respectively.

These variables were finally used together with the operator's sociodemographic characteristics, training, organization of work, near misses as the independent variables in a multiple regression model computed to investigate the effects of these variables on the occurrence of accidents during work history. The analyses were performed using IBM SPSS Statistical Package for Social Science v. 26.

3 Results

The questionnaire was correctly filled out by 86 professional arborists.

Data regarding the sample composition, the level of education, the geographical work area, the working position, the technical and professional training, the training attended, and the coaching received during the work activity were previously published by Bianchini et al. (2021).

Regarding safety, the survey investigated the occurrence of serious accidents (defined as those that have caused an interruption of the work activity longer than one day) and of near misses. It resulted that 24.4% (21 out of 86) of the respondents experienced at least one serious accident during their career, with an average of 2.6 accidents per person (55 accidents for 21 people). The most common causes of the accidents were the use of manual saws (14 events) followed by chainsaw (7), rope swinging (5), insects (5), meteorological conditions (5), falling of objects from above (3) and interference with colleagues at ground (3). The falling for the failure of the anchor and the cutting of ropes registered one event, as well as the problem of miss-communication among operators. No accidents were caused by the tree falling, contact with electrical cables and interference with colleagues operating on the crown. Other causes, not specified by the respondents, triggered ten additional accidents. On average, all these accidents caused 14.1 days off work.

Near misses occurred for 39.5% of the respondents (34 out of 86) with a total of 88 events (equal to an average of 2.6 minor accidents for those persons who experienced at least a minor accidents).

The survey also investigated if respondents were suffering from recurring pains and pathologies while working as a tree-climber. It resulted that 73.2% (63 out of 86) of the respondents feel pain or suffer recurrent pathologies during their work activity. The most frequent pathology was an elbow, wrist, or arm tendonitis (60.3%) followed by backache (39.7%), recurring pain in the hands, wrist, elbow or arm (33.3%), stenosing tenosynovitis of the finger flexors (trigger finger) (19%), and carpal tunnel (7.9%). In addition, a 17.5% indicated other pains (note that this question could be answered by indicating more than one option).

Regarding the organization of work, 44% never alternate work at height and the ground during the same workday and 30% alternate the work rarely; the remaining 26% alternate the work always or often. In most cases (94%) tree-climbers never, or rarely, work with the same team, namely with well-known colleagues.

The ordinary maintenance of PPE (Personal Protective Equipment) is carried out with different periodicity: 29% maintains PPE daily, 15% weekly, 20% monthly and 23% yearly. 7% affirm that they never carried out maintenance while the remaining 6% do this operation every 6 months.

Furthermore, it was asked to comment on what was the most important aspect related to safety, giving a judgment on a scale from 1 to 5 (1 = not at all important 5 = very important). The results showed that arborists consider formation and training the most important factor in safety (average score 4.8), followed by the psycho-physical conditions (4.6), professionalism of the work team (4.5), static conditions of the plant (4.5), personal experience in the field (4.4), maintenance and technical level of the equipment (4.3), organization of the yard (4.3) and weather conditions (3.8).

As for risk perception, the participants were inquired about which is, in their opinion, the most dangerous phase of the work. The EFA yielded two factors, which cumulatively explained 61.95% of the variance (Table 1). The first factor could be labelled "Low-risk activities", since it included items regarding operations, such as the launch of the throwline and the ascending of the tree, that are perceived less dangerous than the others. The second factor, "High-risk activities" included items on activities that exposed the operators to the risk of falls, cutting accidents or others.

The results of the regression analysis are reported in Table 2.

Table 1. Exploratory Factor Analysis (EFA) for risk perception of work operations

Item	Low risk	High risk
Launching of the throwline	.809	
Ascending tree	.868	
Use of the chainsaw at height		.795
Use of the chainsaw at the ground		.757
Descent of the cut logs		.665
Felling		.773
Controlled felling		.768
α	.574	.802
% explained variance	61.953	

Table 2. Regression analysis for the occurrence of accidents in previous work history.

	β	Significativity
Age	−0,001	0,996
Experience	0,160	0,249
Training	−0,228	0,042*
Near misses	0,348	0,002**
Alternance of work at height and at ground	−0,015	0,894
Team formed by well-known colleagues	−0,056	0,615
PPE maintenance	0,127	0,242
Low risk activities	0,015	0,894
High risk activities	0,108	0,333
R^2	.213	

Note. * significant at p <. 05 level; ** significant at p <. 01 level

The analysis showed a significant negative relationship between training and accidents: the more the operators were trained (i.e. participated in a higher number of training events and courses), the less they were involved in accidents. A positive significant association was instead found between the near misses and accidents: when near misses increased, also accidents increased.

4 Discussion and Conclusions

The tree-climbing sector is characterized by a number of safety risks. Despite that, factors affecting the occurrence of accidents are still underinvestigated in this sector. Via a questionnaire submitted to a group of professionale arborists, the study analyzed occupational risks associated with the adoption of tree-climbing techniques to ascend trees. The results of the present study pointed out the significant effects of training and near-misses on the occurrence of serious accidents.

In particular, training showed a negative association with accidents: the more training courses the participants declared to have participated in, the less they reported having being involved in previous accidents. Training is widely acknowledged as a key factor in promoting workers' occupational safety and health among in many different sectors (Burke et al. 2011) and the present results are encouraging, even more if we consider that the tree-climbing sector in Italy is characterized by a high level of operator' education and training (Bianchini et al. 2021). This factor is perceived by the arborists, as they considered the formation and training the most important factors for safety.

The second factor associated with serious accidents were near-misses, which showed a positive relation with the occurrence of accidents. Near misses occur more frequently and are reduced in scale than serious accidents and it is well known that, generally, each major accident is usually preceded by a number of near misses (Phimister et al. 2000). The present result confirms the importance of investigating near misses to prevent more serious accidents (NSC 2013), including in the forestry sector, in which the analysis of near misses have been largely neglected. Operators should be trained to recognize and systematically annotate near misses. In this way, it would be possible to early identify critical aspects of the work system which could lead to a accident and intervene to eliminate or reduce them. Based on Kogler (2015), an effective strategy to pursue this objective could be to increase training in forestry operations, mechanical adaptations, and easy-to-understand and short written operating instructions.

Regarding occupational diseases, the majority of the participants reported to suffer from some pathologies and in particular musculoskeletal disorders due to their work activity. This result is in line with previous evidence and the more recent statistics (Colantoni et al. 2012; Micheletti Cremasco et al. 2019), which recognize that the adoption of repetitive gestures and postures and the use of manual tools increase worker's exposure to biomechanical overload risks.

The evaluation of the biomechanical overload risk in the primary sector activities is more difficult, due to the large variability of the tasks the operators have to perform, depending on crops, cultural operations, the machinery and tools adopted, the weather conditions, sometimes extreme, under which they have to be carried out with, and the daily and seasonal exposure, as well as the lack of a strict standardization of the work in general (Gallo and Mazzetto 2013). To be able to identify all the potential sources of risks and to design effective preventive interventions, the interaction with machines, tools and environment would deserve a particular attention in the application of risk assessment methods conceived for other contexts, or the formulation of specific methods that consider the distinctive characteristics of these activities (European Commission 2011).

Further in-depth studies are desirable to improve the productivity and safety of this type of emerging work dedicated to the correct management of urban green areas. Possible future developments of the research could further explore the relationship between adverse work environment factors and accidents, considering the safety behaviors and coping strategies adopted when dealing with adverse and stressful conditions (Glasscock et al. 2006; Caffaro et al. 2018). More objective techniques of data collection, such as the observation of arborists operating in real work situations, could allow to avoid the typical bias due to self-reporting contained in surveys based on questionnaires.

Acknowledgements. Authors are grateful to the professional arborists belonging to the following firms that actively collaborated in the project: Ecologistica Srl (Viterbo), UTM Soc. Coop. (Oriolo Romano, VT), Vitarborea (Palombara Sabina, RM).

References

Anderson, D.L., Koomjian, W., French, B., Altenhoff, S.R., Luce, J.: Review of rope-based access methods for the forest canopy: safe and unsafe practices in published information sources and a summary of current methods. Methods Ecol. Evol. **6**(8), 865–872 (2015)

Bianchini, L., Cecchini, M., Gallo, P., Biocca, M.: A Survey on rope-based ascending: techniques and materials of professional arborists in Italy. Environ. Sci. Proc. **3**, 23 (2021)

Biocca, M., Bortolini, L.: Macchine e tecniche per il verde urbano. Realizzazione, cura e manutenzione delle aree verdi. Consiglio per la ricerca in agricoltura e l'analisi dell'economia agraria – CREA (Ed.), Roma, p. 234 (2019)

Biocca, M., Gallo, P., Sperandio, G.: Technical and economic evaluation of urban trees pruning by climbing arborists. In: Coppola, A., Di Renzo, G.C., Altieri, G., D'Antonio, P. (eds.) MID-TERM AIIA 2019. LNCE, vol. 67, pp. 653–660. Springer, Cham (2020). https://doi.org/10.1007/978-3-030-39299-4_71

Biocca, M., Gallo, P., Sperandio, G.: Technical and economic analysis of Stone pine (*Pinus pinea* L.) maintenance in urban areas. Trees Forests People **6**, 100162 (2021)

Bortolini, L., Cividino, S.R.S., Gubiani, R., Cecchini, M., Delfanti, L.M.P., Colantoni, A.: Urban green spaces activities: a preparatory groundwork for a safety management system. J. Safety Res. **56**, 75–82 (2016)

Burke, M.J., Salvador, R.O., Smith-Crowe, K., Chan-Serafin, S., Smith, A., Sonesh, S.: The dread factor: how hazards and safety training influence learning and performance. J. Appl. Psychol. **96**, 46–70 (2011)

Caffaro, F., Micheletti Cremasco, M., Roccato, M., Cavallo, E.: It does not occur by chance: a mediation model of the influence of workers' characteristics, work environment factors, and near misses on agricultural machinery-related accidents. Int. J. Occup. Environ. Health **23**(1), 52–59 (2017)

Caffaro, F., Roccato, M., Micheletti Cremasco, M., Cavallo, E.: Falls from agricultural machinery: risk factors related to work experience, worked hours, and operators' behavior. Hum. Factors **60**(1), 20–30 (2018)

Colantoni, A., Marucci, A., Monarca, D., Pagniello, B., Cecchini, M., Bedini, R.: The risk of musculoskeletal disorders due to repetitive movements of upper limbs for workers employed to vegetable grafting. J. Food Agric. Environ. **10**(3–4), 14–18 (2012)

Commission, E.: A Non-binding Guide to Best Practice with a View to Improving the Application of Related Directives on Protecting Health and Safety of Workers in Agriculture, Livestock Farming. Horticulture and Forestry; EU Law and Publications, Brussels (2011)

Gallo, R., Mazzetto, F.: Ergonomic analysis for the assessment of the risk of work-related musculoskeletal disorder in forestry operations. J. Agric. Eng. **44**(s2), 730–735 (2013)

Glasscock, D.J., Rasmussen, K., Carstensen, O., Hansen, O.N.: Psychosocial factors and safety behaviour as predictors of accidental work injuries in farming. Work Stress **20**(2), 173–189 (2006)

Kogler, R., Quendler, E., Boxberger, J.: Near accidents with agricultural vehicles, machinery and equipment in Austria in the year 2013. Agric. Eng. Int. CIGR J. **17**, 141–157 (2015)

Lilley, R., Feyer, A.-M., Kirk, P., Gander, P.: A survey of forest workers in New Zealand Do hours of work, rest, and recovery play a role in accidents and injury? J. Safety Res. **20**, 53–71 (2002)

Lundqvist, P., Gustafsson, B.: Accidents and accident prevention in agriculture a review of selected studies. Int. J. Ind. Ergonom. **10**, 311–319 (1992)

Mazzocchi, F., Cecchini, M., Monarca, D., Colantoni, A., Caruso, L., Leopardi, F.: An overview of risk assessment for tree climber arborists. Contemp. Eng. Sci. **8**(25), 1171–1177 (2015)

Micheletti Cremasco, M., Giustetto, A., Caffaro, F., Colantoni, A., Cavallo, E., Grigolato, S.: Risk assessment for musculoskeletal disorders in forestry: a comparison between RULA and REBA in the manual feeding of a wood-chipper. Int. J. Environ. Res. Public Health **16**(5), 793 (2019)

NSC: Near miss reporting systems. National Safety Council, Itasca (2013)
Phimister, J.R., Oktem, U., Kleindorfer, P.R., et al.: Near-miss system analysis: phase I. Wharton School, Center for Risk Management and Decision Processes, Philadelphia (2000)
Wright, L., Schaaf, T.: Accident versus near miss causation: a critical review of the literature, an empirical test in the UK railway domain, and their implications for other sectors. J. Hazard. Mater. **111**, 105–110 (2004)

Error on Drop Size Measurement Due to Image Analysis Digitisation

Giuseppe Manetto, Emanuele Cerruto(✉), Domenico Longo, and Rita Papa

Department of Agriculture, Food and Environment, University of Catania, Via Santa Sofia, 100, 95123 Catania, Italy
emanuele.cerruto@unict.it

Abstract. For spray pesticide applications, drop size spectrum is one of the most important factors affecting environment and operator safety. Assessment of drop size distribution is based on different techniques, among which those based on image analysis are quite widespread. Errors on drop size measurement when using image analysis systems may depend upon several factors. In the present research the measurement error was evaluated at varying image resolution and drop diameter by means of simulation of Water Sensitive Papers (WSP). The main results showed that lower the image resolution, higher the relative error, and higher the diameter, lower the relative error. Considering an average absolute error on drop diameter lower than 5%, the stain diameter on WSP should be larger than about 300 µm at 300 dpi, larger than about 140 µm at 600 dpi, and larger than about 80 µm at 1200 dpi.

Keywords: Droplet pulverisation · Water sensitive papers · Image resolution · Stain diameter · ImageJ

1 Introduction

Drop size spectrum is one of the most important factors affecting environment and operator safety during pesticide application. In fact, drop size affects environmental pollution in terms of off-target losses due to evaporation, drift, and run-off, as well as operator safety due to its effects on dermal and inhalation exposure (Matthews 2004; Nuyttens et al. 2007; Gil et al. 2014; Cerruto et al. 2018; Lodwik et al. 2020). In addition, it also affects the efficacy of a phytosanitary treatment, as it ensures the correct deposit on the target (Sánchez-Hermosilla et al. 2011; Torrent et al. 2017; Garcerá et al. 2017).

Assessment of drop size distribution is based on different techniques, both intrusive and non-intrusive. Intrusive techniques include the liquid immersion method (Fujimatsu et al. 2014; Kathiravelu et al. 2016; Longo et al. 2020; Cerruto et al. 2021) and the use of Water Sensitive Papers (Cunha et al. 2012; Cerruto et al. 2016; Cerruto et al. 2019). Non-intrusive methods include Phase Doppler Particle Analyzers (Nuyttens et al. 2006), laser diffraction (Fritz and Hoffmann 2016), and imaging principles (De Cock et al. 2016).

M. Biocca et al. (Eds.): SHWA 2020, LNCE 252, pp. 365–374, 2022.
https://doi.org/10.1007/978-3-030-98092-4_37

Each measurement technique produces different results depending on measuring protocol, settings and type of measuring equipment (Sijs et al. 2021), so that the ISO standard 25358 (2018) recommends basing classification of a nozzle on the comparison of its spray spectrum, produced under certain operating conditions, with reference spectra produced by reference nozzles under standard conditions. In addition, when image analysis systems are used, further error sources on drop size measurement come from image features. The main objective of the present research is to evaluate the measurement errors on drop size at varying image resolution and drop diameter, based on WSP simulations.

2 Materials and Methods

The error on drop diameters due to the digitisation process was evaluated by simulating water sensitive paper (WSP) images with resolution ranging from 300 to 1200 dpi, with step of 100 dpi, each image containing 25 not overlapping stains of equal diameter ranging from 20 to 1600 μm (reference values), with step of 10 μm. Overall, 1590 images were produced. The size of each image (white) was 20 × 70 mm and the stains (black) were randomly placed within it, without touching the edge. All the procedures needed for creating the WSP images were implemented by means of custom-made functions using the statistical open source software R (R Core Team 2019). More in detail, image features (size and resolution) were established by using the `png()` function of the `grDevices` package, and stains were drawn by using the `draw.circle()` function of the `plotrix` package (Lemon 2006). The "reference diameter" considered in the experiment is that used as argument of the `draw.circle()` function to draw the stain.

The images simulated were then analysed with the public domain ImageJ processing software (Abramoff et al. 2004), as images of real WSPs after a phytosanitary treatment. It detects all the particles in the image, providing a number of features, among which the stain area A_s (mm^2) was selected. The equivalent diameter of each particle was computed by means of Eq. (1):

$$D_s = \sqrt{\frac{4A_s}{\pi}} \times 1000\ (\mu\text{m}) \tag{1}$$

The computed diameter D_s was compared to the reference one D_s^* and then the error on stain size was computed by means of Eq. (2):

$$E_s = \frac{D_s - D_s^*}{D_s^*}\ (\%). \tag{2}$$

According to the WSP technical data sheet (Syngenta 2021), drop diameter D and stain diameter D_s are correlated by the spread factor S_f (ratio between stain diameter and drop diameter). So, knowing the stain diameter and the spread factor, the drop diameter D can be computed by means of Eq. (3):

$$D = \frac{D_s}{S_f}. \tag{3}$$

The spread factor varies with the physical properties of the spray liquid, such as surface tension, angle of impact, energy of impact, pH, and temperature (Cunha et al. 2012). For water at 20 °C, about 40% of relative humidity and drops reaching WSPs at sedimentation velocity, Syngenta (2021) recommends the spread factor values reported in Table 1. For different working conditions, the spread factor should be measured, especially when accurate drop sizing is required.

Table 1. Stain diameter, spread factor and drop diameter on WSPs (Syngenta 2021).

Stain diameter (μm)	Spread factor	Drop diameter (μm)
100	1.7	59
200	1.8	109
300	1.9	155
400	2.0	200
500	2.1	243
600	2.1	285

Interpolating data in Table 1, the Eq. (4) for spread factor vs. stain diameter was obtained:

$$S_f = 0.9326 \times D_s^{0.1273} \left(\mathrm{R}^2 = 0.9693\right), \tag{4}$$

from which the drop diameter was computed in function of the stain diameter by means of Eq. (5):

$$D = 1.0723 \times D_s^{0.8727}, \tag{5}$$

With reference to the present simulations and considering Eq. (5), when the reference stain diameters ranged from 20 to 1600 μm, the corresponding reference drop diameters D^* ranged from 15 to 671 μm. Reference values were then compared to the measured ones, obtained by means of Eq. (5), and the error on drop diameter was computed according to the Eq. (6):

$$E_D = \frac{D - D^*}{D^*} \times 100\ (\%). \tag{6}$$

Raw data were organised in a Microsoft Excel spreadsheet and then imported in R as a `data.frame` for all subsequent computations and graphical representations.

3 Results and Discussion

Figure 1 is an example of the simulated images, referring to stains with diameter of 500 μm and resolution of 600 dpi. It reports the 25 stains randomly placed, without overlapping and without touching the edge.

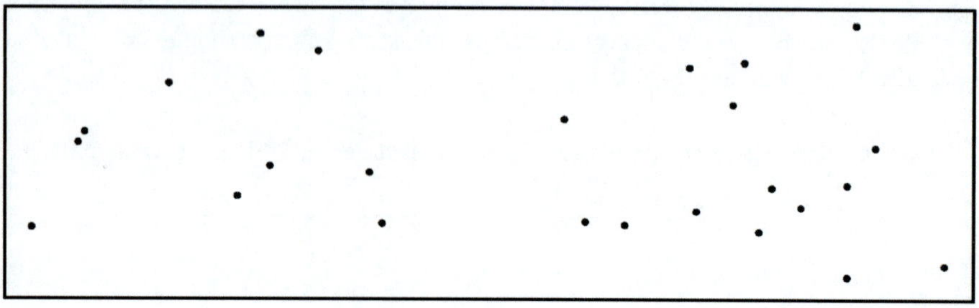

Fig. 1. Example of simulated image (stain diameter = 500 μm, resolution = 600 dpi).

Due to the combined effect of image resolution, stain diameter and random location of the stains inside the image, not always all the 25 stains were present in the image. The percentage of stains correctly reported in the images and identified by ImageJ is shown in Fig. 2.

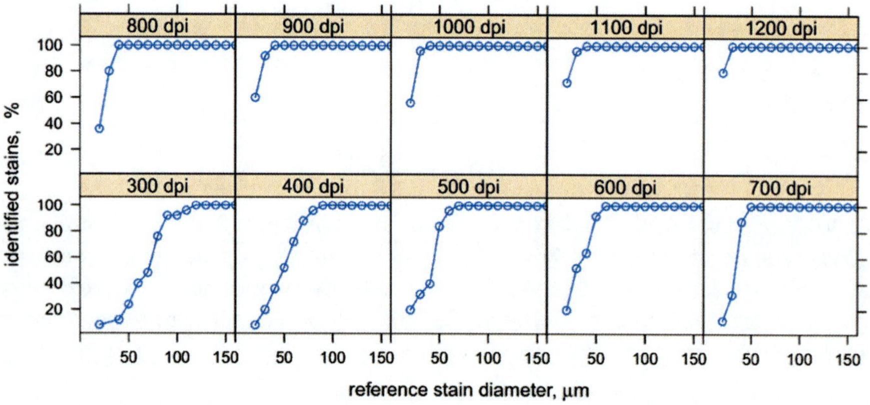

Fig. 2. Percentage of stains reported in the images and identified by ImageJ.

From Fig. 2 it emerges that stains with low diameters were correctly identified only if a proper image resolution was adopted. As an example, at 300 dpi, the percentage of detected stains was only 40% when the reference diameter was 60 μm, 76% with 80 μm, 96% with 110 μm and 100% with diameter greater than 120 μm. At 600 dpi, the percentage was 64% for reference diameter of 40 μm, 92% with 50 μm, and 100%

with diameter greater than 60 μm. At 900 dpi, all stains with reference diameter greater than 40 μm were detected, while the percentage was 60% for reference diameter of 20 μm and 92% with 30 μm. Finally, at 1200 dpi, 80% of 20 μm stains and 100% of stains with diameter greater than 30 μm were detected.

Figures 3 and 4 report the error on stain size in function of reference stain diameter and image resolution, respectively.

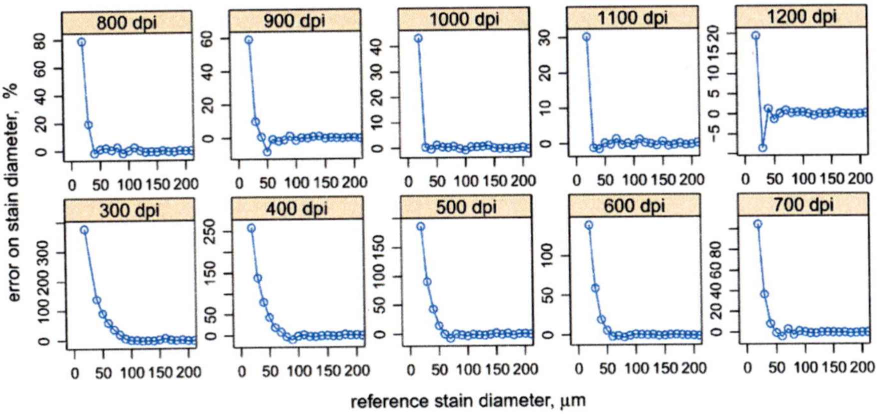

Fig. 3. Error on stain diameter in function of reference stain diameter.

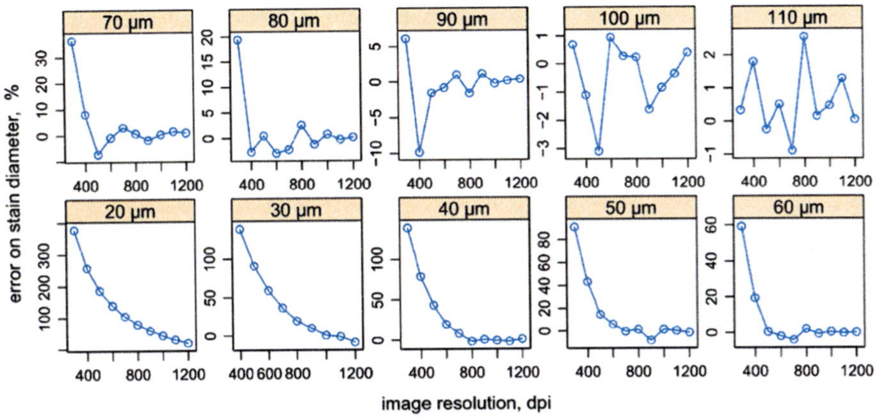

Fig. 4. Error on stain diameter in function of image resolution for some reference stain diameters.

Both figures show a significant effect of image resolution and drop diameter on relative error on stain diameter, especially for small droplets, more subject to drift and more important for worker exposure (stains on simulated WSP with reference diameter lower than 80 μm). As a general result, lower the image resolution, higher the relative error, and higher the diameter, lower the relative error. At 300 dpi, the absolute average relative error decreased from 377.67% (20 μm) to 19.42% (80 μm), at 600 dpi it decreased from 138.83% (20 μm) to 2.95% (80 μm), and at 1200 dpi it decreased from 19.42% (20 μm) to 0.20% (80 μm). When the reference stain diameter was lower than 80 μm, the average relative error was always positive (increasing from 0.98% at 80 μm to 68.30% at 20 μm), whereas with reference stain diameter greater than 100 μm the average relative error was negative (−0.04%, average over all the resolutions).

Figures 5 and 6 report the error on drop diameter in function of reference drop diameter and image resolution, respectively.

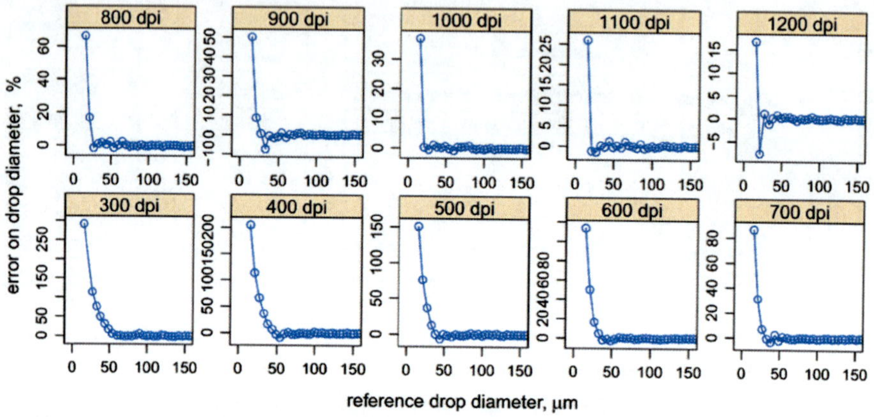

Fig. 5. Error on drop diameter in function of reference drop diameter.

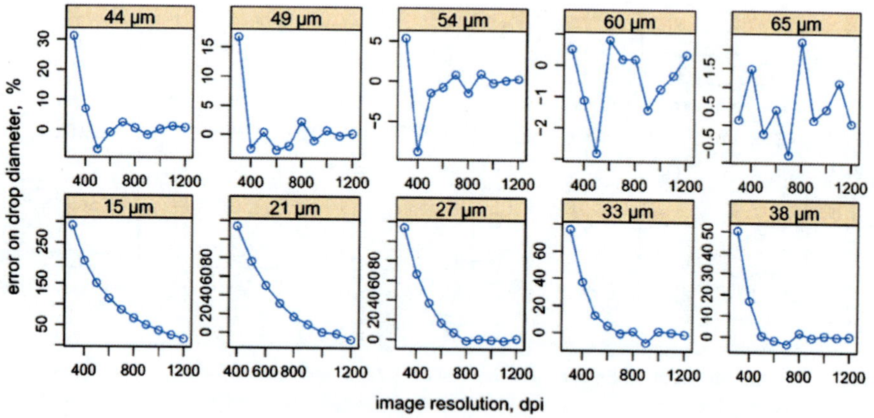

Fig. 6. Error on drop diameter in function of image resolution for some reference drop diameters.

Both figures confirm the results reported in Figs. 3 and 4, but with lower errors. At 300 dpi, the absolute average relative error on drop diameter decreased from 291.45% when the drop diameter was 15 μm to 16.75% (49 μm), at 600 dpi it decreased from 113.78% (15 μm) to 2.75% (49 μm), and at 1200 dpi it decreased from 16.75% (15 μm) to 0.17% (49 μm). When the reference drop diameter was lower than 49 μm (reference stain diameter 80 μm), the average relative error was again always positive (increasing from 0.78% at 49 μm to 56.35% at 15 μm), whereas with reference drop diameter greater than 60 μm it was almost always negative (−0.04%, average over all the resolutions).

Finally, considering that when WSPs are used for drop sizing purposes the starting point is the area of particles measured by ImageJ, the error on drop diameter in function of the measured stain diameter was computed and its average trend is reported in Fig. 7. Each point in the graph refers to a reference drop diameter.

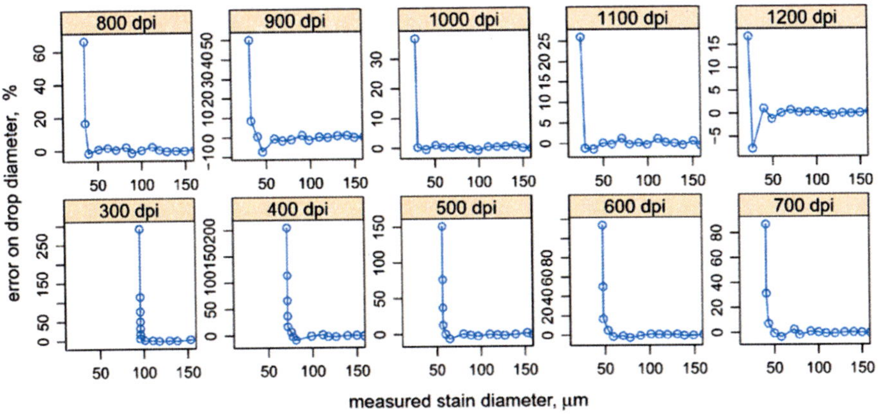

Fig. 7. Error on drop diameter in function of measured stain diameter.

It confirms the general tendency of error on drop diameter at varying stain diameter and image resolution. The greatest errors are at 300 dpi, the lowest at 1200 dpi. Considering that the smallest particle in WSP images is composed by one single pixel, its equivalent diameter at 300 dpi is 96 μm and it was obtained from drops whose diameter ranged from 15 to 85 μm. So, the corresponding error on drop diameter decreased from 291.45% (reference drop diameter 15 μm) to −32.55% (85 μm), with the lowest values at 54–60 μm (5.34% and −3.91%, respectively). Stains with diameter of 135 μm (two contiguous pixels) were obtained from drops with diameter ranging from 60 to 100 μm and then the relative error ranged from 30.03% (60 μm) to −22.15% (100 μm). Analogously, stains with diameter of 191 μm (four contiguous pixels) were obtained from drops with diameter ranging from 75 to 137 μm and then the relative error ranged from 39.95% (60 μm) to −23.57% (137 μm), with minimum values between 104 (0.49%) and 109 μm (−3.91%).

At 600 dpi, resolution widely used for scanning WSPs (Cunha et al. 2012), particles composed by one single pixel (equivalent diameter of 48 μm) were obtained from drops with diameter ranging from 15 to 49 μm, so the corresponding error on drop diameter ranged from 113.78% to −36.24%, with lowest values in the range 27–33 μm. Considering particles composed by four contiguous pixels (equivalent diameter of 96 μm), they were obtained from drops with diameter ranging from 44 to 70 μm, with corresponding error ranging from 31.18% to −18.04%.

Finally, at 1200 dpi, particles composed by one single pixel (equivalent diameter of 24 μm) were obtained from drops with diameter of 15 and 21 μm only, and the corresponding error on drop diameter was 16.75% and -18.04%. Particles composed by four contiguous pixels (equivalent diameter of 48 μm) were obtained from drops with diameter ranging from 27 (error 16.75%) to 38 μm (error −18.04%).

Definitively, the average absolute error on drop diameter was lower than 5% when the stain diameter was larger than 300 μm at 300 dpi, larger than 140 μm at 600 dpi, and larger than 80 μm at 1200 dpi.

4 Conclusions

Measuring drop size from agricultural nozzles is a complex task and several method based upon different principles are used in commercial and custom-made systems. When techniques based on image analysis are used, a possible source of error is due to the digitising process. In this study the error introduced by the use of water sensitive papers has been analysed at varying drop diameter and image resolution. The main results obtained by simulation tests allow drawing the following conclusions:

- Image resolution must be chosen appropriately in function of the lowest particle to be detected: as an example, at 300 dpi only 40% of particles with reference diameter of 60 μm are detected, whereas at 1200 dpi all particles greater than 30 μm are detected.
- Image resolution and drop diameter affect errors on stain and drop diameter: in general, lower the image resolution, higher the relative error, and higher the diameter, lower the relative error.
- Stains with the same area can be obtained from drops with different diameter. To keep the average error on drop diameter lower than prefixed percentage, image resolution must increase when the stain diameter decreases.

Acknowledgements. This research was financed by the University of Catania with the research project "Contributo della meccanica agraria e delle costruzioni rurali per il miglioramento della sostenibilità delle produzioni agricole, zootecniche e agro-industriali. WP1: Impiego sostenibile di macchine irroratrici in serra e in pieno campo", Linea di intervento 2-dotazione ordinaria per attività istituzionali dei dipartimenti-2016-18-II annualità.

References

Abramoff, M.D., Magelhaes, P.J., Ram, S.J.: Image processing with imageJ. Biophotonics Int. **11** (7), 36–42 (2004)

Cerruto, E., Failla, S., Longo, D., Manetto, G.: Simulation of water sensitive papers for spray analysis. Agric. Eng. Int. CIGR J. **18**, 22–29 (2016)

Cerruto, E., Manetto, G., Longo, D., Failla, S., Papa, R.A.: Model to estimate the spray deposit by simulated water sensitive papers. Crop Protection **124**, 104861 (2019)

Cerruto, E., Manetto, G., Papa, R., Longo, D.: Modelling spray pressure effects on droplet size distribution from agricultural nozzles. Appl. Sci. **11**, 9283 (2021)

Cerruto, E., Manetto, G., Santoro, F., Pascuzzi, S.: Operator dermal exposure to pesticides in tomato and strawberry greenhouses from hand-held sprayers. Sustainability **10**(7–2273), 1–21 (2018)

Cunha, M., Carvalho, C., Marcal, A.R.S.: Assessing the ability of image processing software to analyse spray quality on water-sensitive papers used as artificial targets. Biosys. Eng. **111**, 11–23 (2012)

De Cock, N., Massinon, M., Nuyttens, D., Dekeyser, D., Lebeau, F.: Measurements of reference ISO nozzles by high-speed imaging. Crop Prot. **89**, 105–115 (2016)

Fritz, B.K., Hoffmann, W.C.: Measuring spray droplet size from agricultural nozzles using laser diffraction. J. Vis. Exp. (115), e54533 (2016)

Fujimatsu, T., Kito, M., Kondo, K.: Droplet size measurement of liquid atomization by the immersion liquid method (droplet coalescence and solution into the immersion liquid). WIT Trans. Eng. Sci. **82**, 191–202. WIT Press (2014)

Garcerá, C., et al.: Comparison between standard and drift reducing nozzles for pesticide application in citrus: Part II. Effects on canopy spray distribution, control efficacy of *Aonidiella aurantii* (Maskell), beneficial parasitoids and pesticide residues on fruit. Crop Prot. **94**, 83–96 (2017)

Gil, E., et al.: Determination of drift potential of different flat fan nozzles on a boom sprayer using a test bench. Crop Prot. **56**, 58–68 (2014)

Kathiravelu, G., Lucke, T., Nichols, P.: Rain drop measurement techniques: a review. Water **8**, 29 (2016)

Lemon, J.: Plotrix: a package in the red light district of R. R-News **6**(4), 8–12 (2006)

Lodwik, D., Pietrzyk, J., Malesa, W.: Analysis of volume distribution and evaluation of the spraying spectrum in terms of spraying quality. Appl. Sci. **10**, 2395 (2020)

Longo, D., Manetto, G., Papa, R., Cerruto, E.: Design and construction of a low-cost test bench for testing agricultural spray nozzles. Appl. Sci. **10**, 5221 (2020)

Matthews, G.A.: How was the pesticide applied? Crop Prot. **23**, 651–653 (2004)

Nuyttens, D., Baetens, K., De Schampheleire, M., Sonck, B.: PDPA laser based characterisation of agricultural sprays. Agric. Eng. Int. CIGR Ejournal Manuscript PM **8**, 06–024 (2006)

Nuyttens, D., Baetens, K., De Schampheleire, M., Sonck, B.: Effect of nozzle type, size and pressure on spray droplet characteristics. Biosys. Eng. **97**, 333–345 (2007)

R Core Team: R: A Language and Environment for Statistical Computing. R Foundation for Statistical Computing, Vienna, Austria (2019). https://www.R-project.org. Accessed 16 May 2021

Sánchez-Hermosilla, J., Rincón, V.J., Páez, F., Agüera, F., Carvajal, F.: Field evaluation of a self-propelled sprayer and effects of the application rate on spray deposition and losses to the ground in greenhouse tomato crops. Pest Manag. Sci. **67**, 942–947 (2011)

Sijs, R., Kooij, S., Holterman, H.J., van de Zande, J., Bonn, D.: Drop size measurement techniques for sprays: comparison of image analysis, phase Doppler particle analysis, and laser diffraction. AIP Adv. **11**, 015315 (2021)

Syngenta: Water-sensitive paper for monitoring spray distribution. Technical data sheet. https://www.agroconsultasonline.com.ar//ticket.html/Water%20Sensitive%20Paper%20Syngenta%20Agro.pdf?op=d&ticket_id=2388&evento_id=4891. Accessed 10 Aug 2021

Torrent, X., et al.: Comparison between standard and drift reducing nozzles for pesticide application in citrus: Part I. Effects on wind tunnel and field spray drift. Crop Prot. **96**, 130–143 (2017)

Environmental and Health Hazard from Urban Waste Compost: The Role of Information to Overcome Users' Beliefs

Lucia Vigoroso(✉), Giorgia Bagagiolo, Niccolò Pampuro, and Eugenio Cavallo

Institute of Sciences and Technologies for Sustainable Energy and Mobility (STEMS), National Research Council of Italy (CNR), Strada delle Cacce, 73, 10135 Turin, Italy
lucia.vigoroso@stems.cnr.it

Abstract. Composting is an aerobic process that involves the decomposition of the urban waste leading, in accordance with the paradigm of the circular economy, to a stabilized final product rich in fertilizer properties. However, adopting this organic fertilizer may encounter significant barriers since many users still associate compost from organic municipal solid waste (OMSWC) with environmental and health hazards like soil contamination and bad odour. In this context, conveying information about compost use and properties with the appropriate communication channels could help overcome misguided beliefs about OMSWC. The present study investigated which drivers would encourage compost adoption and how the frequency of information through specific sources could influence the interest in using compost among potential users. Three main drivers were identified by surveying 59 potential compost users: providing a certification to the product, receiving more information about the origin of the material, and obtaining evidence about its positive effects on soil. Statistical analysis showed a positive association between frequent use of magazines/advertising as an information source and users' interest in adopting compost. In addition, results pointed out that potential users with higher education levels (with bachelor's degree or above) are more likely to use OMSWC than others. In conclusion, the present study suggests that it would be appropriate to redefine targeted communication campaigns to promote the benefits associated with the compost application for agricultural purposes.

Keywords: Information sources · Interest in use · Recycling waste · Sustainability · Users preferences

1 Introduction

In future years, in accordance with the recent Sustainable Development Goals (SDGs) [1], industrial processes and products will be designed to minimize waste generation and reduce environmental impact [2]. Under this perspective, in the last decades, the linear production model in which wastes are destined to landfills and incinerators has been gradually replaced by the circular economy business model [3]. The circular

M. Biocca et al. (Eds.): SHWA 2020, LNCE 252, pp. 375–382, 2022.
https://doi.org/10.1007/978-3-030-98092-4_38

economy concept discourages the using of new raw materials, promoting sharing, leasing, reusing, repairing, refurbishing, and recycling the existing products and materials as long as possible [4]. Products at the end of their life can be processed and treated, becoming input for new processes and products in a circular metabolism process [5, 6].

The circular economy model can be applied to several different sectors, including agriculture [7]. In this context, urban food waste and urban park residues can be recycled into a stabilized organic material, i.e. compost, free of phytotoxicity and pathogens and with specific humic properties [8]. Moreover, thanks to its agronomic value, compost could be reused as soil fertilizer or conditioner to preserve the soil organic matter content, reduce soil degradation, and substitute the more costly chemical fertilizer [6, 9, 10]. For these reasons, composting represents an efficient alternative for municipal waste management, allowing the reduction of landfill disposal and, at the same time, recycling its agronomic macronutrients. Unfortunately, the adoption of compost made from organic municipal solid waste (OMSWC) may encounter significant barriers since many users still associate this organic product with dirty material and, consequently, environmental and health hazards like soil contamination [11]. Furthermore, though Witherspoon et al., [12] found that the unpleasant and unacceptable odours "per se generate no objective signs of illness, even when people claim symptoms" (p. 10), also the compost odour nuisance was identified as the most critical barrier to using organic waste as fertilizer [13]. In addition, OMSWC, free from phytotoxicity and pathogens after the aerobic stabilization process, does not constitute a hazard to human health or the environment.

For this reason, potential users' perception and intention to adopt recycled material play a relevant role in implementing the circular economy model [14]. In detail, in the present investigation, we considered a potential user as someone whose needs can be satisfied by specific products, but he/she has never bought them. Involving potential users and showing them the benefits of compost adoption can positively influence the increasing market demand for this sustainable material [15]. To increase the adoption of OMSWC on a large scale, and establish a continuous demand market, a good understanding of the factors affecting potential users' attitudes toward this product is required. Potential users' perception and intention to adopt recycled material depends on several factors; the majority of these are demand-side factors (e.g. factors affecting the users' attitudes towards a product or their capacity to use it) [13].

In the present study, besides taking into account socio-demographic aspects, we focused our investigation on the role of information and communication channels. Indeed, to reduce uncertainty and to overcome negative beliefs, as, in the case of OMSWC, information and social interaction play a relevant role in affecting an individual's willingness-to-pay and decision-making behaviour [16]. This may happen both through "impersonal" communication channels, which refer to codified media, such as manuals, documents, magazines, and electronic information repositories [17, 18] and through "interpersonal" communication channels, e.g. with family members, peers and neighbours [18–22]. In the agricultural sector, some previous studies have already emphasized the important role of communication and social interaction in affecting farmers' intentions to use new practices. In this context, Huttel et al. [23], Pham et al. [24], and Weltin et al. [25] noticed that adoption of practices like crop rotation, soil and water conservation, organic fertilizers adoption and use of digital technologies were positively affected by peers' and neighbours' influence. However, to our knowledge, further research is required regarding the role of different information channels in affecting the intention to adopt compost obtained from urban waste, especially among potential compost users. Based on this, the present study aims at: i) investigating which factors could play a role in encouraging potential compost users and ii) understanding which communication channels could be more relevant in affecting their intention to adopt this specific sustainable product.

2 Materials and Methods

2.1 Participants

Participants were recruited exclusively among Italian visitors of the 24th "Ecomondo" Green Technology Expo, which took place in Rimini in November 2019. To be eligible and included in the present investigation, participants were selected among those who do not currently adopt OMSWC but are aware of the existence of this product. In addition, participants involved in the survey were both professionals (farmers and gardeners) and hobbyists.

2.2 Instruments and Procedure

All participants were verbally informed about the study's aim and verbally gave their consent to be included in the research. After obtaining participants' consent, they were asked to fulfill an anonymous paper-and-pencil questionnaire. Moreover, research assistants followed participants when completing the survey asking, as specified in the questionnaire, to focus exclusively on compost from OMSWC.

The questionnaire was made up of three main sections. In the first section, participants were asked to rate on a four-point scale their level of interest in using OMSWC (1 - 'not at all interested'; 2 - 'poorly interested'; 3 - 'interested'; 4 - strongly interested) and their agreement with drivers able to encourage them toward OMSWC adoption (0 - 'no'; 1 - 'yes'). The seven items used to investigate drivers were: i) providing quality certification to the product, ii) receiving information on compost use, iii) receiving more information about the origin of the material, iv) obtaining evidence on compost effectiveness, v) adopting packaging format easier to transport, vi) adopting packaging format suited to user needs, and vii) receiving economic support. In the second section, participants were asked to rate their frequency of use of some information sources to acquire information and/or to keep informed about the OMSWC characteristics, using a four-point scale (1 - 'never'; 2 - 'rarely'; 3 - 'sometimes'; 4 - 'often'). More in detail, information sources investigated were: i) visits other companies, ii) video on the internet, iii) magazines and advertising, iv) courses, v) discussions with experts. The last section focused on participants' socio-demographic characteristics regarding gender, age, education, and profession.

2.3 Data Analysis

Descriptive statistics were computed for all the variables of interest. To analyze and record the relationship between the variables investigated and participants' propensity in using OMSWC, the level of interest in adopting compost was coded as "low" when potential users reported to be poorly interested or not at all interested and "high" when they reported being interested or strongly interested. Then, contingency tables and non-parametric Chi-squared tests were performed. Significance was set at $p < 0.05$. Statistical analysis was performed using SPSS 26.0 (IBM SPSS Inc., Chicago, IL, USA) software.

3 Results

Fifty-nine visitors have been involved in the study. Among these, 47 were male and 12 were female and reported an overall mean age of 39.3 years (SD = 12.4). Concerning their level of education, 25 participants reported a below Bachelor's degree level of education and 34 obtained a Bachelor's degree or above (e.g. Masters post lauream or PhD). Regarding participants' professions, 30 of them were 'professionals', and 29 participants fall under the category of 'hobbyists'.

Concerning the participants' interest in using OMSWC, 62.7% of them reported high interest in using it. Regarding driving factors, three main drivers were detected as more relevant to encourage potential users to adopt compost: providing a certified product (55.9%), receiving more information about the origin of the material (54.2%) and obtaining evidence about its positive effects on soil (40.7%). Percentages for all the seven drivers investigated are reported in Table 1.

Table 1. Percentage of the agreement for drivers able to encourage compost adoption.

Driver	% of participant' agreement	
	Yes	No
Providing quality certification to the product	55.9	44.1
Receiving more information about the origin of the material	54.2	45.8
Obtaining evidence on compost effectiveness	40.7	59.3
Receiving economic support	35.6	64.4
Receiving information on compost use	33.9	66.1
Adopting packaging format suited to user needs	25.4	74.6
Adopting packaging format easier to transport	23.7	76.3

Relationships between the categorical variables are shown in Tables 2 and 3 for participants' socio-demographic characteristics and frequency of use of information sources, respectively. The most frequently used information channels were videos and the internet (66% of potential users reporting a high frequency of use) followed by discussions with experts (63%) and magazines/advertisements (47%). Statistical analysis showed a positive association between frequent use of magazines/advertising as an information source and users' interest in adopting compost ($X^2(1) = 5.732$, $p = .017$). In addition, results pointed out that potential users with higher education levels (with bachelor's degrees or above) are more likely to use compost from organic municipal solid waste than others ($X^2(1) = 6.496$, $p = .011$).

Table 2. Interest use: the count for participants' socio-demographic characteristics.

Interest use	Gender		Age		Profession		Education	
	Female	Male	≤39	≥40	Professional	Hobbyist	Below BSc	Above BSc
Low	2	10	9	13	13	9	14	8
High	10	27	27	14	17	20	11	26

Table 3. Interest use: the count for frequency of use of information sources.

Interest use	Visits		Video/ internet		Magazine /advertising		Courses		Discussion experts	
	Low	High	Low	High	Low	High	Low	High	Low	High
Low	16	6	10	12	16	6	13	9	9	18
High	28	9	10	27	15	22	26	11	13	19

4 Discussion

The present study aimed to understand which factors can affect the intention to adopt compost obtained from organic municipal solid waste among potential users. In particular, the current research considered both personal users' characteristics (such as age, level of education and profession) and different communication channels.

Consistently with present results, also previous studies reported that the intention to adopt composted material had been influenced by the level of education [26–28]. This result may suggest that the more the compost users are educated and trained, the higher is their attitude to comprehend the potential agronomic benefits of compost as well as the best techniques of its application in soils [29].

Regarding the driving factors that may encourage potential users to adopt compost, the study emphasised the importance of information, especially regarding the quality and the composition of the compost; similar results are reported in previous studies. For instance, in a study by Paul et al. [28] no adopter farmers declared that the lack of information concerning compost quality was the main barrier to its adoption. Then, Viaene et al. [15] in their study conducted in Flanders, declared that farmers are uncertain about the compost quality and composition due to the broad definition of the term 'compost'; moreover, no additional distinctions are made based on the target application, e.g., compost for horticulture or agriculture. Similarly, Case et al. [13], investigating the demand-side of organic fertilizers, reported that the lack of quality certifications was ranked among the most significant barriers to accepting manure among surveyed farmers.

Regarding the frequency of use of some information sources about compost, within the investigated context, peer-network (i.e. courses or discussion with experts) is not decisive in the intention of OMSWC adoption and their dissemination as reported in previous studies [30, 31]. However, consistent with present results, some other previous works, mainly describing market survey research [18], showed that there is a positive association between frequent use of magazines/advertising as an information source and users' decision-making process and/or intention to adopt or buy specific technology or product. Focusing on the agricultural context, Viaene et al. [15] reported that farmers pay greater attention to information provided by agricultural magazines. Similarly, Patel & Shekhawat [32] affirmed that farmers who read farm magazines increase their knowledge on agricultural technology compared to non-subscriber farmers. Readers' confidence in magazines could be linked to the fact that conveying information through this type of channel may help consumers "to easily process complex information and comprehend them with little effort" [18] (p. 597). Finally, some limitations of the present study need to be mentioned. First, the investigation is limited in size, so the obtained results can not be generalized, and, in further studies, a representative sample should be involved to confirm these results. Furthermore, in future studies, it could be useful to investigate the time spent per week using mass media such as TV, radio, magazines and newspapers to keep informed, since users who spent more time using these mass media can be less likely to use other social media when gathering information about the products of interest [33].

5 Conclusions and Limitations

Much previous research investigated factors that hinder compost adoption and application in farming contexts. Whereas, the present paper aimed to give a first general overview of some different drivers which may promote the OMSWC adoption among Italian professional and hobbyists users. Overall, results highlighted diverse interesting factors (i.e. users' characteristics, driving factors and information sources) that should be taken into account to support the widespread use of this sustainable product.

Thus, in conclusion, despite its limitations, the present survey allowed us to offer some valuable recommendations for planning advertising campaigns to promote the adoption of OMSWC and that could potentially overcome a number of the previously identified barriers. First, in terms of content, campaigns could overcome users' concerns about compost hazards by focusing on compost certification and providing detailed information on feedstock characteristics, environmental and health aspects, and product effectiveness in improving soil quality. Then, specialized magazines are the most effective media to convey such information and should be able to address all potential users as recipients, reaching users with lower education level who are less likely to adopt OMSWC. Future research and evaluation need to be performed to obtain more insights into the viability of the recommendations proposed and their impact on the organic fertilizers' sector.

References

1. United Nations. Department of Economic and Social Affairs. Sustainable Development. https://sdgs.un.org/goals
2. Sverko Grdic, Z., Krstinic Nizic, M., Rudan, E.: Circular economy concept in the context of economic development in EU countries. Sustainability **12**(7), 3060 (2020)
3. van Kruchten, S., van Eijk, F.: Circular Economy & SDGs. How Circular Economy Practices Help to Achieve the Sustainable Development Goals. https://circulareconomy.europa.eu/platform/sites/default/files/3228_brochure_sdg_-_hch_cmyk_a4_portrait_-_0520-012.pdf
4. European Parliament. Circular Economy: Definition, Importance and Benefits (2021). https://www.europarl.europa.eu/news/en/headlines/economy/20151201STO05603/circular-economy-definition-importance-and-benefits
5. Song, Q., Li, J., Zeng, X.: Minimizing the increasing solid waste through zero waste strategy. J. Clean. Prod. **104**, 199–210 (2015)
6. Vilella, M.: Sustainable Finance for a Zero-Waste Circular Economy (2020). https://zerowasteeurope.eu/wp-content/uploads/2020/11/zero_waste_europe_report_sustainable-finance-for-a-zero-waste-circular-economy_en.pdf
7. Circulareconomy.heart. How the Circular Economy Can Help Realize the Sustainable Development Goals. https://circulareconomy.earth/publications/the-circular-economy-and-the-sdgs
8. Assandri, D., Pampuro, N., Zara, G., Bianco, A., Cavallo, E., Budroni, M.: Co-composting of brewers' spent grain with animal manures and wheat straw: influence of two composting strategies on compost quality. Agronomy **11**, 1349 (2021)
9. Bekchanov, M., Mirzabaev, A.: Circular economy of composting in Sri Lanka: opportunities and challenges for reducing waste related pollution and improving soil health. J. Clean. Prod. **202**, 1107–1119 (2018)
10. Assandri, D., Pampuro, N., Zara, G., Cavallo, E., Budroni, M.: Suitability of composting process for the disposal and valorization of brewer's spent grain. Agriculture **11**, 1–12 (2021)
11. Camacho-Otero, J., Tunn, V.S., Chamberlin, L., Boks, C.: Consumers in the circular economy. In: Handbook of the Circular Economy. Edward Elgar Publishing, Cheltenham (2020)
12. Case, S.D.C., Oelofse, M., Hou, Y., Oenema, O., Jensen, L.S.: Farmer perceptions and use of organic waste products as fertilisers–a survey study of potential benefits and barriers. Agric. Syst. **151**, 84–95 (2017)
13. Witherspoon, J.R., et al.: Water environment research foundation (WERF) anaerobic digestion and related processes, odour and health effects study. Water Sci. Technol. **50**(4), 9–16 (2004)
14. Sijtsema, S.J., Snoek, H.M., Van Haaster-de Winter, M.A., Dagevos, H.: Let's talk about circular economy: a qualitative exploration of consumer perceptions. Sustainability **12**(1), 286 (2020)
15. Viaene, J., et al.: Opportunities and barriers to on-farm composting and compost application: a case study from northwestern Europe. Waste Manag. **48**, 181–192 (2016)
16. Li, C., Shi, Y., Ni, Q., Zhao, M.: Effects of social interactions and information bias on the willingness to pay for transboundary basin ecosystem services. J. Environ. Manag. **296**, 113233 (2021)
17. Shin, D.-H., Choi, M.J.: Ambidextrous information search: linking personal and impersonal search routines with individual performance. Inf. Technol. Manage. **15**(4), 291–304 (2014)

18. Wei, Y.S., Frankwick, G.L., Gao, T.T., Zhou, N.: Consumer adoption intentions toward the Internet in China: the effects of impersonal and interpersonal communication channels. J. Advert. Res. **51**(4), 594–607 (2011)
19. Ajzen, I., Fishbein, M.E.: Understanding Attitudes and Predicting Social Behavior. Prentice Hall, Englewood Cliffs, New York (1980)
20. Welch, E.W., Marc-Aurele, F.J.: Determinants of farmer behavior: adoption of and compliance with best management practices for nonpoint source pollution in the skaneateles lakewatershed. Lake Reserv. Manage. **17**, 233–245 (2001)
21. Mzoughi, N.: Farmers adoption of integrated crop protection and organic farming: do moral and social concerns matter? Ecol. Econ. **70**, 1536–1545 (2011)
22. Greiner, R.: Motivations and attitudes influence farmers' willingness to participate in biodiversity conservation contracts. Agricult. Syst. **137**, 154–165 (2015)
23. Hüttel, S., Leuchten, M.T., Leyer, M.: The importance of social norm on adopting sustainable digital fertilisation methods. Organ. Environ. **35**(1), 79–102 (2020)
24. Pham, H.G., Chuah, S.H., Feeny, S.: Factors affecting the adoption of sustainable agricultural practices: findings from panel data for Vietnam. Ecological Economics **184**, 107000 (2021)
25. Weltin, M., Zasada, I., Hüttel, S.: Relevance of portfolio effects in adopting sustainable farming practices. J. Clean. Prod. **313**, 127809 (2021)
26. Sotamenou, J., Parrot, L.: Sustainable urban agriculture and the adoption of composts in Cameroon. Int. J. Agric. Sustain. **11**(3), 282–295 (2013)
27. Supaporn, P., Kobayashi, T., Supawadee, C.: Factors affecting farmers' decisions on utilization of rice straw compost in Northeastern Thailand. J. Agricult. Rural Dev. Trop. Subtrop. **114**(1), 21–27 (2013)
28. Paul, J., Sierra, J., Causeret, F., Guindé, L., Blazy, J.M.: Factors affecting the adoption of compost use by farmers in small tropical Caribbean islands. J. Clean. Prod. **142**, 1387–1396 (2017)
29. Folefack, A.J.J.: The determinants for the adoption of compost from household waste for crop production by farmers living nearby Yaoundé, Cameroon: descriptive and logit model approaches of analysis. Int. J. Biol. Chem. Sci. **9**(1), 308–328 (2015)
30. Cheng, Z., Chen, Q., Delgado, M.: Understanding the importance of social interactions between small farmers: evidence from farmer planting behavior in rural China (2019)
31. Ćurčin, N., Dulčić, Ž., Mršić, S.P.: Communication channels and innovative farmer behaviour in South Croatia. In: 6th International OFEL Conference on Governance, Management and Entrepreneurship. New Business Models and Institutional Entrepreneurs: Leading Disruptive Change. April 13th–14th, 2018, Dubrovnik, Croatia, pp. 402–416. Governance Research and Development Centre (CIRU), Zagreb (2018)
32. Patel, J.R., Shekhawat, S.S.: Impact of Krushijivan farm magazine. Ind. J. Extn. Educ. RD **22**, 110–113 (2014)
33. Gazal, K.A., Montague, I., Wiedenbeck, J.: Factors affecting social media adoption among wood products consumers. BioProd. Bus. 51–62 (2019)

Environmental Evaluation of Vineyard Airblast Sprayers Through a Comprehensive Spray Mass-Balance Approach

Marco Grella(✉), Paolo Marucco, Gianluca Oggero, Marco Manzone, Fabrizio Stefano Gioelli, and Paolo Balsari

Department of Agricultural, Forest and Food Sciences (DiSAFA), University of Turin (UNITO), Largo Paolo Braccini, 2, 10095 Grugliasco, TO, Italy
marco.grella@unito.it

Abstract. The efficacy and sustainability of pesticide application largely depends on maximizing target coverage, while minimizing off-target losses. It derives that the sprayer efficiency plays a key role for a more sustainable agriculture. The evaluation of sprayer efficiency must consider not only spray off-target losses but also the canopy deposition for a comprehensive assessment. For these purposes an *ad hoc* method designed for the spray mass-balance measurements was applied. Tests were conducted in a vineyard at early and late growth stages, to evaluate the efficiency of two airblast sprayers (conventional vs tower shaped) and the effect of different spray volume rates (conventional full volume vs halved volume) on the efficiency of spray application. For these purposes a spray mass-balance trial layout was used. It consists in sampling the three main spray fractions: i) canopy deposition, ii) ground losses and iii) airborne spray losses, thanks to a frame arch surrounding the sprayed rows and used to hold the collectors exposed to the atmosphere. Results underline the effect of sprayer type on the spray application efficiency and that also in the best cases at early growth stage no more than 13% of spray mixture is retained by canopies. Concurrently the reduction of applied volume had no negative effect on spray application efficiency as the ratio between canopy deposition and off-target losses did not vary when compared with the full volume. It can be concluded that the spray application efficiency largely depends on technical factors mainly related to the sprayer design and its adjustment and to the target morphological characteristics.

Keywords: Airborne losses · Ground losses · Canopy deposition · Environmental contamination · Spray volume rate

1 Introduction

A perfect spraying system aims to deliver and deposit a precise amount of active ingredient uniformly and exclusively upon a target area. Achieving this goal is extremely challenging for agricultural spray applications. Indeed, during spray application in bush/tree crops with airblast sprayers only a fraction of the total applied plant protection products (PPP) is actually deposited on the intended target. Therefore, most

M. Biocca et al. (Eds.): SHWA 2020, LNCE 252, pp. 383–393, 2022.
https://doi.org/10.1007/978-3-030-98092-4_39

of the spray mixture represents undesirable off-target losses. Those, usually known as primary drift [1], are represented by the liquid fraction that can be transported outside the sprayed area by air currents as spray drift; spray ground losses consist in the liquid fraction deposited on the ground either directly in the path of the sprayer tractor, beneath the target tree rows, or indirectly in adjacent areas [2]. An additional loss occurs when pesticides deposited on the target are washed off by rain or volatize and it is known as secondary drift [3]. Thus, during and immediately after a PPP spray application, non-target receptors, including water, plants, animals, and humans can be acutely exposed and may be at risk for adverse effects.

When a pesticide application equipment is evaluated for its efficiency, usually only the off-target losses accounting for the primary drift are considered while the secondary drift is not considered as it is mainly related to the chemical properties of applied PPP, environmental conditions and other variables. Considering that the efficacy and sustainability of the pesticide application largely depends on maximizing target coverage, while minimizing off-target losses, a spray mass-balance approach that includes the evaluation of spray canopy deposition, in-field ground losses and airborne losses as main fractions that compose the total spray mass-balance is a reliable strategy for a comprehensive evaluation of pesticide application equipment efficiency.

To date, few studies have examined all fractions in the spray mass-balance [4]. Typically, measurements consider deposits on plants and soil inside the treated area and spray drift sediments, while airborne drift has been measured in few studies. The reason why a large fraction of the spray is not accounted for when plant and soil deposits are measured and standard values of sedimenting spray drift are used could be methodological problems, such as estimating the LAI correctly and/or recovering pesticide/tracer from leaves and from soil collectors [4]. To overcome some of these difficulties another approach for the calculation of spray mass-balances has been proposed by Balsari et al. [5]. Their idea was to install around the orchard row a sampling structure able to collect spray drift (drift fallout and airborne spray drift) and ground deposits. The proportion of the applied volume deposited on the crop could then be calculated as the difference between the applied volume and the sum of those collected as spray drift and ground deposition. This method was mainly proposed as alternative for potential spray drift measurements and did not consider the quality of spray distribution through a detailed measurement of canopy deposition.

This experimental work proposes a comprehensive spray mass-balance approach aiming to evaluate the efficiency of airblast sprayers in vineyards considering both canopy deposition and off-target losses, as ground and airborne losses. The last fraction accounted for the liquid fraction sampled in the atmosphere just over the sprayed rows that can be potentially blown away outside of the applied area by air currents at the time of application. Two parallel experiments were set up to find out the effects of i) sprayer type and ii) applied liquid volume rate on the spray application efficiency in vineyard for their effect on the relationships between spray canopy deposition and off-target losses.

2 Materials and Methods

2.1 Test Location, Experimental Plot Layout and Sampling System

Field trials were performed in an espalier-trained vineyard (cv: Pinot grigio) located in Fossalta Maggiore di Treviso, Treviso, Italy. The trials were conducted at two growth stages, namely early and late corresponding to average Leaf Wall Area (LWA) 3,549 and 13,078 m^2 ha^{-1}, respectively.

To assess the spray mass-balance three main spray fractions, namely i) spray canopy deposition, ii) spray losses to the ground and iii) spray losses to the air (from this point on airborne spray losses) were sampled transversal in the middle point of the experimental parcel composed by two vineyard rows treated just from one side each one (single sprayer pass) (Fig. 1a). Each trial was replicated five times and replicates were distributed randomly within the vineyard taking care to displace plots so that possible cross contamination was avoided.

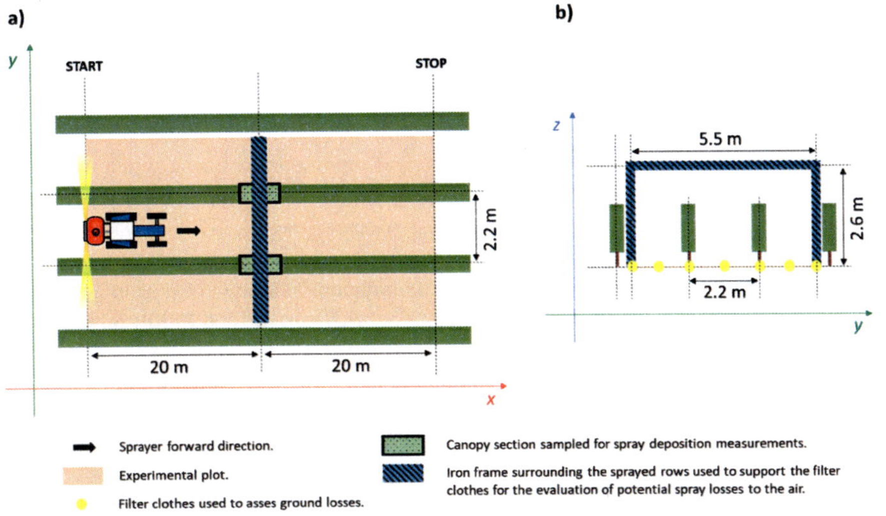

Fig. 1. Schematic of trial layout for the measurements of spray mass-balance: a) aerial and b) frontal views.

An *ad hoc* device, similar to those initially developed [5], was used to hold the artificial collectors for sampling purposes. The device was an iron frame positioned over the two vineyard rows and it consists of two 2.6 m height vertical poles (Fig. 1b), placed beyond the right and left applied rows, and one 5.5 m width horizontal beam placed over the top of the vines (Fig. 1a) to be coupled with the top of the two vertical poles, composing an arch. Along the iron frame a series of supports were distributed every 0.5 m to hold two paired cellulose filter clothes 20 × 10 cm (CAMFIL CM 360) composing an array of collectors all around the two applied vineyard rows.

The spray losses to the ground were measured in line with the iron frame used for the measurements of airborne spray losses (Fig. 1b). Also in this case two cellulose filter clothes (20 × 10 cm) were paired in each sampling position. The collectors were placed under the sprayed vines (two rows) and in the middle of each inter-row open space within the experimental plot (three inter-rows).

The canopy spray deposition was assessed in both applied rows, specifically in the vine canopies placed next to the iron frame used for the measurements of spray airborne losses where collectors for spray ground losses were placed under (Fig. 1a). Once sprayed, all the vine leaves comprised within 1 m row length were collected separately per row (vines were completely defoliated).

2.2 Sprayer Characteristics and Configurations Tested

Two conventional airblast sprayers were tested for comparison. A trailed airblast sprayer Nobili Geo 90-1001T (Nobili S.p.a., Molinella, BO, Italy) equipped with a 1,000 L polyethylene tank (Fig. 2a). A conventional air conveyor with an axial fan (900 mm diameter) generates a nominal airflow rate equal to $51,200\ m^3\ h^{-1}$ setting the tractor Power Take Off (PTO) at 540 rev min^{-1}. Similarly to the sprayer described by Salcedo et al. [6] for the characterization of airflow rate, the fan outlet section was composed by two vertical channels with fixed width where 6 and 4 nozzle holders were equally distant displaced, respectively. This type of sprayer was selected as representative of the most widespread in Italy among small/medium farms where a unique sprayer is typically used by the farmer to spray different crop types like orchards and vineyards. Concurrently, the mounted airblast sprayer Dragone K2 500 (Dragone S.n.c., Castagnole delle Lanze, Asti, Italy), *ad hoc* designed for vineyard spray application, was tested. It was equipped with a 200 L polyethylene tank and six nozzle holders on each side (Fig. 2b). A tower-shaped air conveyor with an axial fan (600 mm diameter) and a two-speeds gearbox enabled the airflow rate to vary from 11,000 to $20,000\ m^3\ h^{-1}$ at 540 rev min^{-1} PTO.

Fig. 2. a) Conventional airblast sprayer Nobili Geo 90-1001T and b) tower shaped airblast sprayer Dragone K2 500 used for the spray mass-balance measurements.

For the comparison of the sprayers their adjustment was based on the common practices used by farmers. In detail, the vertical spray profile was adjusted to the canopy development activating an adequate number of nozzles to completely cover the vegetative strip height. It derives that at early growth stage to reach the intended volumes to be applied, four ATR 80 orange and six ATR 80 yellow hollow cone nozzles (Albuz® CoorsTek, Evreux, France) were used at 1.2 and 1.0 MPa spray pressure combined with forward speeds equal to 1.39 and 1.67 ms $^{-1}$, in order to apply 330 and 285 L ha $^{-1}$ with the conventional and tower-shaped airblast sprayers, respectively. At full growth stage 829 and 715 L ha $^{-1}$ were applied activating for both sprayers five ATR 80 orange hollow cone nozzles (Albuz®) per side. To achieve the intended application rate, the spray pressure was 1.2 and 1.0 MPa and the forward speeds 1.39 and 1.47 m s $^{-1}$ for conventional and tower shaped airblast sprayers, respectively. Concurrently, to evaluate the effect of application rate, the airblast sprayer Dragone K2 500 was tested also applying a halved spray volume compared to those applied for the comparison among sprayer types (from this point on full spray volume). To achieve halved volumes, different size of nozzles and spray pressures were adopted while the forward speed was kept the same as for the application of full spray volume (1.67 m s $^{-1}$). Table 1 summarizes the main parameters adopted for the spray application using the conventional and the tower shaped sprayer at the two different growth stages examined.

Irrespective of growth stage, in all trials a fixed airflow rate was used for both sprayers: the tower shaped airblast sprayer was used with the low fan gear speed activated while the conventional airblast sprayer was used setting the tractor PTO at 400 rev min $^{-1}$ thus reducing the nominal airflow rate produced and achieving values closed to 25,000 m ^{3} h $^{-1}$.

Table 1. Main operating parameters of sprayers tested for the spray mass-balance measurements at different growth stages.

Growth stage	Sprayer	Nozzle[a]	Active nozzles (n°)	Spray pressure (MPa)	Forward speed (m s $^{-1}$)	Fan airflow rate (m ^{3} h $^{-1}$)	Spray volume (L ha $^{-1}$)
Early	Conventional	Orange	4	1.2	1.39	25,000	330
Early	Tower shaped	Yellow	6	1.0	1.67	11,000	285
Early	Tower shaped	Lillac	6	1.1	1.67	11,000	143
Late	Conventional	Orange	10	1.2	1.39	25,000	829
Late	Tower shaped	Orange	10	1.0	1.47	11,000	715
Late	Tower shaped	Brown	10	1.1	1.47	11,000	360

[a]Hollow cone nozzles ATR 80 (Albuz® CoorsTek, Evreux, France)

2.3 Sprayed Mixture and Laboratory Analysis

A solution of water and E-102 Tartrazine yellow dye tracer was applied to measure the collectors' deposit. Both artificial collectors (cellulose filter clothes) and vegetable samples (vine leaves) were afterwards washed with deionized water to extract the tracer. The Tartrazine concentration was determined by measuring the absorbance of the washing solution with a spectrophotometer UV-1600PC (VWR, Radnor, PA, USA) and to compare the results against the calibration curve. Spray deposits on each artificial and leaves collector, expressed per unit area in µL cm^{-2}, was calculated according to ISO 22401:2015. The method is fully detailed in Rincon et al. [7]. A procedure for the characterization of total exposed leaf surface was set up from which the linear fit models that describe the relationships between leaves surface and their weight were obtained and used to calculate the spray deposition per unit area.

2.4 Data Processing and Statistical Analysis

The three fractions that compose the spray mass-balance were calculated from Eqs. (1), (2) and (3) as follows:

$$L_{air} = 100 - ((D_{ground} + D_{canopy}) * 100)/(D_{air} + D_{ground} + D_{canopy}) \quad (1)$$

$$L_{ground} = 100 - ((D_{air} + D_{canopy}) * 100)/(D_{air} + D_{ground} + D_{canopy}) \quad (2)$$

$$C_{dep} = 100 - ((D_{ground} + D_{air}) * 100)/(D_{air} + D_{ground} + D_{canopy}) \quad (3)$$

where: L_{air} is the spray mass-balance fraction referred to airborne spray losses, expressed in %; L_{ground} is the spray mass-balance fraction referred to ground spray losses expressed in %; C_{dep} is the spray mass-balance fraction referred to the spray deposits in the canopy expressed in %; D_{air} is the total amount of deposits measured on artificial collectors placed on the iron frame surrounding the sprayed rows (expressed in µL cm^{-1} of sprayer advancing along the rows); D_{ground} is the total amount of deposits measured on the artificial collectors at ground level (expressed in µL cm^{-1} of sprayer advancing along the rows); D_{canopy} is the amount of spray deposit measured on the vine leaves on both sprayer sides (also expressed as µL cm^{-1} of sprayer advancing along the rows).

To obtain D_{air} and D_{ground} a numerical integration procedure was adopted; it was a variation of those described by Grella et al. [8] for the calculation of sedimenting spray drift.

All statistical analyses were performed using IBM SPSS Statistics (Version 26) predictive analytics software for Windows®. Analyses of Variance (ANOVA) were used to establish the effect of sprayer type (conventional vs tower shaped airblast sprayers), growth stage (early vs late) and applied volume (full vs halved) on spray mass-balance fractions (canopy deposit, ground and airborne losses). The means were compared using a Duncan post-hoc test for multiple comparison ($p < 0.05$). From the relationships between Leaf Wall Area –LWA– (m^2 ha^{-1}) and liquid applied (%) linear fit models were obtained for the two sprayer types and separately according to the spray mass-balance fractions. The linear fit models were compared using Analysis of Covariance (ANCOVA) to evaluate the effect of sprayer type on spray mass-balance fractions while controlling for the LWA (covariate).

3 Results and Discussion

Figure 3 displays the spray mass-balance obtained using conventional and tower shaped airblast sprayers at early and late growth stages.

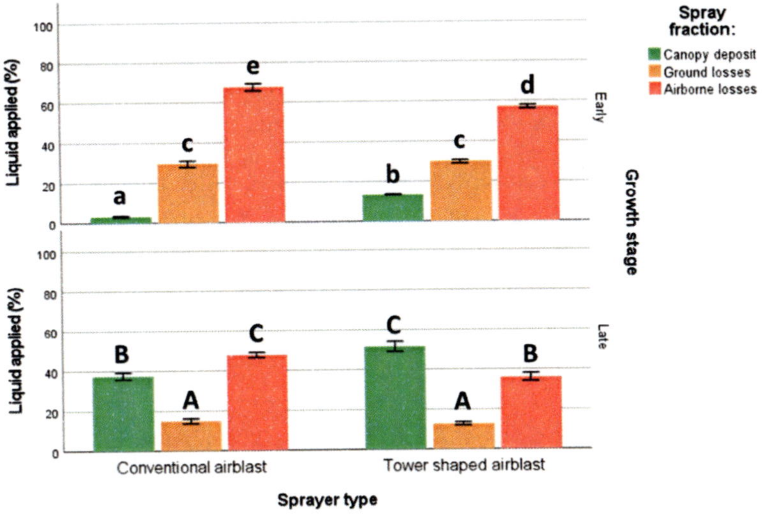

Fig. 3. Comparison of spray liquid applied (%) accounting for the different spray mass-balance fractions obtained using conventional and tower shaped airblast sprayers at early and late growth stages. The bars show the mean ± standard error of the mean. Different letters on the bars, lowercase and capital separately, denote significant differences (Duncan post hoc test, $p < 0.05$).

Concurrently the spray airborne losses were always higher for the conventional airblast sprayer. Interestingly, at early growth stage only 3% and 13% of spray mixture was retained by canopy using conventional and tower shaped airblast sprayers, respectively. At late growth stage the vine canopies density increases substantially trapping 37% and 51% of applied liquid volume using conventional and tower shaped airblast sprayers, respectively. Even if the tower shaped airblast sprayer was selected as most suitable for vineyard spray application, in the best case it generates off target losses equal to 49% of applied liquid volume underlining in general the very low efficiency of airblast sprayers with high potential risk for environment and bystander contamination during spray application. The ground losses were the minor fraction of off-target losses with 29% and 30% at early growth stage and 15% and 13% at late growth stage for the conventional and tower airblast sprayer, respectively. Indeed, while there is no effect of sprayer type on ground losses, at early growth stage the airborne spray losses account for the higher fraction irrespective of sprayer type with values equal to 68% and 57% of total applied spray volume. At full growth stages the airborne losses reach values of 48% and 36% according to the sprayer type, underlining that also when canopies are fully developed using a conventional sprayer almost half of the applied volume does not reach the target and can be potentially blown away as spray drift in case of windy conditions, with consequent implications for the environment and bystanders.

The linear fit models describing the relationships between the Leaf Wall Area – LWA– (m^2 ha^{-1}) of sampled vine canopies and the spray liquid applied (%) accounted by the three spray mass-balance fractions for the two sprayer types are shown in Fig. 4. As expected, the canopy deposit fraction showed the higher r^2 as the particle-trapping capability of vine canopies was fully dependent on their development (canopy dimension and density). Also, the fit models for the off-target losses, namely ground and airborne fractions, showed a good r^2 in all cases higher than 0.871. Concurrently, the line slopes did not differ between sprayers for the three fractions. So, one-way ANCOVA was used to compare sprayer type while controlling for LWA showing that both canopy spray deposits [$F(1,19) = 18.193$, $p = 4.18E{-}04$] and airborne losses [$F(1,19) = 27.702$, $p = 4.40E{-}05$] differed significantly. No statistical differences were detected for ground losses [$F(1,19) = 0.440$, $p = 0.515$]. This underlines the higher efficiency of tower shaped airblast sprayer compared to the conventional one with increased canopy spray deposits corresponding to a decrease of airborne losses irrespective of LWA, selected as a simple parameter to describe the vine canopies development along the growing season.

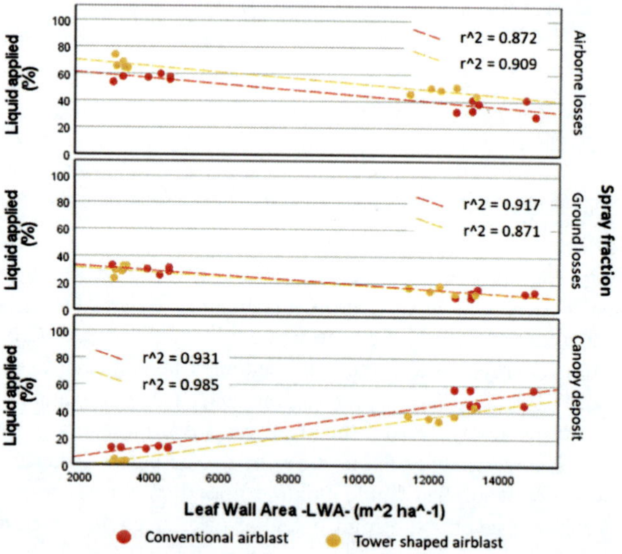

Fig. 4. Relationships between Leaf Wall Area (m^2 ha^{-1}) and liquid applied (%) accounting for the spray mass-balance fractions obtained using conventional and tower shaped airblast sprayers.

The comparison between canopy spray deposition and off-target losses (ground and airborne losses cumulated) obtained applying full and halved spray volumes using the tower shaped airblast vineyard sprayer Dragone K2 500 are displayed in Fig. 5. At early growth stage no statistical differences between full and halved volumes were observed for both canopy spray deposition and off-target losses; the latter were higher than the amount of sprayed liquid trapped by vine leaves, reaching values equal to 87% and 86% for the full and half applied volumes, respectively. When the halved volume (360 L ha^{-1}) was applied at full growth stages the highest canopy spray deposit and lowest off-target losses were found with values equal to 57% and 43% respectively. Intermediate values were measured when the full volume (715 L ha^{-1}) was applied with sprayed liquid almost equally distributed between canopy deposition and off-target losses, namely 51% and 49% with no statistical differences detected.

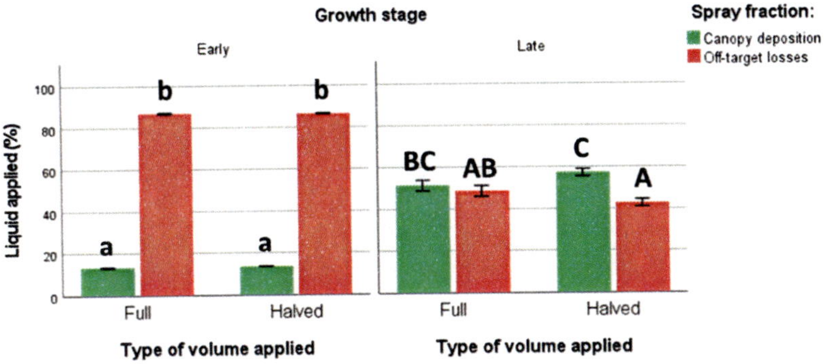

Fig. 5. Comparison of spray liquid applied (%) accounting for spray mass-balance fractions obtained applying the full and halved spray volumes using tower shaped airblast sprayer at early and late growth stages. The bars show the mean ± standard error of the mean. Different letters on the bars, lowercase ad capital separately, denote significant differences (Duncan post hoc test, $p < 0.05$).

These results underline that sprayer adjustments that foresee the reduction of applied volume have not a direct effect on sprayer efficiency with possible benefits for environment thanks to the reduction of off-target losses. Indeed, in conventional farming using a constant liquid rate during application and the dose of plant protection products prescribed by the label irrespective the liquid/water applied, the PPP losses do not vary significantly also halving the liquid applied. Anyway, a slight decrease of off-target losses in favor of canopy deposits was noticed at full growth stage using the halved volume, suggesting that in general reduced spray volumes compared to those generally used by farmers (full volumes) are welcome. These results are also supported by the recent findings described by Grella et al. [9] where spray application volumes in range 330–360 L ha^{-1} were found the most suitable to obtain the optimal spray coverage in trellised vineyard at full growth stage.

4 Conclusions

The applied spray mass-balance method effectively provides a comprehensive and contemporaneous information about the three main fractions that compose the sprayed liquid just after spray application in vineyard. Thus, based on the spray mass-balance assessment method, higher efficiency of tower shaped airblast sprayer compared to the conventional one was detected. In best cases only 13% and 51% of sprayed liquid was trapped by canopies with collateral airborne losses fraction accounting for 57% and 36% at early and late growth stages, respectively. Using the tower shaped sprayer, the reduction of spray volume applied of up to 50% did not affect the efficiency of spray application. A fixed ratio between canopy deposition and off-target losses irrespective of spray volume at both early and late growth stages was indeed observed. Results suggest that, once the sprayer pattern is adjusted to the canopy, the spray application efficiency largely depends on technical factors related to the sprayer design (conventional or tower shaped) and fan settings, as excessive fan airflow speed may reduce deposition and increase spray losses due to canopy compression. Further tests are needed focusing on other parameters like nozzles droplet size and fan air volume that may play a role in improving the deposition on the target. Losses at present are indeed too high and the sprayer efficiency is not in line with the incoming more stringent regulatory requirements set by the European Green Deal, that strives to reduce by 50% the overall use of agrochemicals within 2030.

References

1. Himel, C.M.: Analytical methodology in ULV. In: British Crop Protection Council Monograph no. 11. Pesticide Application by ULV Methods. 112–119 (1974)
2. Grella, M., Miranda-Fuentes, A., Marucco, P., Balsari, P.: Field assessment of a newly-designed pneumatic spout to contain spray drift in vineyards: evaluation of canopy distribution and off-target losses. Pest Manag. Sci. **76**, 4173–4191 (2020). https://doi.org/10.1002/ps.5975
3. Combellack, J.H.: Loss of herbicides from ground sprayers. Weed Res. **22**, 193–204 (1982)
4. Jensen, P.K., Olesen, M.H.: Spray mass balance in pesticide application: a review. Crop Prot. **61**, 23–31 (2014)
5. Balsari, P., Marucco, P., Tamagnone, M.: A system to assess the mass balance of spray applied to tree crops. Trans. ASAE **48**(5), 1689–1694 (2005). https://doi.org/10.13031/2013.19997
6. Salcedo, R., Fonte, A., Grella, M., Garcerá, C., Chueca, P.: Blade pitch and air-outlet width effects on the airflow generated by an airblast sprayer with wireless remote-controlled axial fan. Comput. Electron. Agricult. **190**, 106428 (2021). https://doi.org/10.1016/j.compag.2021.106428
7. Rincón, V.J., Grella, M., Marucco, P., Eloi Alcatrão, L., Sanchez-Hermosilla, J., Balsari, P.: Spray performance assessment of a remote-controlled vehicle prototype for pesticide application in greenhouse tomato crops. Sci. Total Environ. **726**, 138509 (2020). https://doi.org/10.1016/j.scitotenv.2020.138509

8. Grella, M., Marucco, P., Balsari, P.: Toward a new method to classify the airblast sprayers according to their potential drift reduction: comparison of direct and new indirect measurement methods. Pest Manag. Sci. **75**(8), 2219–2235 (2019). https://doi.org/10.1002/ps.5354
9. Grella, M., et al.: Field assessment of a pulse width modulation spray system applying different spray volumes: duty cycle and forward speed effects on vines spray coverage. Precis. Agricult. **23**(1), 219–252 (2022). https://doi.org/10.1007/s11119-021-09835-6

Evaluation of Sustainable Weeding Methods for the Control of Spontaneous Flora in Urban Areas

Lucia Bortolini(✉)

Department of Land, Environment, Agriculture and Forestry, University of Padova, Padova, Italy
lucia.bortolini@unipd.it

Abstract. When not properly managed or eliminated, spontaneous plants may cause more or less serious damage in an urbanized environment. Weed control should be a priority in urban green management, both public and private, but the treatments need more attention towards the environment and people. Following the current Italian laws regarding the use of herbicides in green spaces, a study on alternative methods to the traditional application of glyphosate is proposed. Specifically, the goal was to evaluate the effectiveness of weeds control of four different methods to find the best solution. The experimental tests took place in two paths of the Montecchia Golf course (Selvazzano, Italy). Two bioherbicides and two physical treatments were tested and compared with no treated plots (control): acetic acid (Urban Weed™), pelargonic acid (Finalsan® Plus), flame weeding, and mechanical scraper. After each treatment, photos were taken to evaluate the trend of weed repopulation in each plot. The images were processed using CANOPEO, a Matlab® plugin, to obtain a percentage coverage value for each plot. Considering cost analysis and the effectiveness of weeds control, acetic acid has proved to be the best alternative solution to traditional weed control with glyphosate.

Keywords: Acetic acid · Pelargonic acid · Flame weeding · Mechanical scraper · Cost analysis

1 Introduction

The spontaneous flora can be an element of biodiversity increase in an urbanized environment [1], but the growth of this vegetation must be often kept under control for aesthetic or practical reasons (e.g. in sidewalks, pavements, etc.) or because of its ability to compete with ornamental plant species or due to human health problems [2, 3].

One of the most popular control tools are herbicides. The perception that herbicides are safe for human health and have little impact on the environment is still widespread among users. Among the herbicides, the non-selective, broad-spectrum compound glyphosate (N-Phosphonomethyl glycine) is widely used for the control of wild flora [4]. Microbial degradation of glyphosate produces aminomethyl phosphonic acid (AMPA), the primary glyphosate transformation product [5]. Glyphosate and AMPA have frequently been detected during surface water monitoring in all European

M. Biocca et al. (Eds.): SHWA 2020, LNCE 252, pp. 394–401, 2022.
https://doi.org/10.1007/978-3-030-98092-4_40

countries [6]. Kolpin et al. [7] found that the urban use of glyphosate contributes to glyphosate and AMPA concentrations in streams in the United States, and overall, AMPA was detected much more frequently (67.5%) compared to glyphosate (17.5%). The chemical and toxicological properties of glyphosate and AMPA are well documented [8].

Due to the environmental and health risks in areas frequented by the public or vulnerable groups, current European regulations have banned or limited herbicide use in green areas [9]. In particular, in Italy, the use of plant protection products containing the active substance glyphosate is prohibited in areas frequented by the population, such as: parks, gardens, sports fields and recreational areas, courtyards and green areas within school complexes, children's play areas and areas adjacent to health facilities [10].

As a consequence, many European Member States have required the adoption of alternative strategies to control weeds in public areas. In these areas, the control of spontaneous flora is more applied in sidewalks and driveways and pedestrian paths with paved or gravel surfaces. Different methods of weed removal can be used depending on the surface type, the size of the area and ease of access [11].

The mechanical methods such as strimmers and lawnmowers achieve immediate results in terms of removing the vegetation, but from the basal portion the weeds soon re-grow, so these operations must be repeated several times during the year. On the other hand, these types of equipment cannot be used in gravel paths as the gravel present on the surface would cause serious damage to the blades, and the possible throwing of pebbles can be very dangerous for people and workers. Other mechanical methods more aggressive on weeds include brushes and scrapers.

Thermal methods have become more widespread in recent years and among these, those with hot water (with or without biodegradable foam), water vapour, and direct flame are preferred. For all these systems it is essential to intervene at the initial stage of the weeds [12, 13].

Among non-chemical methods, natural substances that have herbicidal properties are pelargonic acid and acetic acid. These products have the advantage that they can be used using the same equipment as synthetic pesticides.

This experiment aimed to test the effect of different methods for weed control on gravel paths to evaluate the most suitable one as an alternative to glyphosate.

2 Materials and Methods

The experiment was conducted in 2018 at the Montecchia Golf course, located in Selvazzano (Padova, north-eastern Italy) (45°23′09.3″ N; 11°45′46.4″ E).

The trial was carried out on two gravel paths (Fig. 1), both divided into plots 1.80 m wide and 5 m long (9 m^2). In total 30 plots were identified and a different type of weeding treatment was applied to each of them, following a randomized block scheme with five treatments and six replications. Two bioherbicides and two physical treatments were tested and compared with no treated plots (control): acetic acid (Urban Weed™), pelargonic acid (Finalsan® Plus), mechanical scraper, and flame weeding.

Fig. 1. The two gravel paths in the Montecchia Golf course in which the trial was carried out.

Urban Weed contains 82% of acetic acid and it has a significant action of reduction on the transpiration of the vegetation. The product was used with a dilution of 20% with a dosage of about 0,01 L m^{-2}.

Finalsan Plus is a non-selective contact herbicide containing pelargonic acid (also referred to as nonanoic acid) and maleic hydrazide. The dilution was 200 mL L^{-1} of water, with a distribution of about 100 mL m^{-2} of solution.

The distribution for both herbicides was carried out by using a manual backpack sprayer, taking care to cover the vegetation to be treated with a homogeneous layer of product (Fig. 2).

The mechanical scraper was supplied by the company Barbieri (Italy); the working element of the equipment consists of four rotating bars with a vertical axis on whose ends two steel cables are inserted (Fig. 2). During its operation many pebbles were moved so the times for raking and levelling the gravel were also considered to calculate the working time of each treatment.

The equipment for flame weeding was developed by the Emilverde Srl in collaboration with Officine Mingozzi (Italy). The equipment was pulled by a tractor (Fig. 2). Liquefied petroleum gas (LPG) was used by the burner with a working width of 180 cm, perfectly equal to the width of the paths treated. The LPG dose was about 150 kg ha^{-1}.

Fig. 2. Machines used in the experiment for weed control: A) the backpack sprayer used for herbicides distribution; B) the scraper with the four rotating bars and the steel cables; C) the flaming machine.

At the start of the experiment (16 April), the major weeds identified in the paths were *Plantago major L.*, *Plantago lanceolata L.*, *Trifolium repens L.*, *Poa annua L.*, *Cynodon dactylon L.*

Five treatments were done on the following dates: 17 April, 15 May, 11 June, 6 August and 1 October.

Photos were taken to evaluate the trend of weed repopulation in each plot. The surveys were carried out immediately after the treatments and then every one or two weeks, depending on the growth of the weeds, using a camera (Nikon COOLPIX W300) mounted on a trolley and positioned at a height from the ground that allowed to photograph the entire width of each plot. Each parcel required two photographs (1.80 × 2.50 m) to be fully framed. The photos were processed using Adobe Photoshop Lightroom (Adobe Inc., USA) to better highlight the vegetation. The images were then processed using CANOPEO, a MATLAB® software plugin (MathWorks Inc., USA) to obtain the percentages of canopy cover for each parcel [14].

The effectiveness of the different treatments was analysed with parametric (ANOVA, Tukey's HSD, Linear Model with GLS estimation) and non-parametric (Dunn's test) statistical tests.

As the economic analysis, the cost per square meter of each weeding treatment was estimated. Application times, labour and product costs and depreciation of equipment and machines were considered. The time to perform each treatment was calculated considering the actual time of work and any additional times needed to complete the treatment (e.g. to level the gravel). The treatments with the flaming machine were carried out by Emilverde at the cost of 0.14 euro per square meter.

3 Results and Discussion

Starting from the first relief on 16 April to the last one on 24 October, a total of 514,6 mm of rainfall was registered, with an unusual rainy May (123,2 mm) and fairly well distributed rainfall throughout the summer, with a total of about 40 rainy days (>1 mm/d) during the experimental period. The effect of precipitation is evident in the trend of the average values of the vegetation cover measured in the control plots (Fig. 3).

In Fig. 3 it can be noted that the initial condition at the first survey of 16 April is almost homogeneous for all the plots, with average coverage values between 8% and 13%. After each treatment, there was a clear decrease in vegetation cover for all treatments.

In the third survey carried out two weeks after the first treatment, the herbicide effect of the acids and the scraper was maintained and the plots were very clean of weeds, while the flaming did not show the same effect, with a slight increase in weeds, also found in the subsequent survey with a consequent increase in coverage up to an average value of 4%.

In general, the treatment with flame weeding was the one that recorded the most irregular trend, alternating peaks of coverage during the two most favourable periods for the development of weeds, both macrotherms and microtherms. The scraper and pelargonic acid alternated very low coverages in the dates following the treatments, to slightly higher ones when weeding treatments occurred less frequently and during the summer period. For the scraper, it can be noted that, in one of the latest survey (on 25 September), the percentage of canopy cover was very similar to that of the control plots (average values of about 9% vs. 10%). As the scraper performed an action on the soil similar to a light milling, it can be thought that favourable conditions were created for the growing of seed species. In fact, in the previous weeks, abundant rainfall with mild temperatures were registered, ideal conditions for the germination of microthermal species. As regards the flaming machine, the poor results could also be related to the characteristics of the machine, which does not seem to adapt well to too short distances, such as the length of the treatment plots.

Among the treatments, the plots treated with acetic acid maintained a stable trend with a reduced percentage of coverage, which never exceeded 1%. During the entire survey period, it was possible to observe in all the plots that the acetic acid-based preparation was by far the most effective method, both for the control of macrothermal and microthermal weeds.

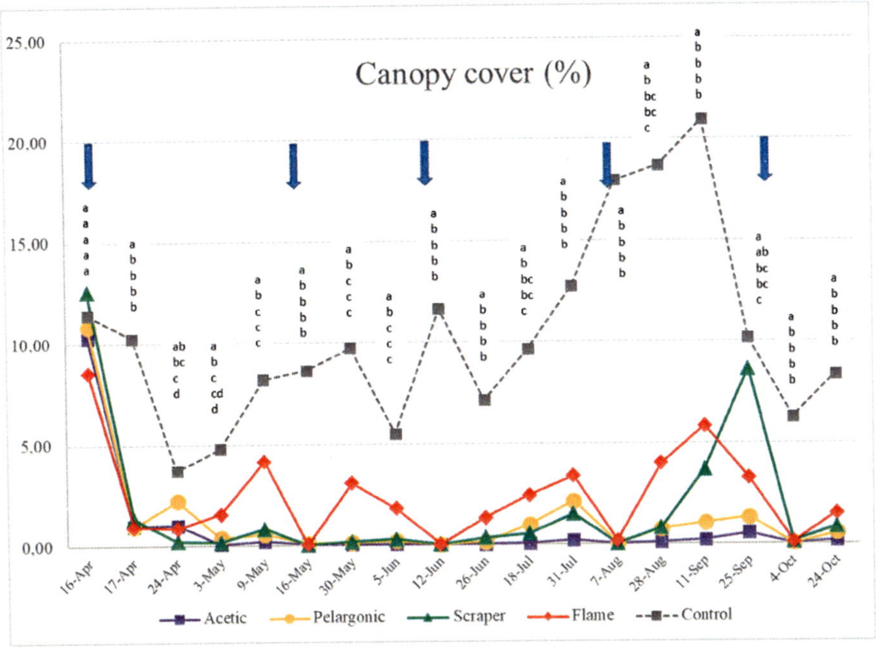

Fig. 3. Trends of the percentage of canopy cover for each treatment during the test period. The arrows indicate the dates when the treatments were performed. At each date, values with the same letter are not different according to Tukey's HSD test ($P \leq 0.05$).

The results are supported by the statistical analysis that showed a substantial difference in efficacy between the treatments and the control. After verifying the homoskedasticity with Levene's test, the application of ANOVA analysis highlighted that acetic acid was the most effective in weed control, due to very low canopy cover values found in all surveys. The application of a linear model with GLS estimation for the heteroskedastic correction also supported the hypothesis that acetic acid was the best of the treatments.

In addition to the effectiveness and environmental sustainability, the economic sustainability of the interventions was also considered. Table 1 summarizes the results relating to the calculation of working times and the total costs for each treatment, referring to square meter of path. The higher cost of the mechanical scraper can be noted, mainly due to the need to level the gravel after each treatment.

Between the two bioherbicides, acetic acid has a cost per square meter that is almost half of pelargonic acid. This cost is the most competitive with respect to the treatment with glyphosate.

Table 1. Working time and total cost for each treatment.

Treatment	Working time (s m^{-2})	Total cost (euro m^{-2})
Acetic acid	6.7	0.11
Pelargonic acid	6.7	0.20
Mech scraper	33.3	0.28
Flame weeding	2.2	0.14
Glyphosate	6.7	0.05

4 Conclusions

Although the efficacy of glyphosate in weed control is well established, its use in areas frequented by the population has been prohibited due to possible negative effects on human health. On the other hand, alternative strategies for the management of urban spaces can bring potential negative impacts on working conditions and local public budgets; therefore, a carefully analysis is necessary to find better sustainable solutions.

The two bioherbicides tested were both effective but pelargonic acid expressed less long-term effectiveness, and the cost treatment is about twice compared to acetic acid.

Among the physical treatments, flame weeding has guaranteed good effects only in the short term, therefore requiring more frequent treatments, even if the treatment is much faster. The mechanical scraper achieved similar results to flame weeding, but its long working time precludes its use on a large scale, in addition to the fact that the cost of the intervention is the highest of the four treatments.

In conclusion, considering cost analysis and the effectiveness of weeds control, among the tested methods the acetic acid has proved to be the best alternative solution to traditional weed control with glyphosate.

References

1. Bonthoux, S., Voisin, L., Bouché-Pillon, S., Chollet, S.: More than weeds: spontaneous vegetation in streets as a neglected element of urban biodiversity. Landsc. Urban Plan. **185**, 163–172 (2019)
2. Benvenuti, S.: Weed dynamics in the Mediterranean urban ecosystem: ecology, biodiversity and management. Weed Res. **44**(5), 341–354 (2004)

3. Fagot, M., De Cauwer, B., Beeldens, A., Boonen, E., Bulcke, R., Reheul, D.: Weed flora in paved areas in relation to environment, pavement characteristics and weed control. Weed Res. **51**(6), 650–660 (2011)
4. Gill, J.P.K., Sethi, N., Mohan, A.: Analysis of the glyphosate herbicide in water, soil and food using derivatising agents. Environ. Chem. Lett. **15**(1), 85–100 (2016)
5. Zhan, H., Feng, Y., Fan, X., Chen, S.: Recent advances in glyphosate biodegradation. Appl. Microbiol. Biotechnol. **102**(12), 5033–5043 (2018)
6. Richmond, M.E.: Glyphosate: a review of its global use, environmental impact, and potential health effects on humans and other species. J. Environ. Stud. Sci. **8**(4), 416–434 (2018)
7. Kolpin, D.W., Thurman, E.M., Lee, E.A., Meyer, M.T., Furlong, E.T., Glassmeyer, S.T.: Urban contributions of glyphosate and its degradate AMPA to streams in the United States. Sci. Total Environ. **354**(2–3), 191–197 (2006)
8. Guyton, K.Z., et al.: Carcinogenicity of tetrachlorvinphos, parathion, malathion, diazinon, and glyphosate. Lancet Oncol. **16**(5), 490–491 (2015). International Agency for Research on Cancer (IARC). Glyphosate 1. IARC Monographs 112 (2016)
9. European Commission. Overview Report on the Implementation of Member States' Measures to Achieve the Sustainable use of Pesticides Under Directive 2009/128/EC (2017)
10. Fogliatto, S., Ferrero, A., Vidotto, F.: Current and future scenarios of glyphosate use in Europe: are there alternatives? Adv. Agron. **163**, 219–278 (2020)
11. Peruzzi, A., Martelloni, L., Frasconi, C., Fontanelli, M., Pirchio, M., Raffaelli, M.: Machines for non-chemical intra-row weed control in narrow and wide-row crops: a review. J. Agricult. Eng. **48**(2), 57–70 (2017)
12. Raffaelli, M., Martelloni, L., Frasconi, C., Fontanelli, M., Peruzzi, A.: Development of machines for flaming weed control on hard surfaces. Appl. Eng. Agric. **29**, 663–673 (2013)
13. Melander, B., et al.: Non-chemical weed management. Thermal weed control. In: Hatcher, P. E., Froud-Williams, R.J., (eds) Weed Research: Expanding Horizons. Wiley, Hoboken (2017)
14. Korhonen, L., Heikkinen, J.: Automated analysis of in situ canopy images for the estimation of forest canopy cover. Forest Sci. **55**, 323–334 (2009)

Analysis of Spray Behavior Using Different Sprayers in Citrus Groves

Souraya Benalia(✉), Laura Violi, Pasquale Barreca, Giuseppe Zimbalatti, and Bruno Bernardi

Dipartimento di Agraria, Università degli Studi Mediterranea di Reggio Calabria, Località Feo di Vito snc, 89122 Reggio Calabria, RC, Italy
soraya.benalia@unirc.it

Abstract. This paper reports the results inherent to technical performances of three different sprayers used in citrus orchards in Calabria (Southern Italy). Accordingly, quantitative and qualitative analyses of the spray were performed. Spray quantitative assessment, related to foliar deposit and losses to the ground, was based on the colorimetric method. While spray qualitative evaluation considered water-sensitive paper (WSP) image analysis. This was performed subsequently to field experimental trials, which consisted in spraying citrus orchards with a dye solution and collecting the required samples, i.e., leaves for foliar deposit, petri dishes for ground losses, and WSP for spray qualitative analysis. Statistically analyzed data showed significant differences between the assessed sprayers considering foliar deposition (Kruskal-Wallis: $\chi^2 = 40.327$, df = 2, p = $1.75e^{-9}$). In addition, some unfavorable aspects were encountered regarding ground losses and qualitative distribution when applying plant protection products with the tested sprayers.

Keywords: Citrus · Foliar deposition · Ground losses · Plant protection equipment · Spray analysis

1 Introduction

Italy is the third largest producer of citrus in the Mediterranean. In this country, this crop is spread over 149,348 ha producing more than 2.89 million tons, according to the data of the National Institute of Statistics related to 2017 campaign [1]. Thanks to the favorable climatic conditions, citrus thrive well in Calabria (Southern Italy), indeed, this district accounts 37,462 ha, representing 25% of the domestic citrus area, making this crop one of the characteristic cultivations of the rural landscape.

Diversely from other fruit crops, citrus is characterized by a dense vegetation canopy. Consequently, this makes it a suitable environment for several pests and diseases, engendering the necessity to intervene with crop protection practices, further than adequate pruning, to guarantee yields from both quantitative and qualitative point of view.

M. Biocca et al. (Eds.): SHWA 2020, LNCE 252, pp. 402–410, 2022.
https://doi.org/10.1007/978-3-030-98092-4_41

Chemical application efficacy technically depends on a series of factors, among which, the maximum vegetation coverage, the maximum evenness, and the optimum penetration of the required amount of the active principle [2]. However, while performing this practice, we often assist to a low deposition of the spray on the target as well as to losses of a great amount of the product in the atmosphere (drift) and to the ground.

This low effectiveness of the application is often attributable to the poor or inadequate calibration and use of the spraying equipment [3]. In fact, depending on the type of treatment, the nature of the formulations and the crop features different operating conditions of the equipment can be set to guarantee effectiveness and sustainability [4]. Following the implementation of the Directive 2009/128/CE [5], which establishes a framework for a sustainable use of pesticides, the present work aims at assessing the performance of crop protection equipment used in Calabrian citrus orchards in terms of spray deposit and ground losses.

2 Material and Methods

2.1 Field and Machinery Description

To assess spray behavior while performing agrochemical application in Calabrian Citrus orchards, three different sprayers were considered: the Maschio mod. UNIGREEN EXPO mist blower (UGE), mounted on NEW HOLLAND TN95F tractor, the DSM Frutteto 1000 (F1000) and the electrostatic EFFE 3 (EF3) mist blowers, both towed by a MASSEY FERGUSON 3435 F tractor. Experimental trials simulating agrochemical application were carried out in private farms in the Province of Reggio Calabria (Fig. 1a, b). Field and main machinery features are reported in the following Table 1. The trials were carried out considering the operative parameters usually employed in each farm.

(a)

(b)

Fig. 1. (a) Application of tartrazine solution in a bergamot orchard using Maschio mod. UNIGREEN EXPO mist blower (UGE). (b) Application of tartrazine solution in an orange orchard using DSM Frutteto 1000 mist blower (F1000).

Table 1. Field and machinery characteristics used in trials

	Unit	UGE	F1000	EF3
Observed area	m^2	1000	896	1512
Crop layout	m	5 × 5	7 × 8	7 × 9
Tank capacity	L	400	1000	1000
Distributed volume	$L \cdot ha^{-1}$	800	1529	165
Advancement speed	$Km \cdot h^{-1}$	1.8	0.3	0.3
Exercise pressure	bar	20	15	15
Nozzle type		Albuz 1.0 Pink HCI60 green	Albuz 1.5 in brass	Anti-drift MGA 8001
Nozzle number		12	16	14

2.2 Quantitative and Qualitative Assessment of Spray Distribution

The quantitative analysis of the spray is based on the estimation of foliar deposit and ground losses expressed in $\mu L \cdot cm^{-2}$. This is performed through the calculation of the intercepted volume by the target (leaves and captors) of a food dye solution applied in the field using the sprayer as to simulate a plant protection product (PPP) application [6, 7]. Therefore, a solution of tartrazine yellow (E102, 85%) was used, as reported by Sánchez-Hermosilla et al. [8] and Sánchez-Hermosilla et al. [9]. Subsequently to spraying operations, samples were collected individually and put in hermetic bags for laboratory analysis as carried out by Bernardi et al. [10]. These included, washing the samples with a known amount of distilled water in order to extract the captured dye by each target. After that, each solution was analyzed using the UV/Visible scanning spectrophotometer Shimadzu UV-1800 at 426 nm, this being the absorbance peak of the tartrazine yellow solution, to determine the solution concentration and subsequently the solution volume that reached each target. Afterward, the leaf area of each sample was measured using a leaf area meter (LI-COR 3100, LICOR, USA), to reference the intercepted volume of tartrazine solution to the relative area and express it in $\mu L\ cm^{-2}$.

To assess foliar deposit evenness on the tree canopy, different foliage heights were considered for leaf sampling. Particularly, in the first trial, when implementing the sprayer UGE, leaves were sampled at about 1 to 1.2 m for lower foliage, at 2 m for intermediate foliage and above 2.5 m for upper foliage. However, int the second and third trials, using respectively the sprayers F1000 and EF3, sampling was performed taking into account just the lower and upper foliage, as tree dimensional characteristics were different from those of the first trial. Hence a total of 96 natural samples (leaves) were collected. For ground losses, Petri dishes used as artificial captors, were randomly positioned between the rows interested by tartrazine yellow application. A minimum of 10 captors were used in each trial, for a total of 36. In addition, water-sensitive papers (WSP) were randomly set amongst the vegetation (Fig. 2a, b), to assess spray distribution from the qualitative point of view as carried out by Bernardi et al. [10].

(a)

(b)

Fig. 2. Water-sensitive paper used for qualitative assessment of foliar deposit.

As the distributed volume was not equal in all the trials, the foliar deposit and ground losses data were first normalized referring them to a volume of 1000 $l \cdot ha^{-1}$.

The obtained data were statistically analyzed using R software (version 3.3.1). First, data normal distribution was checked by performing the Shapiro-Wilk normality test. After that one-way analysis of variance (ANOVA) was applied when data normality was verified, otherwise, the non-parametric Kruskal-Wallis was chosen to highlight any difference between the treatments.

Water-sensitive papers were analyzed with ImageJ 1.45s software using Java (1.6.0-20). This included the conversion of their scanned images into grayscale and then to a binary process that permitted particle (or spot) counting.

3 Results and Discussion

Normal distribution of foliar deposit data was checked by performing the Shapiro-Wilk normality test. The results showed that the null-hypothesis is rejected (W = 0.78758, p-value = $1.867e^{-10}$). Hence, the Kruskal–Wallis test was used to compare the foliar deposit generated by each tested equipment. Such a test showed a significant difference between the performances of the tested sprayers (Kruskal-Wallis $\chi 2 = 40.327$, df = 2, p = $1.75e^{-9}$). The mean and median values corresponding to the normalized foliar deposit generated by these sprayers are reported in the following Table 2.

Table 2. Mean and Median values of the normalized foliar deposit ($\mu L \cdot cm^{-2}$) for each sprayer type.

	UGE	F1000	EF3
Mean values ($\mu L \cdot cm^{-2}$)	4.366	3.662	10.180
Median values ($\mu L \cdot cm^{-2}$)	3.9330	3.6985	9.0240

The One-Way Anova performed on the foliar deposit obtained by UGE in the first trial did not show any significant difference according to foliage height (F = 0.434; df (2; 21), Pr (>F) = 0.654) meaning that vertical distribution evenness is quite guaranteed when using this sprayer with the previously mentioned parameters (Fig. 3).

Fig. 3. Normalized foliar deposit mean values obtained by UGE in the first trial according to foliage height (μL·cm^{-2}).

The same situation was verified for the sprayer F1000 ($F = 0.246$; df (1; 32), Pr (>F) = 0.623) as shshown in the following Fig. 4. While for the sprayer EF3, normal foliar deposit was significantly affected by sampling height (Kruskal-Wallis $\chi 2 = 8.2696$, df = 1, p = 0.004032). Mean values are reported in Fig. 5.

Fig. 4. Normalized foliar deposit mean values obtained by F1000 in the second trial according to foliage height (μL·cm^{-2}).

Fig. 5. Normalized foliar deposit mean values obtained by EF3 in the third trial according to foliage height (μL·cm^{-2}).

The foliar deposit generated by a determined spraying machine represents the resultant of the adjustment of its operative parameters and constitutes an important factor for the evaluation of spray quantitative and qualitative assessment. In fact, the study conducted by Garcerá et al. [11], allowed the modelling of the impact area coverage as well as of the efficiency of two oily formulations according to spray deposit in laboratory conditions. The results they obtained were subsequently validated in the field considering the cochineal of the citrus fruits *Aonidiella aurantii Maskell* (CRS) as a reference pest. Deposits of 1.01 μL·cm^{-2} for the organophosphorus formulations and 3.41 to 4.72 μL·cm^{-2} for the mineral oils, were necessary to obtain the efficiency of 90% of a generation with prevalence of juvenile stages. The authors believe that deposits of 3.41 μL·cm^{-2} for the first category and 4.72 μL·cm^{-2} for the second category are adequate for the control of subsequent pest generations. The sprayers UGE and F1000 generated spray deposit like those recommended in this study. Even though spray deposition generated by the sprayer EF3 is relatively high, if compared to the two other sprayers, it remains lower than that obtained by De Araújo et al. [12] (12.42 μL·cm^{-2}) who assessed spray deposit according to spray volume, fruit growth and rainfall in citrus black spot control.

Qualitative analysis carried out by analyzing water-sensitive paper images (Fig. 6a–c) showed that drop foliar coverage reached 46.364% with drop dimensions varying from 0.001 cm^2 to 0.04 cm^2 when using the sprayer UGE. The average foliar coverage obtained by the sprayer F1000 was 38.23%, while for the sprayer EF3, it was of 37.84% with droplet dimensions varying from 0.002 cm^2 to 0.013 cm^2 for both.

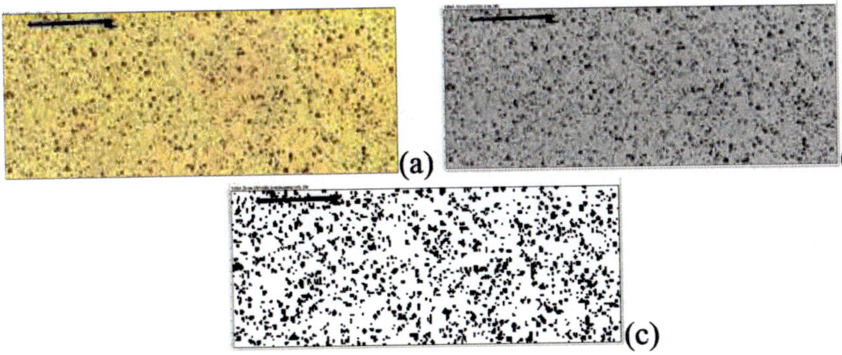

Fig. 6. Conversion and analysis of water sensitive paper image for spray qualitative analysis: (a) original image, (b) gray-scale image, (c) application of a binary process.

Also for ground losses, a significant difference was found between the analyzed equipments (Kruskal-Wallis $\chi 2 = 20.481$, df = 2, p = $3.57e^{-05}$). The mean and median values of the normalized ground losses produced by these sprayers are reported in Table 3. These data show the high amount of losses generated by both sprayers F1000 and EF3, indicating that these two systems need adjustment of operative parameters.

Table 3. Mean and Median values of the normalized ground losses ($\mu L \cdot cm^{-2}$) for each sprayer type.

	UGE	F1000	EF 3
Mean values ($\mu L \cdot cm^{-2}$)	1.504	5.836	8.341
Median values ($\mu L \cdot cm^{-2}$)	1.04300	6.01514	5.40900

4 Conclusions

The present work constitutes a preliminary evaluation of the equipment employed in citrus crop protection in Calabria. The obtained outputs clearly reveal that this practice presents critical issues. Although somehow the foliar deposit was homogeneous considering foliage height, the qualitative analysis of the spray as well as the ground losses confirmed that two of the three tested sprayers need serious adjustments of the operative parameters before implementing them in plant protection practices. Further experimental trials are needed in view to enhance this practice by increasing agrochemical application efficiency and significantly reducing the losses.

Acknowledgment. The present research was realized in the framework of the National Operative Project PON Ricerca e Competitività 2007–2013, PON03PE_00090_3 MODELLI SOSTENIBILI E NUOVE TECNOLOGIE PER LA VALORIZZAZIONE DELLE FILIERE VEGETALI MEDITERRANEE” funded by the Italian Ministry of Education, University and Research.

References

1. ISTAT. http://agri.istat.it/sag_is_pdwout/jsp/introduzione.jsp?id=15a|21a|31a
2. Grella, M., et al.: Field assessment of a pulse width modulation (PWM) spray system applying different spray volumes: duty cycle and forward speed effects on vines spray coverage. Precis. Agricult. (2021). https://doi.org/10.1007/s11119-021-09835-6
3. Balsari, P., Marucco, P., Tamagnone, M.: A system to assess the mass balance of spray applied to tree crops. Trans. ASAE **48**(5), 1689–1694 (2005). https://doi.org/10.13031/2013.19997
4. ENAMA. https://www.enama.it/userfiles/sfogliabili/Pubblicazione/monografie/enama_irroratrici/files/assets/common/downloads/ENAMA%20-%20Attivit.pdf
5. EU (European Parliament, Council of the European Union): Directive 2009/128/CE of the European Parliament and of the Council of 21 October 2009 Establishing a Framework for Community Action to Achieve the Sustainable use of Pesticides. Off. J. Eur. U. L. **309** 71–82 (2009)
6. Cerruto, E., Manetto, G., Emma, G.: Ottimizzazione dei Depositi nei Trattamenti su Pomodoro in Ambiente Protetto Siciliano. Macchine e Loro Regolazioni per una Difesa Sostenibile delle Colture Protette, 1st edn, pp. 45–78. Tipo-litografia FIORDO, Galliate (2008)
7. Guarella, P., Pascuzzi, S., Guarella, A.: Tecnologie per la distribuzione dei fitofarmaci alla fragola in coltura protetta. Macchine e Loro Regolazioni per una Difesa Sostenibile delle Colture Protette, 1st edn, pp. 79–102. Tipo-litografia FIORDO, Galliate (2008)
8. Sánchez-Hermosilla, J., Rincón, V.J., Páez, F., Fernández, M.: Comparative spray deposits by manually pulled trolley sprayer and a spray gun in greenhouse tomato crops. Crop Prot. **31**(1), 119–124 (2012). https://doi.org/10.1016/j.cropro.2011.10.007
9. Sánchez-Hermosilla, J., Páez, F., Rincón, V.J., Callejón, A.J.: Evaluation of a fog cooling system for applying plant-protection products in a greenhouse tomato. Crop. Prot. **48**, 76–81 (2013). https://doi.org/10.1016/j.cropro.2013.02.018
10. Bernardi, B., Benalia, S., Panuccio, M.R., Zimbalatti, G.: Assessing the "Special-Serre" sprayer for pesticide application to a greenhouse chrysanthemum crop. Acta Hortic. **1170**, 603–610 (2017). https://doi.org/10.17660/ActaHortic.2017.1170.75
11. Garcerá, C., Moltó, E., Zarzo, M., Chueca, P.: Modelling the spray deposition and efficacy of two mineral oil-based products for the control of California red scale, Aonidiella Aurantii (Maskell). Crop. Prot. **31**, 78–84 (2012). https://doi.org/10.1016/j.cropro.2011.10.004
12. De Araújo, D., Raetano, C.G., Ramos, H.U., Ribeiro da Rocha, D.S., Prado, E.P., Aguiar V. A.: Interference of spray volume, fruit growth and rainfall on spray deposits in citrus black spot control periods. Cienc. Rural. **46**(5), 825–831 (2016). https://doi.org/10.1590/0103-8478cr20150944

A Contribution to Environmental Protection Through the Valorisation of Kitchen Biowaste

Carlo Greco[1], Antonio Comparetti[1], Santo Orlando[1](✉), and Michele Massimo Mammano[2]

[1] Department of Agricultural, Food and Forest Sciences, University of Palermo, Palermo, Italy
Santo.orlando@unipa.it
[2] Council for Agricultural Research and Economics (CREA), Research Centre Protection and Certification, Bagheria (Palermo), Italy

Abstract. The worldwide interest in Renewable Energy Sources (RES) is gathering momentum. Anaerobic Digestion (AD) process is steadily growing, as more people are setting up bioreactors in order to produce biogas and digestate. AD process is the fermentation of solid biowaste in the absence of oxygen and is zero emission, i.e. it is CO_2 neutral. The aim of this work is to test HomeBiogas 2.0, i.e. a revolutionary equipment that used the kitchen biowaste from the canteen of the agricultural institute of Marsala (Trapani, Italy). This equipment works as a continuous flow bioreactor: the waste is fed in one end, while biogas and digestate come out from the other. The process continuously took place at an average temperature of 18 °C ca. The liquid digestate was sanitised by running through three chlorine tablets, in order to reduce the amount of active bacteria in this effluent. Therefore, the digestate was immediately used as a liquid biofertiliser for the soilless cultivation of tomato, strawberry, aubergine and pepper plants, inside a greenhouse. HomeBiogas 2.0 had a disposal capacity of 1.7 t of organic waste per year, equal to approximately 4.8 kg per day. During the testing period, this equipment produced 525 L of biogas and 7.78 L of liquid digestate per day. Sicily and the other islands of Mediterranean Sea are optimal environments for the operation of the tested equipment, that is an example of circular bioeconomy implementation.

Keywords: Anaerobic Digestion · Biogas · Circular bioeconomy · Digestate · Organic waste · Renewable Energy Sources

1 Introduction

The organic waste from livestock (manure and slurry) and the crop residues from agricultural activities, as well as the food industry by-products and the Organic Fraction of Municipal Solid Waste (OFMSW) from the separate waste collection (including garden and kitchen waste), are Renewable Energy Sources (RES), that have been only partially exploited in Sicily and in the other islands of Mediterranean Sea [1, 2].

In the separate waste collection, the possibility of energy valorisation of the bio-waste through Anaerobic Digestion (AD) process, producing biogas (constituted, above all, by biomethane) and digestate, is now mandatory, according to EU legislation on

M. Biocca et al. (Eds.): SHWA 2020, LNCE 252, pp. 411–420, 2022.
https://doi.org/10.1007/978-3-030-98092-4_42

waste [2–9]. Therefore, there is a need for convergence among various stakeholders who are directly or indirectly involved in the valorisation of this resource [10].

AD is a fermentation process in the absence of oxygen and a feasible model of development of the agri-food supply chain, according to the principles of circular bioeconomy (CBE) [7, 11–16]: its implementation is a valuable opportunity for the current society, which must face complex and important challenges, such as food security, competition for natural resources, dependence from fossil sources and climate change. Therefore, the link between food waste management and sustainable biomass production is a key element of circular bioeconomy [17].

AD process is a solution for both implementing a correct biowaste management and drastically reduce the fee for waste disposal due by citizens and operators in the agri-food supply chain.

Innovative technological, financial and legislative solutions, that add further value to the already virtuous supply chain of biomethane from OFMSW and food industry by-products are a paramount opportunity for Sicily [2, 3].

AD process has the following advantages:

- it reduces the emission of dangerous amounts of methane from landfill waste to the atmosphere;
- it reduces the excessive dependence from fossil fuels (e.g. oil and coal) for energy generation;
- it ensures the ecological disposal of organic waste;
- it is CO_2 neutral, as it produces biogas, including an amount of CH_4, i.e. a greenhouse gas (GHG), that is equal to this GHG that would be emitted from the fermentation of waste materials in landfills [3];
- it uses compounds contained in the organic waste for generating bioenergy;
- it meets sanitation requirements, as it deactivates pathogens and parasites and, therefore, reduces the incidence of waterborne diseases;
- it produces high quality stabilised liquid fertiliser (digestate);
- it is carried out inside a bioreactor, having no wear part;
- it is not followed by odour emission;
- it integrates an eventually existing composting process;
- it has a high reliability, as shown by many thousands of bioreactors operational all over the world.

Biogas properties are shown in Table 1.

Table 1. Biogas properties: the amounts of the gases included inside biogas depend on the type and quantity of the waste fed in the bioreactor.

Compound	Formula	Amount inside biogas (%)
Methane	CH_4	55–65
Carbon dioxide	CO_2	35–45
Nitrogen	N_2	0–10
Hydrogen	H_2	0–1
Oxygen	O_2	0–0.5
Hydrogen sulphide	H_2S	0.00001

Biogas is used for:

- combustion inside boilers;
- production of biomethane, to be used as biofuel for vehicles or fed into the natural gas grid;
- co-generation (of electrical and thermal energy) inside CHP (Combined Heat and Power) plants or gas turbines.

As far as the calorific value of biogas, 1 Nm^3 of biogas produces 6.5–7.2 MJ of electrical energy and 7.2–10.8 MJ of thermal energy.

AD is a process for converting biowaste, such as waste from livestock (manure and slurry) and crop residues from agricultural activities, food industry by-products, OFMSW (including garden and kitchen waste), into biogas and digestate [2–8, 18].

AD allows the valorisation of OFMSW, to be treated together with livestock waste and food industry by-products, at the farms located in the areas of waste production [4, 6, 19].

Furthermore, the short distance between the place of biowaste production and that of its disposal considerably reduces the cost of waste management and the atmospheric pollution caused by its transport to the landfill.

Farms producing clean energy from biowaste have adopted sustainable production methods, by offering products having higher value to markets that are increasingly sensitive to environmental issues [10, 20–22].

Therefore, the aim of this work is to test HomeBiogas 2.0, i.e. a revolutionary equipment that used the kitchen biowaste from the canteen of the agricultural institute of Marsala (Trapani, Italy).

2 Materials and Methods

HomeBiogas 2.0 project was carried out by three Sicilian citizens, Vittorio Nocera, Gianfranco Maltese and Mario Ragusa, by means of the start-up "I Nuovi Mille" of Marsala (Trapani), aimed at promoting the green economy of Sicily, through innovative researches in energy and agriculture sectors.

HomeBiogas 2.0 equipment (Fig. 1) allows:

- up to 30% reduction of waste taxes;
- reduction of gas bills and energy dependence;
- reduction of CO_2 emission without sending waste to landfills;
- production of biofertiliser, to be locally used or sold;
- application of the circular bioeconomy principles.

Fig. 1. HomeBiogas 2.0 equipment, using the kitchen biowaste from the canteen of the agricultural institute of Marsala (Trapani).

Moreover, this equipment, whose technical specifications are shown in Table 2, is an educational tool for school pupils and families.

Table 2. Technical specifications of HomeBiogas 2.0 equipment.

Parameters	Value
Inputs	Maximum amount/day
Kitchen waste	12 L
Livestock waste (manure and slurry)	36 L
Outputs	Maximum amount/day
Biogas	700 L
Digestate	12 L
Physical characteristics	Specifications
Size	115 × 205 × 125 cm
Biogas tank volume	700 L
Outlet valve pressure	1 kPa

HomeBiogas 2.0 works as a continuous flow bioreactor: the waste is fed in one end, while biogas and digestate come out from the other. In fact, this small bioreactor is constituted by a funnel for charging organic matter, a 1200 L tank containing water and bacteria, an expandable vessel for collecting biogas, a valve with a filter for biogas supply and a tap for pouring the digestate. HomeBiogas 2.0 includes a single cooker, three chlorine tablets, pipes and connectors.

The organic matter can be charged into the bioreactor, so that bacteria can digest the biowaste, by releasing biogas, which is sent to the outlet valve. It is possible to store in the vessel up to 700 L per day of produced biogas, that can be used by means of the cooker, able to work for two hours. The connection with conventional hobs may require an adaptor.

The produced biogas is temporarily stored in the bioreactor vessel at a relatively low pressure, so that, in case of accidental fire, there is a negligible risk of explosion. Moreover, any excess of biogas is released into the atmosphere by a security valve.

In order to pour the digestate, the appropriate tap must be opened.

This equipment is provided with detailed instructions and everything needed for allowing to assemble the bioreactor in two hours, without the help of specialised technical staff. When it is used for the first time, the activation of bacteria can take up to 2–4 weeks, after which HomeBiogas 2.0 autonomously produces biogas for 24 h a day. Its operation requires only the charging of waste and the collection of digestate. For an optimal operation, its installation in an area having a good sun exposure and within 20 m from the cooker is recommended.

This equipment can use the following organic waste:

- food (e.g. kitchen) waste;
- vegetable residues;
- livestock waste (manure and slurry);
- cut grass.

It is needed to carry out the following ordinary maintenance operations:

- to replace the filter for hydrogen sulphide once a year;
- to remove the residual digestate that accumulates on the bottom of the tank once every three years, in order to use it as biofertiliser;
- to replace the chlorine tablets when they are completely dissolved.

HomeBiogas 2.0 is certified according to CE, ISO 9000 and ISO 14000 standards.

From the 1st November to the 31st December 2019, a test was carried out for measuring the production of biogas and digestate from the kitchen biowaste deriving from the school canteen.

This approach is the basis of the concept of circular bioeconomy, which keeps the added value inside products as long as possible and removes waste [11].

The waste was fed into the bioreactor, where an active and stable bacterial culture was previously introduced. If the waste is shredded before feeding it into the bioreactor, it is possible to increase the efficiency of bacteria. During this step, hydrolysis and self-heating took place, so that the fermentation is triggered.

In the tightly closed bioreactor, working in an anaerobic environment, i.e. in absence of oxygen, bacteria transformed the biowaste into biogas and digestate.

Biogas is constituted, above all, by methane, that can be used for cooking or heating.

Methane has a molecular weight lower rather than air, so that it tends to go to the atmosphere.

Biogas includes also, H_2S, equal to about 300 ppm, that is generally is reduced, through filtration, to less than 0.1 ppm. The smell of H_2S, similar to that of hard boiled eggs, is detectable by the human nose at a level of 0.01 ppm or higher.

The produced biogas can be transferred to a CHP (Combined Heat and Power) plant, in order to produce electrical energy, that can be locally consumed, and thermal energy, that can contribute to heat a greenhouse. This equipment can produce about 65,700 MJ of electrical energy per year. This electrical energy can be introduced into the grid, in order to assure an income of € 4927 ca. (price of € 0.27 per kWh).

Differently from many commercial fertilisers, the digestate produced by means of HomeBiogas 2.0 is a high quality biofertiliser having the following properties:

- it contains many macro- and micronutrients, as solute for being quickly absorbed by plants, in order to accelerate their growth and increase their resistance to diseases;
- it is liquid, so that it can be quickly and easily applied through root drenching, i.e. pouring it on the soil, or foliar spraying, by means of a bottle sprayer.

Before using this biofertiliser a dilution ratio of 5:1 water/fertiliser must be prepared.

The following rates per week are suggested for different plant species:

- 3.5 L/plant for flower species;
- 15–20 L/plant; for trees;
- 5–10 L/plant for some vegetable species;
- 3–5 L/plant for other vegetable species.

As the digestate comes from the bioreactor tank, it contains a high number of bacteria. Therefore, the chlorine tablets must be inserted in the pipes used for fertilisation for at least one day, in order to sanitise it, especially if this biofertiliser is used for foliar spraying.

3 Results and Discussion

During the testing period, the minimum temperature resulted 10.75 °C, while the maximum temperature was 18.20 °C, so that a mean temperature of 14.95 °C was recorded.

AD process continuously took place at an average temperature of 18.28 °C.

HomeBiogas 2.0 had a disposal capacity of 1.7 t of organic waste per year, equal to approximately 4.8 kg per day.

During the testing period, this equipment produced 525 L of biogas and 7.78 L of liquid digestate per day.

The digestate produced, whose properties are shown in Table 3, was sanitised by means of three chlorine tablets.

Table 3. Digestate properties: the values can vary depending on the type of biowaste charged in the equipment; cfu means colony forming units.

Parameters	Values
Nitrogen	260 mg/L
Phosphorus	20 mg/L
Potassium	275 mg/L
Micronutrients	Ca, Mg, S, B, Cu, Fe, Mn, Zn
Escherichia coli	<10 cfu/100 mL
Fecal coliforms	<10 cfu/100 mL
Escherichia coli 0157:H7	None
Listeria monocytogenes	None
Salmonella	None
Vibrio cholera	None

Therefore, it was immediately used as a biofertiliser for the soilless cultivation of tomato, strawberry, aubergine and pepper plants, inside a greenhouse (Fig. 2).

Fig. 2. Soilless cultivation of strawberry plants, using the digestate produced as biofertiliser, inside a greenhouse.

4 Conclusions

AD process, that can be carried out by means of the tested HomeBiogas 2.0 equipment, is a smart choice for both the environment and the economy: it produces biogas, that is a green Renewable Energy Source (RES), and digestate, that is a bio-fertiliser [4, 6].

Moreover, this process can contribute to reduce soil, water and air pollution. In fact, the biomethane included in the produced biogas does not go to the air, so that there is zero GHG emission in the atmosphere, while the digestate used as biofertiliser can replace the chemical ones, able to cause water pollution and, sometimes, eutrophisation.

Therefore, AD process can contribute to mitigate the effects of climate change, e.g. global warming.

Furthermore, biowaste collection and management are significantly more effective in areas where there are AD plants, so that environment, sanitation and hygiene are improved. In fact, AD plants can considerably reduce the amount of waste disposed in landfills, that are harmful for spreading foul smells and draining toxic liquids into underground water sources.

HomeBiogas 2.0 is a small, user-friendly and low-cost equipment, that can be used at home for valorising kitchen waste and pays for itself after a short period [18, 23].

Similarly to other RES (e.g. solar and wind energy), AD is also affected by weather parameters: the optimal temperature needed by bacteria for digesting biowaste is about 37 °C. However, during the testing winter period (November-December), the process continuously took place at a mean temperature of 18.28 °C. Therefore, Sicily and the other islands of Mediterranean Sea, where the cost of energy and waste man-agement are much higher than elsewhere in EU, are optimal environments for the operation of HomeBiogas 2.0 equipment. Instead, in environments characterised by cold climate, AD process requires thermal energy for keeping a constant biogas and digestate production.

Therefore, the tested equipment is an example of circular bioeconomy implementation and, at the same time, an educational tool for school pupils and families [24, 25].

References

1. Attard, G., Comparetti, A., Febo, P., Greco, C., Mammano, M.M., Orlando, S.: Case study of potential production of renewable energy sources (RES) from livestock wastes in Mediterranean Islands. Chem. Eng. Trans. **58**, 1–6 (2017). https://doi.org/10.3303/CET1758093
2. Comparetti, A., Febo, P., Greco, C., Orlando, S.: Italian potential biogas and biomethane production from OFMSW. In: IV International Conference Ragusa SHWA on "Safety, Health and Welfare in Agriculture, Agro-food and Forestry Systems", Lodi, Italy, 8–11 September 2015, pp. 206–215
3. Comparetti, A., et al.: Assessment of organic waste management methods through energy balance. Am. J. Appl. Sci. **11**(9), 1631–1644 (2014). https://doi.org/10.3844/ajassp.2014.1631.1644
4. Comparetti, A., Greco, C., Navickas, K., Orlando, S., Venslauskas, K.: Life cycle impact assessment applied to cactus pear crop production for generating bioenergy and biofertiliser. Riv. Studi Sulla Sostenib. **2**, 315–329 (2019). https://doi.org/10.3280/RISS2019-002-S1020
5. Comparetti, A., Greco, C., Mammano, M.M., Navickas, K., Orlando, S., Venslauskas, K.: Valorisation of urban green areas for producing renewable energy and biochar as growing substrate of Sicilian aromatic and nutraceutical species in a circular economy. Riv. Studi Sulla Sostenib. **2**, 299–314 (2019). https://doi.org/10.3280/RISS2019-002-S1019

6. Greco, C., Comparetti, A., Febo, P., Navickas, K., Orlando, S., Venslauskas, K.: LCA applied to an anaerobic digestion plant for biomethane and digestate production. In: 9th International Scientific Conference "Rural Development 2019: Research and Innovation for Bioeconomy", Akademija, Kaunas district, Lithuania, 26–28 September 2019, pp. 1–7 (2019). https://doi.org/10.15544/RD.2019.038
7. Greco, C., Agnello, A., La Placa, G., Mammano, M.M., Navickas, K.: Biowaste in a circular bioeconomy in Mediterranean area: a case study of compost and vermicompost as growing substrates alternative to peat. Riv. Studi Sulla Sostenib. **2**, 345–362 (2019). https://doi.org/10.3280/RISS2019-002-S1022
8. Orlando, S., Greco, C., Tuttolomondo, T., Leto, C., Cammalleri, I., La Bella, S.: Identification of energy hubs for the exploitation of residual biomass in an area of western sicily. In: EUBCE 2017, Online Conference Proceedings, pp. 64–69. ETA srl, Firenze (2017)
9. Hanif, M., Hesam, N.M., Akhiar, A., Fazril, I., Zamri, M.F.M.A., Shamsuddin, A.H.: Economic feasibility of smart city power generation from biogas produced by food waste in Malaysia via techno-economic analysis. IOP Conf. Ser., Earth Environ. Sci. **476**, 012076 (2020). https://doi.org/10.1088/1755-1315/476/1/012076
10. Chen, L., Zhao, L., Ren, C., Wang, F.: The progress and prospects of rural biogas production in China. Energy Policy **51**, 58–63 (2012)
11. Greco, C., Comparetti, A., Mammano, M.M., Orlando, S.: Sustainable, circular and innovative value chains using growing substrata alternative to peat for the cultivation of nutraceutical species. In: 9th International Scientific Conference "Rural Development 2019: Research and Innovation for Bioeconomy", Akademija, Kaunas district, Lithuania, 26–28 September 2019, pp. 1–8 (2019). https://doi.org/10.15544/RD.2019.040
12. Alexander, S., Harris, P., Bernadette, K., McCabe, B.K.: Biogas in the suburbs: an untapped source of clean energy? J. Clean. Prod. **215**, 1025–1035 (2019). https://doi.org/10.1016/j.jclepro.2019.01.118
13. Miller, K.E., Grossman, E., Stuart, B.J., Davis, S.C.: Pilot-scale biogas production in a temperate climate using variable food waste. Biomass Bioenergy **138**, 105568 (2020). https://doi.org/10.1016/j.biombioe.2020.105568
14. Ellacuriaga, M., García-Cascallana, J., Gómez, X.: Biogas production from organic wastes: integrating concepts of circular economy. Fuels **2**, 144–167 (2021). https://doi.org/10.3390/fuels2020009
15. Loizia, P., Neofytou, N., Zorpas, A.A.: The concept of circular economy strategy in food waste management for the optimization of energy production through anaerobic digestion. Environ. Sci. Pollut. Res. **26**(15), 14766–14773 (2018). https://doi.org/10.1007/s11356-018-3519-4
16. Porselvam, S., Mahendra, B., Srinivasan, S.V., Ravindranath, E., Suthanthararajan, R.: Enhanced biogas production from co-digestion of intestine waste from slaughterhouse and food waste. Energy Fuels **31**, 12133–12140 (2017). https://doi.org/10.1021/acs.energyfuels.7b01764
17. Paul, S., Dutta, A., Defersha, F., Dubey, B.: Municipal food waste to biomethane and biofertilizer: a circular economy concept. Waste Biomass Valoriz. **9**(4), 601–611 (2017). https://doi.org/10.1007/s12649-017-0014-y
18. Malik, W., Mohan, C., Annachhatre, A.P.: Community based biogas plant utilizing food waste and cow dung. In: Materials Today: Proceedings, vol. 28, Part 3, pp. 1910–1915 (2020). https://doi.org/10.1016/j.matpr.2020.05.312
19. Gosens, J., Lu, Y., He, G., Bluemling, B., Beckers, T.A.M.: Sustainability effects of household-scale biogas in rural China. Energy Policy **54**, 273–287 (2013). https://doi.org/10.1016/j.enpol.2012.11.032

20. Meeks, R., Sims, K.R.E., Thompson, H.: Waste not: can household biogas deliver sustainable development? Environ. Resource Econ. **72**(3), 763–794 (2018). https://doi.org/10.1007/s10640-018-0224-1
21. Bond, T., Templeton, M.R.: History and future of domestic biogas plants in the developing world. Energy Sustain. Dev. **15**(4), 347–354 (2011). https://doi.org/10.1016/j.esd.2011.09.003
22. Christiaensen, L., Heltberg, R.: Greening China's rural energy: new insights on the potential of smallholder biogas. Environ. Dev. Econ. 19(Special Issue 1), 8–29 (2014). https://doi.org/10.1017/S1355770X13000375
23. Petracchini, F., et al.: A novel pilot scale multistage semidry anaerobic digestion reactor to treat food waste and cow manure. Int. J. Environ. Sci. Technol. **15**(9), 1999–2008 (2017). https://doi.org/10.1007/s13762-017-1572-z
24. Greco, C., Comparetti, A., Febo, P., La Placa, G., Mammano, M.M., Orlando, S.: Sustainable valorisation of biowaste for soilless cultivation of Salvia officinalis in a circular bioeconomy. Agronomy **10**(1158), 1–13 (2020). https://doi.org/10.3390/agronomy10081158
25. Greco, C., et al.: Effects of vermicompost, compost and digestate as commercial alternative peat-based substrates on qualitative parameters of Salvia officinalis. Agronomy **11**(98), 1–11 (2021). https://doi.org/10.3390/agronomy11010098

ROPS and Stability Research

"Ask Me What I Need": Investigating Users' Training Needs and Design Requirements to Encourage the Correct Operation of Foldable Rollover Protective Structures

Federica Caffaro[1(✉)], Lucia Vigoroso[2], Margherita Micheletti Cremasco[3], Giuseppe Paletto[2], and Eugenio Cavallo[2]

[1] Department of Education, University of Roma Tre, Via del Castro Pretorio, 20, 00185 Rome, Italy
federica.caffaro@uniroma3.it
[2] Institute of Sciences and Technologies for Sustainable Energy and Mobility (STEMS), National Research Council of Italy (CNR), Strada delle Cacce, 73, 10135 Torino, Italy
[3] Department of Life Sciences and Systems Biology, University of Torino, via Accademia Albertina, 13, 10123 Torino, Italy

Abstract. To facilitate tractor operation in case of overhead obstacles, Foldable ROPS (FROPS) have been developed. However, many tractor rollover fatalities occur as frequently the operators leave the FROPS in a folded-down position. In the present study, we performed an ergonomic user-centred analysis of FROPS operation, to identify critical issues in training, provision of information and machine design to promote the correct use of the protective structure. Twenty male farm operators (Mage = 49.4 years, $SD = 11.8$) were interviewed while manually rising and folding down the rear-mounted FROPS fitted on their tractor. Participants were asked about: 1) the frequency of FROPS operation, 2) information available on the user manuals and training received from the dealers about the correct FROPS handling behaviour, 3) discomfort in use and 4) suggestions for a FROPS re-design to overcome the reported sources of discomfort. The results showed that operator manuals lack specific information regarding FROPS manual handling and they are often left unread. Some training is provided by the dealers but often the operators learn how to operate the protective structure by themselves. Discomfort in grasping the foldable frame was reported by the participants as discouraging the correct FROPS operation. Many participants would have welcomed some FROPS re-design solutions to increase folding frame reachability and to have more stable support for the feet. The results highlighted the need for actions at multiple levels to promote the correct use of rear-mounted FROPS among agricultural operators.

Keywords: Ergonomics · Tractor safety · Education · User-centered design

M. Biocca et al. (Eds.): SHWA 2020, LNCE 252, pp. 423–430, 2022.
https://doi.org/10.1007/978-3-030-98092-4_43

1 Introduction

The Rollover Protective Structure (ROPS) fitted on tractors avoids severe injuries to the driver during machinery rollover [1]. To facilitate tractor operation in low overhead clearance zones and confined spaces, Foldable ROPS (FROPS) has been developed [2]. FROPS has been designed to be manually folded frontward, or rearward when it is fitted on the rear of the driving seat. However, many tractor rollover fatalities occur since the operators often leave the FROPS in a folded-down position [3]. For instance, 40% of serious injuries and deaths during tractor rollovers occurred in the European Union member states when a FROPS was not deployed into its protective position [4]. In Italy, in particular, in 2016, 90 out of 114 fatal accidents involving tractors were rollovers, and about 19% of these fatalities resulted from FROPS in the folded-down position [3]. In previous studies, it was noticed that this unsafe behaviour was related to issues in FROPSs manual handling, which is considered a time-consuming and/or arduous operation [5, 6]. Regarding this last issue, previous studies already demonstrated criticalities concerning the human-tractor interaction quality. In particular, issues in FROPS reachability and its comfort in use, unnatural gestures and awkward postures of both upper and lower limbs, and unsafe behaviours in FROPS operation were previously observed [7, 8].

To ensure workers' safety and health, the safety hierarchy protocol should be applied in the design of machinery and equipment [9, 10]. In detail, the last step of this protocol specifies that when residual risks on machinery cannot be eliminated by design or using the adoption of Personal Protective Equipment (PPE), users need to be warned against these risks [9, 10]. Thus, safety training courses and/or safety manuals should be easily used to convey these safety messages. Regarding training, even though these issues are discussed during safety training courses in different industrialized countries [11, 12], the available statistics on FROPS adoption and its correct use seem to suggest the need for further interventions. As regards the manuals, previous studies [13, 14] showed that the number of information contained in these manuals does not encourage the final users to read it. In this way, the users improvise a series of actions, gestures and movements that can be hazardous and uncomfortable for their health and safety [8].

To develop targeted and effective solutions the user-centred design (UCD) approach should be adopted, to better understand real users' needs and behaviours and design systems able to satisfy these needs thus being more effective and satisfactory [15, 16]. The UCD approach is an iterative process, based on the users' involvement in all the design phases of a system, from the definition of the requirements the system should satisfy, to the evaluation of the prototypes developed based on these requirements. The approach has been previously applied in different sectors, including the agricultural context [17–20]. For instance, thanks to the collaboration of farmworkers, ergonomic solutions to reduce oil palm workers' discomfort and awkward postures were proposed [21] and users' characteristics, needs and limits were considered to reach users' satisfaction, ease of use and acceptability in the design of agricultural machinery [19]. Safety training materials have been also developed adopting the UCD approach [20, 22] to increase safety knowledge, comprehensibility, and satisfaction and to reduce the perceived effort during the training lectures. Based on this previous evidence and the ergonomic literature, we can expect that adopting this approach could provide tailored suggestions to improve FROPS manual handling.

Based on the above considerations, in the present study, we performed an ergonomic user-centred analysis of FROPS operation to identify critical issues in training, provision of information and machinery design that would benefit from targeted interventions, to promote the correct use of the protective structure.

2 Materials and Methods

2.1 Participants

Twenty male farm operators took part in the study. They had a mean age of 49.4 years (SD = 11.8) and a mean work experience in farming of 25.7 years (SD = 17.3). The study was performed in Cuneo, Piedmont region (Northwestern Italy), one of the Italian regions with the higher number of fatal overturning accidents involving tractors [23]. In particular, the province of Cuneo reports the highest number of tractor-related accidents within the region maybe because it has a large percentage of hilly and mountainous agricultural areas [24].

2.2 Instruments

Data was collected through a structured interview. Four main topics were investigated:

1) the frequency of FROPS operation;
2) training received about the correct FROPS handling behaviour and the information available on the user manual;
3) discomfort in use;
4) suggestions for a FROPS re-design to overcome the reported sources of discomfort.

Data on the sociodemographic characteristics (i.e., age and farming experience) of the participants were also collected. The interviews were recorded and then were transcribed verbatim.

2.3 Procedure

To be included in the study, farmers had to own a tractor satisfying two inclusion criteria: a) track width larger than 1150 mm, and b) fitted with rear-mounted two-pillar FROPS. A list of possible participants was provided by the dealers of various agricultural machinery brands in the province of Cuneo. Respondents were contacted by telephone and, if willing to participate, were met at their farm. Each visit lasted about 20 min. During the visit, participants were asked to raise and fold down the rear-mounted FROPS fitted on their tractor. After each task, the researchers proceeded with the interview. Moreover, regarding the information available on the user manual, farmers were also asked if they had the operation and maintenance manual of the machine and if it was possible to consult it. The participation was voluntary, and all the farmers gave their written informed consent before their inclusion in the study.

3 Results

3.1 Information and Training

Regarding the frequency of FROPS operation, 8 participants declared to keep the device always in the upright position, while 12 participants reported seasonal handling of the device. Regarding the training received and information available on the user manual, participants' answers can be synthesized as follows:

- 11 participants received some explanation from the dealer on how to operate the FROPS; moreover, 5 of them were also shown the correct handling behaviour;
- 8 participants were taught the importance of correct use of FROPS during tractor driving certification training, but the proper handling behaviour was not shown; Some participants stated: "*The dealer told me that I have to unscrew screws and pins*" (S12), "*The dealer showed it [how to handle the FROPS] to me when they delivered it to me [the tractor]*" (S17), "*Trainers explained it during the training for the tractor license, but they provide general information, they didn't tell me how to handle it. and I didn't discuss it with the other colleagues*" (S8), "*The mechanic showed it to me, how to allocate the bolts and the pins. Instead, during the training courses, trainers talked about the folding frames, but they did not provide either specific information or practical demonstration*" (S15).
- only 1 participant declared to have read the handling information on the user manual:
  "*Yes, the technical manual explains that I can move it [FROPS] by removing the screws and pins, moving the mirror shaft, but it is not written or indicated how to perform the gestures...*" (S17), "*No [I do not read the manual], because it's so intuitive that I can handle the FROPS without using the manual*" (S1), "*It could be written on the manual, but I didn't look. I know that I should do it ...*" (S5), "*Yes, in general, I read the instructions on the manual if they are written in Italian, when I search something specific to understand what I have to do ...*" (S8), "*No, I never read the technical manual. I don't even know if it explains how to do it*" (S12).

When participants were asked to show the researchers the technical manual, the researchers took some pictures of the FROPS handling instruction pages. Some examples are reported in Fig. 1.

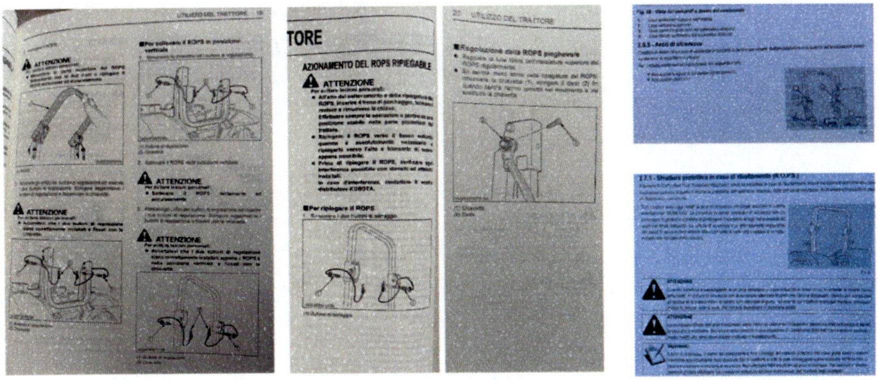

a) b) c)

Fig. 1. Pictures taken from the user manuals of different tractors to check the content and the page layout of the sections dedicated to FROPS operation.

3.2 Discomfort in Use

Overall, the participants reported some discomfort in grasping the foldable frame, regarding reachability issues and weight of the folding frame: "*I find FROPS handling uncomfortable, hard to pull, push, folding down and raise it, it's not really easy. It is hard to remove the pins and it takes time*" (S17), "*Folding down the FROPS from the central part of the folding bar is heavier, maybe a handle on the side would be better*" (S16), "*The effort is not so much in folding the FROPS down, rather in removing the pins and screws first, the only fatigue is that the bar is high*", (S14), "*It is too complex. A grip is missing (...) and if I lose my grip, and it falls it can end up on my hand or my head*" (S12), "*It's comfortable even though it is heavy*" (S18), "*When I fold the bar I have to stretch a little forward, and it seems that you are falling [when folding the FROPS standing on the seat]*" (S15).

Also, the unstable support for the feet was noted as a source of discomfort and possible risk of falling: "*The main issue is about where to place your feet, with or without tools*" (S9), "*You must have a platform for the feet, something that allows you to stand safely*" (S8).

3.3 Redesign Suggestions

Regarding the redesign suggestions to make FROPS handling more comfortable, 13 participants suggested providing some kind of handle to ease FROPS operation: "*A handle on the right side of the FROPS would be useful*" or "*A rope that makes it easier to pull it down or raise it*" or "*a friction on the screws to prevent it from rapid falling*" (S17), "*A platform that I can attach to the tractor and that raises me of about 20 cm to grip the upper part of FROPS easier*" (S16), "*A handle that does not interfere or hinder the other parts of the tractor when the FROPS is down. Maybe it can be retractable*" (S16).

A few other comments were generic or related to other parts of the machinery to improve the FROPS reachability: "*Maybe lowering the lift arms or putting them further back because they are too high now*" (S18), "*...to develop something that makes it easier to handle the FROPS*" (S20), "*Maybe a third supporting point for the feet*" (S11), "*Maybe making it telescopic*" (S7).

4 Discussion and Conclusions

The present study performed a user-centred ergonomic analysis of FROPS handling, to identify critical areas regarding training, information and machinery design that can hinder the correct handling behaviour and use of FROPS. The results highlighted the need for actions at multiple levels to promote the proper use of rear-mounted FROPS among agricultural operators. In particular:

- Most of the participants did not receive either useful information or training about the correct and safe handling behaviour. Therefore, this issue deserves specific attention and behavioural training during training courses, to become routine behaviour.
- Manuals are poorly used as a source of safety information. The recommended grasping areas and places to stand for the operator may be embossed or identified through coloured labels on the FROPS and the machine itself, acting as an affordance capable of suggesting the correct behaviour to the user. The same information may be integrated and reinforced by being reported with simple drawings in the user manual.
- Many participants would have welcomed some FROPS re-design solutions to increase the folding frame reachability and to have more stable support for the feet. In addition, handles and platforms could be designed and tested with users to find the best solution fitting technical and user requirements.

We acknowledge that the present results cannot be considered as conclusive due to the small number of participants involved and the fact that data on training received and manual consultation was self-reported; however, we believe that the study provides a useful contribution toward a better quality of human-agricultural machinery interaction by reflecting on the importance of bringing real users' needs to light and collecting their suggestions for an improved machinery redesign and tailored training and information materials and activities.

References

1. Organization for Economic Co-operation and Development. OECD Standard Code for the Official Testing of Agricultural and Forestry Tractors. Author, Paris (2017)
2. NIOSH. Preventing Death and Injury in Tractor Overturns with Roll-Over Protective Structures. National Institute for Occupational Safe and Health. https://blogs.cdc.gov/niosh-science-blog/2009/01/05/rops/. Accessed 20 Jan 2022

3. Fargnoli, M., Lombarbi, M., Haber, N., Puri, D.: The impact of human error in the use of agricultural tractors: a case study research in Vineyard Cultivation in Italy. Agriculture **8** (82), 21 (2018)
4. Hoy, R.M.: Farm tractor rollover protection: why simply getting rollover protective structures installed on all tractors is not sufficient. J. Agric. Saf. Health **15**, 3–4 (2009)
5. Khorsandi, F., Ayers, P.D.: The effect of friction on actuation torques of foldable rollover protective structures. J. Agric. Saf. Health **24**, 227–242 (2018)
6. Caffaro, F., Cremasco, M.M., Giustetto, A., Vigoroso, L., Paletto, G., Cavallo, E.: Ergonomics in agriculture: critical postures, gestures, and perceived effort in handling foldable roll-over protective structures (ROPS) fitted on tractors. In: Bagnara, S., Tartaglia, R., Albolino, S., Alexander, T., Fujita, Y. (eds.) IEA 2018. AISC, vol. 819, pp. 194–202. Springer, Cham (2019). https://doi.org/10.1007/978-3-319-96089-0_22
7. Ferrari, E., Cavallo, E.: Operators' perception of comfort in two tractor cabs. J. Agric. Saf. Health **19**, 3–18 (2013)
8. Micheletti Cremasco, M., Caffaro, F., Giustetto, A., Vigoroso, L., Paletto, G., Cavallo, E.: Tractor rollover protection: is the incorrect use of foldable rollover protective structures due to human or to technical issues? Hum. Factors **62**(1), 64–76 (2020)
9. Caputo, A.C., Pelagagge, P.M., Salini, P.: AHP-based methodology for selecting safety devices of industrial machinery. Saf. Sci. **53**, 202–218 (2013)
10. Purschwitz, M.A.: Personal protective equipment and safety engineering of machinery. In: Lessenger, J.E. (ed) Agricultural Medicine. A Practical Guide, pp. 53–69. Springer, New York (2006)
11. Murphy, D.J., Myers, J., McKenzie, E.A., Jr., Cavaletto, R., May, J., Sorensen, J.: Tractors and rollover protection in the United States. J. Agromedicine **15**(3), 249–263 (2010)
12. Cavallo, E., Langle, T., Bueno, D., Tsukamoto, S., Görücü, S., Murphy, D.: Rollover protective structure (ROPS) retrofitting on agricultural tractors: goals and approaches in different countries. J. Agromedicine **19**(2), 208–209 (2014)
13. Tebeaux, E.: Improving tractor safety warnings: readability is missing. J. Agric. Saf. Health **16**(3), 181–205 (2010)
14. Tebeaux, E.: Safety warnings in tractor operation manuals, 1920–1980: manuals and warnings don't always work. J. Tech. Writ. Commun. **40**(1), 3–28 (2010)
15. Bevan N, Curson I: Planning and implementing user-centred design. In: Conference Summary on Human Factors in Computing Systems (CHI 1998), pp. 111–112 (1998)
16. ISO. ISO 9241-210: 2010 Ergonomics of human-system interaction. Part 210: Human-Centred Design for Interactive Systems. International Organization for Standardization, Geneva (2010)
17. Rossi, V., Salinari, F., Poni, S., Caffi, T., Bettati, T.: Addressing the implementation problem in agricultural decision support systems: the example of vite.net®. Comput. Electron. Agricult. **100**, 88–99 (2014)
18. Walisadeera, A.I., Ginige, A., Wikramanayake, G.N.: Ontology evaluation approaches: a case study from agriculture domain. In: Gervasi, O., et al. (eds.) ICCSA 2016. LNCS, vol. 9789, pp. 318–333. Springer, Cham (2016). https://doi.org/10.1007/978-3-319-42089-9_23
19. Haapala, H.: User-centred design and multi-actor approach in agricultural innovations - case: combi-drill design. Agricult. Mach. Technol. **13**, 15–19 (2019)
20. Vigoroso, L., Caffaro, F., Cavallo, E.: Occupational safety and visual communication: user-centred design of safety training material for migrant farmworkers in Italy. Saf. Sci. **121**, 562–572 (2020)

21. Mohd Yusoff, I.S., Mohd Tamrin, S.B., Mat Said, A.M., Ng, Y.G., Ippei, M.: Oil palm workers: designing ergonomics harvesting tool using user centered design approach to reducing awkward body posture by CATIA simulation. Iran. J. Public Health **43**, 72–80 (2014)
22. Cutini, M., Forte, G., Maietta, M., Mazzenga, M., Mastrangelo, S., Bisaglia, C.: Safety-critical manuals for agricultural tractor drivers: a method to improve their usability. Agriculture **7**, 1–20 (2017)
23. Pessina, D., Facchinetti, D.: A survey on fatal accidents for overturning of agricultural tractors in Italy. Chem. Eng. Trans. **58**, 79–84 (2017)
24. Basso, M., Libener, M., Miotti, F., Pasqualini, O., Ruvolo, M.: Sistema di sorveglianza sugli infortuni mortali della Regione Piemonte. Rapporto Sulla Ricostruzione Degli Infortuni Mortali In Regione Piemonte, Anni 2009–2010 (2010)

Can Digital Games Be Adopted as a Learning Method to Engage Farmworkers in Training Sessions? Perspectives and Possible Applications

Lucia Vigoroso[1(✉)], Federica Caffaro[2], Margherita Micheletti Cremasco[3], and Eugenio Cavallo[1]

[1] Institute of Sciences and Technologies for Sustainable Energy and Mobility (STEMS), National Research Council of Italy (CNR), Strada delle Cacce, 73-10135 Torino, Italy
lucia.vigoroso@stems.cnr.it

[2] Department of Education, University of Roma Tre, Via del Castro Pretorio, 20-00185 Rome, Italy

[3] Department of Life Sciences and Systems Biology, University of Torino, via Accademia Albertina, 13-10123 Torino, Italy

Abstract. Safety training programs help to increase farmworkers' knowledge on the risks and the correct operational behavior in the workplace. To be more effective, training interventions should be engaging and make trainees emotionally involved. A game-based training can achieve this objective. However, few studies used games as a safety training method in the agricultural context. The present study aimed to show what is available in the literature concerning safety game-based training in different industries and which elements can be used to proceed with the "Safety! Is a serious game" ongoing project, which intends to develop an engaging and usable safety training game addressing farmworkers. The literature review allowed to identify the available technologies, the most used game mechanics and the addressed topics, considering the relationships between the worker and the components of the working system as defined by the SHELL-model (Software-Hardware-Environment-Liveware). The results showed that previous interventions focused on eight macro-topics regarding the interaction with machinery/equipment, work procedures, environment and other workers, that can be encountered also during farmworking. Although different game mechanics exist, "points", "levels", "challenges" and "discovery" were preferred; whereas computer-game was detected as the most adopted, followed by virtual and augmented reality. Moreover, to be realistic and satisfying the game should recreate the decision making processes that operators encounter in hazardous situations. Proceeding with the project and adopting a participatory approach, the game mechanics, degrees of complexity and goals leading to greater players' engagement, satisfaction and knowledge will be identified and evaluated with the users.

Keywords: Agriculture · Engagement · Gamification · Human-machine interaction · Safety training

M. Biocca et al. (Eds.): SHWA 2020, LNCE 252, pp. 431–441, 2022.
https://doi.org/10.1007/978-3-030-98092-4_44

1 Introduction

In different industries, workers are often exposed to on-site hazards which can result in serious injuries and fatalities [1]. Adequate safety training can increase workers' risk perception and improve their safe behaviors. Previous studies showed that safety and health training is provide through different methods which can be more or less engaging. In details, lectures, videos and written materials are considered among the less engaging training methods, whereas computer-based instructions and hand-on demonstrations are considered among the most engaging [2]. More recently, thanks to the technological development, new safety training methods are emerging, providing realistic simulations of real-life experiences: digital games, by means of a simulated environment, have the advantage to create an interactive and more engaging learning experience. Within these "safe" virtual environments workers can focus their attention on specific tasks, can learn how to use the personal protective equipment during the working hours and they can also learn what they have to do and which behaviors to avoid in case of hazardous situations and injuries [3]. Moreover, digital games train workers to operate dangerous machinery allowing them to make mistakes which would hurt the individual in real-life situations. The scientific literature defines these simulation tools as "serious games" or "gamified tools" in which game thinking, game design elements and game mechanics as rewards and incentives in forms of points and badges are used to increase players' motivation to find effective problem-solving strategies and encourage behavioral change. Moreover, game mechanics can be defined as the rules and the control mechanisms that guide the player's actions [4]. Several previous studies showed that digital games have been already applied to safety training in different sectors, such as the automotive manufacturing, the firefighting, construction and mining [1, 5, 6]. Despite agriculture being one of the most hazardous industries, digital games were introduced as a training method in this sector mainly with the aim to teach workers how to operate specific smart farming technologies or with agro-ecology learning purposes [7]. To our knowledge, very few studies used games as a safety training method in agriculture. The available studies deal with agricultural machinery driving and All-Terrain Vehicles (ATVs) and the need to push operators toward safer behaviors [8]. However, agriculture exposes workers not only to risks related with driving activities but, as in other occupational sector includes different potential hazardous interaction with workers, procedures, instruments and environment, which need of targeted training intervention. Indeed, farm work, like the other work systems, includes "one or more workers and work equipment acting together to perform the system function, in the workspace, in the work environment, under the conditions imposed by the work tasks" [9] (p. 2) and each interaction between these components can be a source of risk. Considering the few studies available on game based-training in the agricultural context, surveying how these risks are addressed in gamified training interventions in sectors other than agriculture, could be useful to a gamified tool specifically addressing in the primary sector. Based on all these considerations, the present contribution aims at identifying and synthesizing the results currently available in the literature about the game-based safety training adopted in different occupational sectors. Analysing the interactions between the worker and the other components of the working system, the present will help to understand which risks have been considered in previous gamified training interventions, and then which game mechanics and technologies are able to offer a satisfactory and effective experience; subsequently, starting from these results, insights for future game-based safety training development for workers employed in agriculture will be discussed.

2 Materials and Methods

This contribution represents an update of the scoping review by Vigoroso et al. [10] and it adopted a different theoretical framework to analyze the results. The literature search was carried in July 2021 on WEB OF SCIENCE, SCOPUS, PUBMED and PsycINFO databases. Two groups of keywords searches were performed to identify the relevant articles: 1) "gamification" AND "safety" AND "training" OR "education" and 2) "serious game" AND "safety" AND "training" OR "education". To be considered, research articles had to: (1) focused on a game-based training intervention; (2) the game developed had to face risks in the interaction with machinery or equipment, work procedures, workplace environment or other workers, (3) it can be possible to deduce information on the game mechanics adopted and (4) the full-text available should be published in English. For each paper retained for the analysis, the year in which the article was published, the working sector in which the game was applied, the technology and the game mechanics used, the risks addressed and similarities with agriculture were recorded. Then, the results were analyzed referring to the SHELL-model [11] to comprehend which of the interactions between the individual and the elements of the working system able to influence the individual's behavior were considered in previous studies (Fig. 1): Software (S – i.e. the rules, procedures, written documents), Hardware (H – i.e. controls and surfaces, displays), Environment (E – i.e. the situation in which the L-H-S system must function, the social and the natural environment) and Liveware (L – i.e. the human beings, workers and administration people).

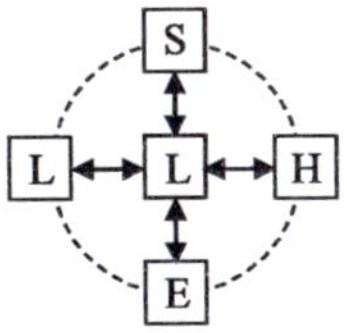

Fig. 1. The building-block representation of the SHELL conceptual model. The model takes its name from the initial letters of its four components.

3 Results

A total of 985 articles were identified. Three hundred and sixty-three articles remained after the removal of duplicates and they were screened on the basis of the inclusion criteria applied to the titles, abstracts, and full texts. Forty-seven publications were included in the final analysis. Regarding the year of publication, articles ranged from

2007 to 2021 (Fig. 2) but the majority of the articles (57.5%) were published between 2018 and 2021 (Fig. 2a). Regarding the working sectors in which the game training was applied, they can be summarized as: industrial sector (which includes also automotive and construction industries), mining, fire and firefighting, aviation, maritime and business (Fig. 2b). The computer games, virtual reality (VR) and augmented reality (AR, which also includes mixed-reality – MR) were detected as the three main technologies used (Fig. 2c).

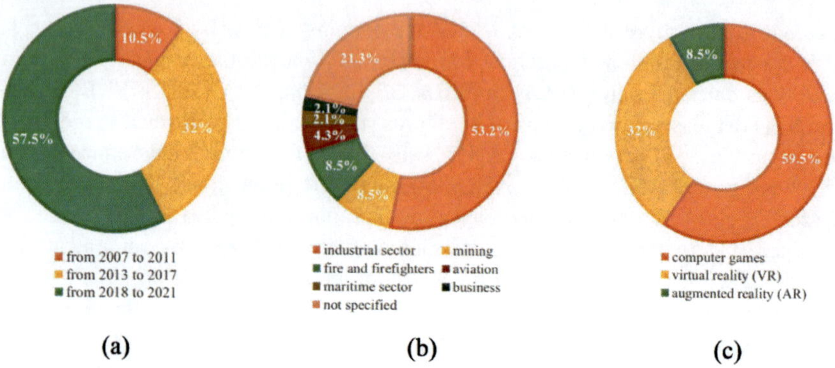

Fig. 2. Information recorded in the papers retained: (a) year in which the article was published, (b) the working sector considered and (c) the technology used.

When considering the game mechanics, in some studies more than one was applied. In details, five game mechanics were recorded: challenges (n = 25), points (n = 21), levels (n = 14), discovery (n = 18) and virtual goods (n = 2). Regarding the risks addressed, some studies dealt with more than one risk. It is possible to reduce these risks to eight macro-topics were recorded, which correspond to those suggested for training in the agricultural sector by the International Convention and the International Labour Office (ILO) codes of practice [12]: agricultural installation (which includes fire risk, electrical risk and falling risk n = 20), vehicle driving (n = 8), machinery operation (n = 10) biomechanical overload (n = 4), personal protective equipment adoption (PPE, n = 11), chemical risk (n = 6), management (n = 6) and general hazards (n = 6) (Table 1). Reading these risks and topics through the theoretical lens of the SHELL-model, it can be noted that also in this case many studies dealt with more than one component (Table 1). Trying to disentangle all the components and analyze each interaction, it is possible to make the following considerations:

- The interaction between Liveware and Software was identified in 37 studies and can be classified in four groups: learn a step by step operation (21%), learn operation using the correct PPE (5%), how to act in case of emergency or danger (27%) and recognize hazards at the workplace (47%);
- The interaction between Liveware and Hardware was identified in 18 studies and can be classified in four groups: simulation of the use of an object (19%), simulation

with tracker on the real object (19%), simulation of machine operations (14%) and driving simulation (48%);

- The interaction between Liveware and Environment was identified in 13 studies and can be classified in three groups: weather conditions (such as rain, fog, fire, smoke etc. 23%), different road conditions (e.g. sand, concrete, slope etc. 30.8%) and physical working environment (46.2%);
- The interaction between Liveware and Liveware was identified in 12 studies. It has to be noticed that almost all the games were designed to be played by a single player. Thus, the role of the Non-Player Characters (NPCs) in the game, i.e., those characters in a game which are not controlled by the player but by the computer, was analyzed. The NPCs could be either not very interactive with the player (33%) having very limited number of actions and reactions to the player (e.g. they provide few pre-set information only or follow pre-established paths in the virtual environment), or more interactive (67%), meaning that their actions are influenced by the player's actions (e.g. in case of time penalties, poor decisions and/or procrastination, NPCs can become increasingly stressed, irritated and can culminate in the their abandoning of the team work team, causing the end of the game).

Table 1. Parallels with training topics in the agricultural sector in the reviewed articles and the corresponding SHELL-model component considered.

Source	Risks addressed and similarities with agriculture	SHELL component
Albert et al., 2014 [13]	Chemical risk; agricultural installation (falling risk)	S
All et al., 2017 [14]	Agricultural installation (fire risk)	S
Alshiar et al., 2019 [15]	PPE adoption; operate on machinery; agricultural installation (falling risk)	S
Amal et al., 2016 [16]	Operate on machinery	S + H
Brown et al., 2018 [17]	Agricultural installation (fire risk, falling risk)	S + E + L
Bukru et al., 2020 [18]	General hazards; PPE adoption; operate on machinery	S + H + E + L
Burt et al., 2018 [19]	Drive vehicle	S + H
Cai et al., 2017 [20]	Agricultural installation (fire risk); Drive vehicle	H + E + L
Cavalvanti et al., 2021 [21]	Agricultural installation (fire and falling risk)	S + E
Chodan et al., 2017 [22]	Management	S
Dawood et al., 2013 [23]	Chemical risk; agricultural installation (falling risk)	S + E
Dawood et al., 2014 [24]	Chemical risk; agricultural installation (falling risk)	S
Dholakiya et al., 2018 [25]	Chemical risk; PPE adoption	S

(*continued*)

Table 1. (*continued*)

Source	Risks addressed and similarities with agriculture	SHELL component
Dickinson et al., 2011 [26]	Operate on machinery	S + E
Din et al., 2019 [27]	Agricultural installation (electrical and falling risk)	S + E
Eller et al., 2018 [28]	Agricultural installation (fire risk)	H + L
Gilotta et al., 2018 [29]	Operate on machinery; PPE adoption	S
Golovina et al., 2019 [30]	Operate on machinery; agricultural installation (electrical risk); chemical risk	H
Gonzales et al., 2009 [31]	Drive vehicle	H + E
Gounaridou et al., 2021 [32]	Drive vehicle; PPE adoption	S + H
Greuter et al., 2013 [33]	Agricultural installation (fire risk); PPE adoption; management	S
Hafsia et al., 2018 [34]	PPE adoption	S + H
Hall et al., 2016 [35]	Agricultural installation (fire risk); Management	S + L
Harper et al., 2018 [36]	Operate on machinery; PPE adoption	S + H
Hulme et al., 2021 [37]	Drive vehicle;	H + E
Jiang et al., 2016 [38]	Drive vehicle	S + H + E
Kamkuimo et al., 2020 [39]	Chemical risk; PPE adoption	S
Kuindersma et al., 2017 [40]	Drive vehicle	S + H + L
Kwegyir-Afful et al., 2020 [41]	Operate on machinery	H
Lanzotti et al., 2019 [42]	Biomechanical overload; PPE adoption	S
Leong et al., 2013 [43]	General hazards	S
Liang et al., 2019 [6]	Agricultural installation (falling risk)	S + E + L
Lovreglio et al., 2020 [5]	Agricultural installation (fire risk)	S + H
Mallet et al., 2008 [44]	PPE adoption	S + E + L
Matsas et al., 2015 [45]	Operate on machinery	H
Mayer et al., 2013 [46]	Agricultural installation (fire and falling risk)	S + L
Mazur et al., 2014 [8]	General hazards	S
Morelot et al., 2021 [47]	Agricultural installation (fire risk);	H
Pietrafesa et al., 2020 [48]	General hazards; management	S
Proctor et al. 2007 [49]	Drive vehicle	H + E
Rapp et al., 2019 [50]	Management	L
Rahouti et al., 2021 [51]	Agricultural installation (fire risk);	S + L
Regent et al., 2013 [52]	Agricultural installation (electrical risk)	S + L
Somerkoski et al., 2020 [53]	Agricultural installation (fire risk);	S
Stothard et al., 2010 [54]	Agricultural installation (falling risk)	S
Torres et al., 2020 [55]	General hazards	S
Wang et al., 2020 [56]	Use of machinery	S

4 Discussion and Conclusions

The literature analysis performed demonstrates that there is a recent and increasing interest in the adoption of game-based training in several working sectors. Moreover, gaming has been previously applied to train workers on many types of risks and avoidance behaviors that may also be encountered in agriculture. Regarding the components of the working system, the Software was detected as the main investigated component, whereas Environment and Liveware (introduced as environmental details and NPCs, respectively) have been relatively less considered and in only five studies they were developed together [6, 17, 18, 20, 44]. Considering that the design of NPCs, the simulation of specific weather conditions and the addition of physical elements (e.g., smoke) represent successful factors in the development of more enjoyable and immersive training sessions [17], it may be particularly relevant to introduce these detailed elements in future educational games. Even though actions, dialogues and interactions with NPCs are preset, they can properly simulate the mental pressure exerted by supervisors on workers, help workers to establish an emotional link between the player, the other characters and the environment and help them learn and complete more complex tasks [17]. Moreover, the game environment has to suitably designed to allow the players to become more aware of their working environment, and it should be also able to recreate the real-life decision making processes to improve operators' capability to react to hazards in real situations [17, 24, 45]. In light of the current results and to create an engaging digital game for agricultural operators, some requirements able to influence the player motivation should be taken into account. Regarding the technology to be adopted, the computer game was identified as the most used and "practicable", especially if we consider that, in some users, VR technology may cause dizziness in case of extended use [6]. Regarding the game mechanics, a limited set of core game mechanics and interface elements are recommended, based on the analyzed studies. However, even though these game made extensive use of points, challenges and levels as game mechanics, future projects may implement some few other and more complex game mechanics solutions or some variations in tasks to sustains the workers' motivation, such as allowing the player to learn how to use some features or tools under the influences of external factors. As a final remark, it has to pointed out that due to the COVID-19 pandemic, the use of digital platforms has doubled and proved to be an effective instrument to train workers: the opportunities offered by these new training methods should be further investigated in agriculture and gaming platforms specifically addressing agricultural operators developed.

References

1. Gao, Y., Gonzalez, V.A., Yiu, T,W.: The effectiveness of traditional tools and computer-aided technologies for health and safety training in the construction sector: a systematic review. Comput. Educ. **138**, 101–115 (2019)
2. Burke, M.J., Sarpy, S.A., Smith-Crowe, K., Chan-Serafin, S., Salvador, R.O., Islam, G.: Relative effectiveness of worker safety and health training methods. Am. J. Public Health **96**, 315–324 (2006)

3. Leba, M., Ionică, A., Apostu, D.: Educational software based on gamification techniques for medical students. Recent Adv. Comput. Eng. Commun. Inf. Technol. 225–230 (2014)
4. Zichermann, G., Cunningham, C.: Gamification by design. O'Reilly Media, Inc., Sebastopol (2011)
5. Lovreglio, R., Duan, X., Rahouti, A., Phipps, R., Nilsson, D.: Comparing the effectiveness of fire extinguisher virtual reality and video training. Virtual Reality **25**(1), 133–145 (2020). https://doi.org/10.1007/s10055-020-00447-5
6. Liang, Z., Zhou, K., Gao, K.: Development of virtual reality serious game for underground rock-related hazards safety training. IEEE Access **7**, 118639–118649 (2019)
7. Fountas, S., et al.: A serious game for smart farming technologies. In: 12th EFITA International Conference, Rhodes Island, Greece, pp. 240–245 (2019)
8. Mazur, J.M., Meister, J., Hoagland, R.: Hazard Ridge: a serious gaming intervention for preventing injury to rural farm youth. In: Proceedings of CGAMES 2014 USA - 19th International Conference on Computer Games AI, Animation, Mobile, Multimedia, Educational and Serious Games, pp 118–120 (2014)
9. International Organization for Standardization (ISO): Ergonomics principles in the design of work systems. ISO 6385:2016(E), vol. 2016. Vernier, Geneva (2016)
10. Vigoroso, L., Caffaro, F., Micheletti Cremasco, M., Cavallo, E.: Innovating occupational safety training: a scoping review on digital games and possible applications in agriculture. Int. J. Environ. Res. Public Health **18**(4), 1868 (2021)
11. Hawkins, F.H.: Human Factors in Flight, 2nd edn. Routledge (1987)
12. International Labour Organization. (ILO): Safety and Health in Agriculture. Geneva, Switzerland: International Labour Organization (ILO) (2011)
13. Albert, A., Hallowell, M.R., Kleiner, B., Chen, A., Golparvar-fard. M.: Enhancing construction hazard recognition with high-fidelity augmented virtuality. J. Constr. Eng. Manag. **140**, 1–11 (2014)
14. All, A., Plovie, B., Nunez Castellar, E.P., Looy, J. Van.: Pre-test influences on the effectiveness of digital-game based learning: a case study of a fire safety game. Comput. Educ. **114**, 24–37 (2017)
15. Alshiar, M., Holtkamp, B., Biediger, D., Wilson, M., Yun, C., Kim, K.: SMACK: subjective measure of applied contextual knowledge. In: 2019 IEEE Games, Entertainment & Media Conference. IEEE, pp. 1–10 (2019)
16. Amal, G.K., Mclain, M.L., Bijlani, K., Jayakrishnan, R., Bhavani, R.R.: A novel approach for training crane operators. In: IEEE Eighth International Conference on Technology For Education, IEEE pp. 116–119 (2016)
17. Brown, L.D., Poulton, M.M.: Improving safety training through gamification: an analysis of gaming attributes and design prototypes. Adv. Intell. Syst. Comput. **780**, 392–403 (2018)
18. Bükrü, S., Wolf, M., Böhm, B., König, M., Teizer, J.: Augmented virtuality in construction safety education and training. In: 27th EG-ICE International Workshop on Intelligent Computing in Engineering 2020. Universitatsverlag Der TU, Berlin, pp. 115–124 (2020)
19. Burt, C., Crowe, L., Thomas, K.: Validation of a gamified measure of safety behavior: the SBT. In: Al, H. et al. (eds.) SBT Safety and Reliability–Safe Societies in a Changing World. Taylor & Francis Group, London, pp. 263–270 (2018)
20. Cai, P., Chandrasekaran, I., Cai, Y., Chen, Y., Wu, X.: Simulation-enabled vocational training for heavy crane operations. In: Cai, Y., Goei, S.L., Trooster, W. (eds.) Simulation and Serious Games for Education. GMSE, pp. 47–59. Springer, Singapore (2017). https://doi.org/10.1007/978-981-10-0861-0_4
21. Cavalcanti, J., Valls, V., Contero, M.: Fonseca, D: Gamification and hazard communication in virtual reality: a qualitative study. Sensors **21**(14), 4663 (2021)

22. Chodan, C., Mirza-Babaei, P., Sankaranarayanan, K.: Safety does not happen by accident, can gaming help improve occupational health and safety in organizations?. In: Duffy, V. (ed.) Digital Human Modeling. Applications in Health, Safety, Ergonomics, and Risk Management: Health and Safety. DHM 2017. Lecture Notes in Computer Science, vol. 10287. Springer, Cham. https://doi.org/10.1007/978-3-319-58466-9_29
23. Dawood, N., Miller, G., Pataca, J., Kassem, M.: Construction health and safety training: the utilisation of 4D enabled serious games. Spec. Issue BIM Cloud-Based Technol. AEC Sect.: Present Status Future Trends **19**, 326–335 (2013)
24. Dawood, N., Miller, G., Patacas, J., Kassem, M.: Combining serious games and 4D modelling for construction health and safety training Nashwan. Computing in Civil Building Engineering, pp. 2087–2094 (2014)
25. Dholakiya, N.D., Ferjencik, M., Schofield, D., Kubík, J.: Virtual learning for safety, why not a smartphone? Process Safety Prog. **38**, e12005 (2019)
26. Dickinson, J.K., Woodard, P., Canas, R., Ahamed, S., Lockston, D.: Game-based trench safety education: development and lessons learned. Electr. J. Inf. Technol. Constr. **16**, 118–132 (2011)
27. Din, Z.U., Gibson, G.E.: Serious games for learning prevention through design concepts: an experimental study. Saf. Sci. **115**, 176–187 (2019)
28. Eller, C., Bittner, T., Dombois, M., Rüppel, U.: Collaborative immersive planning and training scenarios in VR. In: Smith, I., Domer, B. (eds.) Advanced Computing Strategies for Engineering. EG-ICE 2018. Lecture Notes in Computer Science, vol. 10863, pp. 164–185 Springer, Cham (2018). https://doi.org/10.1007/978-3-319-91635-4_9
29. Gilotta, S., Spada, S., Ghibaudo, L., Isoardi, M.: A technology corner for operator training in manufacturing tasks. In: Bagnara, S., Tartaglia, R., Albolino, S., Alexander, T., Fujita, Y. (eds.) IEA 2018. AISC, vol. 824, pp. 935–943. Springer, Cham (2019). https://doi.org/10.1007/978-3-319-96071-5_96
30. Golovina, O., Kazanci, C., Teizer, J., König, M.: Using serious games in virtual reality for automated close call and contact collision analysis in construction safety. In: 36th International Symposium on Automation and Robotics in Construction ISARC, pp. 967–974 (2019)
31. González, N., Kalyakin, I., Lyytinen, H.: RACER: a non-commercial driving game which became a serious tool in the research of driver fatigue. Des Use Seri. Games **2010**, 171–184 (2008)
32. Gounaridou, A., Siamtanidou, E., Dimoulas, C.: A serious game for mediated education on traffic behavior and safety awareness. Educ. Sci. **11**(3), 127 (2021)
33. Greuter, S., Tepe, S.: Engaging students in OH&S hazard identification through a game. In: DiGRA 2013 - Proceedings of the 2013 DiGRA International Conference DeFragging Game Studies 2013, vol. 6 (2013)
34. Hafsia, M., Monacelli, E., Martin, H.: Virtual reality simulator for construction workers. In: Proceedings of the Virtual Reality International Conference - Laval Virtual, pp. 1–7 (2018)
35. Hall, J., Wyeth, P., Johnson, D.: Creating authentic experiences within a serious game context: evaluation of engagement and learning. In: Marsh, T., Ma, M., Oliveira, M.F., Baalsrud Hauge, J., Göbel, S. (eds.) JCSG 2016. LNCS, vol. 9894, pp. 55–66. Springer, Cham (2016). https://doi.org/10.1007/978-3-319-45841-0_5
36. Harper, S., Sivanathan, A., Lim, T., Mcgibbon, S., Ritchie, J.: Development of a mixed reality game for simulation based education. In: European Conference on Games Based Learning, Academic Conferences International Limited, pp. 212–217 (2008)
37. Hulme, K.F., Schiferle, M., Lim, R.S.A., Estes, A., Schmid, M.: Incorporation of modeling, simulation, and game-based learning in engineering dynamics education towards improving vehicle design and driver safety. Safety **7**(2), 30 (2021)

38. Jiang, M., Chang, J., Dodwell, M., Jekins, J., Yang, H.J., Zhang, J.J.: Game prototype for understanding safety issues of life boat launching process. In: 8th International Conference on Games and Virtual Worlds for Serious Applications (VS-GAMES) (2016)
39. Kamkuimo, S.A.K., Girard, B., Lapointe, P., Menelas, B.-A.: Design and implementation of a serious game to make construction workers aware of exposure to silica dust in the workplace. In: Ma, M., Fletcher, B., Göbel, S., Baalsrud Hauge, J., Marsh, T. (eds.) JCSG 2020. LNCS, vol. 12434, pp. 85–98. Springer, Cham (2020). https://doi.org/10.1007/978-3-030-61814-8_7
40. Kuindersma, E., van der Pal, J., van den Herik, J., Plaat, A.: Building a game to build competencies. In: Dias, J., Santos, P.A., Veltkamp, R.C. (eds.) GALA 2017. LNCS, vol. 10653, pp. 14–24. Springer, Cham (2017). https://doi.org/10.1007/978-3-319-71940-5_2
41. Kwegyir-Afful, E., Kantola, J.: Simulation-based safety training for plant maintenance in virtual reality. In: Cassenti, D., Scataglini, S., Rajulu, S., Wright, J. (eds.) Advances in Simulation and Digital Human Modeling. AHFE 2020. Advances in Intelligent Systems and Computing, vol. 1206, pp. 167–173. Springer, Cham (2021). https://doi.org/10.1007/978-3-030-51064-0_22
42. Lanzotti, A., et al.: Interactive tools for safety 4.0: virtual ergonomics and serious games in real working contexts. Ergonomics **63**, 324–333 (2020)
43. Leong, P., Goh, V.: REAPSG: work safety and health games for construction sector. 2013 IEEE International Games Innovation Conference. IEEE, pp. 134–137 (2013)
44. Mallett, L., Orr, T.J.: Working in the classroom - a vision of miner training in the 21st Century. In: First International Future Mining Conference and Exhibition 2008, Proceedings, pp. 83–89 (2008)
45. Matsas, E., Vosniakos, G.-C.: Design of a virtual reality training system for human–robot collaboration in manufacturing tasks. Int. J. Interact. Des. Manufact. (IJIDeM) **11**(2), 139–153 (2015). https://doi.org/10.1007/s12008-015-0259-2
46. Mayer, I., Wolff, A., Wenzler, I.: Learning efficacy of the 'hazard recognition' serious game. In: Ma, M., Oliveira, M.F., Petersen, S., Hauge, J.B. (eds.) Serious Games Development and Applications. SGDA 2013. Lecture Notes in Computer Science, vol. 8101. Springer, Heidelberg (2013). https://doi.org/10.1007/978-3-642-40790-1_12
47. Morélot, S., Garrigou, A., Dedieu, J., N'Kaoua, B.: Virtual reality for fire safety training: Influence of immersion and sense of presence on conceptual and procedural acquisition. Comput. Educ. **166** (2021)
48. Pietrafesa, E., Bentivenga, R., Lalli, P., Capelli, C., Farina, G., Stabile, S.: Becoming safe: A serious game for occupational safety and health training in a WBL Italian experience. In: Advances in Intelligent Systems and Computing 1236 AISC, pp. 264–271 (2021)
49. Proctor, M.D., Bauer, M., Lucario, T.: Helicopter flight training through serious aviation gaming. J. Defense Model. Simul. **4**, 277–294 (2007)
50. Rapp, J., Rose, J., Narciss, S., Kapp, F.: How to set the game characteristics, design the instructional content and the didactical setting for a serious game for health prevention in the workplace. In: Gentile, M., Allegra, M., Söbke, H. (eds.) GALA 2018. LNCS, vol. 11385, pp. 210–220. Springer, Cham (2019). https://doi.org/10.1007/978-3-030-11548-7_20
51. Rahouti, A., Lovreglio, R., Datoussaïd, S., Descamps, T.: Prototyping and validating a non-immersive virtual reality serious game for healthcare fire safety training. Fire Technol. **57**(6), 3041–3078 (2021). https://doi.org/10.1007/s10694-021-01098-x
52. Regent, P., Raimbault, P., Marty, A.: "Serious game" serving the professionalization of elecricity technicians. In: 22nd International Conference and Exhibition on Electricity Distribution, Stockholm, vol. 1485, pp. 5–7 (2013)

53. Somerkoski, B, Oliva, D, Tarkkanen, K, Luimula, M.: Digital learning environments - constructing augmented and virtual reality in fire safety. In: ACM 11th International Conference on E-Education, E-Business, E-Management, and E-Learning, pp. 103–108 (2020)
54. Stothard, P., Hengel, A. Van Den: Development of serious computer game based training module and its integration into working at heights mine site induction — Part I. Min. Technol. **119**, 68–78 (2010)
55. Torres, A., et al.: A 360 video editor framework for interactive training. In: 2020 IEEE 8th International Conference on Serious Games and Applications for Health, pp. 1–7 (2020)
56. Wang, T., Ahmad, R.: Design of a new game for teaching assembly process. In: Hernandez, E.E., Keshtkar, S., Valdez, S.I. (eds.) LASIRS 2019. MMS, vol. 86, pp. 44–52. Springer, Cham (2020). https://doi.org/10.1007/978-3-030-45402-9_6

Track Width for a Safe Travelling of a Timber Tractor-Trailer Combination

Gianfranco Airoldi, Mario Tamagnone, Marco Manzone, and Angela Calvo(✉)

DISAFA (Department of Agricultural, Forest and Food Sciences and Technologies), Largo P. Braccini 2, 10095 Grugliasco, Turin, Italy
angela.calvo@unito.it

Abstract. In Italian Alpine Regions, wood transportation from the forest to local terminals is commonly practiced using low powered tractors with trailers (equipped with single axis or two axes) on steep downhill tracks, that must be wide enough when cornering. Aim of the work was to study and to define a calculation model to detect the width of the turning carriageway as a function of the turning radius of the vehicle used in the forestry sector for timber transportation.

When a tractor-trailer system turns, four trajectories are plotted, drawn by the wheels of the tractor and by the wheels of the trailer. The overall width of the carriageway to make the turn possible is given by parameters depending on some vehicle characteristics, and it can be summarized as the difference between the radius of the circumference described by the external front wheel of the tractor and the radius of the circumference traced by the innermost wheel of the trailer. The track width is therefore function of two parameters: the distance between the hitch and the rear axle of the tractor, and the distance between the towing eye and the trailer axles.

The calculation model studied in this work identifies the internal radius of the curve and the width of the carriageway necessary for the transit of a tractor-trailer system. A correct design of the forest track accordingly to the dimensional needs of the vehicles is essential for a safe driving, especially in the case of the downhill timber transportation.

Keywords: Turnout width · Timber transportation · Driver safety · Forestry road

1 Introduction

In the Italian hilly and mountain forest areas, the wood transportation from the trees felling points to the local terminal is commonly practiced using low power European 4 WD farm tractors with front wheels with a smaller diameter than the rear wheels and with an Ackermann steering system (named FWA in the USA). The Ackermann steering system provides for the presence on the steering axle of a kinematics - between the actuator connected to the driving system (steering wheel) and the wheels of the steering axle - which allows a progressive reduction of the steering angle of the outer wheel compared to the inner wheel. As a consequence, the wheels can travel (in the

M. Biocca et al. (Eds.): SHWA 2020, LNCE 252, pp. 442–448, 2022.
https://doi.org/10.1007/978-3-030-98092-4_45

direction changes) concentric circumferences keeping the position of the steering axle constant with respect to the vehicle chassis.

These machines tow short log trailers downhill along steep and rough forest roads. Even though they mostly are low-volume traffic roads, these routes are the only access to the woodland.

Forest management needs moreover an effective road network. It not only provides the access to the forest for the timber harvesting, but it permits also the necessary operations for the fire protection, the grazing and the land management. On the other hand, forestry roads may change the environment they cross and, if they are not properly constructed and maintained, may be cause of soil erosions and landslips [1].

It is possible to build different woodland roads, the surface shapes and the characteristics depending on the terrain and the management objectives [2].

Concerning the management objectives, the characteristics of the timber machines that travels these routes are also important. Giummarra and Blanksby [3] provided practical guidelines to address requests for tractor-trailer access to forestry roads, providing some criteria to match specific woodland roads to vehicle types.

If machineries are an important cost item to define the volume of timber that can be extracted, the woodland roads used to transport the timber from the forest to the asphalt roads are also an important part of the total costs [4]. It is possible to reduce the cost of the woodland road construction by an effective design, which must include slopes, distances, soil removal, as well as a correct curve width required by the transporting machines [2].

The design of a correct curve width is an important point for the roadway safety, because an incorrect road design that not consider the turning width may result in dangerous off-tracking wheels of the travelling vehicles [1]. In case of too narrow curves, other than the machine overturning risk [5], the off-tracking wheels may moreover erode the road shoulders, weakening the road.

It is therefore necessary to know if the road is wide enough when cornering, and this fact may be critic when the machine is an agricultural tractor-trailer vehicle.

The turning radius of the machine and the steering angle of the wheels are for this purpose important information, often written in the machine manual. The steering of an agricultural tractor is an Ackermann type, where the higher steering wheel is the front inner, and the other front wheel is bound by a higher turning radius (the steering angle in this case is consequently lower).

The four wheels describe on the ground four trajectories, which depend on the construction characteristics of the tractor (length, width, steering) and on the driver's behavior (curve setting, driver skill, Fig. 1).

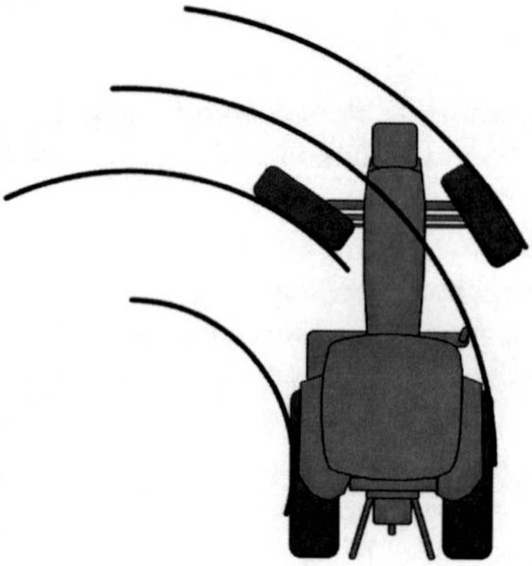

Fig. 1. Trajectories described on the ground by the wheels of an agricultural tractor when steering.

The situation is different for the tractor-trailer combination, because the manuals do not give information about the steering of the tractor-trailer system.

It is here necessary to consider many parameters: the trailer connection system (hitch), the trailer coupling system (fifth wheel or tow bar), the different drivers' behaviors. Forestry trailers may have one (without fifth wheel), two or three axles, but in the Italian alpine forest road framework they may be one axle or tandem, as they are better suited for the transportation of medium-sized logs even on steep forestry tracks.

The preliminary objective of this work was to study and to define a theoretical calculation model for the determination of the width of the turning carriageway as a function of the turning radius of a tractor-trailer combination used in the Alpine forestry sector for the timber transportation.

2 Material and Methodology

The width of the track when cornering depends on the size and on the steering of the crossing vehicle.

To describe the method, it is therefore necessary to start from the features of the tractor-trailer system and from the behavior of the vehicle when turning.

2.1 Curves Travelled by an Agricultural Tractor

Analyzing the right-angled triangles formed by the radii of the circles tangents to each wheel and the wheelbase of the tractor, it is possible to model the turning radius (the radius of the largest turning circle - the path of the outer front wheel) in function of the tractor dimensions and steering angle of the inner front wheels (Table 1 and Fig. 2).

The steering angle (α), the wheelbase (P) and overall width (L) of an agricultural tractor are the parameters necessary for the calculation of the turning radius of the outer front wheel (Tr) of an agricultural tractor with an Ackermann steering (Fig. 2).

Table 1. Requested parameters for modelling the turning radius with a tractor crossing.

Parameter	Measurement unit
Maximum steering angle described by the inner front wheel	°
Wheelbase	m
Overall width (tyres included)	m

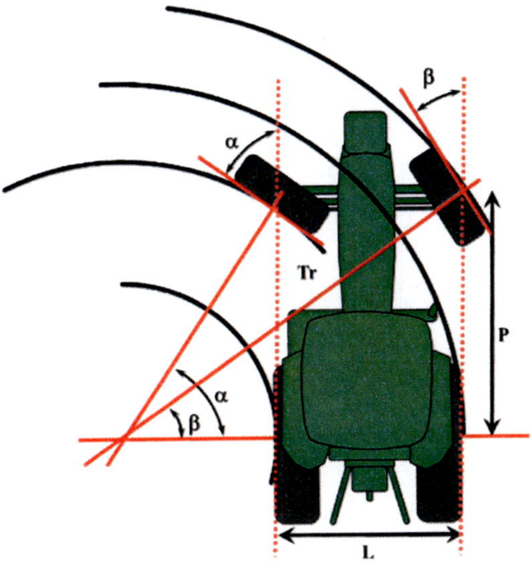

Fig. 2. Steering angle (α), wheelbase (P) and total width (L) of an agricultural tractor for the calculation of the turning radius (Tr) of an agricultural tractor with an Ackermann steering.

2.2 Curves Travelled by an Agricultural Tractor Linked to a Single-Axle or a Tandem Trailer

The tractor's maximum steering capacity is however never used when towing a trailer.

When a tractor-trailer system turns, six trajectories are plotted, depending on the tractor and trailer track values.

The overall width of the carriageway to make the turn possible and safe is here given by the difference between the radius of the circumference described by the external front wheel of the tractor and the radius of the circumference traced by the innermost wheel of the trailer.

To calculate the radius of this last circumference, two parameters are necessary: the distance between the hitch and the rear axle of the tractor, and the distance between the towing eye and the trailer axle (or the midpoint between the trailer axles in the case of a tandem trailer) (Fig. 3).

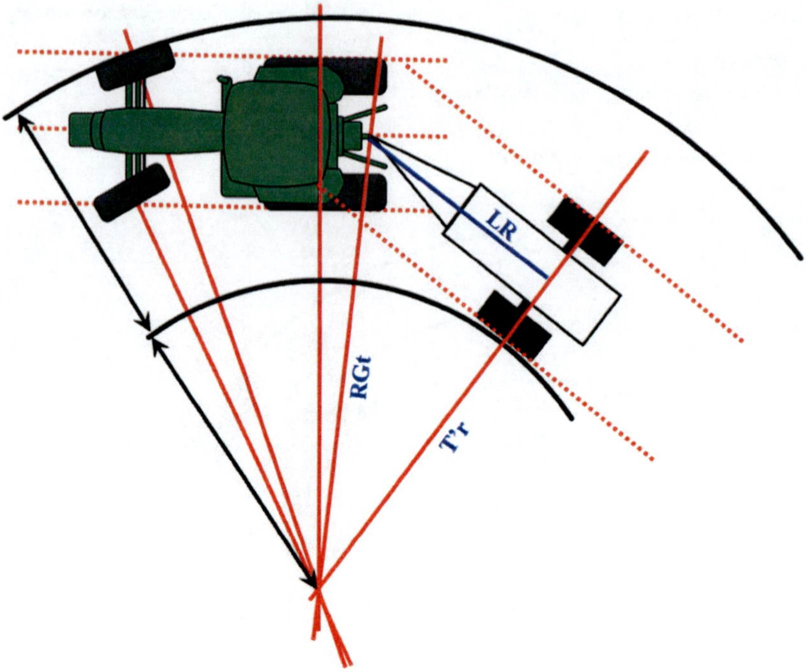

Fig. 3. Curves travelled by a tractor-trailer system (trailer with single axle)

3 Results and Discussion

The first modelled system regards the turning radius of the tractor. Analyzing the triangles formed by the tangents to the steering wheel and by the parameters of the tractor (Fig. 2), it is possible to model the minimum turning radius of the road when the agricultural tractor steers (Eq. 1).

$$Tr = \sqrt{\frac{P^2}{sin^2 a} + L^2 + 2LPcotga} \tag{1}$$

where:

Tr: turning radius (m).

P: wheelbase (m).

L. total width (m).

α: steering angle (°).

For example, with $\alpha = 55°$, using a 96 kW farm tractor (Fendt 700 Vario), P = 2.7 m, L = 2.5 m, the turning radius Tr is 5.17 m.

In the case of a tractor-trailer vehicle, the overall width of the carriageway to make the turn possible is here given by the difference between the radius of the circumference described by the external front wheel of the tractor and the radius of the circumference traced by the innermost wheel of the trailer.

To calculate the radius of this last circumference, two parameters are necessary: the turning radius followed by the hitch, and the distance between the towing eye and the trailer axle (or the midpoint between the trailer axles in the case of a tandem trailer).

The equation is therefore the following (Eq. 2).

$$T'r = \sqrt{RGt^2 - LR^2} - \frac{L}{2} \tag{2}$$

where:

T′r: turning radius of the inner wheel of the trailer (m).

RGt: turning radius followed by the hitch (m).

LR: distance between the towing eye and the trailer axle (m).

L: total width of the vehicle (m).

For example, with the same previous tractor and a Terpa trailer with RGt = 6 m, LR = 3.4 m and L = 2.5 m, we obtain a turning radius T'r equal to 3.7 m. To have a turning radius of the hitch of 6 m, the tractor must curve with a steering angle of 30 degrees and therefore with a turning radius of 7.7 m. The dynamic width of the combination is 4 m (7.7 m–3.7 m), with an increase in the width of the carriageway with respect of the total width of the vehicle of 1.5 m (4 m–2.5 m).

Equations 1 and 2 may work if there are perfect boundary conditions. As a matter of fact, things may also be done differently. For example, the variability of the position of the tractor-trailer system during the drive along the curve should be considered.

The movements of the drawbar, the towing eye and the steering, as well as the tire-ground interaction and the slippage due to the load on the trailer should be taken into account. Moreover, starting with a certain steering angle and going straight with the tractor, it may take a lot of space before the trailer is aligned.

The operator's behaviour, his experience, know-how and skill should be taken into account as well.

All these aspects need to be verified in a future work.

4 Conclusions

Traveling along forest roads during timber transportation increases the risk of accidents for the driver, especially along sloped and narrow curves. A correct design of the forest track according to the dimensional needs of the vehicles is essential for safe driving, especially in the case of timber transportation.

Forest roads do not provide size limit signs, and the driver has the responsibility to decide if it is possible to travel with his convoy (tractor and trailer) in function of the road characteristics. The driver must safely travel especially in proximity of the curves, where the trailer could have problems with trajectories and bearing capacity of the ground. It is therefore important to know the turning radius of all the hairpin bends of the road and compare the smaller with the convoy characteristics.

The starting models suggested in this work permit to calculate the width of the curve, in function on the construction characteristics of the tractor-trailer combination and on the different steering angles.

The proposed models consider only the ideal trajectory of the convoy and do not take into account possible drifts due to any poor tire grip. In order to improve the models, experimental tests are planned to verify the influence of other variables (tire grip, trailer drive axles, frame structure of the trailer, and non-conventional tractor).

References

1. Schiess, P., Whitaker, C.A.: Road Design and Construction in Sensitive Watersheds. FAO, Rome (1986)
2. Bowers, S. Adams, P.: Managing woodland roads: a field guide, p. 104. Oregon State University PNW 641 (2017)
3. Giummarra, G., Blanksby, C.: Engineering Requirements for logging truck operations on forest roads (Research report) Paperback. ARRB Group Ltd. (2006)
4. Epstein, R., et al.: A combinatorial heuristic approach for solving real-size machinery location and road design problems in forestry planning. Oper. Res. **54**(6), 1017–1027 (2006)
5. Facchinetti, D., Santoro, S., Galli, L.E.: Pessina, D: Tractor roll-over related fatalities in italy: results from a 12 years analysis. Sustainability **13**, 4536 (2021)

ROPS Design Guidelines for the Effective Driver's Protection in Case of Tractor Overturning

Domenico Pessina(✉), Lavinia Eleonora Galli, and Davide Facchinetti

Department of Agricultural and Environmental Sciences - Production, Landscape, Agroenergy, University of Milan, Via Celoria 2, 20133 Milan, Italy
domenico.pessina@unimi.it

Abstract. For over 60 years, the Roll Over Protective Structure (ROPS) has been the most diffused means to reduce the operator's risk in case of overturning of the agricultural and forestry tractors. The most important requirement provided is the protection of a given volume (named "clearance zone" or "deflection-limiting volume", depending on the standard considered), located at the driver's place, provided that he/she is properly fastened with a seat belt. The manufacturers design the ROPS in such a way to satisfy the shown requirement playing on two factors, i.e. the strength of the ROPS and its overall dimensions. In other words, the required energy could be absorbed mostly in terms of strength or mostly playing on the deflection. The first case represents a forced solution for the narrow tractors, where the available volume is restricted, and then it is necessary to build a very rigid ROPS to assure limited deflections. The second option is more frequently adopted on open-field tractor models, on which generally there are no serious problems in respecting the clearance zone, even if the deflection is remarkable.

The ROPS test results of the last 15 years activity of the Italy-Milan Testing Station were investigated, to examine the elastic and plastic deflection values, to fix the manufacturers tendency when designing the ROPS. Moreover, once the clearance zone location was defined, its effective protection was verified in terms of "margin", i.e. the minimum distance of the ROPS structural elements from the clearance zone contour.

The results highlighted that ROPS show generally a good balance between the elastic (temporary) and the plastic (permanent) deflection values when they both range between 40 and 60% of the total deflection, even if, in some cases, very rigid ROPS models were detected, showing a very low elastic deflection.

Keywords: OECD codes · Clearance zone · Strength · Deflection

1 Introduction

Operators' safety is one of the most important items to be considered in agriculture: despite the overall risk level recently decreasing, the agricultural sector still suffers at present several serious and even fatal injuries. Agricultural machinery, especially

M. Biocca et al. (Eds.): SHWA 2020, LNCE 252, pp. 449–456, 2022.
https://doi.org/10.1007/978-3-030-98092-4_46

tractors, represent important occupational hazard factors, causing very severe injuries. Statistics show that tractor drivers are exposed to grip and dragging from motion transmission elements, falls to the ground getting on or off from the driving place, crushing while hooking up or disconnecting tools, inhalation of exhaust gases, burns for hot surface contact, high levels of noise and vibrations and incongruous postures. But the most serious hazard is the **tractor overturning** (lateral, longitudinal or both, combined) due to vehicle overload and/or very high pull force and/or excessive ground slope [1].

ROPS (Roll Over Protective Structure) is still now the most diffused device to reduce the operator's risk in the case of agricultural and forestry tractor overturning [2]. ROPS became popular in the 50s, starting their application in Scandinavian countries, where dedicated studies established the energy levels to be applied to simulate the mechanical stress produced in case of a rollover accident occurred in well-defined conditions. This information represented later the basis for the most popular International Standards (issued by OECD, EU, ISO and other international organisations) currently in force to test ROPS strength. The first official Standard Code for the official testing of protective structures on agricultural tractors was approved on 21st April 1959 by the OEEC Council, then turned in OECD (Organization for Economic Co-operation Development). Since then, the various OECD codes were also applied to forestry tractors, to determine other performance characteristics of safety and noise [3, 4]. All ROPS testing methods are based on a rollover simulation through blows (with a mass equipped with pendular motion) or horizontal loadings (applied with a hydraulic cylinder) and crushing tests. The resulting energy absorbed by the ROPS is quantified vs the mass of the tractor on which the protective structure is fitted.

In Italy, since 1st January 1974, each new open-field wheeled tractor must be fitted with a ROPS (cab or frame, front or rear roll-bar), being it a structure of welded and bolted steel tubes, plates and sheets of various dimensions and thickness, fitted around the driving place, tested and then homologated according to dedicated international Standards. In the following years, further specific standards were issued for specialized tractors (i.e. those narrow tracked, mainly used in vineyard and orchards); at a later time, also tracklaying tractors were equipped with tested ROPS. To comply with the technical legislation, ROPS manufacturers have to check their prototypes through strength tests, to be carried out in the officially qualified and accredited laboratories. At present, the static tests procedure is that mostly applied worldwide (Fig. 1); in this case, to be considered positively, the only requirement for a ROPS is the compliance with a safety volume (clearance zone) arranged around the seat and the steering wheel, in case of overturning.

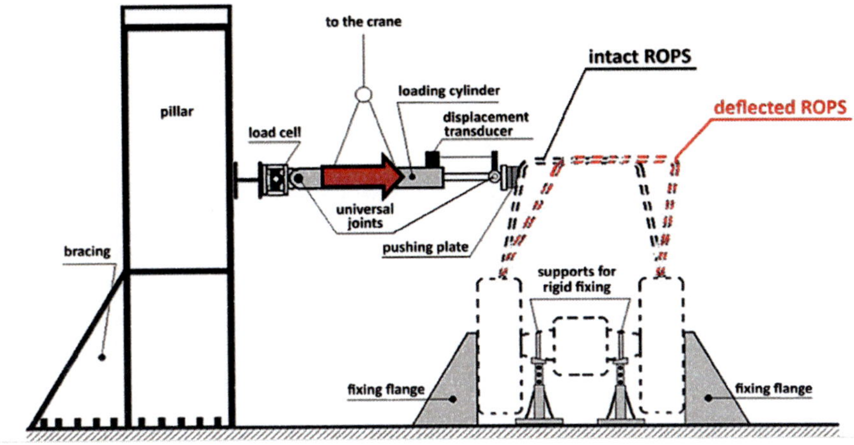

Fig. 1. Sketch of the execution of a ROPS static (lateral) test.

The primary purpose of the present study is to examine the total, elastic and plastic deflection values recorded in a series of strength tests, to fix the trend followed by the manufacturers when designing the ROPS. Moreover, once defined the clearance zone location, its effective protection was verified in terms of "margin", i.e. the minimum distance of the ROPS structural elements from the clearance zone contour.

2 Materials and Method

The most important requirement provided for a ROPS is the complete protection of a given volume (named "clearance zone" or "deflection-limiting volume"), located at the driver's place, provided that he/she is properly fastened with a seat belt (Fig. 2).

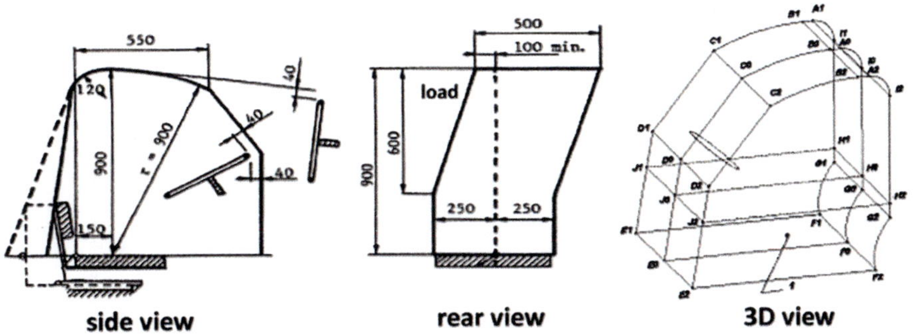

Fig. 2. Views of the "clearance zone" provided by the OECD Code 4, for the testing of ROPS fitted on open-field tractors.

Each longitudinal test can be considered as positive if, having absorbed the entire energy defined by dedicated formulae (the energy is resulting from the integral of the curve derived from the loading force and the total deflection data), the volume of the clearance zone is respected, i.e. the structural members do not enter into the volume. In this condition, the volume does not remain unprotected in respect to the ground in case of tractor overturning.

To satisfy this requirement, the manufacturers are playing both on the strength of the ROPS and its overall dimensions [5, 6]. *In fact, the required energy of the tests could be absorbed mostly in terms of strength or mostly playing on the deflection of the ROPS.*

The results obtained from lateral loading of some strength tests on ROPS (cabs) carried out in the last 15 years activity of the Italy-Milan Testing Station were investigated, to examine the Elastic (De) and Plastic (Dp) deflection values, to fix the manufacturers tendency when designing the ROPS, in terms of the "**elasticity ratio**" (**De/Dp**) (Fig. 3).

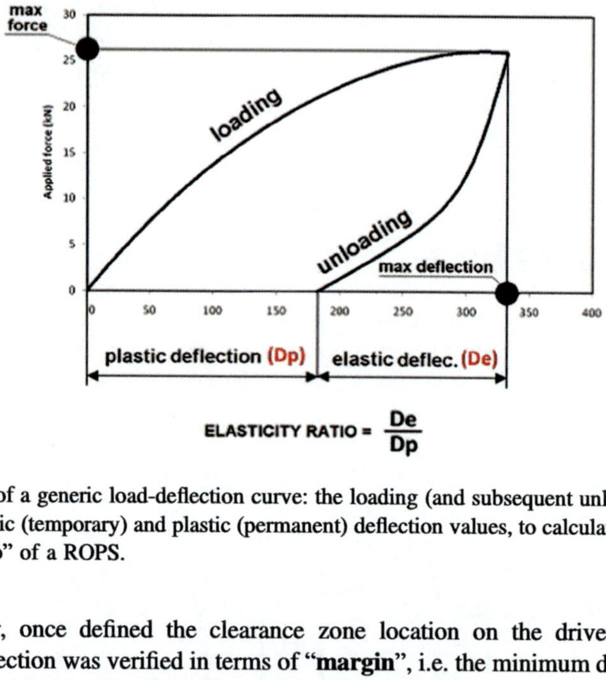

Fig. 3. View of a generic load-deflection curve: the loading (and subsequent unloading) curves define the elastic (temporary) and plastic (permanent) deflection values, to calculate the so-called "elasticity ratio" of a ROPS.

Moreover, once defined the clearance zone location on the driver's place, its effective protection was verified in terms of "**margin**", i.e. the minimum distance of the ROPS structural elements from the clearance zone contour, to be measured at the maximum deflection, i.e. usually at the end of the loading (Fig. 4). Both the results of elasticity ratio and margin values were classified into two different tractor categories on which each ROPS was fitted, named "open fields" and "narrow track". The technical feature to classify the models taken into consideration was referred to what stated into the OECD Codes, i.e. the tractor minimum track value. If the tractor has at least the minimum track of one of the two axles below 1150 mm, it is defined as "narrow" (so typically used in vineyards/orchards); viceversa, it is included in the open field category if has both axle tracks exceeding the limit.

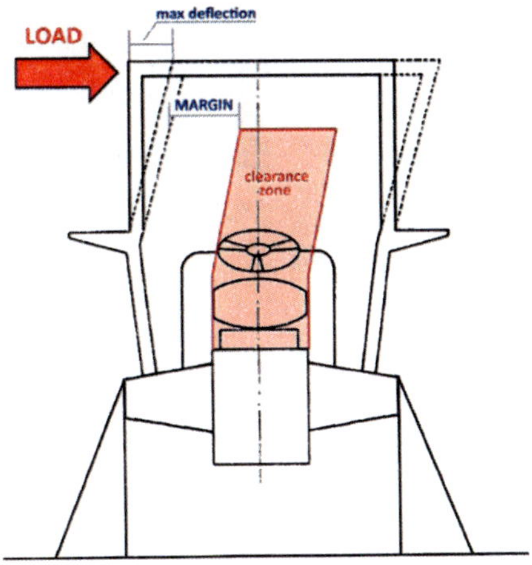

Fig. 4. The "margin" is defined as the minimum distance of the ROPS structural elements from the clearance zone contour, at the maximum deflection recorded at the end of the loading curve.

3 Results and Discussion

3.1 Elasticity Ratio

In Figs. 5 and 6 the results of Dp/De ratio are shown, divided for open-field and narrow-track tractors vs their mass. This tractor classification was considered at a first extent significant, considering that on narrow tractors (those used in orchards and vineyards) the limited volume available forces the designers to create very robust cabs, able to absorb energy mostly in terms of force rather than deflection. Conversely, on open field tractors the volume is higher, and therefore they can play better on the deflection rather than on the strength (i.e. the resistance of the ROPS).

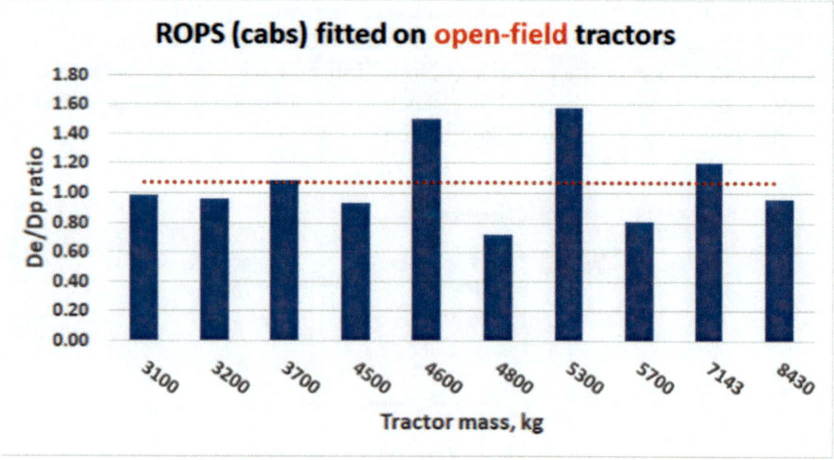

Fig. 5. Elasticity ratio for the tested cabs fitted on open-field tractors.

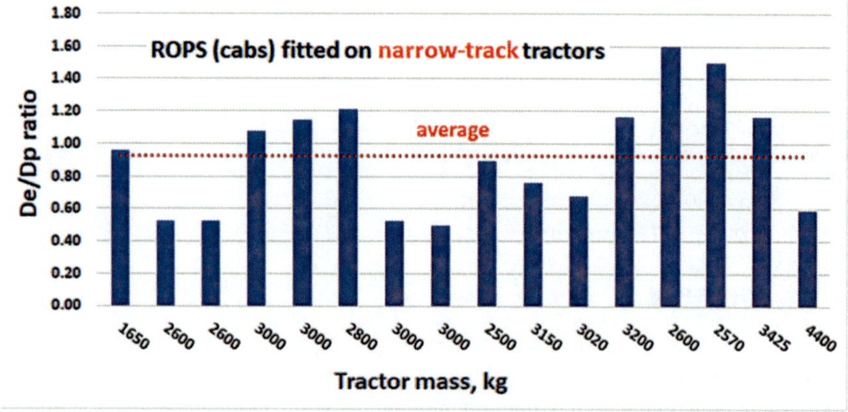

Fig. 6. Elasticity ratio for the tested cabs fitted on narrow-track tractors.

In general, the De/Dp elasticity ratio in the side loading showed a wide variation. In detail:

- no relationship was found considering the tractor mass;
- for open-field tractors, the elasticity ratio ranged from 0.72 to 1.58, with an average of 1.07;
- for narrow-track (vineyard and orchard) tractors the elasticity ratio ranged from 0.50 to 1.60, with an average of 0.93.

On the open-field tractors, the average of the elasticity ratio is higher than that of the narrow-track category; in other words, in proportion, the elastic (temporary) deflection is higher than the plastic (permanent) deflection. This is confirming the hypothesis, i.e. the higher overall width of the open-field tractors, and therefore of the cabs fitted, is allowing a higher elastic deflection of the ROPS, at the full load, without infringing the clearance zone.

3.2 Margin

As expected, the "margin" values recorded (in the side loading) for conventional tractors are remarkably higher than those of narrow-track tractors, so confirming that the width of the volume available for the driver on the open-field models is higher (Fig. 7).

Conversely, in some cases, the margin of some cabs fitted on narrow-track tractor is minimum (few millimeters), so evidencing a non optimal situation about the driver's protection in case of serious overturning.

In these critical cases, the stiffness of the ROPS has to be remarkably increased, to assure effective protection of the clearance zone. This lead to a strong challenge for the ROPS designers, that often have to take into account also the use of high quality steel, which offers high stress-strain values.

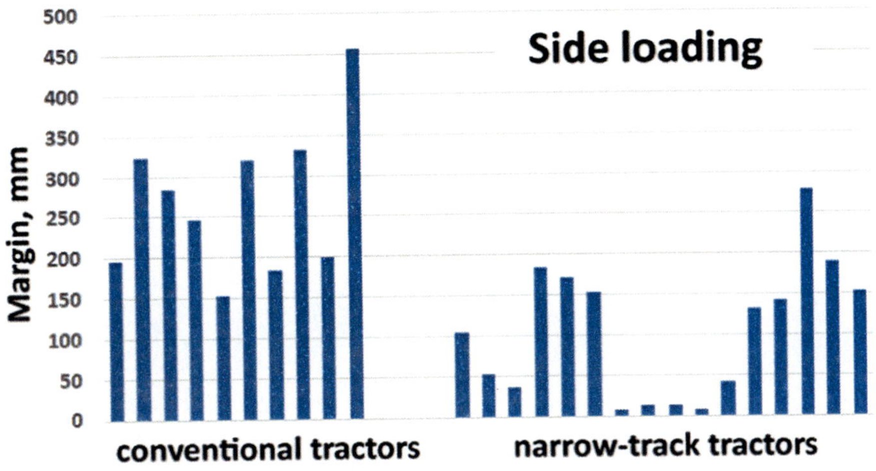

Fig. 7. "Margin" values recorded in the side loading of some strength tests carried out both on open-field (conventional) and narrow-track tractors.

4 Conclusions

The overall size of the driver's place, particularly its width, is affecting the design criteria of the ROPS (cabs) fitted on agricultural tractors. When the tractor is narrow, to assure the respect of the clearance zone the cab has to be very stiff, so absorbing the energy defined by the standards most in terms of force (resistance) rather than in terms of total deflection (elastic plus plastic). On the contrary, for open-field tractors, a less stringent constraint could be applied, so playing more on the deflection rather than the force.

In this view, to evaluate the possible impact of structural modifications on the strength of a ROPS, and therefore their role in changing the force/deflection curves data, the Finite Element Analysis (FEA) technique has been by time profitably adopted, then using as a reference the most popular software tools, (such as Ansys, Nastran, Msc Pal 2, etc.).

References

1. Pessina, D., Facchinetti, D., Belli, M.: Definition of unambiguous criteria to evaluate tractor rops equivalence. In: CIGR XVIIth World Congress, Quebec (Canada), June 2010
2. Reynolds, S.J., Groves, W.: Effectiveness of roll-over protective structures in reducing farm tractor fatalities. Am. J. Prev. Med. **18**(4 Suppl.), 63–69 (2000). https://doi.org/10.1016/s0749-3797(00)00142-2
3. OECD Code 4: OECD Standard Code for the official testing of protective structures on agricultural and forestry tractors (static test), OECD, Paris (2021)
4. OECD Code 7: OECD Standard Code for the official testing of front mounted roll-over protective structures on narrow-track agricultural and forestry tractors, OECD, Paris (2021)
5. Harris, J.R.: Predicting the performance characteristics of assembled rollover protective structure designs for tractors. Graduate theses, Dissertations, and Problem Reports. 2838 (2008). https://researchrepository.wvu.edu/etd/2838
6. Chetwin, J.M.: Approved code of practice for Roll Over Protective Structures on tractors in agricultural operations. Occupational Safety and Health Service Department of Labour, Wellington, New Zealand (2001)

Survey on the Lateral Stability of Wheeled Narrow-Track Tractors

Domenico Pessina[1(✉)], Davide Facchinetti[1], Lavinia Eleonora Galli[1], Carlo Bisaglia[2], and Maurizio Cutini[2]

[1] Department of Agricultural and Environmental Sciences - Production, Landscape, Agroenergy, University of Milan, Via Celoria 2, 20133 Milan, Italy
domenico.pessina@unimi.it

[2] CREA Research Centre for Engineering and Agro-Food Processing, Via Milano 43, 24047 Treviglio, BG, Italy

Abstract. Narrow-track tractors are spread worldwide, being used in important agricultural sectors such as vineyards, orchards, horticultural cultivation, greenhouses and nursery. The architecture of these tractors and their basic structure have remained practically unchanged in the last decades. Despite this, their construction can differ considering the origin of the design: (1) models coming from the traditional open-field tractors, having rigid body and front and rear wheels of different dimensions; (2) models derived from the evolution of the walking tractor, having both rigid or articulated body and iso-diametric wheels. In addition to the design, in case of lateral overturning the behaviour of a narrow tractor could be affected by many other features, such as the wheels dimension and inflation pressure and the fitting of important accessories. This paper is aimed to show the limit angles of the static lateral stability of some narrow-track tractors, adopting different testing procedures.

Keywords: Safety · Overturning · Stability limit angle · OECD Code 6

1 Introduction

Agricultural work is one of the most hazardous jobs, as it ranks among the top in work injury statistics. The fatality rate of such injuries is six times higher than the rate of all industries combined. In addition, concern about the growing number of leisure-related farm injuries is arising as well. Recently (2013) figures concerning the burden of these injuries in the EU countries (EU15) showed an average mortality rate of 13 deaths per 100,000 farm workers; this also occurs in the United States, where an average rate of 22 deaths per 100,000 workers is recorded. In both Regions, peaks of more than 30 deaths per 100,000 workers were recorded as well [1, 2]. Statistics referring to Italy reported tractors as the main cause both of injuries and fatal accidents; as a matter of fact, 56.5% of the total number of accidents in agriculture and forestry has been related to tractor tasks. Within this framework and considering only the fatal accidents in agriculture and forestry operations, concern arises as 51% of these happened while workers were riding tractors (75% located on field and 25% when driving on roads). As

M. Biocca et al. (Eds.): SHWA 2020, LNCE 252, pp. 457–467, 2022.
https://doi.org/10.1007/978-3-030-98092-4_47

far as accident dynamics are concerned, machine overturning represents the 77% of the accidents, while those involving the PTO shaft account for 0.7% [3, 4].

The reason for the tractors overturning involves several factors, as the location of additional loads (ballast, implements), poor visibility and/or awareness of the terrain on which is operating, and/or burrows or ditches. All these factors may jeopardize the machine's stability even when standing still or moving slowly. The loss of machine stability cannot be effectively prevented and the possible driver's error in tilt angle estimate could make any recommendations about the maximum admissible operating gradient vain [5, 6]. The literature contains many studies on the stability of agricultural vehicles and related risks for operators [7–16]. These studies have been mainly focused on the determination of the risks and of the methods for testing the lateral stability. Most of them related to vehicles able to lift objects or workers, but very few address a method to characterize tractor's stability. Moreover, the problem of the tractor stability – especially lateral – is accentuated in the so-called "narrow tractors", generally designed to work between rows of tree crops and vineyards. For this reason, they are similarly considered as specialized: in detail, a tractor can be considered as "narrow" when at least one of the two tracks is not exceeding 1150 mm [17, 18], so making them very susceptible to side tipping.

At present, before the strength test, the OECD Code 6 ("OECD Standard Code for the Official Testing of Front Mounted Roll-Over Protective Structures on Narrow-Track Agricultural and Forestry Tractors) [17] provides two preliminary tests:

- the static lateral stability on a 38° slope test;
- the non-continuous rolling on a 1/1.5 gradient test (normally assessed through a computer simulation).

For the first preliminary test, at point 3.1.3.2, the Code 6 states: *"Using a jack or a hoist, tilt the part of the tractor which is rigidly connected to the axle that bears more than 50% of the tractor's weight, while constantly measuring the angle of inclination. This angle must be at least 38° at the moment when the tractor is resting in a state of unstable equilibrium on the wheels or tracks touching the ground. Perform the test once with the steering wheel turned to full right lock and once with the steering wheel turned to full left lock"*. Thus, Code 6 allows the adoption of both methods to ascertain the static stability of the tested tractor. In the "jack" method, one of the two wheels of the tractor axle that bears more than 50% of its weight is placed on a tilting platform; then, this axle is inclined up to an angle of 38° in respect to the horizontal. The tractor must remain stable at this slope, i.e. the upstream wheel must remain in touch with the platform surface. The test has to be repeated in two conditions, i.e. with the steering wheel turned both to full left and right lock; for carrying out the "hoist" method, one of the wheels of the axle carrying more than 50% of the total mass of the tractor is raised using a crane; the tractor must remain stable at an angle of 38° of the raised axle, i.e. the downstream wheels maintain an effective contact with the supporting surface and do not skid on the ground. Also in this case, the test must be repeated in two conditions, i.e. with the steering wheel turned both to full left and right lock. The details of the two test methods are summarized in Fig. 1.

The history of this preliminary test is controversial. At the beginning of the '80 of the last century, an EU technical Commission studied the problem, but did not share a

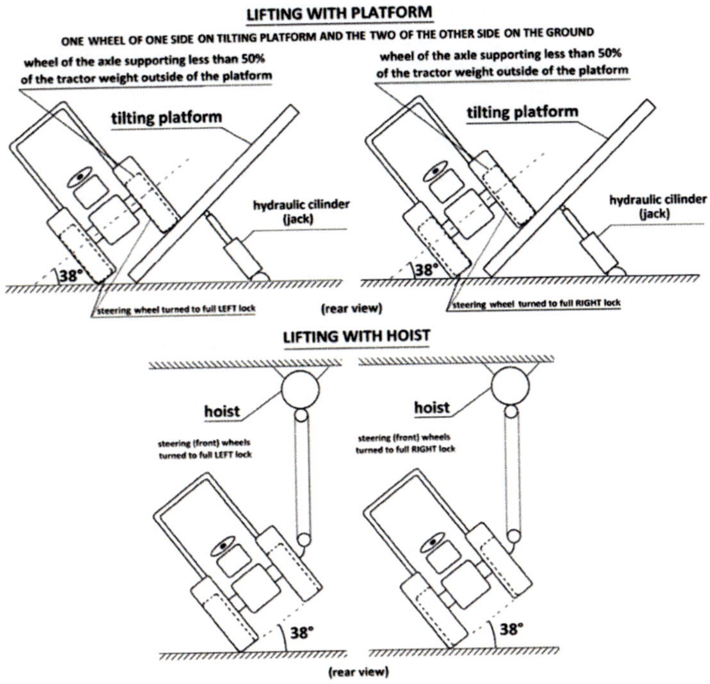

Fig. 1. Details of application of the "jack" (top) and "hoist" (bottom) methods for the ascertainment of the 38° lateral static stability provided in the OECD Code 6.

well-defined method, leaving the possibility to select the most appropriate test method in accordance with the opinion of the Testing Station engineers. In that situation, the EEC 87/402 Directive [19], and later the OECD Code 6 were issued, both including the two test methods.

Moreover, it has to be also considered that in other Standards (still in force) provided for similar check (i.e. static stability; UNI EN 1459-1:2017 Title: Rough-terrain trucks - Safety requirements and verification - Part 1: Variable-reach trucks) [20], but dedicated for vehicles used in professional sectors different from agriculture (i.e. load handling; ISO 22915-1:2016 Industrial trucks – Verification of stability – Part 1: General) [21], the machine has to be placed in such a way that all the wheels are lying on the tiltable platform (the so called "Integral Platform Method, IPM, Fig. 2).

Coming back to the agricultural and forestry tractors, some detailed comparative tests were then carried out in the mid'80 of the last century by the former Institutes of Agricultural Mechanics of the Universities of Milano and Bologna (Italy) to ascertain the possible differences of the static lateral stability limit angles of some tractor models on the market at that time.

Eleven narrow-track tractor models were tested on a tilting platform (with all wheels lying the platform, i.e. the described IPM) or using a crane. For some of them, a direct comparison of the two methods was carried out (Table 1).

On the tilting platform, the lateral static limit angles ranged from about 25° and 48°.

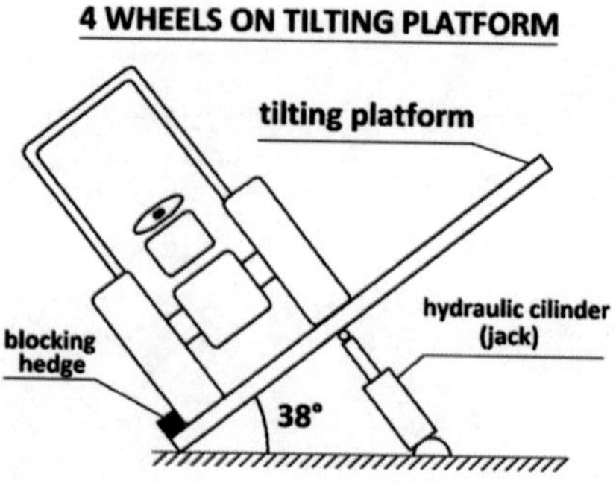

Fig. 2. In the so called "Integral Platform Method" (IPM), the tractor must be placed in such a way that all the wheels are lying on the tiltable platform.

Table 1. Lateral static stability limit angles of some tractor models on the market in Italy on the '80 of the last century (TR = traditional; IA = iso-diametric articulated; IS = iso-diam. steering wheels).

Tractor no.	TR; IA; IS	Minimum track, mm	Tyres of the raised axle	Roll-over angle, deg	
				Tilting platform	Crane
1	IA	685	6.50–15	37.45	—
2	IS	997	9.50–20	41.00	—
3	IA	860	7.50–16	39.00	48.30
4	IA	980	10.50–20	41.30	—
5	IS	1010	7.50–16	48.30	—
6	IA	690	8.25–16	32.00	—
7	IA	685	8.25–16	34.00	—
8	TR	997	11.2–24	29.20	40.00
9	TR	680	9.50–24	25.00	—
10	IA	790	8.25–16	—	43.33
11	IA	730	6.00–16	—	44.00

Using the crane, all the 4 tractor models tested highlighted a result higher than 38°. The comparisons carried out between the crane (hoist method) and the tilting platform with all wheels lying the platform (IPM) evidenced remarkable differences, ranging from about 9 to 11°. This paper is aimed to ascertain the possible difference in maximum static lateral stability angles adopting different evaluation methods. In this way, the results of the survey will provide a preliminary database on tractors' stability limit angle. This study aimed to help address tractor safety issues by developing a procedure for assessing the lateral stability of these widely used agriculture vehicles under different working conditions.

2 Material and Methods

2.1 Term and Definitions

Following ISO 16231 part 1 [22], the experimental test regarded the determination of the Static Overturning Angle (SOA), defined as the angle of tilt on which the vertical projection of the Center of Gravity (CoG) falls beyond the area of stability.

Moreover, the standard introduces the difference between:

- **rollover**, defined as the loss of machine stability characterized by a clockwise or counterclockwise rotation of more than 90° around either both the longitudinal of lateral axis of the machine, and.
- **tip-over**, defined as the loss of machine stability characterized by a clockwise or counter clockwise rotation of no more than 90° around either both the longitudinal of lateral axis of the machine.

  In this work this difference has not been considered and the term "roll-over" was used to indicate the loss of stability. Always ISO 16231 part 2 [23] reports that most self-propelled machines are equipped with a swivel angle limiting device on the swiveling axle, which, when hit during lateral rollover, restricts further swiveling of the axle prior to the complete overturn of the machine. The wheel of the fixed axle loses contact with the ground and lifts up. The body of the machine rolls and stops when the swiveling axle hits the stroke limiting device. At that point, the stability line is formed by the contact points of a front and a rear tyre. The stroke limiting device is effective only when the angle of the swiveling axle keeps the vertical projection on the inside of the stability line formed by the tyres, to absorb the dynamic effects of rolling. The state of the art does not allow to assess whether the inertia of the machine, caused by rolling is reaching a level which results in a complete tip-over. This last sentence is fundamental for the research and conducted to define two roll-over angles:
- RR: at the moment when only rear wheel starts to overturn;
- RB: at the moment when both wheels start to overturn.

These captures have been adopted for describing the test conditions and for discussing the results.

2.2 Tested Tractors

Six specialized tractors, mainly narrow, were tested at the CREA facility of Treviglio, (BG) - Italy. The tractors, had less than 73 kW nominal power and a mean unladen mass of 2700 kg, a mean width-track of 1100 mm and a mean Center of Gravity (CoG) height located approx. 600 mm above the ground. On each surveyed tractor, the CoG location coordinates were ascertained previously, in accordance with ISO 789-6:2019 standard (Agricultural tractors – Test procedures – Part 6: Centre of gravity). The main geometric parameters, mass and the height of the CoG are reported in Fig. 3.

2.3 Test Method

The Standards applied for the survey regarded the determination of the stability of agricultural or industrial machines. In detail:

- Industrial Trucks. Verification of Stability. Part 1: General; UNI ISO 22915–1/1:2008.
- Industrial Trucks. Verification of Stability. Part 2: Counterbalanced Trucks with Mast; UNI ISO 22915–1/2:2008;
- Safety of Industrial Trucks—Self Propelled Variable Reach Trucks.9–6 (Agricultural Tractors—Test Procedures— Part 6: Centre of gravity); UNI EN 1459:2010;
- Agricultural Tractors—Test Procedures, Part 6: Centre of Gravity Standard; ISO 789–6; 2019 [24].
- Self-Propelled Agricultural Machinery—Assessment of Stability—Part 1: Principles; ISO 16231–1:2013;
- Self-Propelled Agricultural Machinery—Assessment of Stability—Part 2: Determination of Static Stability and Test Procedures; ISO 16231–2:2015

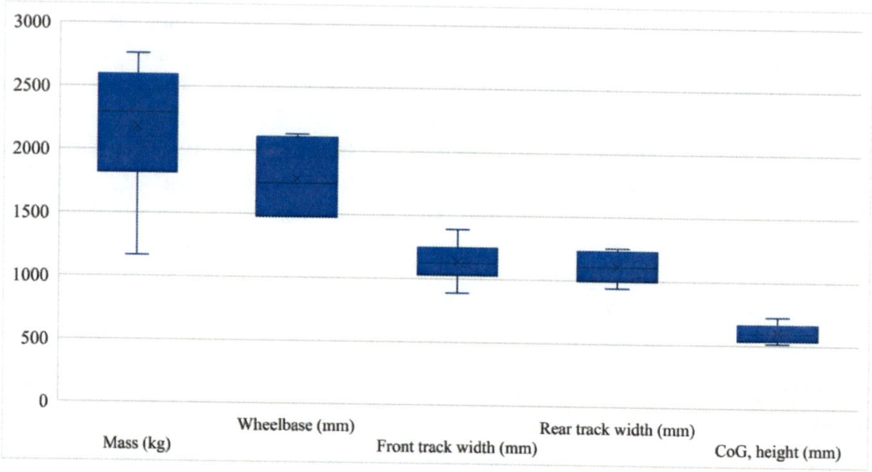

Fig. 3. Main characteristics of the 6 tested tractors (highest and lowest data value, median, first and third quartile).

The survey has been developed considering:

1. number of wheels placed on lifting platform: 1-wheel (1W), 2-wheel (2W), 4-wheel (4W) (Table 2).
2. The position of the tractor on the platform in the 4-wheel lifting method:
   a. parallel to the slope (OECD Code 6)
   b. not parallel to the slope (ISO 22915).
3. The tyre pressure (on 2 tractors):

   a. low (80 kPa)
   b. middle (160 kPa)
   c. high (240 kPa)
4. The direction of steering wheel in the 1-wheel method:
   a. left;
   b. right.
5. Number of repetitions: 3

Table 2. Lateral static stability tests layout

Scheme	Test code	Test layout	Experimental test example
1-wheel (OECD Code 6)	1W	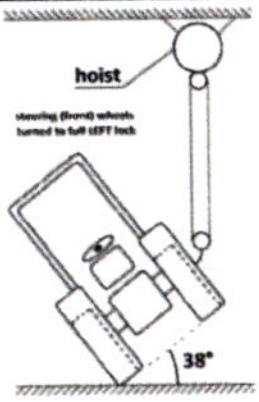	
2-wheel (OECD Code 6)	2W_RR 2W_RB	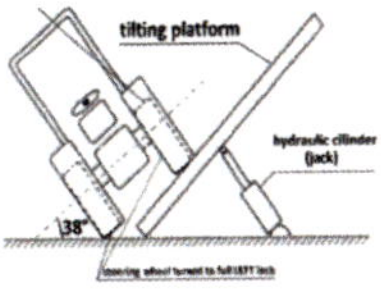	
4-wheel lateral (OECD Code 6)	4W_L_RR 4W_L_RB	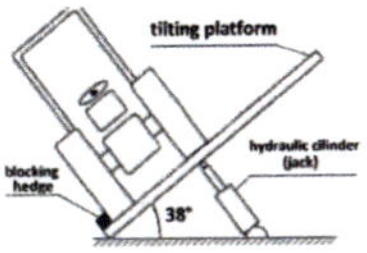	
4-wheel (ISO 22915)	4W_ISO_RR 4W_ISO_RB	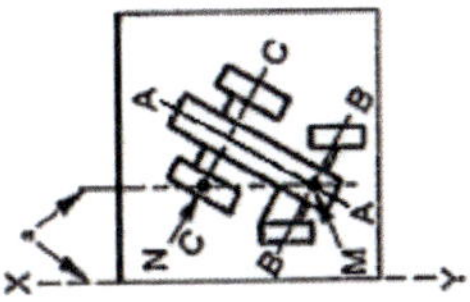	

3 Results and Discussion

According to the different methods applied, the results of the tests carried out on the 6 tractors are shown in Fig. 4.

Apart of course the vehicle architecture, in case of side overturning the behaviour of the tractors could be affected by some further features, such as the tyres dimension and their front/rear combination, as well as the inflation pressure.

The following remarks on the method and the results have been highlighted:

- the four methods applied (1-wheel, 2-wheel, 4-wheel on platform and ISO 22915) showed significantly different results;
- when using hoist, the tractors do not overturn. This method should be applied only to evaluate the requirement of static lateral stability at a slope of 38°, not for ascertaining the roll-over limit angle;
- when 2-wheel and 4-wheel on platform methods are used, the rear and front wheels are overturning sequentially; in the 1-wheel method the behaviour of the tractor is different, i.e. the two wheels touching the ground start to overturn at the same time;
- on tractors having a certain architecture, the RR and the RB roll-over dynamics could coincide, but this particular behaviour could be ascertained only with the 2-wheel and 4-wheel methods;
- in case of 1-wheel and 2-wheels methods, the slope angle is measured at the tractor axis. Conversely, in 4-wheels method the angle is measured at the platform;
- the mean standard deviation has been very low, of 0.1 deg in 1-wheel test and 0.2 deg in 4-wheel test.

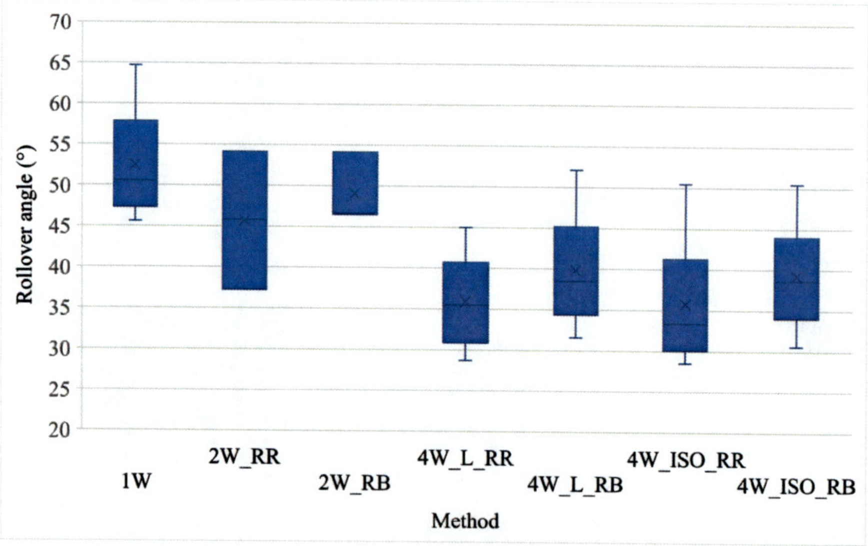

Fig. 4. Rollover limit angle of the 6 tractors tested, obtained with the different test settings (highest and lowest data value, median, first and third quartile). (1W = one wheel on platform; 2W = two wheels on platform; 4W = four wheels on platform; L = tractor parallel to the slope; ISO = tractor not parallel to the slope; RR = angle measured when only the upstream rear wheel starts to overturn; RB = angle measured when both front and rear wheels start to overturn)

Concerning the roll-over "RB", the 4W_L_RB and the 4W_ISO_RB methods showed quite close roll-over angle values, with a mean difference of 0.8°.

Moreover, for the results obtained from the six tractor models tested an interesting correlation was found, between the 1-wheel and the 4-wheel methods. In detail, the roll-over limit angle in the 4-wheel method resulted about 78% of the 1-wheel method (regression equation:"4W_L_RB" = 0.781* "1W"; R^2 = 99.89%).

4 Conclusions

Among the different methods applied to ascertain the limit lateral static rollover angle, the integral platform method was the only able to give comparable results. Moreover, considering the positioning of the tractor on the platform, the solution "4W_L_RB" make clear the tractor set-up (i.e. with the machine placed parallel to the slope, with all its wheels lying on the platform and the two wheels downstream leaning to a blocking beam, to prevent side skid). On the contrary, it's not the same for the 4W_ISO_RB, because the tractor is not placed parallel to the slope, and therefore a possible side skid could be not avoided easily.

So, at this stage of the project the most suitable method able to give an unambiguous indication for stability properties of the tractor seem to be the 4W_L_RB.

At the moment, the OECD Code 6 allows the application of different methods to ascertain the 38° static lateral stability requirement: they are the so called "jack" or "hoist" methods, but for the same tractor the results are different.

However, the tests carried out in this survey revealed a correlation between the 4-wheel (i.e. integral platform method) and 1-wheel (hoist method). So, in case a tilting platform it unavailable, the hoist method could still be adopted, taking into account a suitable correction factor.

Acknowledgments. The authors wish to thank Ivan Carminati, Gianluigi Rozzoni, Alex Filisetti and Elia Premoli for their valuable help in carrying out the experimental trials.

Funding. This study was funded by the Organization for Economic Co-operation and Development (OECD), Tractor Codes Programme. The opinions expressed and arguments shown are solely those of the authors and do not necessarily reflect the official views of the OECD or of its member countries.

Conflicts of Interest. The authors declare no conflict of interest. The funding sponsor had no role in the design of the study; in the collection, analyses, or interpretation of data; in the writing of the manuscript, or in the decision to publish the results.

References

1. European Commission (EC): The Magnitude and Spectrum of Farm Injuries in the European Union Countries. EC, Athens, Greece (2004)
2. Health and Safety Authority (HSA): Farm Safety Action Plan 2013–2015; Health and Safety Authority (HSA), Dublin, Ireland (2013)

3. Osservatorio INAIL Sugli Infortuni Nel Settore Agricolo e Forestale. In Report Annuale Sugli Infortuni Mortali e Con Feriti Gravi Verificatisi Nel 2014 Nel Settore Agricolo e Forestale; Dipartimento Innovazioni Tecnologiche e sicurezza degli Impianti, Prodotti e Insediamenti Antropici, Rome, Italy (2015)
4. Facchinetti, D., Santoro, S., Galli, L.E., Pessina, D.: Agricultural tractor roll-over related fatalities in Italy: Results from a 12 years analysis. Sustainability **13**(82), Article number 4536 (2021). https://doi.org/10.3390/su13084536
5. Cavallo, E., Görücü, S., Murphy, D.J.: Perception of side rollover hazards in a Pennsylvania rural population while operating an all-terrain vehicle (ATV). Work **51**, 281–288 (2015)
6. Görücü, S., Cavallo, E., Murphy, D.J.: Perceptions of tilt angles of an agricultural tractor. J. Agromed. **19**, 5–14 (2019)
7. Franceschetti, B., Lenain, R., Rondelli, V.: Comparison between a rollover tractor dynamic model and actual lateral tests. Biosyst. Eng. **127**, 79–91 (2014)
8. Vidoni, R., Bietresato, M., Gasparetto, A., Mazzetto, F.: Evaluation and stability comparison of different vehicle configurations for robotic agricultural operations on side-slopes. Biosyst. Eng. **129**, 197–211 (2015)
9. Wang, W., Wu, T., Hohimer, C.J., Mo, C., Zhang, Q.: Stability analysis for orchard wearable robotic system. IFAC PapersOnline **49**, 61–65 (2016)
10. Jung, D., Jeong, J., Woo, S.M., Jang, E., Park, K., Son, J.: A study on the stability of a vehicle with lifting utility. Adv. Mater. Res. **753–755**, 1169–1174 (2013)
11. Myers, M.L.: Ride-On Lawnmowers. The hazards of overturning. Prof. Saf. **54**, 52–63 (2009). www.asse.org. Accessed 7 June 2017
12. Liu, J., Ayers, P.D.: Off-road vehicle rollover and field testing of stability index. J. Agric. Saf. Health **5**, 59–72 (1999)
13. Molari, G., Badodi, M., Guarnieri, A., Mattetti, M.: Structural strength evaluation of driver's protective structures for self-propelled agricultural machines. J. Agric. Saf. Health **20**, 165–174 (2014)
14. Shulruf, B., Balemi, A.: Risk and preventive factors for fatalities in all-terrain vehicle accidents in New Zealand. Accid. Anal. Prev. **42**, 612–618 (2010)
15. Cutini, M., et al.: A study of the lateral stability of self-propelled fruit harvesters. Agriculture **7**(11), Article number 92 (2017). https://doi.org/10.3390/agriculture7110092
16. Cutini, M., Forte, G., Maietta, M., Mazzenga, M., Mastrangelo, S., Bisaglia, C. Safety-critical manuals for agricultural tractor drivers: a method to improve their usability Agriculture **7**(84), Article number 67 (2017). https://doi.org/10.3390/agriculture7080067
17. OECD Code 6: Standard Code for the official testing of front mounted roll-over protective structures on narrow track agricultural and forestry tractors, Paris, France (2019)
18. Molari, G., Rondelli, V. On the Definition of narrow-track wheeled agricultural tractors. Biosyst. Eng. **88**(1), 75–80 (2004). https://doi.org/10.1016/j.biosystemseng.2004.02.001
19. Council Directive 87/402/EEC of 25 June 1987 on roll- over protection structures mounted in front of the driver's seat on narrow-track wheeled agricultural and forestry tractors (1987)
20. UNI EN 1459-1:2017 Title: Rough-terrain trucks - Safety requirements and verification - Part 1: Variable-reach trucks), Milan, Italy (2017)
21. UNI ISO 22915-1/2:2008: Industrial Trucks. Verification of Stability. Part 2: Counterbalanced Trucks with Mast; International Organization for Standardization: Geneva, Switzerland (2008)
22. ISO 16231-1:2013: Self-Propelled Agricultural Machinery—Assessment of Stability—Part 1: Principles; International Organization for Standardization: Geneva, Switzerland, (2013)

23. ISO 16231-2:2015: Self-Propelled Agricultural Machinery—Assessment of Stability—Part 2: Determination of Static Stability and Test Procedures; International Organization for Standardization: Geneva, Switzerland (2015)
24. Agricultural Tractors—Test Procedures, Part 6: Centre of Gravity Standard; ISO 789-6:2019; International Organization for Standardization: Geneva, Switzerland (2019)

Design, Manufacturing, and Strength Test of a 4-post ROPS Fitted on a Very Low-Profile Tractor (TRACLAS Project)

Domenico Pessina[1(✉)], Davide Facchinetti[1], Francesco Santoro[2], Pierluigi Febo[3], Santo Orlando[3], Danilo Monarca[4], Massimo Cecchini[4], Maurizio Cutini[5], Davide Gattamelata[6], Vincenzo Laurendi[6], and Simone Pascuzzi[2]

[1] Department of Agricultural and Environmental Sciences - Production, Landscape, Agroenergy, University of Milan, Via Celoria 2, 20133 Milan, Italy
domenico.pessina@unimi.it

[2] Department Agricultural and Environmental Science, University of Bari, Via G. Amendola 165/a, 70126 Bari, Italy
simone.pascuzzi@uniba.it

[3] Department of Agricultural, Food and Forest Sciences, University of Palermo, Viale delle Scienze, 90128 Palermo, Italy

[4] Department of Agriculture and Forest Sciences, University of Tuscia, Via S. Camillo de Lellis, 01100 Viterbo, Italy

[5] CREA-ING - Consiglio per la ricerca in agricoltura e l'analisi dell'economia agraria, Via Milano, 43, 27047 Treviglio (Bg), Italy

[6] INAIL - National Insurance Institute for Occupational Accidents, P.le Pastore, 6, 00144 Roma, Italy

Abstract. To improve the safety level when agricultural tractors work on slope under canopy, INAIL (the Italian National Institute for Insurance against Accidents at Work) funded a research project, named TRACLAS, devoted to set-up a prototype of low-profile compact tractor, not exceeding 1.60 m overall height, equipped with a 4-post ROPS (RollOver Protective Structure).

After the complete set-up of the prototype, some field and lab tests were carried out to compare its performance with those of some comparable tractor models already on the market, used in vineyards, orchards and greenhouses.

A basic step of the TRACLAS project was the execution of the strength test of the customized 4-post frame (the ROPS) carried out in compliance with the OECD Code 4. The main features concerning the design, the manufacturing and the validation test of the ROPS are reported in this paper. The low-profile compact tractor prototype was obtained starting from a Goldoni E100 commercial tractor. The 4-post ROPS was tested successfully, being able to fully satisfy the desired requirement, i.e., the protection of the clearance zone, located with reference to the tractor driving place (operator's seat and steering-wheel) during all the loads provided by the official procedure.

Keywords: Agricultural work safety · Tractor overturning · Roll-Over Protective Structure (ROPS) · Vineyard · Orchard and greenhouse

All the authors contributed equally to this paper.

M. Biocca et al. (Eds.): SHWA 2020, LNCE 252, pp. 468–476, 2022.
https://doi.org/10.1007/978-3-030-98092-4_48

1 Introduction

The main cause of accidents in agriculture concerns the tractor loss of control and the consequent its overturning (Facchinetti et al. 2021; Cecchini et al. 2013). Considering the working conditions and the typical shape of agricultural tractors, the most frequent rollovers occur on the lateral axis (about 75% of cases) (Inail 2017, Inail 2016). Furthermore, the highest number of fatal accidents, resulting in death or permanent disability involving the operators of agricultural tractors, demonstrates that the current safety devices mounted on tractors, especially on narrow-track tractors, used particularly in specialized crops (orchards) need to be improved (Bambach et al. 2013). The mainly adopted solution to reduce the consequences of the overturning of tractors is made up of passive protection devices (Rondelli et al. 2018). The combination of a RollOver Protective Structure (ROPS) with a restraint system (seat belt) creates and guarantees, around the operator, a "safety volume", named also "clearance zone" (Pascuzzi 2015; Kim et al. 2011). Mainly on narrow-track tractors, ROPS are folding type, to make easy the execution of the garaging and some working routines under canopy. Nevertheless, the folding of the ROPS could lead to an improper use (Mattetti et al. 2017). So, several fatal accidents are caused by the tractor overturning, resulting in hitting and crushing the driver, due to the missing of the re-positioning in safety condition of the ROPS (Facchinetti et al. 2021; Melvin et al. 2009; Molari and Rondelli 2007). This is a quite important question, on which much work has to be still carried out, to improve the overall tractor safety, by setting-up more efficient solutions able to protect suitably the operator (Myers 2010).

To improve the safety level when tractors work on slope under canopy, and solve the described problem, with the collaboration of some research units belonging to university and research bodies, INAIL (the Italian National Institute for Insurance against Accidents at Work) funded a research project, named TRACLAS (messa a punto di un TRattore Agricolo Compatto per le LAvorazioni Sottochioma). The project was devoted to set-up a compact tractor (called like the project, i.e. "TRACLAS"), complementary and competitive in terms of performance to the models at present available on the market, but equipped with a not-folding ROPS.

In addition to compactness, another important feature was the very high lateral and longitudinal stability, as well as an excellent manoeuvrability. A further added value has been the fitting of a 4-post ROPS (or, alternatively, a cab), obviously coupled with a safety belt of the driving seat, to assure an effective driver protection in case of overturning.

2 Materials and Methods

2.1 Farms Survey

Taking into account 20 farms located all over Italy, found concerning the most popular crops in the reference areas, the first step of the project involved a series of detailed static (planting layout) and dynamic (cultivation methods) surveys of the main specialized italian open field crops (grape, olive, hazel, almond, kiwi, etc.). The selection of these crops was connected in particular to the difficulty of working with tractors under canopy. In each farm, the main characteristics of the tractors and implements used were carefully analysed. In detail, the data investigated concerned: methods and frequency of execution, assessment of the boundary conditions (e.g. state of the soil in different cultivation periods, vegetation, stage of growth of the main crop, etc.); size and particularity of the equipment used; power and dimensions of coupled tractors; plant layout; working speeds, theoretical and real work capacity, including the accessory transfer routines.

In the investigated farms, the operators were interviewed about the critical issues encountered in their activity with tractor, the desired stability of the machine and its most suitable overall dimensions taking into account the non damage of the canopy and the production. Furthermore, they were asked about the desirable design and performance features of a new compact tractor for specialized crops.

The further step concerned an ad-hoc study of the canopy volumes of the investigated crops, in the period of their maximum growing, to assess the possible interferences during the travelling of specific tractor models currently operating in the farms. Also the overall dimensions of a theoretical tractor, passing in the inter-rows with no interference were properly defined (Pascuzzi et al. 2020).

2.2 Tractor Prototype

The results of this study led to the definition of the main technical requirements for the TRACLAS. The prototype was set-up by modifying a Goldoni E100 (Goldoni, Migliarina di Carpi, Modena - Italy) narrow-track, isodiametric wheels tractor, equipped as standard with a 2-post front foldable ROPS, homologated according to OECD Code 6.

As part of the modifications introduced on the Goldoni E100, the 2-post front rollbar rollbar was replaced with a 4-post frame. Moreover, to reduce the overall height of the machine, the tyres 320/70R20 were replaced with tyres 31.5–15, thus satisfying this limit provided by the project, i.e. overall height not exceeding 1600 mm. Due to the modified set-up (in particular for the fitting of very wide tyres), the minimum track width of the front and rear axles exceeded the conventional limit of 1150 mm; as a consequence, the tractor was included in the "conventional" category instead of the "narrow-track" category.

One of the most important goals of the project was to ascertain the safety requirements of the new ROPS fitted on the prototype. Thus, the main goal of this paper is to report the study carried out to design and manufacture the 4-post frame mounted on the TRACLAS, as well as the tests carried out according to the OECD Code 4.

2.3 The 4-post Frame

The new customized 4-post frame is fixed rigidly on the tractor body, with no silent blocks. Usually, silent-blocks reduce the transmission of high frequency vibrations, that are producing negative effects, such as a noise resonance in case of fitting of a closed cab as well as fatigue phenomena on the main structure.

The four posts of the frame are made with steel square tubes 50 × 50 × 5 mm, welded to an upper ring also made by square tubes 50 × 50 × 5 mm, reinforced with some crossbars (always made by square tubes 50 × 50 × 5 mm) and plates, to increase the structure stiffness. Steps were then taken to optimize the ROPS stiffness, concerning the optimization of the overall dimensions, by reducing the so called "margin", i.e. the free space remaining between the safety zone volume and the ROPS structural members. Therefore, this is the reason why several reinforcements were added (in terms of crossbars and corner plates), to reduce at a minimum the elastic and plastic deflections resulting from the loads application, thus ensuring the respect of the clearance zone, avoiding the intrusion of structural elements (Figs. 1 and 2).

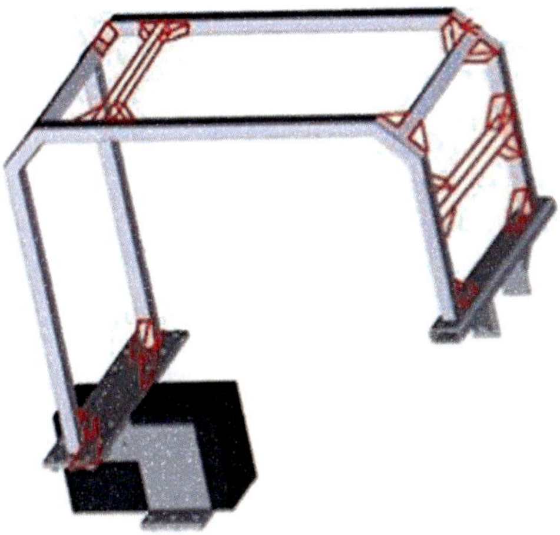

Fig. 1. The 4-post frame installed on the compact agricultural tractor (TRACLAS). In void lines, thecrossbars and corner plates added as stiffening elements.

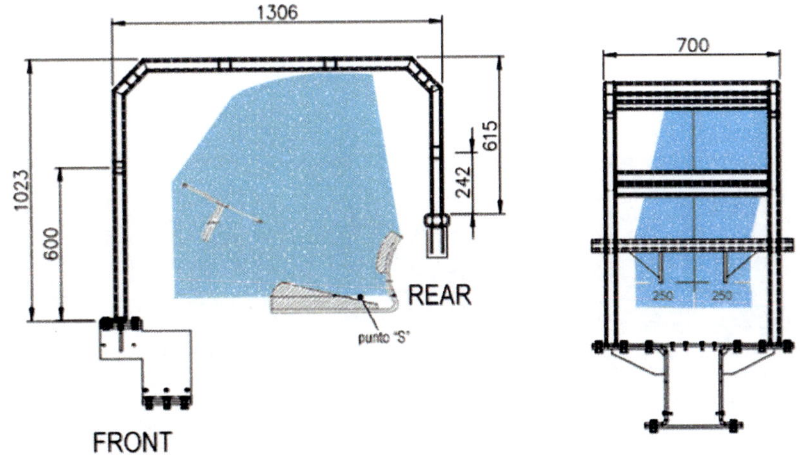

Fig. 2. Main dimensions of the 4-post ROPS fitted on TRACLAS, and location of the clearance zone (according OECD Code 4).

2.4 The ROPS Tests

The test of the 4-post frame installed on the compact profile agricultural tractor prototype was carried out according to OECD Code 4, that provides a given sequence of horizontal and vertical loads. More in detail, in the horizontal (longitudinal and side) loads, a given energy is to be absorbed, derived from the integral of the load-deflection curve, having the so-called "reference mass" of the tractor as a reference (Mrif.). Conversely, for the vertical loads a given crushing force has to be attained, always connected to the reference mass. Each test is passed if during its execution and at the end no part of the frame shall enter the clearance zone and furthermore the clearance zone shall not be outside the protection of the protective structure (OECD 2021).

Considering the fitting of the 4-post frame and all the tanks full, the compact tractor reached an overall mass of 1950 kg, burdening for the 65.4% (1275 kg) on the front axle and the 34.6% (680 kg) on the rear axle. The "reference mass" (i.e. the value used for the calculation of the values of energy to be absorbed and the force to be attained) (OECD 2021) was therefore set at 2000 kg, a little bit more than the measured mass, to take into account some accessories that could be fitted at a later extent.

The longitudinal load (energy to be absorbed, to be calculated with the formula (*not an equation*): E (J) = 1.4 * Mrif = 2800 J) was carried out from the front (because this is the part of the tractor bearing the major part of the mass) by means of a hydraulic cylinder, operated in a quasi-static mode (max speed: 5 mm/s). During the test, the incremental values of force and deflection were recorded (Fig. 3a). The energy absorbed E_{tot} (J) by the ROPS was calculated in accordance of the following formula:

$$E_{tot} = \sum(F_i \cdot \Delta d_i)$$

where:

F_i = measured loading force (N);

Δd_i = incremental deformation in the loading direction (single step 0.1 mm).

The front vertical load (max force to be attained, to be calculated with the formula: F (N) = 20 * Mrif = 40 kN) was carried out by means of a steel beam, connected at the ends to two hydraulic cylinders, in turn fixed to the plant base (Fig. 3b).

Fig. 3. The tractor ROPS at the end of the front loads: longitudinal (a) and vertical (b).

The side load provided an energy absorption of 3500 J (E = 1.75 * Mrif), and was executed placing the TRACLAS laterally in respect to the loading hydraulic cylinder (Fig. 4a). Finally, the rear vertical test (again the max force to be attained was 40 kN) was carried out in a similar way to the aforesaid front compression test, but with the horizontal beam placed on the rear area of the safety frame (Fig. 4b).

Fig. 4. The tractor ROPS at the end of the side (a) and rear vertical (b) loads.

3 Results and Discussion

3.1 The Manufactured Compact Tractor Fitted with the 4-post ROPS

As results of previous specific tests, carried out in accordance with ISO 789, ISO 22915 and OECD Code 6 standards, the TRACLAS highlighted a remarkable lateral static stability, combined with excellent manoeuvrability (Figs. 5 and 6).

The fitting of the non-foldable 4-post ROPS assured a constant and effective protection to the driver in the event of overturning, provided he/she is properly fastened to the seat with a safety belt.

Table 1 reports the main features of the compact tractor.

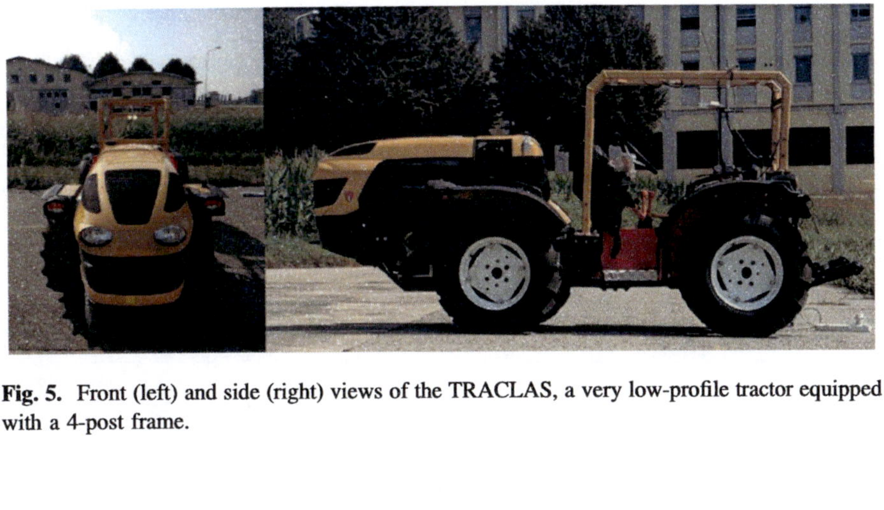

Fig. 5. Front (left) and side (right) views of the TRACLAS, a very low-profile tractor equipped with a 4-post frame.

Fig. 6. Lateral static stability test: hoist (left) and platform (right) method, carried out in accordance with OECD Code 6.

Table 1. Main features of the TRACLAS

Track, mm	Front	Rear
	1350	1290
Tyres	31.5–15	31.5–15
Mass, kg	1275	680
Overall width, mm	1710	

3.2 Results of Strength Test

In Fig. 7 are shown the load-deflection curves of the front and side strength tests, as well as the elastic, plastic and total deflection values recorded at the end of each test.

Table 2 shows the final deflection values measured at the end of the sequence of tests. The remarkable robustness of the ROPS in fully confirmed, because the overall deflections have been very limited, especially in the longitudinal and vertical direction.

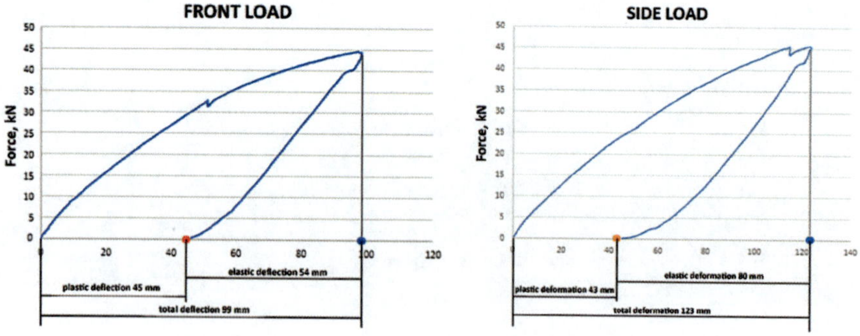

Fig. 7. Load-deflection curves of the front and side strength tests.

Table 2. ROPS deflection values measured at the end of the test sequence.

		Deflection, mm
Longitudinal (to the rear)	Right	4
	Left	4
Side	Front	100
	Rear	100
Vertical (front)	Right	5
	Left	3
Vertical (rear)	Right	2
	Left	23

The side deflection recorded has been higher than those measured in the other directions, but this is justified by the severity of the correspondent load, for the highest energy amount to be absorbed by the ROPS. In general, the deflection values have been much lower than those commonly recorded in the safety frames installed on other tractors comparable to the TRACLAS.

Moreover, to validate the suitability in carrying out the typical tasks of specialized crops, the TRACLAS was tested at work either inside a hazelnut orchard and an olive grove. Also considering the cultivation season, a shredder make Facma model TR 200 BVE has been coupled to the compact tractor to execute an inter-row grass cutting at a travelling speed of 5 km/h, obtaining a quite good working capacity, of about 1 ha/h.

4 Conclusions

All the tests required by the OECD Code 4, to which the 4-post frame fitted on the TRACLAS was subjected, were successfully passed. The safety zone, representing the volume occupied by the driver normally placed at the seat with the torso erect and his hands on the steering wheel and suitably fastened with a seat belt, has been always protected throughout and at the end of each loading test.

Indeed, this is the only requirement provided by the applied standard but, in this specific case, it was revealed particularly critical, considering the severe constraint of the ROPS design, i.e. the compact dimensions especially in height (max 1600 mm above the ground).

The design and manufacturing of a very rigid ROPS (significantly oversized compared to commercial ROPS) was very challenging, because it was mandatory the absorption of the energy especially in terms of resistance (i.e. high strength) rather than deformation (i.e. high elasticity). This goal was fully achieved, thus confirming the design forewords, being those the protection of a certain safety volume on a compact vehicle.

The TRACLAS was easy to be used, highlighting good reliability and trim stability. Above all, due to its extremely low overall height, the compact tractor evidenced no interference at all with the canopy when passing in the inter-rows, between the plants.

Acknowledgements. The study presented in this paper is part of a project funded by INAIL (the Italian National Institute for Insurance against Accidents at Work) entitled "*Messa a punto di un trattore agricolo a profilo compatto per le lavorazioni nelle coltivazioni arboree specializzate*".

References

Bambach, M.R., Grzebieta, R.H., McIntosh, A.S.: Thoracic injuries to contained and restrained occupants in single-vehicle pure rollover crashes. Accid. Anal. Prev. **50**, 115–121 (2013)

Cecchini, M., Monarca, D., Colantoni, A., Di Giacinto, S., Menghini, G., Longo, L.: Study on the possibility of application of a compact roll over protective structure for agricultural wheeled narrow track tractors. J. Agric. Eng. **44**(1S) (2013). 10th Conference of the Italian Society of Agricultural Engineering, Viterbo

Facchinetti, D., Galli, L.E., Santoro, S., Pessina, D.: Agricultural tractor roll-over related fatalities in italy: results from a 12 years analysis. Sustainability **13**(8), 4536 (2021). https://doi.org/10.3390/su13084536

Kim, H.J., et al.: A study on improving the tractor ROPS and seat belt use of Korean farmers. In: American Society of Agricultural and Biological Engineers Annual International Meeting (2011)

Mattetti, M., Molari, G., Sereni, E.: Damage evaluation of driving events for agricultural tractors. Comput. Electron. Agric. **135**, 328–337 (2017)

Molari, G., Rondelli, V.: Evaluation criteria for the anchorage resistance of safety belts on agricultural tractors. Biosys. Eng. **97**, 163–169 (2007)

Melvin, L.M., Henry, P.C., Susan, C.W.: Injury severity related to overturn characteristics of tractors. J. Saf. Res. **40**, 165–170 (2009)

Myers, J.R.: Prevalence of roll-over protective structure (ROPS)-equipped tractors on hispanic-operated farms in the United States. J. Agromed. **15**, 137–147 (2010)

Myers, J., Cole, H., Westneat, S.: Seatbelt use during tractor overturns. J. Agric. Saf. Health **12** (1), 46 (2006)

OECD: OECD Code 4, standard code for the official testing of protective structures on agricultural and forestry tractors (static test) (2021)

INAIL: Osservatorio INAIL sugli infortuni nel settore agricolo e forestale. Report annuale sugli infortuni mortali e con feriti verificatisi nel 2017 nel settore agricolo e forestale. Dipartimento Innovazioni Tecnologiche e Sicurezza degli Impianti, Prodotti e Insediamenti Antropici dell'INAIL (2017)

INAIL: Osservatorio INAIL sugli infortuni nel settore agricolo e forestale. Report annuale sugli infortuni mortali e con feriti verificatisi nel 2016 nel settore agricolo e forestale. Dipartimento Innovazioni Tecnologiche e Sicurezza degli Impianti, Prodotti e Insediamenti Antropici dell'INAIL (2016)

Pascuzzi, S.: A multibody approach applied to the study of driver injuries due to a narrow-track wheeled tractor rollover. J. Agric. Eng. **46**, 105–114 (2015)

Pascuzzi, S., Anifantis, A.S., Santoro, F.: The concept of a compact profile agricultural tractor suitable for use on specialised tree crops. Agriculture **10**, 123 (2020). https://doi.org/10.3390/agriculture10040123

Rondelli, V., Casazza, C., Martelli, R.: Tractor rollover fatalities, analysing accident scenario. J. Saf. Res. **67**, 99–106 (2018)

Food Safety

Effect of Nano Particles of Pomegranate Peel Extract on Shelf Life of Some Fruit and Vegetable Products

Domenico Rongai[1(✉)], Francesca Baldassarre[2], Carlo Di Marco[1], and Giuseppe Ciccarella[2]

[1] CREA Research Centre for Engineering and Agro-Food Processing, C.da Bucceri Cepagatti, Pescara, Italy
domenico.rongai@crea.gov.it
[2] Department of Biological Sciences and Technologies and Environmental, University of Salento & UdR INSTM Salento, via Monteroni, 73100 Lecce, Italy

Abstract. Natural anti-microbial compounds are a valid and safe alternative to agrochemicals to control plants pathologies and post-harvest fruit diseases. Safety and efficacy of these new biopesticides could be enhanced through micro/nanoencapsulation methods and nanomaterials exploitation. In this study, *in vitro* and *in vivo* tests were performed to evaluate the effects of pomegranate peel extract-loaded nanoparticles on *Botrytis cinerea.* $CaCO_3$ nanocrystals were selected as carriers and the pomegranate peel extract loading was obtained through physical adsorption. Fruit and vegetable products were obtained from a commercial market, while *B. cinerea* was a strain of CRA-DC Roma collection. *In vitro* tests were achieved by measuring radial growth on Petri dishes. *In vivo* tests were performed immersing fruit and vegetable products for 30 s in a volume of the respective solutions. Disease severity (DS), disease incidence (DI) and FT-NIR spectra were recorded. $CaCO_3$ nanocrystals effectively absorb extract producing pomegranate (*Punica granatum*) peel extract-loaded nanoparticles (PCN). *In vitro* tests showed the ability of PCN to control *B. cinerea.* The effectiveness was also established by the treatment on plant products to extend the shelf life. Our study suggests that PCN could be an alternative to synthetic products.

Keywords: Gray mould · Punicalagin · Postharvest diseases · $CaCO_3$ · Nanocrystals

1 Introduction

In the past the use of pesticides to control crop disease has been useful to increased production of food worldwide. Unfortunately, the abuse of their use in crop protection had a serious environmental impact. Pesticides also pose serious health risks within our food chain and have been linked to the increased incidence of several types of cancer. Increased resistance in fungal populations, on the other hand, accelerates the development of new formulations that are both effective and environmentally friendly.

M. Biocca et al. (Eds.): SHWA 2020, LNCE 252, pp. 479–485, 2022.
https://doi.org/10.1007/978-3-030-98092-4_49

The fungus *Botrytis cinerea* infects various plants and in different ways. But the main damage is the decline of ripe fruit after harvest. It represents the most significant disease in strawberries production, causing major quality and quantity losses in postharvest storage. Treatments are normally applied by during flowering and strawberries are harvested in the early morning to control the fungus spread into the crop and to extend the shelf life. Indeed, strawberry shelf life is one day shorter when harvested after 10 AM [1]. Some authors have investigated the antifungal activity of environmentally friendly natural formulations, which could be suitably applied in place of fungicides in the treatment of strawberry. Some of these contain glutathione, oligosaccharides, laminarine and chitosan, substances capable of inhibiting postharvest decay by direct inhibition of the fungus, or by stimulating natural defense mechanisms [2].

Recent research has been focused on *Punica granatum* peel extract that showed high efficiency as natural pesticide to control phytopathogens in the crops and postharvest fruits diseases [3, 4]. These bioactive substances are contained in the peel of the fruit and could be used to obtain innovative materials for active packaging [5, 6]. Natural anti-microbial compounds are a valid and safe alternative to agrochemicals to control plants pathologies and post-harvest fruit diseases. Safety and efficacy of these new biopesticides could be enhanced through micro/nanoencapsulation methods and nanomaterials exploitation [7–9]. Gradual release reduces dosages and frequency of applications improving the sustainability of the system.

In this study, the pomegranate peel extract from *P. granatum, var. dente di cavallo* [10], has been efficiently adsorbed on $CaCO_3$ nanocrystals (nano$CaCO_3$) and its bioactivity against gray mould has been improved thanks to nanotechnologies application. The effects of pomegranate (*Punica granatum*) peel extract-loaded nanoparticles have been investigated against the gray mould caused by *B. cinerea*. Nano$CaCO_3$were selected as carriers thanks to their high bio and eco compatibility and ability in drug delivery [11, 12]. $CaCO_3$ nanocrystals effectively absorbed different organic chemicals ranging from anticancer drugs to phytochemicals [11–13]. Previous studies have been proposed and supported $CaCO_3$ nanocarriers as delivery systems to control the phytopathogen *Xylella fastidiosa* [14]. PAE metabolites have been adsorbed on nano$CaCO_3$ porous surface with a great efficiency of 50%. Therefore, we have performed *in vitro* and *in vivo* tests to study PAE-loaded nanocrystals (PCN) activity on gray mould causal agent, *B. cinerea*.

2 Methods

2.1 Plants Materials, Fungal Strains and Pomegranate Extracts Used

Strawberry fruits (*Fragaria x ananassa Duch.*) of uniform size, without defects and with the same degree of ripeness, were obtained from a commercial market (Rome, Italy) of organic products. The fungus *B. cinerea* was a CRA-DC Rome strain (collection number 1623) maintained on potato dextrose agar (PDA, OXOID CM 0139, Basingstoke, UK) and stored at 4 °C. The isolate was grown for 8 days on PDA in the dark at 25 ± 2 °C, before use.

Samples of *P. granatum, var. dente di cavallo*, provided by the botanical garden at "CRA-DC", Rome, Italy, were used. The powder of pomegranate peel hydro alcoholic extract (PAE) was obtained according to the method previously described [10].

2.2 Preparation of PAE-Adsorbed $CaCO_3$ Nanocrystals

The powder of nano$CaCO_3$@PAE (PCN) was obtained through physical adsorption as previous described for other organic species, such as phospholipids and polyphenols [13, 14]. The adsorption of PAE on nano$CaCO_3$ was performed adding, drop by drop, a solution of PAE 3% W/V to nano$CaCO_3$ (100 mg) under vigorous agitation for 12 h. The quantification of PAE in supernatant (not loaded) was obtained through spectrophotometric analysis, recording UV-Vis absorption spectra at 260 nm by a Varian-Cary 500 spectrophotometer. The selected wavelength of 260 nm is the peak of PAE that corresponds to the abundant punicalagins adsorption peak. The unknown concentration was obtained referring to a standard curve using PAE solutions at known concentration in the range of 0.01–10% W/V and fitting the line through Origin software (Abs values have been multiplied by dilution factor).

2.3 Fungus and *in vitro* Assays

In vitro tests were achieved by measuring radial growth on Petri dishes containing 20 mL of Potato Dextrose Agar (PDA). In treated plates, a total of 5, 10, or 20 mg of PCN was added to PDA such that the concentration corresponded to 0.25, 0.50, or 1% (w/v), respectively. In untreated plates (negative control), samples contained PDA without extract. A plate containing a standard fungicide (Marisan 50 PB, Dichloran 60%, SIAPA, Milan, Italy) was used at the recommended concentration as a positive control. A 5 mm diameter plug of inoculum was taken from the actively growing margin of 8-day-old cultures of *B. cinerea* and placed face up in the centre of each Petri dish. Plates were incubated at 24 ± 2 °C, radial growth was measured at 4 days after inoculation. Inhibition was calculated by the formula: % inhibition = 100 × (colony diameter (mm) of control – colony diameter (mm) of sample)/colony diameter (mm) of control. Each treatment was replicated four times and the test was repeated twice.

2.4 Fruits Protection Assay

In vivo, strawberry fruits were immersed for 30 s in a 2 L volume of the respective solutions. Fruits soaked in deionized water were used as the control. The fruits were then dried in air for 1 h and arranged in groups of 18 in plastic boxes containing three absorbent layers on the bottom. After treatments, strawberries were stored for 2 days at 0 ± 1 °C, 95–98% of relative humidity, and then at 20 ± 1 °C, 95–98% of relative humidity for another 4 days (shelf life). Each treatment consisted of 60 fruits (15 fruits × 4 replicates) and included: a) non treated control; b) fruits treated with PAE at the concentration of 1%; c) fruits treated with PCN at the concentrations of 0.25; 0.50 and 1%.

Disease severity (DS) and disease incidence (DI) were recorded. An empirical scale was designed for this experiment. DS was recorded on a 0–5 scale where: 0 = no symptoms; 1 = 1–20% fruit surface infected; 2 = 21–40% fruit surface infected; 3 = 41–60% fruit surface infected; 4 = 61–80% fruit surface infected; 5 = more than 81% of the strawberry surface infected and with sporulation. The percentage disease severity of each treatment was calculated by the formula: DS % = (sum of all disease ratings × 100)/(total number of ratings × maximum disease grade). Disease incidence was determined according to the formula: DI % = (No. of infected plants × 100)/Total n. of plants assessed.

Mycelial growth inhibition, disease severity, and disease incidence were analysed by ANOVA with means separation by Fisher's protected LSD test at $\alpha = 0.05$. Because the results from two independent experiments were consistent and there were no interactions between experiment run and treatments, data from two independent experiments were combined and the experiment was treated as a block term in the analysis. To correct for heterogeneity of variance, the data were arcsine-transformed prior to analysis. SigmaPlot V10 (London, UK) was used to create graphics and Sigma Stat V3.5 (London, UK) was used for ANOVA.

3 Results and Discussion

3.1 *In vitro* Assays

In vitro tests showed the ability of PCN and PAE to control *B. cinerea*. PCN sample has been obtained as just described in materials and methods section. $CaCO_3$ nanocrystals adsorbed PAE phytochemicals with a great loading efficiency of 50%. Four days after inoculum, the mycelia growth was 62 mm in the control plates while it significantly decreased to 44.7 (PCN at 0.25%) or 21.0 mm in plates with PCN at concentration of 1%. There were no significant differences in mycelium growth between 1% PCN and fungicide, proving that PAE-adsorbed $CaCO_3$ nanocrystals can replace chemical fungicides in the control of strawberry gray mold. In the plates treated with PCN (0.5%) the growth of the mycelium was 36%, a lower value even if statistically not different from the 39.5% recorded in PAE (1%) (Table 1).

Table 1. *In vitro* mycelial growth of *B. cinerea* at 4th day by increasing doses of nano$CaCO_3$@PAE (PCN). The fungicide (Marisan 50 PB) and PAE (pomegranate peel hydro alcoholic extract), were also used. Values with different letters for each group are statistically different (LSD test, $P \leq 0.05$). Standard deviations of the means are indicated.

Treatment	Dose	Mycelial growth of *B. cinerea* mm	
Control		62.0	± 2.70a
Fungicide	(0.15%)	18.8	± 1.92c
PCN	(0.25%)	44.7	± 2.80b
PCN	(0.5%)	36.0	± 2.92b
PCN	(1%)	21.0	± 1.83c
PAE	(1%)	39.5	± 2.50b
			F = 556.88 $P \leq 0.05$

Therefore, although the PAE used pure may inhibit *B. cinerea* [15], when formulated with calcium carbonate nanocrystals (PCN), its efficiency increases. These results have been provided the potential utilization of PCN formulation to control *B. cinerea* postharvest disease; so, we have conducted a validation test through *in vivo* assay on strawberry.

3.2 Fruits Protection Assay

Both PCN and PAE can control the fungus *B. cinerea.* PCN reduced disease severity and the incidence of gray mold on strawberries, even at the lowest dose (0.25%). By increasing the concentration of the PCN, both disease severity (DS) and disease incidence (DI) decreased: in strawberries treated with PCN 0.25% DS and DI were 35 and 84.42% respectively, statistically lower values than 48 and 100% obtained in the untreated control. No statistically significant difference was detected between PCN 1% and fungicide. Good antifungal activity was also detected in Petri dishes containing PCN at the 0.5% concentration and PAE at 1% (Fig. 1).

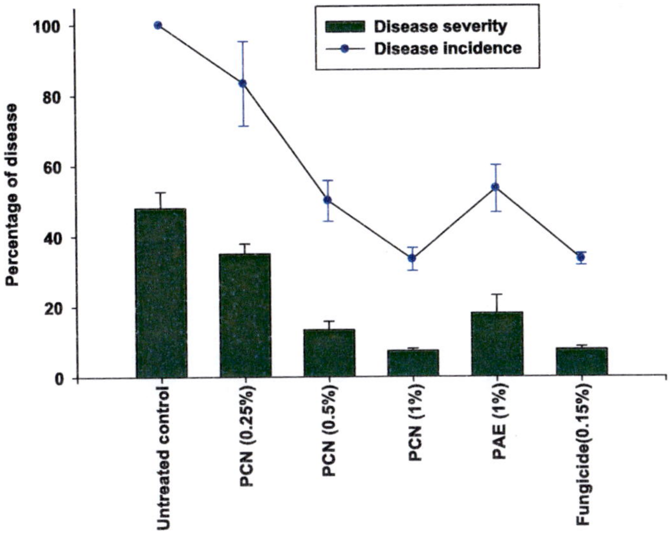

Fig. 1. Disease severity and incidence of gray mould (*Botrytis cinerea*) on strawberry *versus* increasing doses of PCN (0.25–1% W/V). Free PAE (1% W/V) and commercial fungicide (0.15%W/V) are used as positive control.

The antifungal activity of the PCN is also confirmed by the direct observation of the fruits. Samples treated with the highest dose were apparently free of mold compared to the corresponding untreated controls. Indeed, some untreated samples were invaded by gray mold, while those treated with PCN appeared healthier (Fig. 2).

Fig. 2. Photos of strawberry fruits infected with *B. cinerea* after 7 days of treatments: fungicide (Marisan 50 PB) at 0.15% W/V, PCN at 0.25%, 0.50 and 1% and free PAE at 1% W/V.

4 Conclusion

The obtained results have demonstrated the ability of PCN to contain the development of *B. cinerea* on strawberry and therefore increase its shelf-life. This research promotes the use of PCN as natural fungicide to control the decay of post-harvest strawberries. Replacing synthetic pesticides with plant molecules reduces the amount of chemicals in the environment and undesirable effects such as contamination of the food chain. Since the $CaCO_3$ nano crystals-based formulation allowed to a simple and effective enhancement of PAE bioactivity as demonstrated by *in vitro* and *in vivo* tests. This research suggests that PCN could be a safe alternative to agrochemicals to control post-harvest spoilage and improve fruit quality while avoiding the impact on human health.

References

1. Romanazzi, G., Smilanick, J.L., Feliziani, E., Droby, S.: Integrated management of postharvest gray mold on fruits crops. Postharvest Biol. Technol. **113**, 69–76 (2016)
2. Landi, L., Feliziani, E., Romanazzi, G.: Expression of defense genes in strawberry fruits treated with different resistance inducers. J. Agric. Food Chem. **62**, 3047–3056 (2014)
3. Quattrucci, A., Ovidi, E., Tiezzi, A., Vinciguerra, V., Balestra, G.: Biological control of tomato bacterial speck using *Punica granatum* fruit peel extract. Crop Prot. **46**, 18–22 (2013)
4. Regaieg, H., et al.: Evaluation of pomegranate (*Punica granatum L. var. Gabsi*) peel extract for control of root-knot nematode Meloidogyne javanica on tomato. Arch. Phytopathol. Plant Prot. **50**(2), 1–11 (2017)
5. Cui, H., Surendhiran, D., Li, C., Lin, L.: Biodegradable zein active film containing chitosan nanoparticle encapsulated with pomegranate peel extract for food packaging. Food Packag. Shelf Life **24** (2020). https://doi.org/10.1016/j.fpsl.2020.100511
6. Surendhiran, D., Li, C., Cui, H., Lin, L.: Fabrication of high stability active nanofibers encapsulated with pomegranate peel extract using chitosan/PEO for meat preservation. Food Packag. Shelf Life **23** (2020). https://doi.org/10.1016/j.fpsl.2019.100439
7. Liolios, C., Gortzi, O., Lalas, S., Tsaknis, J., Chinou, I.: Liposomal incorporation of carvacrol and thymol isolated from the essential oil of Origanum dictamnus L. and in vitro antimicrobial activity. Food Chem. **112**, 77–83 (2009)

8. Pinho, E., Grootveld, M., Soares, G., Henriques, M.: Cyclodextrins as encapsulation agents for plant bioactive compounds. Carbohydr. Polym. **101**, 121–135 (2014)
9. Conte, R., Marturano, V., Peluso, G., Calarco, A., Cerruti, P.: Recent advances in nanoparticle-mediated delivery of anti-inflammatory phytocompounds. Int. J. Mol. Sci. **18** (2017). https://doi.org/10.3390/ijms18040709
10. Rongai, D., Pulcini, P., Pesce, B., Milano, F.: Antifungal activity of pomegranate peel extract against fusarium wilt of tomato. Eur. J. Plant Pathol. **147**(1), 229–238 (2016). https://doi.org/10.1007/s10658-016-0994-7
11. Vergaro, V., et al.: Scalable production of calcite nanocrystals by atomization process: Synthesis, characterization, and biological interactions study. Adv. Powder Technol. **28**(10), 2445–2455 (2017)
12. Vergaro, V., et al.: Cell-penetrating $CaCO_3$ nanocrystals for improved transport of NVP-BEZ235 across membrane barrier in T-cell lymphoma. Cancers **10** (2018). https://doi.org/10.3390/cancers10020031
13. Baldassarre, F., et al.: Biocatalytic synthesis of phospholipids and their application as coating agents for $CaCO_3$ Nano–crystals: characterization and intracellular localization analysis. Chem. Sel. **1**(20), 6507–6514 (2016)
14. Baldassarre, F., et al.: Application of calcium carbonate nanocarriers for controlled release of phytodrugs against *Xylella fastidiosa* pathogen. Pure Appl. Chem. (2020). https://doi.org/10.1515/pac-2018-1223
15. Rongai, D., Sabatini, N., Pulcini, P., Di Marco, C., Storchi, L., Marrone, A.: Effect of pomegranate peel extract on shelf life of strawberries: computational chemistry approaches to assess antifungal mechanisms involved. J. Food Sci. Technol. **55**(7), 2702–2711 (2018). https://doi.org/10.1007/s13197-018-3192-0

Ozone and Sanodyna® Treatments to Increase Safety and Sustainability in Extra Virgin Olive Oil Production Chain

Alessio Cappelli(✉), Piernicola Masella, Alessandro Parenti, and Enrico Cini

Department of Agriculture, Food, Environment and Forestry (DAGRI), University of Florence, Piazzale delle cascine 16, 50144 Florence, Italy
alessio.cappelli@unifi.it

Abstract. The aim of this work is to assess the effectiveness of ozone and sanodyna® treatments in olive washing stage to reduce microbial load, water consumptions, and environmental impacts, safeguarding the extra virgin olive oil characteristics. With respect to ozonization treatments, 12 samples of olives were tested (six treated insufflating ozone and 6 (not treated) used as control). Three of the treated samples were processed to produce olive oil within 24 h, while the others three were processed after seven days. The same apply for the six untreated samples. In this case, the effect on olives microbiological load and on the quality of extra virgin olive oil were assessed. Sanodyna® was sprayed directly to olives (at 0.05%) and was also tested at three different concentrations (4, 8, 12%) in the olives washing water. Notably, the effects on the microbial load, on washing water characteristics, and on the obtained oil were assessed. The ozonization treatment did not reduce the olive's microbial load. In addition, ozone had detrimental effects on the total phenolic content of the olive oil, highlighting its unsuitability to this application. The direct treatment with sanodyna® did not decrease the microbial load. On the other hand, the application of higher dosages in the washing stage showed promising results. In particular, sanodyna® treatments reduced the microbial load and extended the duration of washing water. In conclusion, despite the effects on the olive oils need to be evaluated, the latter application seems to be effective in reducing microbiological load and environmental impacts.

Keywords: EVOO · Hypochlorous acid (HOCl) · Olive washing water · Safety · Environmental pressures

1 Introduction

The production and consumption of extra virgin olive oil (EVOO) have a considerable importance in Europe, with both economic and cultural relevance, especially in Italy [1]. The extra virgin olive oil, considered worldwide as the standard bearer of the Mediterranean diet, is constantly gaining interest, and its production is continuously increasing [2]. The European Commission report [3] highlighted that 3 million tons of olive oil are approximately produced each year in the world. In particular, two million

M. Biocca et al. (Eds.): SHWA 2020, LNCE 252, pp. 486–495, 2022.
https://doi.org/10.1007/978-3-030-98092-4_50

tons are produced in the EU, mainly by four member states (i.e. Spain (66% of EU production), Italy (15%), Greece (13%), and Portugal (5%)) [3]. The quality and the characteristics of EVOO's are influenced by the management of several key unit operations such as, harvesting, olives washing, olives crushing, paste malaxing, centrifugation, and filtration [2, 4]. Certain unit operations in EVOO production, such as harvest and post-harvest [5], malaxing [6], centrifugation [7], and filtration [8], have been examined in depth, leading to a significant improvement of the EVOO quality. On the other hand, others unit operations, like the washing stage, have been slightly neglected or leastwise did not experience the same involvement in terms of research efforts [4].

The washing stage represent a crucial steps for ensuring the safety and the quality of the EVOO [4, 9, 10]. Moreover, the improvement of this stage could significantly reduce water consumptions and environmental impacts [4, 9]. Nevertheless, only few works are available in the literature [4, 9–11]. Some authors suggested a continuous recycling approach of exhausted washing water through a closed circuit to reduce both the microbial load and the environmental impact, but with unsatisfactory results [11]. However, with respect to the improvement of the olive washing phase, Cappelli et al. (2019) [4] developed an innovative washing machine based on usability engineering. The results presented by the latter authors showed that the innovative washing machine was able to significantly reduce water consumption (60 L per hour) and, additionally, proved to be able to improve the olive washing stage. Nevertheless, given the lack of studies providing innovations and improvement strategies aimed to advance the washing stage, further investigations are necessary, thus motivating this work.

As highlighted by Espadas-Aldana et al. (2019) [12], approximately eighty percent of olive mass is composed by pulp, stones, and water; therefore, the production process of EVOO generates four times more waste than quantity of oil [2, 12]. Consequently, the disposal of olive mill wastes (i.e. olive washing water, olive mill wastewater, olive seed, pomace, and olive tree leaves) represent one of the most critical environmental problem in the Mediterranean basin [2, 12]. For this reason, innovations, improvement strategies, and solutions for the washing phase (but also for all the other unit operations in EVOO production process) must mandatorily consider the environmental concerns. Therefore, the aim of this work is to assess the effectiveness of ozone and sanodyna® treatments in olive washing stage in order to reduce microbial load, water consumptions, and environmental impacts, safeguarding the extra virgin olive oil characteristics.

2 Materials and Methods

2.1 Raw Materials, Equipment, and Sanitizing Products

Fresh olives were kindly provided by Giacomo Grassi farm (Greve in Chianti, Tuscany, Italy). With respect to ozonization treatment, ozone was produced using an ozone generator machine (model QLA 3G- 220 V/50 Hz) kindly provided by Innovazioni e tecnologie Ltd. (Florence, Italy). Successively, ozone was insufflated into three plastic containers to perform the ozonization treatment. After the ozonization treatment, treated and untreated olives samples were processed to produce extra virgin olive oil in our laboratory-scale olive oil mill plant. The oil microextraction plant consisted of a crusher, a lab-scale malaxator, and a laboratory centrifuge (Model NEYA8, Neya centrifuges, Modena, Italy). Regarding sanodyna® treatments, six 0.5 L bottles with spray trigger containing 0.05% of hypochlorous acid (HOCl) and 99.95% of electrolyzed water for direct applications on the olives, were kindly provided by Nuova Control System Ltd (Prato, Italy). The latter company also provided sanodyna® with higher percentages of hypochlorous acid (i.e. 4, 8, and 12%) for the treatment of the exhausted washing water of the olive oil mill. In the case of direct application of sanodyna® to the olives, treated and untreated olives samples were processed to produce extra virgin olive oil in our laboratory-scale olive oil mill plant.

2.2 Ozonization Treatment and Experimental Design

Ozonization treatment was performed using fresh olives which were divided into 12 samples of 1.5 kg each. Successively, six samples were treated insufflating ozone and the other six were not treated and used as control. Three of the treated samples were processed to produce olive oil within 24 h, while the others three were processed after seven days to assess the effects of storage time in addition to the ozonization treatment. The same apply for the six untreated samples. The ozonization treatment was performed using an ozone generator machine connected with a three-liter plastic container (especially modified) containing 1.5 kg of sample to be treated. The treatment was carried out insufflating 333 mg of ozone per Kg of olives in 10 min. 60 g of olives, for each sample, were collected before processing for microbiological analyses.

From treated and untreated olive samples, 12 oils were produced in our laboratory-scale olive oil mill plant. After crushing, the olive paste was malaxed in a hermetically-sealed cylindrical malaxing machine for 20 min at a controlled temperature of 27 °C. Successively, the oil was separated from the vegetation water and from the solid fractions by centrifugation for 10 min at 6500 rpm. Consecutively, the oil was recovered with a separatory glass funnel and stored in glass bottles. Finally, the 12 oils were analyzed for acidity, peroxide index, and total polyphenols content. All the tests were carried out in three replicates.

2.3 Sanodyna® Treatments and Experimental Design

Sanodyna® treatments were performed directly on the fresh olives and on the exhausted washing water collected at the end of the washing stage in an olive oil mill. With respect to direct application, fresh olives were divided into 12 samples of 20 kg each. Successively, six samples were treated using a sanodyna® solution with 0.05% of Hypochlorous acid (HOCl) and 99.95% of electrolyzed water. In particular, 0.5 L of solution were directly sprayed on each sample (0.0125 ml of HOCl per kg of olives). The other six samples were used as control. Three of the treated and untreated samples were processed to produce olive oil within 24 h, while the others six were processed after seven days to assess the effects of storage time. 60 g of olives, for each sample, were collected before processing for microbiological analyses. From treated and untreated olive samples, 12 oils were produced in our laboratory-scale olive oil mill plant with the same procedure described in Sect. 2.2. Finally, the 12 oils were analyzed for acidity, peroxide Index, and total polyphenols content. All the tests were carried out in three replicates.

Regarding the treatment of exhausted washing water with sanodyna®, four levels of HOCL per liter of exhausted washing water were tested: 0% (Control); 4% (0.020 mL/L); 8% (0.040 mL/L); and 12% (0.060 mL/L). The exhausted washing water was collected directly from the washing machine of the olive oil mill at the end of the washing stage and treated by Valoritalia laboratory (Florence, Italy) within 12 h. Treated and untreated samples of washing water were analyzed by the latter laboratory as described in Sect. 2.4. All the tests were carried out in three replicates.

2.4 Olives, Oils, and Washing Water Analyses

The analyses of treated and untreated samples of olives, olive oil, and exhausted washing water, were carried out by Valoritalia laboratory (Florence, Italy), following approved, official methods [13]. With respect to olives samples, total bacterial count (log CFU/g), molds (log CFU/g), yeasts (log CFU/g), and *Aureobasidium spp.* (log CFU/g) were investigated [13]. Moreover, the olive oil samples, obtained from our laboratory-scale mill trials, were analyzed for free fatty acids (% oleic acid), peroxide value (*mEq* of O_2 per kg of oil), UV spectroscopic indices (K_{232}, K_{270}, and ΔK), and total polyphenols content (mg/Kg), according to official methods [14]. Finally, with respect to exhausted washing water, turbidity (NTU), Biochemical Oxygen Demand (BOD) (mg O_2/L), Chemical Oxygen Demand (COD) (mg O_2/L), total bacterial count (CFU/mL), *Escherichia coli* (CFU/mL), intestinal *Enterococci* (CFU/mL), and *Clostridium perfringens* (CFU/mL) were evaluated.

2.5 Statistical Analyses

The effects of storage time and ozonization treatment were assessed using a fixed effects model, as described by Montgomery (2017) [15]. The same apply for the sanodyna® treatments in direct application to the olives with 0.05% of HOCl. In particular, in both cases, a two-way ANOVA for storage time, treatment (ozone or sanodyna®), and their interaction was performed [15]. Significance was set at $p < 0.05$. When the significance level was reached, a post-hoc Tukey HSD test was run [15].

With respect to the effects of sanodyna® treatments with higher percentages of HOCl on the exhausted washing water of the olive oil mill, a one-way ANOVA was performed. Four levels of HOCl were tested: 0% (control), 4%, 8%, and 12%. Significance was set at $p < 0.05$. In cases of statistically significant results ($p < 0.05$) a Tukey HSD post hoc test was performed.

3 Results and Discussion

3.1 Ozonization Treatment

The results of ozonization treatment and storage time on the microbiological load of the olives are summarized in Table 1.

Table 1. Results of microbiological analyses (log CFU/g) expressed as the mean of the three replicates ± SD.

Sample	Total bacterial count (log CFU/g)	Molds (log CFU/g)	Yeasts (log CFU/g)	Aureobasidium spp. (log CFU/g)
Untreated (0 week storage)	2.22 ± 0.08	2.16 ± 0.15	1.95 ± 0.05	2.45 ± 0.05
Ozonized (0 week storage)	2.19 ± 0.20	2.03 ± 0.05	1.96 ± 0.04	1.98 ± 0.03
Untreated (1 week storage)	2.33 ± 0.70	3.26 ± 0.19	2.64 ± 0.73	4.61 ± 0.69
Ozonized (1 week storage)	1.95 ± 0.05	3.19 ± 0.27	2.27 ± 0.59	4.40 ± 0.26

With respect to total bacterial count, molds, yeasts, and *Aureobasidium*, the two-way ANOVA found that the effects of storage time is statistically-significant, while this was not the case for ozonization treatment and for their interaction. As highlighted in Table 1, with the increases of storage time the microbiological load of the olives significantly increases, as expected. However, no statistically-significant effects for ozone treatment were found. This is supported by Bravo et al. (2007) [16] which highlighted that the treatment with ozone did not lead to the production of any inhibitory (toxic) compound for microbial growth and was always favorable for the development of both yeasts and lactobacilli. However, on the contrary, other authors found interesting applications of ozone in reducing the microbiological load in table olives processing [17].

Concerning the effects of ozone treatment and storage time on EVOO characteristics, Table 2 summarizes the results on acidity, peroxide value, UV spectroscopic indices (K_{232}, K_{270}, and ΔK), and total polyphenols content.

Table 2. Results of oils characterization expressed as the mean of the three replicates ± SD.

Sample	Acidity (%)	Peroxide index (mEq O2/kg)	K232	K 268	Delta-K	Total polyphenols content (mg/Kg)
Untreated (0 week storage)	0.48 ± 0.07	5.27 ± 0.65	1.92 ± 0.04	0.17 ± 0.04	0.004 ± 0.002	962.00 ± 52.37
Ozonized (0 week storage)	0.54 ± 0.16	6.50 ± 0.89	1.86 ± 0.02	0.14 ± 0.01	0.004 ± 0.001	922.33 ± 67.93
Untreated (1 week storage)	0.44 ± 0.03	7.30 ± 0.62	1.91 ± 0.05	0.15 ± 0.01	0.004 ± 0.000	661.67 ± 10.60
Ozonized (1 week storage)	0.39 ± 0.03	7.37 ± 0.67	1.88 ± 0.03	0.14 ± 0.01	0.004 ± 0.000	593.00 ± 9.64

Regarding acidity and UV spectroscopic indices (K_{232}, K_{270}, and ΔK), the two-way ANOVA did not find any statistically significant differences for storage time, ozonization treatment, and their interaction. However, with respect to peroxide index, the two-way ANOVA found individual effects for storage time, while this is not observed for ozonization treatment and the factor interaction. In particular, storage time significantly increases peroxide index by 25%, highlighting a negative effect on this key parameter. Moreover, regarding total polyphenols content, the two-way ANOVA found individual effects for storage time and ozonization treatment, while this was not the case for their interaction. In particular, storage time significantly reduced total polyphenols content by 30%, showing a detrimental effects. Additionally, ozonization treatment generates a statistically-significant reduction of the total polyphenols content which were reduced by 7%. The latter finding is supported by several authors [16–18]. In particular, Chedeville et al. (2009) [18] developed a very efficient method based on ozonization treatment to remove up to 80% of the phenolic compounds, highlighting the unsuitability of ozonization treatments to improve the EVOO production chain.

3.2 Sanodyna® Treatment on Olives

The results of sanodyna® treatment and storage time on the microbiological load of the olives are summarized in Table 3.

Table 3. Results of microbiological analyses (log CFU/g) expressed as the mean of the three replicates ± SD.

Sample	Total bacterial count (log CFU/g)	Molds and yeasts (log CFU/g)
Untreated (0 week storage)	2.40 ± 0.14	1.06 ± 0.06
Treated with Sanodyna at 0.05% (0 week storage)	2.42 ± 0.12	1.04 ± 0.04
Untreated (1 week storage)	2.81 ± 0.29	1.99 ± 0.32
Treated with Sanodyna at 0.05% (1 week storage)	2.96 ± 0.07	2.21 ± 0.20

With respect to microbiological analyses, the two-way ANOVA found that the effects of storage time is statistically-significant, while this was not the case for sanodyna® treatment and for the factor interaction. As shown in Table 3, storage time significantly increase total bacterial count, molds, and yeasts, as found for the ozonization treatment (Sect. 3.1). However, no statistically-significant effects for sanodyna® treatment were found. This might be due to the low concentration of Hypochlorous acid tested in the trials.

Regarding the effects of sanodyna® treatment and storage time on EVOO characteristics, Table 4 summarizes the results on acidity, peroxide value, UV spectroscopic indices (K232, K270, and ΔK), and total polyphenols content.

Table 4. Results of oils characterization expressed as the mean of the three replicates ± SD.

Sample	Acidity (%)	Peroxide index (mEq O2/kg)	K232	K 268	Delta-K	Total polyphenols content (mg/Kg)
Untreated (0 week storage)	0.19 ± 0.02	2.67 ± 0.80	1.24 ± 0.15	0.08 ± 0.02	0.002 ± 0.001	444.00 ± 48.33
Treated with Sanodyna at 0.05% (0 week storage)	0.15 ± 0.02	2.60 ± 0.83	1.29 ± 0.15	0.08 ± 0.02	0.003 ± 0.001	402.67 ± 44.33
Untreated (1 week storage)	0.20 ± 0.02	4.07 ± 0.80	1.36 ± 0.15	0.07 ± 0.03	0.000 ± 0.000	440.67 ± 47.67
Treated with Sanodyna at 0.05% (1 week storage)	0.21 ± 0.02	5.50 ± 0.90	1.35 ± 0.15	0.08 ± 0.02	0.000 ± 0.000	472.00 ± 51.00

Concerning the EVOO characteristics (acidity, peroxide value, UV spectroscopic indices, and total polyphenols content), the two-way ANOVA did not find any statistically significant differences for storage time, sanodyna® treatment, and their interaction. Contrarily to ozonization treatment, sanodyna® did not show any harmful effects neither on the oils quality nor on the total polyphenols content. However, the concentration of 0.05% of HOCl proved to be infective in reducing the microbiological load of the olives, motivating the increase of HOCl percentages in the treatment of exhausted washing water (Sect. 3.3).

3.3 Sanodyna® Treatments on Exhausted Washing Water

The results of sanodyna® treatments on the microbiological load and on the physical and chemical characteristics of the exhausted washing water are summarized in Table 5.

Table 5. Results of exhausted washing water characterization expressed as the mean of the three replicates ± SD.

Sample	Turbidity (NTU)	Dry residues at 180 °C (mg/L)	Biochemical oxygen demand (mg O_2/L)	Chemical oxygen demand (mg O_2/L)	Total bacterial count (CFU/mL)	*Escherichia coli* (CFU/mL)	*Intestinal Enterococci (CFU/mL)*	*Clostridium perfringens (CFU/mL)*
0% Control	387.00 ± 23.30	588.00 ± 15.00	140.00 ± 20.10	378.00 ± 20.60	390.00 ± 8.33	1600.00 ± 3.33	870.00 ± 10.33	1500.00 ± 12.00
Treated with 4% Sanodyna	362.00 ± 18.20	750.00 ± 13.25	131.00 ± 18.40	355.00 ± 14.20	350.00 ± 6.00	4.00 ± 0.50	45.00 ± 3.33	160.00 ± 4.00
Treated with 8% Sanodyna	190.00 ± 12.00	884.00 ± 18.00	119.00 ± 12.00	314.00 ± 12.85	210.00 ± 5.33	0.00 ± 0.00	5.00 ± 2.00	26.00 ± 2.66
Treated with 12% Sanodyna	148.00 ± 10.60	936.00 ± 15.20	114.00 ± 16.00	299.00 ± 10.00	12.00 ± 2.66	0.00 ± 0.00	0.00 ± 0.00	0.00 ± 0.00

With respect to turbidity, the one-way ANOVA found a significant main effect for the factor "percentage of HOCl" (Table 5). In particular, turbidity values obtained with the addition of 8% and 12% of HOCl were significantly lower compared to the control. The statistically-significant reduction of water turbidity is mainly related to the powerful antibacterial activity of HOCl, consistently with other authors [19, 20]. Furthermore, this is confirmed by the statistically-significant reduction of total bacterial count, *Escherichia coli*, Intestinal *Enterococci*, and *Clostridium perfringens* in exhausted washing water treated with increased percentages of HOCl [21–23]. In particular, the treatment with 12% HOCl proved to be able to eliminate *Escherichia coli*, Intestinal *Enterococci*, and *Clostridium perfringens*, increasing the safety of the treated water. Finally, with regard to BOD and COD, the one-way ANOVA did not found any statistically significant differences.

4 Conclusions

Notwithstanding the great strides made by the scientific research in improving the most important unit operations of the EVOO production chain, the washing stage has been slightly neglected, highlighting the need for innovations and improvement strategies. Our results highlighted that the tested ozone treatment was not effective in reducing the microbiological load of the olives. In addition, a negative effect on the total phenolic content of EVOO was found. The treatment with Sanodyna® at 0.05% directly sprayed on the olives did not show any statistically-significant effects neither on the microbiological load of the olives nor on the EVOO quality.

However, the treatment of the exhausted washing water of the olive oil mill with higher percentages of Sanodyna® proved to be effective in reducing microbiological load and water turbidity in the treated washing water samples. In particular, the treatment with 12% of Hypochlorous acid completely eliminated *Escherichia coli*, *Enterococci*, and *Clostridium perfringens*, increasing the safety of the treated water. In future researches the effects of Sanodyna® treatment with 12% of Hypochlorous acid on the EVOO quality will be carefully investigated. However, the application of Sanodyna® at high levels of hypochlorous acid (in particular at 12%) proved to be very effective in the sanitization of olive washing water with a subsequent reduction of water consumptions.

Funding. This research did not receive any specific grant from funding agencies in the public, commercial, or not-for-profit sectors.

Declarations of Interest. The authors declare no conflict of interests.

References

1. Altieri, R., Esposito, A.: Olive orchard amended with two experimental olive mill wastes mixtures: effects on soil organic carbon, plant growth and yield. Biores. Technol. **99**, 8390–8393 (2008)

2. Cappelli, A., Menditto, N., Cini, E.: Innovative olive tree leaves shredder prototype for the valorization of wasted leaves: an application to high-quality compost production. Sustainability **13**(16), 9421 (2021)
3. European Commission: Olive Oil: An Overview of the Production and Marketing of Olive Oil in the EU (2020). https://ec.europa.eu/info/food-farming-fisheries/plants-and-plant-products/plant-products/olive-oil
4. Cappelli, A., Parretti, C., Cini, E., Citti, P.: Development of a new washing machine in olive oil extraction plant: a first application of usability-based approach. J. Agric. Eng. **50**(3), 134–142 (2019)
5. Masella, P., et al.: An artificial neural network model to predict olive mechanical harvesting: a first approach based on metadata. In: VI International Symposium on Applications of Modelling as an Innovative Technology in the Horticultural Supply Chain Model-IT, vol. 1311, pp. 355–359 (2021). Acta Horticulturae
6. Tamborrino, A., et al.: Industrial trials on coadjuvants in olive oil extraction process: effect on rheological properties, energy consumption, oil yield and olive oil characteristics. J. Food Eng. **205**, 34–46 (2017)
7. Boncinelli, P., Daou, M., Cini, E., Catalano, P.: A simplified model for designing and regulating centrifugal decanters for olive oil production. Trans. ASABE **52**, 1961–1968 (2009)
8. Bottino, A., et al.: Application of membrane processes for the filtration of extra virgin olive oil. J. Food Eng. **65**(2), 303–309 (2004)
9. Maza-Márquez, P., González-Martínez, A., Rodelas, B., González-López, J.: Full-scale photobioreactor for biotreatment of olive washing water: structure and diversity of the microalgae bacteria consortium. Biores. Technol. **238**, 389–398 (2017)
10. Di Giovacchino, L., Sestili, S., Di Vincenzo, D.: Influence of olive processing on virgin olive oil quality. Eur. J. Lipid Sci. Technol. **104**, 587–601 (2002)
11. Vichi, S., Boynuegri, P., Caixach, J., Romero, A.: Quality losses in virgin olive oil due to washing and short-term storage before olive milling. Eur. J. Lipid Sci. Technol. **117**, 12 (2015)
12. Espadas-Aldana, G., Vialle, C., Belaud, J.P., Vaca-Garcia, C., Sablayrolles, C.: Analysis and trends for Life Cycle Assessment of olive oil production. Sustain. Prod. Consump. **19**, 216–230 (2019)
13. AOAC: Official Methods of Analysis, 19th edn. Association of Official Analytical Chemists (2007)
14. EC, 2008: Commission Regulation (EC) No 640/2008 of 4 July 2008 amending Regulation (EEC) No 2568/91 on the characteristics of olive oil and olive-residue oil and on the relevant methods of analysis. Off. J. Eur. Union **L178**, 11–16 (2008)
15. Montgomery, D.C.: Design and Analysis of Experiments. Wiley, Hoboken (2017)
16. Bravo, K.S., López, F.A., García, P.G., Quintana, M.D., Fernández, A.G.: Treatment of green table olive solutions with ozone. Effect on their polyphenol content and on Lactobacillus pentosus and Saccharomyces cerevisiae growth. Int. J. Food Microbiol. **114** (1), 60–68 (2007)
17. García-García, P., Arroyo-López, F.N., Rodríguez-Gómez, F.: Partial purification of iron solutions from ripe table olive processing using ozone and electro-coagulation. Sep. Purif. Technol. **133**, 227–235 (2014)
18. Chedeville, O., Debacq, M., Porte, C.: Removal of phenolic compounds present in olive mill wastewaters by ozonation. Desalination **249**(2), 865–869 (2009)
19. Ghernaout, D., Elboughdiri, N.: On the other side of viruses in the background of water disinfection. Open Access Libr. J. **7**(5), 1–29 (2020)

20. Ghernaout, D.: Water treatment chlorination: an updated mechanistic insight review. Chem. Res. J. **2**, 125–138 (2017)
21. Pullar, J.M., Vissers, M.C., Winterbourn, C.C.: Living with a killer: the effects of hypochlorous acid on mammalian cells. IUBMB Life **50**(4–5), 259–266 (2000)
22. Fang, F.C.: Antimicrobial reactive oxygen and nitrogen species: concepts and controversies. Nat. Rev. Microbiol. **2**(10), 820–832 (2004)
23. Block, M.S., Rowan, B.G.: Hypochlorous acid: a review. J. Oral Maxillofac. Surg. **78**(9), 1461–1466 (2020)

Food Process Safety Investigated Through NIR Water Absorption Variations

Tiziana Maria Piera Cattaneo[1,2(✉)], Laura Marinoni[1], Alessandro Cammerata[4], Annamaria Stellari[3], Massimo Brambilla[3], and Elio Romano[3]

[1] Research Centre for Engineering and Agro-Food Processing, CREA, Via G. Venezian, 26, 20133 Milan, Italy
tiziana.cattaneo@crea.gov.it
[2] DAFNE, Tuscia University, Viterbo, Italy
[3] Research Centre for Engineering and Agro-Food Processing, CREA, Via Milano, 43, 24047 Treviglio, BG, Italy
[4] Research Centre for Engineering and Agro-Food Processing, CREA, Via Manziana, 30, 00189 Roma, Italy

Abstract. This paper aimed to study the Aquaphotomics application, as NIR water absorption variations, to monitor and address food dehydration processes for food safety and preservation.

A solar drying micro-system with a fan always active at the bottom was used. Two matrices were tested: eggplant and apple slices. The water activity (a_w) content was determined at the end of the drying process. The NIR spectra were collected in reflectance mode from 900 to 1600 nm using a MicroNIR 1700 (VIAVI Solutions, Italy). The pre-treated and averaged absorbance spectra were displayed on radial axes defined by the activated water absorbance bands. PCA, and MBM (Moving Block Model) were then built up.

Some critical points were identified: a) the positioning and the isolation of the NIR probe; b) the working temperature; c) the sample size; d) the identification of the process end. The applicability of Aquaphotomics for the continuous monitoring of food drying processes using an adequate NIR probe can be a good opportunity to understand product safety better. Aquagrams can help to identify the actual endpoint of the process. The PCA loadings confirmed the critical contribution of the water absorption independently from the food matrix. The selection of few wavelengths to study food processes and quality can optimise the transfer of instrumental data to IT platforms and networks. Further investigations are in progress to check the possibility of transfer the obtained results on a pilot scale for future applicability of these applications.

Keywords: Food · Water absorption pattern · NIR sensors · Food storage

1 Introduction

Drying is the oldest method of food preservation involving the removal of moisture using thermal energy. Dehydration primary purpose is to extend the food products shelf-life by reducing water activity (a_w), thus inhibiting microbial growth and

M. Biocca et al. (Eds.): SHWA 2020, LNCE 252, pp. 496–505, 2022.
https://doi.org/10.1007/978-3-030-98092-4_51

enzymatic activity. Several drying methods are commercially available, based on different principles of operation, operating pressure, and mode of heat transfer. However, such methods are expensive, high-energy demanding, and greenhouse gases producing [1, 2].

Some of the quality changes that may occur in food during drying include loss of aroma and flavours, changes in colour and texture, and a decrease in its nutritional value [3]. Solar drying is an economical and sustainable alternative to traditional techniques, being a low-cost and straightforward technology, both in terms of initial capital and management. No skilled labour is required, and large quantities of crops can be dried using renewable energy [4]. Sun drying, without specific equipment, is the most widely used food processing operation globally, and more than 250,000,000 tons of fruit, vegetables, and cereals are dried every year using solar energy [4]. In some Countries, foods are simply laid out on the field or roofs and regularly turned until completely dry. More sophisticated methods use types of equipment to pick up solar energy and heat the air that will be used in the process.

The transformation process influences the organisation of water molecules (free water, dimers, trimers, solution shells) in the food tissues and produces changes in the water absorbance pattern.

When an aqueous system, such as fruit, is measured by Near InfraRed (NIR) spectroscopy, the light absorbance reflects the water molecules vibrations [5].

The Aquaphotomics approach bases on the high sensitivity of water's hydrogen bonds which can reflect any change of the aqueous system highlighting differences that can be observed, measured, analyzed, and interpreted.

The past two decades have seen much progress in water science with the establishment of a new common platform that can provide integration of knowledge.

Several groups of researchers have investigated the application of this holistic approach to the food sector. The usefulness of this discipline in understanding the role of water in food matrices has been proved [6, 7].

This study aims to evaluate the Aquaphotomics application, as NIR water absorption variations, to monitor and address food dehydration processes for food safety and preservation.

2 Materials and Methods

2.1 Dehydration Processes

Drying experiments were performed at the Research Centre for Engineering and Agro-Food Processing in Treviglio (BG, Italy) using eggplants as a preliminary activity and apples as a validation experiment.

Samples were purchased from local markets in Treviglio. To prevent physiological and chemical changes due to the respiration process, samples were stored at 4 ± 1 °C before the experiments. Fruits were mechanically cut into slices of uniform thickness (0.3–0.5 cm) to ensure a uniform drying and placed on the drier trays close to each other without overlapping.

The dehydration process was performed with a solar micro drying plant equipped with a continuously working fan at the bottom for the air extraction (Fig. 1).

Data for weight, external and internal temperature, and relative humidity were recorded during the drying process using Arduino sensors (ATmega32U4 Arduino Leonardo AVR® ATmega AVR MCU 8 bit; Arduino UNO Rev3 orig. HIGH QUALITY; Yun Rev 2 Arduino).

(a) (b)

Fig. 1. Solar micro drying plant employed in the experimentation: front (a) and back (b) sides.

2.2 Chemical Analyses

The water activity (a_w) content was determined at the end of the drying process. The measurements were performed with the AquaLab 4TE (METER Group, Inc. USA) benchtop water activity meter, based on the chilled-mirror dew point technique [8]. To obtain accurate values, the analysis was carried out in customized mode: 3 consecutive readings of the same sample were performed within a range of $\pm$ 0.005 a_w. The output a_w value is the average of the 3 readings. For each sample, 3 independent measurements were performed.

2.3 NIR Spectra Acquisition

NIR spectra were collected in reflectance mode using the MicroNIR1700™ (VIAVI Solutions Italia S.r.l., Monza, Italy) portable spectrometer with the MicroNIR Pro ES 1700 software. Spectra were acquired continuously throughout the process by placing the NIR probe over a sample slice, positioned on the intermediate shelf. The probe was thermally insulated to avoid temperature variations. In the eggplant preliminary process, the probe was wrapped in aluminium insulating fabric (Fig. 2a), while for the apple process, a commercial extended probe housing was used (Fig. 2b). For the eggplant process, spectra were collected for 50 h, one spectrum every two hours, for a total of 26 spectra; for the apple process, spectra were collected for 61 h, one spectrum every hour, for a total of 61 spectra. Before the analysis, the instrument underwent calibration for the black, in the air, and for the white on the supplied standard tile.

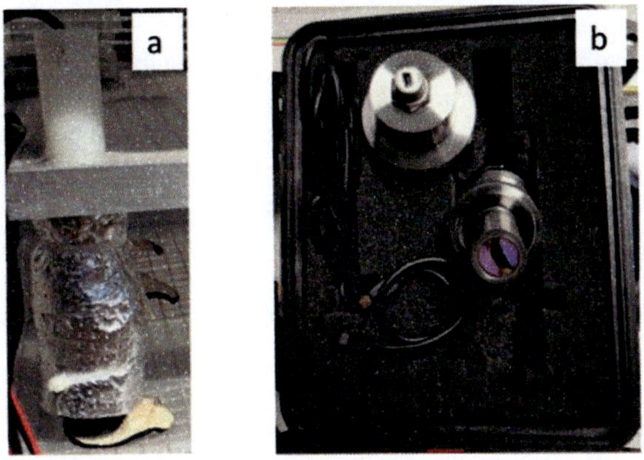

Fig. 2. NIR probe conditioning: aluminium insulating fabric for the eggplant dehydration process (a); commercial extended probe housing for the apple process (b).

2.4 Data Analyses

Spectra were mathematically prepreprocessed according to Tsenkova et al. (2018) [9]. In particular, the 2nd derivative Savitzky-Golay filter (2nd order polynomial fit and 21 points) and multiplicative scatter correction (MSC) were applied to remove potential scatter effects. Transformed spectra underwent normalization by applying the following formula:

$$(A\lambda - \mu\lambda)/\sigma\lambda$$

where $A\lambda$ is the transformed absorbance, $\mu\lambda$ is the mean value of all spectra and $\sigma\lambda$ is the standard deviation of all spectra at wavelength λ. Finally, Aquagrams were calculated from the spectral data. Aquagram is a spider-shaped diagram that shows the normalized and averaged absorbances at the selected wavelengths according to specific water absorbance bands.

Principal Component Analysis (PCA, 95% confidence level) were performed on 1st derivative spectral data (Savitzky-Golay filter, 2nd order polynomial fit and 11 points) by using The Unscrambler software (v 9.7, Camo, Inondhcim, Norway).

The Moving Block mean Model (MBM) was applied to the data using a block size of 3 for eggplant process and 7 for apple process. Moving Block Models are calibration-free, unsupervised methods which allow to monitor the process with no need for prior knowledge of NIR spectra of the analysed matrix. Thus, they are a valuable tool for assessing the endpoints of evolving processes such as powder blending or drying operations. They are based on the summarized variance in blocks of spectra collected over the time course of the process. A block is a predefined number of spectra previously demonstrated to provide enough information about the changes within the evolving process [10].

3 Results and Discussion

The eggplant dehydration process was used as preliminary experimentation to assess and optimize the dehydration process conditions to be validated with the apple process. In this first activity, some critical points were identified: a) the positioning of the NIR probe, which must remain in contact with the product throughout the process; b) the working temperature and the NIR probe insulation to avoid that temperature variations could have a detrimental effect on NIR data; c) the sample size; d) the identification of the process end. Figure 3 shows the system's setup with the chosen configuration of the critical elements: NIR probe positioning over an eggplant slice at the beginning (a) and the end (b) of the dehydration process.

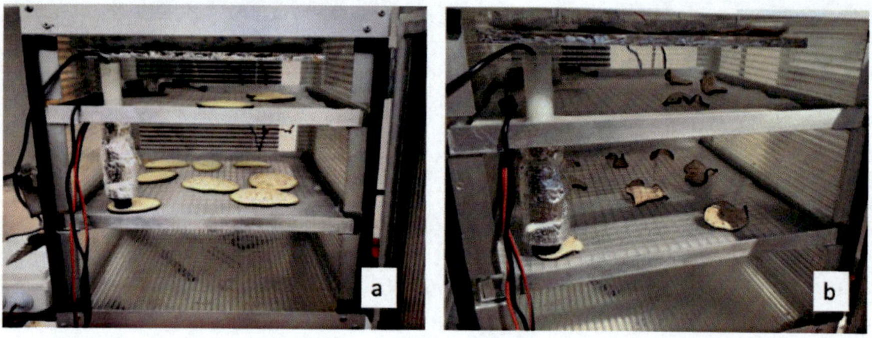

Fig. 3. NIR probe positioning at the beginning (a) and at the end (b) of the dehydration process of eggplants.

The drying process lasted 50 h and it was stopped when the samples reached a constant weight, considering a difference lower than 1% between two consecutive weightings. At the end of the process, it was verified that the probe was still in the starting position and that it was still in contact with the slice of product. This ending point corresponded to an a_w value of the dried product of 0.388. This value is well below the microbiological safety threshold identified as 0.6 [11] and indicates that the safety of the product was ensured.

Spectra acquired during the eggplant dehydration are shown in Fig. 4.

The raw spectra (Fig. 4a) showed broad and overlapped absorption bands detecting predominant water absorption bands at 960–990 (2nd overtone -OH stretching), 1150–1190 (Combination of 1st overtone -OH stretching and OH bending), and 1400–1435 nm (1st overtone -OH stretching) [9]. Applying the 1st derivative calculation as pretreatment (Fig. 4b) allowed better visualization of the changes in the spectral profile during the process. In particular, great variations were recorded in the spectral region

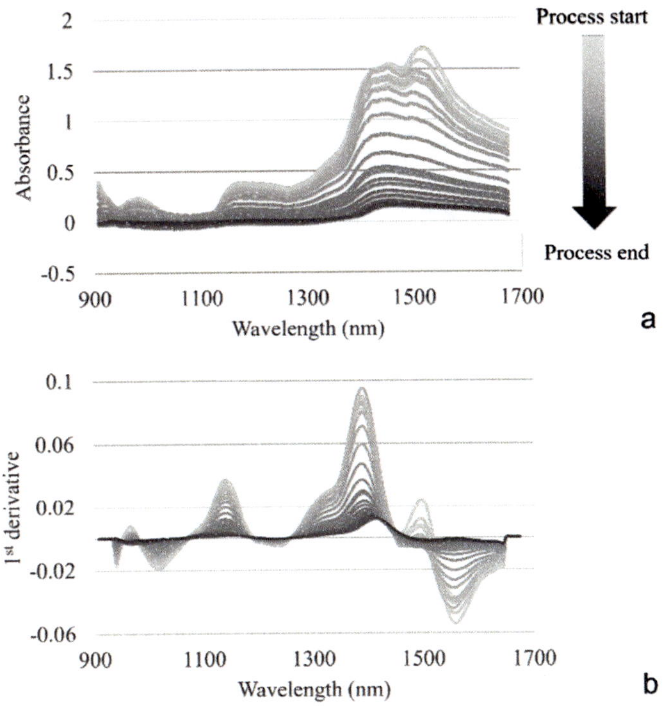

Fig. 4. Raw (a) and 1st derivative transformed (b) NIR spectra of the eggplant dehydration process.

between 1300 and 1600 nm, associated with the first overtone of water, indicating the possibility of using the Aquaphotomic approach to follow the process.

The first two principal components of the PCA performed on the pretreated spectra could explain 99% of the total variance. The trend of the scores defines the positioning of the samples in the multivariate space according to the progress of the process (Fig. 5a). PCA score plot loadings confirmed significant variations in the first overtone of the water spectral region (Fig. 5b).

Figure 6a provides the results of MBM applied to the acquired data using a block of 3 spectra, which means that the first point represents 1–3 spectra, the second point represents 2–4 spectra and so on. The trajectory of the MBM showed a continuous decrease during the process and tended to a constant value when repeated low values of variability were recorded, greatly helping in identifying the process endpoint. According to the PCA loadings, it can be assumed that the MBM trajectory reflects the modifications of the water component during the process.

Based on the water absorbance bands found in pretreated spectra according to Tsenkova et al. (2018) [9], 12 wavelengths were selected for representation of spectral patterns of water samples using Aquagrams: 1342, 1366, 1373, 1410, 1428, 1441, 1453, 1459, 1478, 1490, 1515 nm (Fig. 6b). Each wavelength corresponds to the absorption of specific water molecular species [9]. A shift from left to right was

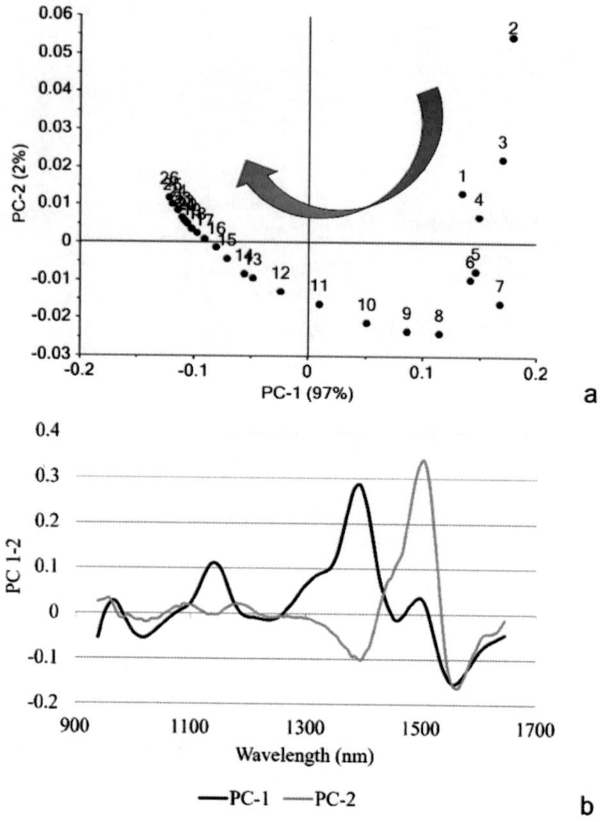

Fig. 5. PCA scores (a) and loadings (b) plot performed on eggplant pretreated spectra.

observable during the dehydration process, which means moving from a system characterized by the prevalence of bonded water forms to a situation in which free water molecules prevail due to the evaporation process.

To improve the process and confirm these first results, the apple dehydration process was carried out. The drying process lasted 61 h, and it was stopped when the samples reached a constant weight as above specified. At the end of the process, the a_w value of the dried product was equal to 0.329.

For this process, a commercial thermally insulated NIR probe housing was used, and the number of acquired spectra was increased by increasing the acquisition frequency to one spectrum per hour. The PCA on these spectra confirmed the involvement of the water absorption bands even if a different matrix was used [data not shown].

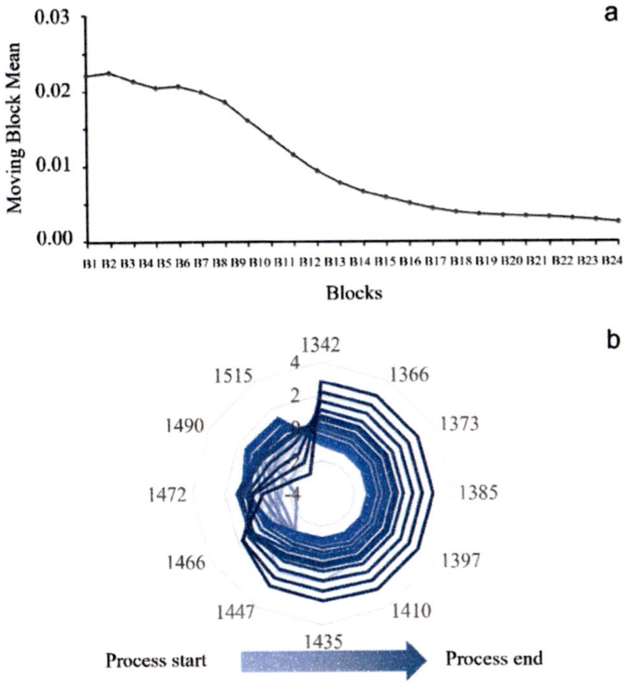

Fig. 6. Eggplant Moving Block mean trajectory (a); Aquagrams (b).

For this trial, MBM blocks of 7 spectra were used to build up the chart and establish the endpoint criteria. On the basis of MBM trajectory, lower and upper limits of 0.0365 and 0.0368 were tentatively set (Fig. 7a), to verify the MBM suitability in identifying the process end. According to these limits, the spectra corresponding to the block B43 should belong to completely dehydrated samples. The same spectra were represented as yellow lines in the Aquagram of Fig. 7b. These spectra did not correspond to the completely dried samples. They were located in the right side of the Aquagram (Fig. 7b) and characterized by a water pattern with a prevalence of free water molecules. However, they did not fit with the sample on the far right of the chart. Therefore, by combining the information deriving from the two approaches, we should identify the actual end of the process corresponding to a block within the range defined by the endpoint criteria beyond the B43 block.

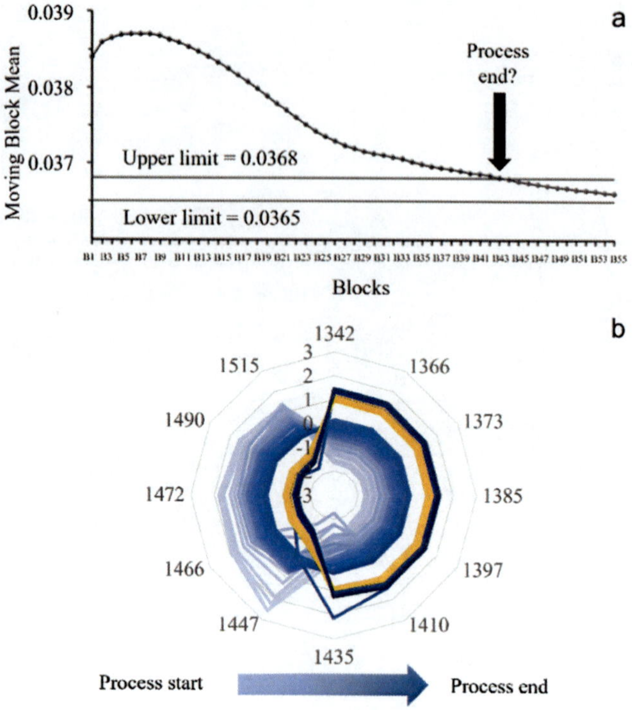

Fig. 7. Apple Moving Block mean trajectory (a); Aquagrams (b).

4 Conclusions

The obtained results suggested the applicability of Aquaphotomics for the continuous monitoring of food drying processes using an adequate NIR probe.

PCA loadings confirmed the important contribution of the water absorption in the wavelength range 1300–1600 nm during the dehydration processes independently from the food matrix. The great variations recorded in the spectral region of the first overtone of water, suggested the possibility of using the Aquaphotomic approach to follow the dehydration process. Thus, Aquagrams can help to identify the actual endpoint of the process, due to the detection and the monitoring of the different water forms (free water, dimers, aggregates, clusters, etc.…). The combination of information from the two chemometric instruments, MBM and Aquagrams, provided a valuable tool for defining and predicting the end of the dehydration process. The choice of the proper size of blocks is crucial in order to gain a good predictive power of the MBM.

The selection of few wavelengths to study food processes and quality can optimise the transfer of instrumental data to IT platforms and networks.

Further investigations are in progress to check the possibility of transfer the obtained results on a pilot scale for future applicability of these applications.

Aknowledgements. Scientific and technical support: Maurizio Cutini, Alex Filisetti, Carlo Bisaglia, Ivan Carminati, Gianluigi Rozzoni, CREA-IT, Treviglio, Bergamo.

Funder: Italian Ministry of Agriculture, Agridigit project, sub-project Agrofiliere (D.M. 36503/7305/2018, 20/12/2018).

PhD course: DAFNE-EEE, Tuscia University, Viterbo, ITALY.

References

1. Mohammed, S., Edna, M., Siraj, K.: The effect of traditional and improved solar drying methods on the sensory quality and nutritional composition of fruits: a case of mangoes and pineapples. Heliyon **6**(6), e04163 (2020). https://doi.org/10.1016/j.heliyon.2020.e04163
2. Motevali, A., Koloor, R.T.: A comparison between pollutants and greenhouse gas emissions fromoperation of different dryers based on energy consumption of powerplants. J. Clean. Prod. **154**, 445–461 (2017). https://doi.org/10.1016/j.jclepro.2017.03.219
3. Araya-Farias, M., Ratti, C.: Dehydration of foods: general concepts advances in food dehydration. In: Ratti, C. (ed.) Advances in Food Dehydration, pp. 1–36. CRC Press, Taylor & Francis Group, LLC (2009)
4. Fellows, P.J.: Food Processing Technology - Principles and Practice, 2nd edn. CRC Press, Taylor & Francis, UK (2000)
5. Tsenkova, R.: Visible-Near Infrared perturbation spectroscopy: water in action seen as a source of information. In: Burling-Claridge, G.R., Holroyd, S.E., Sumner, R.M.W. (eds) "NIR in Action – Making a Difference", Proceedings of the 12th International Conference on Near-infrared Spectroscopy, Auckland, New Zealand, pp. 514–519. New Zealand Near Infrared Spectroscopy Society Inc., Hamilton (2007). ISBN:978-0-473-11746-7
6. Tsenkova, R.: Aquaphotomics: dynamic spectroscopy of aqueous and biological systems describes peculiarities of water. J. Near Infrared Spectrosc. **17**, 303–313 (2009)
7. Muncan, J., Tsenkova, R.: Review: aquaphotomics - from innovative knowledge to integrative platform in science and technology. Molecules Open Access **24**, 2742 (2019). https://doi.org/10.3390/molecules24152742
8. Decagon Device. Aqualab Water Activity Meter Operator's Manual. Decagon Devices Inc (2008)
9. Tsenkova, R., Muncan, J., Pollner, B., Kovacs, Z.: Essentials of aquaphotomics and its chemometrics approaches. Front. Chem. Open Access **6**, 363, 1–25 (2018) https://doi.org/10.3389/fchem.2018.00363
10. VIAVI Solutions. VIAVI MicroNIR™ Pro Software Suite v3.0. User Manual. VIAVI Solutions Inc. (2019)
11. Galli Volonterio, A.: Microbiologia Degli Alimenti. Casa Editrice Ambrosiana, Rome (2005)

CO_2 Use and Energy Efficiency in Closed Plant Production System by Means of Mini-air Handling Unit

Claudio Perone[1], Michela Orsino[2](✉), Giovanna La Fianza[2], Lucio Brunetti[2], Ferruccio Giametta[2], and Pasquale Catalano[2]

[1] Department of the Science of Agriculture, Food and Environment, University of Foggia, 71100 Foggia, Italy

[2] Department of Agricultural, Environmental and Food Sciences, University of Molise, 86100 Campobasso, Italy

m.orsino2@studenti.unimol.it

Abstract. The concept of resource use efficiency is strictly related to crops production in closed plant production system (CPPS). In fact, it is essential to maximize the crop yield with the minimum employment of resources. Among the main ones there are the CO_2, to replace that absorbed for photosynthetic activity, and primary energy for the conditioning of the closed environment. In this work the CO_2 level inside Controlled Environment Agriculture (CEA) is controlled by means of a mini-air handling unit (MAHU) with heat recovery, which renews the air to keep constant the indoor CO_2 concentration. The MAHU is installed in a local firm, and it is equipped as an IoT system and remote control. Both sensors, actuators and controller are not directly connected to the Unit but by mean of a remote unit located in the University laboratory. The overall performance of the connection was evaluated. Tests were conducted in an experimental CPPS for the lettuce production. The air conditioned by the handling unit was supplied in the environment by a perforated duct. CO_2 probes allowed to measure the concentration inside the supply duct, exhaust duct, and indoor environment. In addition, a set of probes to measure both physic and energy parameters was installed on MAHU to calculate the Energy Efficiency Ratio (EER) of the system.

The performed tests make it possible to evaluate the efficiency of the MAHU both in terms of overall connection performance and in controlling the CO_2 inside CPPS. The continuous replacement of the indoor air allows keeping almost constant its concentration, leading to an important saving in terms of CO_2 enrichment. Remarkable energy performances were also highlighted. The proposed IoT system well suites the Industry 4.0 paradigm as it allows installation/maintenance of the different parts of the whole system without any other operation on the remaining parts.

Keywords: CO_2 Use Efficiency · Energy efficiency · Controlled Environment Agriculture · Coefficient of performance

M. Biocca et al. (Eds.): SHWA 2020, LNCE 252, pp. 506–515, 2022.
https://doi.org/10.1007/978-3-030-98092-4_52

Nomenclature		*Subscripts*	
CPPS	Closed Plant Production System	a	Air
CUE	Carbon dioxide Use Efficiency	e	Electric
E.E.	External Environment	E	Exhaust
EER	Energy efficiency ratio [Wth We^{-1}]	O	Outdoor
MAHU	Modular Handling unit	ren	Renewal
P	Power [W]	s	System
rps	Revolutions per second	S	Supply
T	Temperature [° C]	SET_calc	Calculated Set-point
V	Volumetric flow rate		
Greek symbols			
η	Temperature ratio		
ρ	Density		

1 Introduction

The increasing demand for food supply, the increasing problems linked to climate change and the lack of natural resources and energy has made urgent the need for more sustainable production technologies [1].

The growing expansion of cities has reviewed traditional farming methods in favour of new urban farming systems: in this regard Closed Plant Production System (CPPS) represents an innovative product solution. In fact, thanks to a high level of environmental control, these types of plant factories can reduce the time production up to four times compared to traditional outdoor cultivation [2], optimizing yields. At higher levels of yields corresponds to the need of an increasing sustainability during the production as a more efficient resources usage (as water, CO_2, light energy, inorganic nutrients) [3], that can be higher compared to the traditional greenhouses [4]. This system becomes productive and closed when each of the resources is projected to produce at its maximum level, minimizing the emission of environmental pollutants and resources consumption [5, 6].

Starting from resources use efficiency, the carbon dioxide represents an important factor to consider. In fact, carbon dioxide is fundamental for the efficiency of the photosynthetic process [7, 8]. Hence, it is clear that the accurate control of CO_2 contributes in the final crop yield and quality, as well as reducing environmental impact.

In a confined agriculture environment the enrichment of carbon dioxide is essential not only for the major productivity, but also for the independence from external environment [9]. Thus, it becomes an important cost to consider in the entire cultivation system. Finding the best method for generating and supplying CO_2 becomes relevant for efficiency and profits. CO_2 sources used can be summarized from literature as reported: natural or forced ventilation, compost, exhaust gas from fossil fuels and from renewable energy, and pure liquefied CO_2 [10].

The development of sustainable technologies solutions in the food production processes must take into account also a careful evaluation of the energy saving aspects during the crops growing cycle [11–13]. In particular, this is an aspect to consider when the replacement of CO_2 is made through mechanical ventilation. In addition to the CO_2 concentration, mechanical ventilation permits the management of indoor humidity and temperature [14]. Among these parameters that are essential for the well-being of crops and their growth, in particular, temperature control in the confined environment is essential for fast plant growth and accumulation of organic matter [15–17]. Thus, it becomes interesting to evaluate the performance of the system in terms of energy efficiency ratio (EER) [18, 19].

In this study the CO_2 level inside a confined cultivation room was controlled by means of a mini-air handling unit (MAHU) with heat recovery. The unit renews the air to keep constant the indoor CO_2 concentration. The CO_2 supply rate and the overall performance inside the working environment were evaluated.

2 Materials and Methods

2.1 Tested CPPS

Tests were carried out in a laboratory, where an empty experimental CPPS for successive lettuce production was reproduced. The workspace is given by a 40 m^3 inner volume of a closed environment with dimensions of 4.1 × 3.6 × 2.7 m (Width × Depth × Height), well insulated on the wall and on the roof with sandwich panels made of poly- urethane foam.

As showed in Fig. 1, the experimental CPPS (Confined Environment) is in a laboratory that consists also of: an External Environment (E.E.), where are reproduced external climatic conditions; an Engine Room, where the mini-air handling unit (MAHU) and the control system are placed; a Duct Room, where an insulated duct connects the E.E. room to the suction side of recovery unit; a Supervisory Room, to control and manage the entire implant system.

The conditioned air provided by a handling unit of 400 $m^3 h^{-1}$ was supplied inside the CPPS by a perforated duct installed on one side at a height of 2.30 m. On the opposite side of the supply duct, located in the corner, there is a recovery grid at a height of about 60 cm.

Carbon dioxide (CO_2) probes were located on the recovery duct at level of recovery grid and on the insulated duct in the Duct Room. CO_2 probes were used to control on demand the airflow in the mechanical ventilation system.

A set of NTC (Negative Temperature Coefficient) temperature probes were installed around the heat exchanger (TS supply air temperature probe, TE exhaust air temperature probe, TO outdoor air temperature probe, TDIS_ST discharge air temperature probe downstream the static recovery and TS_ST supply air temperature probe downstream the static recovery) to evaluate the heat trans fer between supply and exhaust airstream.

In addition, an energy meter is installed to measure the global electric demand of the whole recovery system, considering also the electric absorption due to ventilation

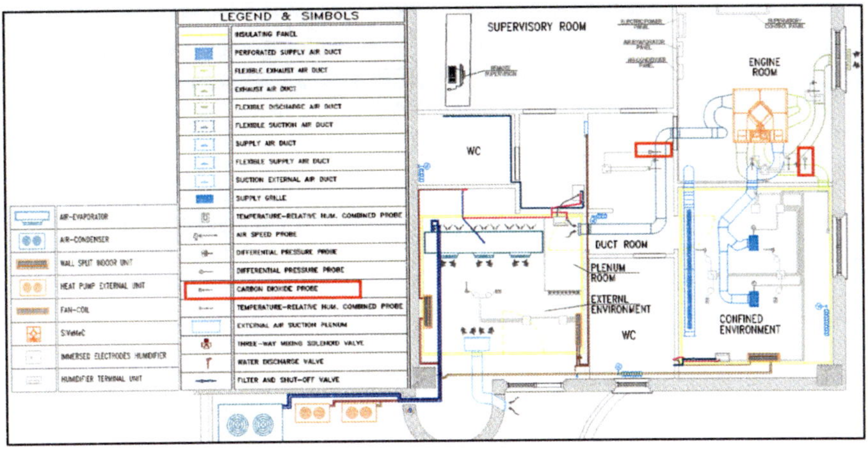

Fig. 1. CPPS configuration environment and CO_2 probes location.

(fans). In the Supervisory Room a supervision software allowed to acquire and historicize all measured data thanks a series of probes (Table 1).

Table 1. Probes' characteristics used for experimental tests.

Probes	Accuracy
Humidity and temperature room sensors Range 5–95% −30–70 °C, output 0–10 Vcc	±0.2 K at 25 °C ±3% at 25 °C 30–70% rh ±5% at 25 °C 10–30% 70–90% rh ±10% at 25 °C 5–10% 90–95% rh
Duct CO_2 transmitter. Output. 0–10 Vcc	+2%
Duct humidity-temperature sensor. Range 5–95% 30–70 °C. Output 0–10 Vcc	±0.3 K at 25 °C ±2.5% at 20 °C 10–95% rh
Transmitter of differential pressure 0–500 Pa	±5%

2.2 Experimental Procedures

The work target is the next lettuce cultivation, thus the microclimatic parameters were properly set. The air handling unit was tested in cooling mode and in static recovery mode by reproducing 32 °C outdoors on the exhaust side. The setpoint temperature was 24 °C as day temperature, with a dead band of ±1 °C. Once achieved the setpoint, MAHU worked modulating the engine compressor speed, expressed in rotations per second (rps), increasing in this way the whole efficiency. When the system reached the steady-state, all the chosen data were acquired with an acquisition time of 40 min. The sampling time for each parameter was set to 5 s.

In this work the energy efficiency ratio (EER) of the unit was evaluated. EER is a ratio that depends on the net cooling capacity of the unit and on the total electric power usage, including compressor and fans [20]. EER ratio is calculated as follows:

$$\mathrm{EER} = \frac{\mathrm{P_{fr}}}{\mathrm{P_e}}$$

where, *Pe* is the global absorbed electric power [kW], *Pfr* represents the cooling power of the system [kW] and is calculated as follows:

$$\mathrm{P_{fr}} = \mathrm{V_{ren}} \rho_a c_{p_a} (\mathrm{T_s} - \mathrm{T_O}) \quad (2)$$

where, $\dot{V}ren$ is the flow rate of the renewal air to be cooled [m^3 s^{-1}], *ρa* is the air density [kg m^{-3}], *cp* is the air specific heat at constant pressure [kJ kg^{-1} K^{-1}], *TS* is the supply air temperature [°C], *TO* is the outdoor air temperature [°C], *Pe* is measured by an energy meter. It is important to highlight that in the electric absorption *P* value is also considered the power consumption due to ventilation (i.e. fans).

Regarding the CO_2 flow inside the controlled environment, this is an important parameter measured in the culture room in μmol m^{-2} h^{-1}. In particular, in this paper we focused on the CO_2 supply rate CS, calculated starting from the CO_2 Use Efficiency ratio [4]:

$$CUE = \frac{C_P}{C_S - C_R} = \frac{C_S - C_L}{C_R + C_S} \quad (3)$$

where CP is the net photosynthetic rate (μmol m^{-2} h^{-1}), CS represents the CO_2 supply rate (μmol m^{-2} h^{-1}), CR is the respiration rate of workers in the culture room (μmol m^{-2} h^{-1}), and CL is the rate of loss of CO_2 to the outside due to air infiltration (μmol m^{-2} h^{-1}). In this work CR, CP, CL were not taken into account because, respectively: the automatized cultivation room does not need the employment of workers; there are not crops physically for having the real acquisition of net photosynthetic rate, so are considered values based on literature that for lettuce is about 20 μmol mol^{-1} [8, 21]; air infiltrations are negligible because the income and the outcome of air are balanced, so the environment is neither in overpressure nor in depression. CS is calculated as follows:

$$\mathrm{CS} = \frac{1}{kc}\frac{\dot{V}_{ren}}{F}\mathrm{CO_{2in-out}} \quad (4)$$

where, kC is the conversion factor from mol to volume (0.0244 m^3 mol^{-1} at 24 °C setpoint temperature), $\dot{V}ren$ is the renewed air volume given by the number of air exchanges per hour in the cultivation room (h^{-1}) and the air volume in the culture room (m^3), F is the floor area of the cultivation room (m^2), $CO_{2in\text{-}out}$ is given by the difference between CO_2 concentration inside and outside the culture room (μmol mol^{-1}).

3 Results and Discussion

In the steady-state, reasonably, the physical conditions were almost constant and can be asserted that the temporal mean of each data-set is an estimation of each parameter. The standard deviation is assumed as its uncertainty. In Table 2 are shown average values and standard deviation of measured parameters at outdoor air temperature of 32 °C.

Table 2. Average values and standard deviation of measured parameters at outdoor temperature of 32 °C.

		I sampling	II sampling
TS [°C]	$\bar{x}$	21.08	21.16
	σ	1.81	1.79
TE [°C]	$\bar{x}$	23.17	23.34
	σ	0.72	0.78
$\dot{V}$ ren	$\bar{x}$	404.97	408.92
	σ	9.88	12.61
η [%]	$\bar{x}$	92.74	92.33
	σ	0.88	0.98
EER [Wth We^{-1}]	$\bar{x}$	3.49	3.44
	σ	0.73	0.58
EERS [Wth We^{-1}]	$\bar{x}$	3.46	
	σ	0.45	

From the table is clear that we set the system to work at 24 °C setpoint temperature and that the system had a mean supply temperature TS of 21.12 ± 1.27 °C, with a mean value of temperature measured on the exhaust side of recovery unit TE of 23.25 ± 0.53 °C. In addition, the system worked at an average of renewable air flow of about 406.47 ± 7.78 m^3/h.

The test results are shown in Figs. 2a, 2b, and 3a, 3b. Figure 2a and 3a show the time trend of the measured temperatures. In both cases, TS oscillates around about 21 °C because the system is working in attenuation phase as described in [18] and as Fig. 2b and 3b show: the attenuation phase occurred when the compressor operated at the minimum allowed speed (30 rps). The system, in fact, after start with the power up phase, reached the regime (60 rps for a minimum time of 120 s); then the compressor speed was reduced to 30 rps, since the set point was reached, and as a result an increasing of EER was observed. The system showed also a good passive recovery efficiency (η = (TO − TS_ST)/(TO − TE) equals to 92.53% at 32 °C) that make it possible the compressor to operate at attenuation phase (30 rps). In these conditions the mean EER is 3.46 ± 0.45. Looking at the two samplings, the system proved to be very reliable. In fact, all the measured parameters are consistent in their errors and also their temporal evolution is similar as a direct comparison between Figs. 2a and 3a, and Figs. 2b and 3b demonstrates.

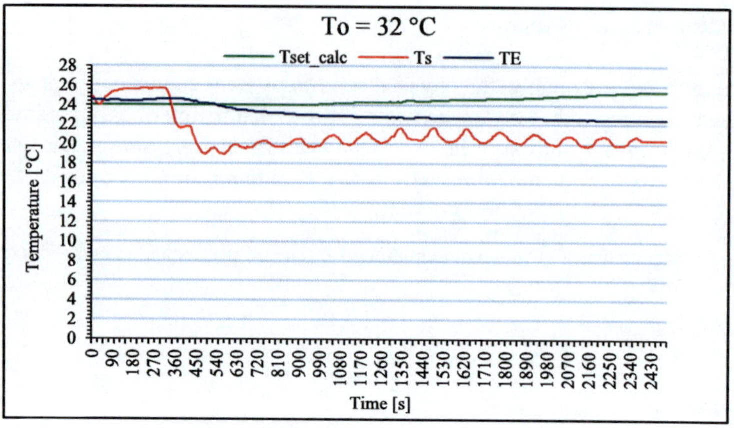

Fig. 2a. First test: time trend of the exhaust air temperature (TE), supply air temperature (TS), and calculated set point temperature (TSET_calc).

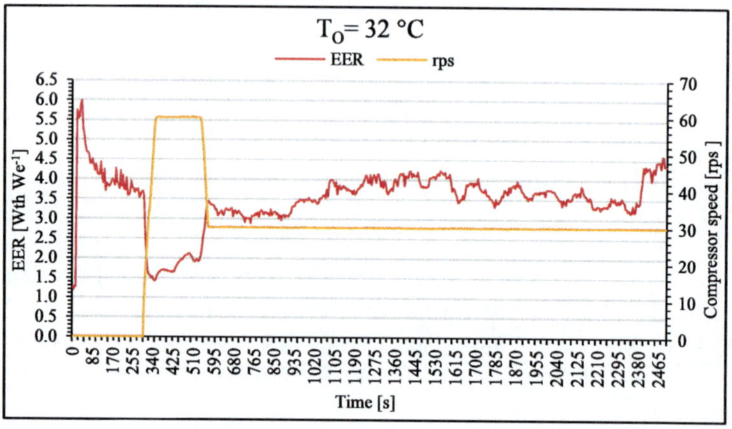

Fig. 2b. First test: energy efficiency ratio (EER) in function of the compressor speed (rps).

Regarding the supplied CO_2 in the environment, this is an important parameter because permits to understand for each square meter how many μmol of CO_2 are released into the environment per hour.

The following Tables 3 and 4 show, respectively, the average values and standard deviation of $CO_{2in\text{-}out}$ and the of CO_2 supply rate.

From an average value of 552.38 ppm, converted in μmol mol^{-1} [22], it was obtained a CS mean value of all tests equal to 0.63 μmol m^{-2} h^{-1}, with a standard deviation of 0.02. The obtained value is useful because it helps to take a first step forward in terms of CO_2 supply enrichment. In fact, the obtained CS supplied by the unit represents a useful aid to reduce the CO_2 that should be supplied by an external source. This contributes to saving energy and resources, considering that CO_2

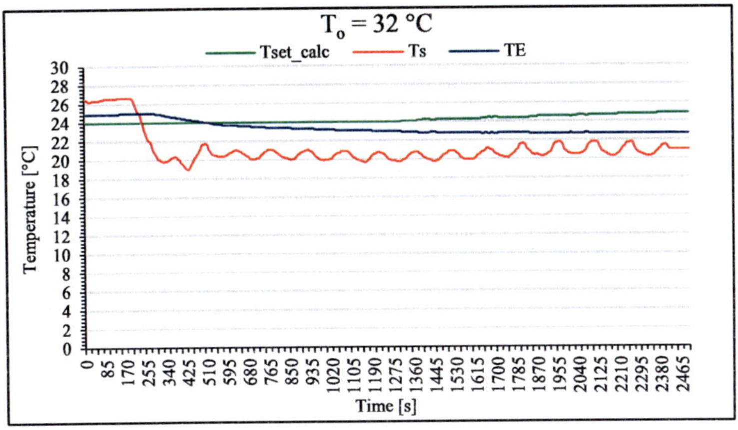

Fig. 3a. Second test: time trend of the exhaust air temperature (TE), supply air temperature (TS), and calculated set point temperature (TSET_calc).

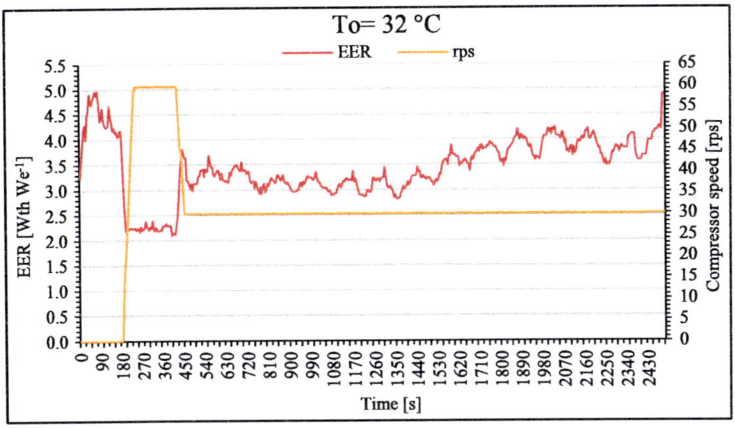

Fig. 3b. Second test: energy efficiency ratio (EER) in function of the compressor speed (rps).

Table 3. Average values and standard deviation of $CO_{2in\text{-}out}$.

		I sampling	II sampling
CO_2	$\bar{x}$	552.1	552.64
	σ	4.71	8.33

enrichment in CPPS is generally high (around 1000–2000 μmol mol^{-1}). This study represents a preliminary investigation to design a future CPPS configuration, in order to reduce both energy and CO_2 use.

Table 4. Average values and standard deviation of CS supply rate.

		I sampling	II sampling
CS	$\bar{x}$	0.63	0.63
	σ	0.02	0.02

4 Conclusion

The resource use efficiency in confined cultivation production is increasingly a key issue. In particular, in CPPS the CO_2 enrichment is essential for constant and efficient productivity. The controlled environment conditions were set optimally, with a 24 °C setpoint temperature, a mean supply temperature of 21.12 ± 1.27 °C and a mean value of temperature measured on the exhaust side of the recovery unit of 23.25 ± 0.53 °C. The system showed a good passive recovery efficiency (η is equal to 92.53% at 32 °C) that make it possible for the compressor to operate at the attenuation phase (30 rps). This system ensures a good EER for microclimatic conditions set in the environment. The continuous replacement of the indoor air allowed to keep almost constant its concentration leading to a CO_2 enrichment sufficient to have a first saving in terms of carbon dioxide, essential for optimal photosynthetic activity of lettuce crops.

There will be much to do in terms of CO_2 saving using mechanical ventilation and in the future will be investigated a better configuration of the system to increase the rate of carbon dioxide supply in the environment from outdoor.

References

1. Fedoroff, N.V.: Food in a future of 10 billion. Agric. Food Secur. **4**(1), 1–10 (2015)
2. Luna-Maldonado, A.I., Vidales-Contreras, J.A., Rodríguez-Fuentes, H.: Advances and trends in development of plant factories. Front. Plant Sci. **7**, 1848 (2016)
3. Hu, M.C., Chen, Y.H., Huang, L.C.: A sustainable vegetable supply chain using plant factories in Taiwanese markets: a Nash-Cournot model. Int. J. Prod. Econ. **152**, 49–56 (2014)
4. Kozai, T.: Resource use efficiency of closed plant production system with artificial light: concept, estimation and application to plant factory. Proc. Jpn. Acad. Ser. B **89**(10), 447–461 (2013)
5. Kozai, T., Niu, G., Takagaki, M. (eds.): Plant Factory: An Indoor Vertical Farming System for Efficient Quality Food Production. Academic Press, San Diego (2019)
6. Zhang, R., Liu, T., Ma, J.: Plant factory: a new method for reducing carbon emissions. In AIP Conference Proceedings, vol. 1820(1), p. 040016. AIP Publishing LLC, March 2017
7. Zelitch, I.: Improving the efficiency of photosynthesis. Science **188**(4188), 626–633 (1975)
8. Zhou, J., Wang, J.Z., Hang, T., Li, P.P.: Photosynthetic characteristics and growth performance of lettuce (Lactuca sativa L.) under different light/dark cycles in mini plant factories. Photosynthetica **58**(3), 740–747 (2020)
9. Kozai, T., Niu, G.: Role of the plant factory with artificial lighting (PFAL) in urban areas. In: Plant Factory, pp. 7–34. Academic Press (2020)

10. Li, Y., Ding, Y., Li, D., Miao, Z.: Automatic carbon dioxide enrichment strategies in the greenhouse: a review. Biosys. Eng. **171**, 101–119 (2018)
11. Perone, C., et al.: Experimental study of a mechanical ventilation system in a greenhouse. Chem. Eng. Trans. **58**, 811–816 (2017)
12. Perone, C., Catalano, P., Giametta, F., Brunetti, L., Bianchi, B.: Controlled mechanical ventilation to reduce primary energy consumption in air conditioning of greenhouses. In: Coppola, A., Di Renzo, G., Altieri, G., D'Antonio, P. (eds.) Innovative Biosystems Engineering for Sustainable Agriculture, Forestry and Food Production. MID-TERM AIIA 2019. LNCE, vol. 67, pp. 399–407. Springer, Cham (2020). https://doi.org/10.1007/978-3-030-39299-4_45
13. Perone, C., Orsino, M., La Fianza, G., Giametta, F., Catalano, P.: Study of a mechanical ventilation system with heat recovery to control temperature in a monitored agricultural environ ment under Summer conditions. J. Build. Eng. **43**, 102745 (2021)
14. Orsino, M., Perone, C., La Fianza, G., Brunetti, L., Giametta, F., Catalano, P.: Microclimatic control in confined agricultural environment for plants cultivation. Chem. Eng. Trans. **87**, 229–234 (2021)
15. Max, J.F., Horst, W.J., Mutwiwa, U.N., Tantau, H.J.: Effects of greenhouse cooling method on growth, fruit yield and quality of tomato (Solanum lycopersicum L.) in a tropical climate. Scientia Horticulturae **122**(2), 179–186 (2009)
16. Hatfield, J.L., Prueger, J.H.: Temperature extremes: effect on plant growth and development. Weather Clim. Extremes **10**, 4–10 (2015)
17. Kaiser, E., Morales, A., Harbinson, J., Kromdijk, J., Heuvelink, E., Marcelis, L.F.: Dynamic photosynthesis in different environmental conditions. J. Exp. Bot. **66**(9), 2415–2426 (2015)
18. Perone, C., Orsino, M., La Fianza, G., Giametta, F., Catalano, P.: Study of a mechanical ventilation system with heat recovery to control temperature in a monitored agricultural environment under Summer conditions. J. Build. Eng. **43**, 102745 (2021)
19. Engler, N., Krarti, M.: Review of energy efficiency in controlled environment agriculture. Renew. Sustain. Energy Rev. **141**, 110786 (2021)
20. ANSR/AHRI Standard: Standard for performance rating of unitary air-conditioning & air-source heat pump equipment, ANSI/AHRI Standard 210/240. 240 (2008)
21. Jishi, T., Matsuda, R., Fujiwara, K.: A kinetic model for estimating net photosynthetic rates of cos lettuce leaves under pulsed light. Photosynth. Res. **124**(1), 107–116 (2015). https://doi.org/10.1007/s11120-015-0107-z
22. Kimball, B.A.: Impact of carbon dioxide, trace gases, and climate change on global agriculture. In: Proceedings of a Symposium. American Society of Agronomy (1990)

Author Index

M. Biocca et al. (Eds.): SHWA 2020, LNCE 252, pp. 517–519, 2022.
https://doi.org/10.1007/978-3-030-98092-4